Climate Variability and Change in the 21th Century

Climate Variability and Change in the 21th Century

Editors

Stefanos Stefanidis
Konstantia (Dia) Tolika

MDPI • Basel • Beijing • Wuhan • Barcelona • Belgrade • Manchester • Tokyo • Cluj • Tianjin

Editors
Stefanos Stefanidis
Aristotle University of Thessaloniki
Greece

Konstantia (Dia) Tolika
Aristotle University of Thessaloniki
Greece

Editorial Office
MDPI
St. Alban-Anlage 66
4052 Basel, Switzerland

This is a reprint of articles from the Special Issue published online in the open access journal *Climate* (ISSN 2225-1154) (available at: https://www.mdpi.com/journal/climate/special_issues/climate_variability_21).

For citation purposes, cite each article independently as indicated on the article page online and as indicated below:

LastName, A.A.; LastName, B.B.; LastName, C.C. Article Title. *Journal Name* **Year**, *Volume Number*, Page Range.

ISBN 978-3-0365-0108-6 (Hbk)
ISBN 978-3-0365-0109-3 (PDF)

Contents

About the Editors

Stefanos Stefanidis has a basic degree in Forestry (AUTh) and has done postgraduate studies in the Laboratory of Mountainous Water Management and Control (AUTh). He also completed his Ph.D. on the rational mountainous watershed management under climate change. Since 2018, he has been a teaching assistant in the Faculty of Forestry (AUTh). His main scientific interests are flash floods, water resources management, hydrometeorology and climate change effects.

Konstantia (Dia) Tolika is an Assistant Professor in the Department of Meteorology and Climatology at AUTh University. She has a basic degree in Physics (AUTh) and has done postgraduate studies in the Department of Meteorology and Climatology (AUTh). She also finished her Ph.D. on the development of climate scenarios for the changes in rainfall in the Greek region. This was followed by her post-doc studies on climate change and extreme weather phenomena. Since 2009, she has held an academic position in the Department of Meteorology and Climatology. Her main scientific interests are climatology, especially in the Mediterranean region, as well as climate models, climate change scenarios and extreme weather events.

Article

Study on Temporal Variations of Surface Temperature and Rainfall at Conakry Airport, Guinea: 1960–2016

René Tato Loua [1,2,3,*], Hassan Bencherif [1,4], Nkanyiso Mbatha [5], Nelson Bègue [1], Alain Hauchecorne [6], Zoumana Bamba [2] and Venkataraman Sivakumar [4]

[1] Laboratoire de l'Atmosphère et des Cyclones, UMR 8105, CNRS, Université de La Réunion, Météo-France, 97490 Réunion, France
[2] Centre de Recherche Scientifique de Conakry Rogbane, Conakry 1615, Guinée
[3] Direction Nationale de la Météorologie de Guinée, Conakry 566, Guinée
[4] School of Chemistry and Physics, University of KwaZulu Natal, Durban 4000, South Africa
[5] Department of Geography, University of Zululand, KwaDlangezwa 3886, South Africa
[6] Laboratoire Atmosphère, Milieux, Observations Spatiales/Institut Pierre-Simon-Laplace, UVSQ Université Paris-Saclay, Sorbonne Université, CNRS, 78280 Guyancourt, France
[*] Correspondence: rene-tato.loua@univ-reunion.fr or lrenetatometeo@gmail.com

Received: 15 March 2019; Accepted: 3 July 2019; Published: 18 July 2019

Abstract: The monthly averaged data time series of temperatures and rainfall without interruption of Conakry Airport (9.34° N 13.37° W, Guinea) from 1960 to 2016 were used. Inter-annual and annual changes in temperature and rainfall were investigated. Then, different models: Mann-Kendall Test, Multi-Linear-Regression analysis, Theil-Sen's slope estimates and wavelet analysis where used for trend analysis and the dependency with these climate forcings. Results showed an increase in temperature with semi-annual and annual cycles. A sharp and abrupt rise in the temperature in 1998 was found. The results of study have shown increasing trends for temperature (about 0.21°/year). A decrease in rainfall (about −8.14 mm/year) is found since the end of 1960s and annual cycle with a maximum value of about 1118.3 mm recorded in August in average. The coherence between the two parameters and climate indices: El Niño 3.4, Atlantic Meridional Mode, Tropical Northern Atlantic and Atlantic Niño, were investigated. Thus, there is a clear need for increased and integrated research efforts in climate parameters variations to improve knowledge in climate change.

Keywords: temperature; rainfall; climate indices; wavelet; trend analysis; climate change and Conakry

---

1. Introduction

Climate is naturally variable as evidenced by the irregularity of the seasons from one year to another. Long-term climate variability is of great importance for the estimation of its impact on human activities and for predicting the future climate [1]. The need to develop science programs that, in addition to exploring long-term climate change, can meet the more immediate needs of people and organizations to begin factoring climate risks into planning and management processes [2]. Over the twentieth century, west African region continues to receive a lot of unusual disasters at unexpected moments and areas. This might be a consequence of climate change and then that change is generated on the one hand by anthropogenic activities and on the other hand by natural variation. That's why some previous studies done on West Africa regions highlight the variability of temperature and rainfall and their relationship with climate indices, as Zerbo et al. [3] who studied the relationship between the solar cycle and meteorological fluctuations in West Africa and found that temperature and rainfall are influenced by solar activity. Schulte et al. [4] analysed the influence of climate modes on streamflow in the Mid-Atlantic region of the United States. Many studies also have been done on the intra-seasonal and inter-annual variability of temperature and rainfall [5–9] over West Africa areas. Previous studies have shown the crucial role of sea surface temperature (SST) anomalies in the tropical Atlantic region. For instance, SST induces forcing on the summer monsoon rainfall over sub-Saharan West Africa [10]. Vizy and Cook [8] highlighted that warm sea surface temperature anomalies influences positively the increase in rainfall along the Guinean coast. In their study on variability of summer rainfall over tropical north Africa during the 1906–1992 period, Rowell et al. [11] showed that the global SST variation are responsible for most of the variability of seasonal (July-August-September) rainfall from 1949 to 1990. Indeed, the annual cycle of rainfall over West Africa depends greatly on SSTs in the Gulf of Guinea [11].

However, any climatological study over West Africa could take into account at least West African Monsoon (WAM) and Inter-Tropical Convergence Zone (ITCZ). It is for the reason aforementioned that several studies have been done on the WAM influence on annual climatic variability in West Africa, [6,12] and its dynamic and onset [13–15]. Furthermore, it was also reported by Nicholson [16,17], that a major role of the WAM system is to transport moisture into West Africa from the Atlantic. In response to the onset of the African monsoon, the upwelling cooling is strongest in the east both because of the strong acceleration of the southerly winds and because the thermocline is shallow there [11].

The inter-annual variability of the WAM is mainly explained by the surface of ocean. It is worthy to note that the surface temperature of the inter-tropical Atlantic can be analysed efficiently. It constitutes an important climatic parameter, in the event of a strong anomaly, in all the coastal areas subjected to the direct impact of the WAM [18]. It is for this reason that Joly and Voldoire. Ref. [12] reported that SST anomalies are maximum in June–July, and are associated with a convective anomaly in the marine ITCZ with a spread over the Guinean coast. ITCZ is the major synoptic-scale system controlling seasonal rainfall [19]. It is well known that the distribution of temperature and rainfall through Earth surface is not homogeneous. Espinoza Villar et al. [20] pointed out the impact of mountain ranges on rainfall and specified that the long-term variability with a decreasing rainfall since the 1980s prevails in June-July-August and September-October-November in the Amazon Basin countries.

Our study area is localized in West Africa, enclosing the three major West African climate zones: Guinean zone (approximately 6–8° N); Soudanian zone (approximately 8–12° N) and Sahelian zone (approximately 12–16° N) [21]. It may be stated that the region of Conakry is part of the Soudanian zone (see Figure 1a). The station of Conakry is located at the international airport of Conakry at 9.34° N and 13.37° W, at 26 m height above the sea level (sl). Given that Conakry is a coastal zone that lies between the Atlantic Ocean and the Kakoulima Mountain range, which forms a barrier and promotes the Foehn phenomenon (see Figure 1b). This feature seems to be the reason that makes it the rainiest area compared to other parts of the country. This coastal site is the national socioeconomic development centre of Guinea, but is always threatened by heavy precipitations and strong heat waves causing significant economic and sanitary damages and loss of lives.

The absolute poverty of a large proportion of the African continent's people renders them highly vulnerable to changes in climate [22]. According to the increasing impact of the climate change in this area and the geo-climatic and environmental factors influences mentioned above, the purpose of our approach is to investigate with the keenest interest the climate variability as well as the forcing led by some climate indices on the temperature and rainfall at Conakry during 57 years. The aim of our study is to improve the understanding and strengthen the knowledge on the climate variability in this region of Guinea through a climatological approach coupled with a digital tool of analysis. After the station dataset description and methodology, obtained results are presented and discussed.

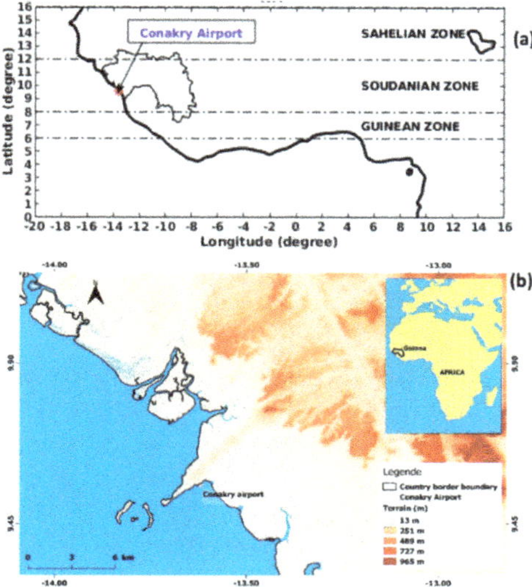

Figure 1. Geo-localisation of Conakry station in West Africa, map showing the three major west african climate zones (**a**), map showing Conakry Airport between Atlantic Ocean and Kakoulima Mountain range (**b**).

2. Materials and Methods

2.1. Data

Monthly averages of temperature and rainfall time-series are used in this study for the 1960–2016 period. They were obtained from continuous measurements at the synoptic weather station of Conakry in Guinea. A set of 684 monthly average temperature measurement during 57 years were used, for rainfall, the same data number were used too. The location of this synoptic station at the international airport of Conakry makes the data set uninterrupted and of good quality. The daily mean temperatures were calculated by averaging the daily minimum and maximum temperatures. The monthly and yearly temperature averages were calculated from the daily and monthly averages, respectively, for the complete study period. The histogram of monthly mean temperature peaks at 26 °C with 133 occurrences (Figure 2a).

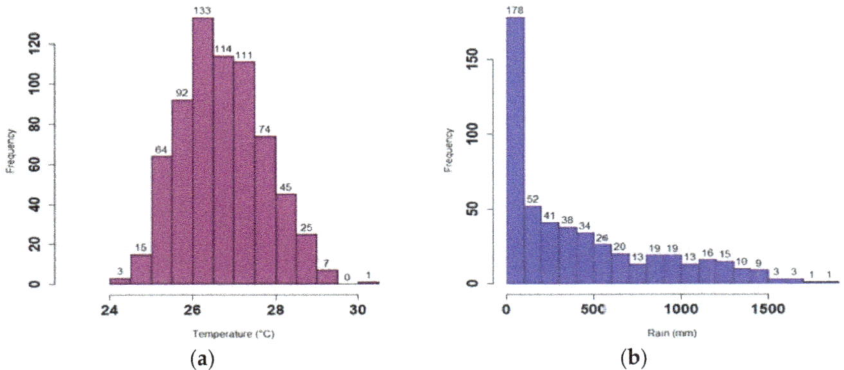

Figure 2. Histograms of monthly temperature frequency, the monthly temperature mean value of 26 °C has higher frequency of observation (**a**) and monthly rainfall (**b**) overall frequency showing that the monthly rainfall value of 0.1–100 mm has higher frequency of observation at Conakry Airport station.

The monthly rainfall is the accumulated based on daily rainfall obtained for a particular month. The overall annual rainfall is calculated as the sum of monthly rainfall. However, measured rainfall commonly consists of discrete series of rainfall events with different durations and time intervals [23]. It is noteworthy that rainfall is a discontinuous parameter, thus, monthly rainfall accumulated values used in our study oscillate between 0.1 and 1839.3 mm. Rstudio and Matlab software were used to perform all computational tasks. The histogram of monthly accumulated rainfall peaks at 0.1–100 mm with 178 occurrences, such as values above 1500 mm, have a lower occurrence (<10), but are very quantitatively significant from disaster (flood, landslide) point of view (Figure 2b).

To achieve a better understanding of the forcing that may influence temperature and rainfall of Conakry, four climate indices were used:

1. Niño3.4 monthly mean time series from 1960 to 2016 (684 measurements) were downloaded from the National Oceanic and Atmospheric Administration (NOAA) website (https://www.esrl. noaa.gov/psd_wgsp/Timeseries). The Niño3.4 index is calculated by taking the area-averaged sea-surface temperature (SST) within the Niño3.4 region, which extends from 5° N to 5° S in latitude and from 120° W to 170° W in longitude (in the Pacific Ocean). We use Niño3.4 averages calculated from the HadISST SST dataset, which is given by 1° in latitude–longitude.
2. Atlantic Meridional Mode (AMM) SST index from 1960 to 2016 (684 data) were downloaded from the National Oceanic and Atmospheric Administration (NOAA) website (https://www.esrl.noaa. gov/psd/data/timeseries/monthly). The AMM time series is calculated by projecting SST on to the spatial structure resulting from Maximum Covariance Analysis (MCA) to sea surface temperature (SST) over the region of 21° S–32° N, 74° W–15° E.
3. Tropical Northern Atlantic index (TNA) is the anomaly of the monthly averaged SST values. The TNA SST index is defined as region-averaged SST anomalies in the domain (0°–20° N, 60° W–20° E) [24]. TNA monthly mean time series from 1960 to 2016 (684 data) were downloaded from the NOAA website (https://www.esrl.noaa.gov/psd/data/climateindices/list/#TNA).
4. The Atlantic Niño (AN) or Atlantic Equatorial Mode is a quasiperiodic interannual climate pattern of the equatorial Atlantic Ocean. The term Atlantic Niño comes from its close similarity with the El Niño-Southern Oscillation (ENSO) that dominates the tropical Pacific basin [25]. The Atlantic Niño (AN) index is defined as the SST anomaly in the central-eastern tropical Atlantic: (3° S–3° N; 20° W–0° E) [26]. The AN monthly mean time series from 1960 to 2016 (684 measurements) were download from KNMI website (https://climexp.knmi.nl/start.cgi). The equatorial warming and cooling events associated with the Atlantic Niño are known to be strongly related to atmospheric climate anomalies, especially in African countries bordering the Gulf of Guinea [27]. As Conakry is a coastal region in west Africa, we used AN in our study.

In this work, the monthly average temperature (684 measurements) and monthly rainfall (684 measurements) as well as the monthly mean (684 measurements) of 4 climatic indices (Nñio 3.4, AMM, TNA and AN) were used as input data for our investigation during the period 1960–2016.

2.2. Method

According to the World Meteorological Organization [28,29] (WMO), climatological standard normals are defined as the averages calculated for uniform and relatively long period including at least three consecutive periods of ten years. The climatological standard normals are hence the averages of the climatological data calculated for the consecutive periods of 30 years. Our climatological study is based on the standard normal calculated for the 1961–1990 period. The obtained arithmetic mean was calculated (for temperature and rainfall) by using the following formula:

$$X = \frac{1}{n} \sum_{i=1}^{1} (x_i), \tag{1}$$

where, n = year number.

Over the Conakry site, the climatological normal calculated for temperature and rainfall on the 1961–1990 period corresponds to the values of 26.5 °C and 3806.8 mm, respectively. These values are used to calculate corresponding anomalies.

2.2.1. Mann-Kendal Test

It is always essential to work out monotonic trends in the time series of any geophysical data before any further use. In this study, the Mann-Kendall test [30–32] was employed to detect the trends that exist in both the temperature and rainfall time series. This method is defined as a non-parametric, rank-based method which is commonly used to extract monotonic trends in the time series of climate data, environmental data or hydrological data. The Mann-Kendall test statistic gives information about the trend of the total time series and its significance. However, it is important to investigate how the trend varies with respect to time. Therefore, the calculation of the forward/progressive ($u(t)$) and backward/retrograde ($u'(t)$) values of the Mann-Kendal test statistic is essential in order to investigate both the potential trend turning points and the general variability of trends in respect to time. This method is called the Sequential Mann-Kendall (SQ-MK), and it is well explained by [33] and other authors [31,34]. This method has been found to perform very well in trend analyses of stream flow and precipitation [35] and also in the field of earth remote sensing [36].

2.2.2. Multilinear Regression

One of the primaries aims of this study is to identify the relationship between the studied time-series (temperature and rainfall at Conakry station in the present paper) and climate indices such as TNA, Niño3.4, AMM and AN. The multi-linear regression (MLR) is a method that is frequently used to explain the relationship between one continuous dependent and two or more independent variables (climatic indexes in this case). The MLR model output y_i based on a number "n" of observations can be expressed as follow:

$$y_i = \beta_0 + \beta_1 x_{i2} + \cdots + \beta_p x_{ip} + \varepsilon_i, \tag{2}$$

where

$$i = 1, 2, 3, \ldots, n, \tag{3}$$

where in this case, y_i is the dependent variable x_{ip} represents the independent variables, β_0 is the intercept, and $\beta_1, \beta_2, \ldots \beta_p$ are the x's coefficients. The final term (ε_i) represents the residual term which the model should always keep its contribution as minimum as possible.

2.2.3. Theil-Sen's Estimator

Theil-Sen slope estimate method were used to analyse the long-term trend in the data and the seasonality. The Theil-Sen estimator (TSE) is fairly resistant to outliers and is robust with a high breakdown point of 29.3% [37,38]. TSE method was first outlined by Theil [39] and later expanded upon by Sen [40]. The determination of trend slope of n-pair of data is given by the formula:

$$T_i = \frac{x_j - x_i}{j - i}, \tag{4}$$

where, x_j and x_i presents as data values at time j and i ($j > i$) respectively [37].

2.2.4. Wavelet Analysis

The present study employed the Morlet wavelet which provides a good balance between time and frequency localization [41], especially for geophysical data. Wavelet analysis includes different wavelet functions such as the windowed Fourier transform, wavelet transform, normalization, wavelet power spectrum, etc. The main advantage of the wavelet analysis in comparison with other techniques is that it analyses localized variations of power within a time series. By decomposing a time-series into time-frequency space, one is able to determine the dominant modes of variability and their variation with time [42]. Wavelet transform coherence (WTC) is a good method for analyzing the coherence and phase lag between two time-series as a function of both time and frequency [43]. Therefore, we adopted the Monte Carlo wavelet and coherence analysis to quantify the relationships between climate forcing and the two data sets (rainfall and temperature) recorded at Conakry. More details about wavelets and wavelet coherence and phase are given by Torrence and Compo [42], Grinsted et al. [41] and others.

Basically, from a climatological point of view, the 1961–1990 normal (30 years) was used for this study. The models used thus show complementarity in the sense that the Mann-Kendall test gives information about the trend of the total time series and its significance. In addition, SQ-MK is important to determining both the trend variability in time and the trend change points in the time series. However, it is important to identify the relationship between the studied time-series and climate indices, for that purpose, the MLR and Wavelet are used. But the difference between these two models is that the multi-linear regression (MLR) helps to explain the relationship between one continuous dependent and two or more independent variables. The Wavelet analysis method helps to determine the dominant modes of variability and their variation with time, in addition it helps to quantify the relationships between climate forcings and the two data sets indicating the period when the correlation is significant as well. Furthermore, it also specifies whether the parameters are correlated or not and if so, whether they are in-phase or out-of-phase or if the causal relationship is identified or if there is simultaneity. The results from this methodology are then discussed in the following sections and some figures are plotted according to that done by Bilbao et al. [44].

3. Results

3.1. Climatology and Seasonality of Temperature and Rainfall

3.1.1. Inter-Annual Variation of Temperature and Rainfall

Figure 3a shows the month versus year evolution of the monthly averaged temperatures recorded at Conakry station from 1960 to 2016. This figure indicates that Conakry is experiencing an increasing temperature, found to be significant since 1970s. It is clearly shown on this figure that 1998 is the year with the highest temperature (30 °C in mean recorded on April). The 1997/98 El Niño phenomenon, which started in March 1997 and lasted until mid-1998, had resulted in severe flooding and drought in several parts of the world [45,46].

By analysing the evolution of the temperature in two different periods, 1960–1998 and 1998–2016 (Figure 3c), we have found an increase of 0.8 °C from the first period to the 2nd one. In average, the temperature ranges from 26.5 °C to 27.8 °C from one period to the other. The annual averaged temperature of 1998 is 28.1 °C. Similar analysis for another Guinean station located at 7.74° N; 8.82° W, 900 km far from Conakry is reported by Loua et al. [9]. They highlighted a warming due to the increase in evaporation. We assume that the 1998 warming observed at Conakry seem to be linked to the 1998's strong El Niño. Angell et al. [47], shown that the record global warmth in 1998, particularly in the 850–300 mb layer, is partly, if not mostly, due to the very strong El Niño of 1997–1998. Strong El Niño event made 1998 relatively hot at the surface and in the atmosphere. The exceptionally warm El Niño year of 1998 was an outlier from the continuing temperature trend. Previous works have also pointed out the influence of the large tropical explosive volcanic eruptions and ENSO on precipitation and temperature changes over West Africa [48–50]. However, these studies reveal that thus far no consensus has been reached on either the sign or physical mechanism of El Niño response to volcanism. Based on the use of the Fifth Coupled Model Intercomparison (CMIP5), Khodori. [49] showed that large tropical volcanic explosions favour an El Niño within 2 years following the eruption. They demonstrated that volcanically induced cooling in tropical Africa weakens the West African monsoon and the resulting atmospheric Kelvin wave drives equatorial westerly wind anomalies over the western Pacific. This wind anomaly is further amplified by air–sea interactions in the Pacific, favouring an El Niño-like response. This analysis was found in agreement with the study reported by

Liu et al. [50]. Through the use of the Community Earth System Model (CESM1), they shown that volcanic eruptions are efficient in reducing the monsoon precipitation. In addition to reduce moisture heavily, the volcanic eruptions can affect the circulation field much [50].

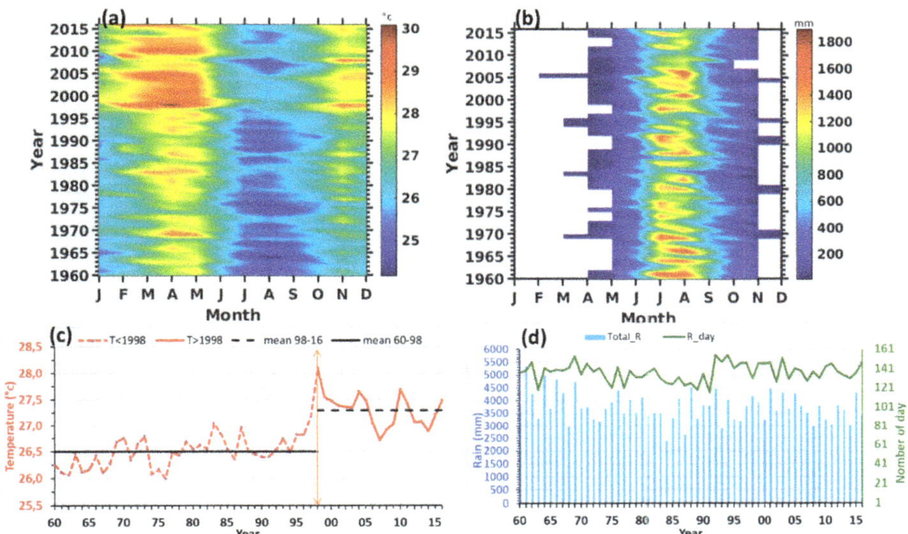

Figure 3. Yearly/monthly evolution of temperature (**a**) and rainfall with a pick during June-July-August (**b**); interannual evolution of temperature, dotted vertical line is the year 1998 (**c**) and interannual evolution of rainfall (blue bars) with rain day (black solid line) (**d**).

For the 1998 warming, Wang Shaowu et al. [51] explained that it is evident the annual temperature of 1998 set the highest record for the past century in China. Foster and Rahmstorf [52] reported the strong influence of known forcings on short-term variations in global temperature, including El Niño–Southern Oscillation (ENSO), and to a lesser degree, solar cycles. It so happens that 1982–1983 and 1997–1998 were the times of two biggest El Niño on record, and it is well established that a mini global warming occurs at the latter stages of an El Niño as heat comes out of the upper ocean and contributes to a warmer atmosphere and surface, but resulting in a cooler ocean [53].

The Figure 3b shows the yearly/monthly evolution of rainfall. There are climate conditions where one summer may be sunny, dry, and warm, whereas another may be cool, cloudy, and wet. Globally, the biggest cause of such regimes that last several seasons is the ENSO phenomenon [54]. For the specific case of Conakry, we remark that the evolution of temperature and rainfall through Hovmöller representation (year/month) shows an interseasonality for rainfall, and an increase in temperature. Additionally, for each year, the monthly maximum values of rainfall (>1500 mm) are recorded during the period June-July-August-September. Smallest amount of annual overall rainfall was recorded on 1984 (Figure 3d) and seem to be linked to the 1980s severe drought [55,56] that West Africa has experienced. On the one hand, the increase in temperature observed may be a response to global warming, and it is therefore consistent to diagnose whether warm years coincide with the occurrence of some geophysical phenomena such as El Niño. On the other hand, the remarkable interseasonality can be associated not only to the climatic warming but also to the irregularity in the intensification of the WAM and the dynamics of the ITCZ.

From this general overview of the interannual evolution of these meteorological parameters, we therefore proceeded to the analysis of the monthly climatology in the following section. This analysis allowed us to better understand the variability of the monthly climatological average of each parameter during the year. It will therefore be necessary to highlight the different seasons to which this region is subject.

3.1.2. Monthly Climatological Variations

Figure 4a depicts the variation of monthly climatology of temperature during the year over the whole period of observation at Conakry. Climatologically, the variation of monthly mean temperature shows clearly that the semi-annual cycle is dominant than the annual cycle. During the year, the temperature means oscillate between 24.2 °C and 30.1 °C with an annual mean of 26.8 °C. During the winter season, the monthly mean temperature may reach a peak in November (27.3 °C) and a second one in April (28.1 °C). In summer (June–October), it decreases in August (25.4 °C). This abrupt decrease in temperature that starts in may corresponds to the beginning of the rainy season, remarkably, a strong shift appears in June which seems to be due to the onset of WAM. Sylla et al. [57] reported that the beginning of rainy season in the West Africa Region can be associated with the northward migration of ITCZ from 4° N to 10° N, and the onset of the West African summer monsoon in the second half of June.

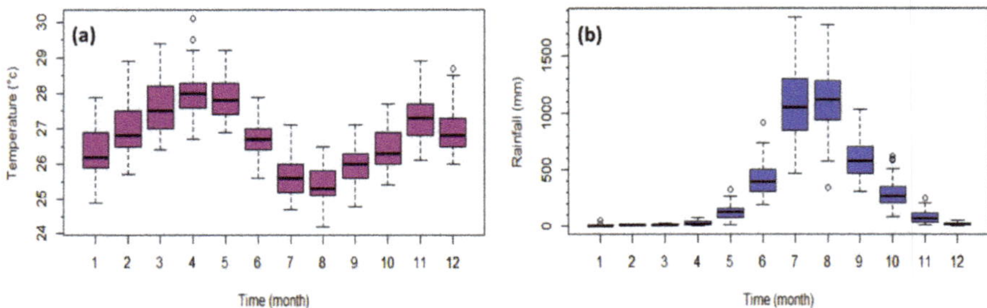

Figure 4. Climatology of monthly temperature showing two picks on April and November (**a**) and monthly rainfall showing a pick on July (**b**) as derived from ground observations at Conakry station from 1960 to 2016.

While during summer, under a cloudy sky, or overcast and rainiest, there is less solar radiation that reaches the Earth's surface. The temperature remains relatively low, resulting in a small thermal amplitude. The equatorial cooling intensifies the southerly monsoon in the Gulf of Guinea and pushes the continental rain band inland from the Guinean coast [11].

The Figure 4b shows the evolution of monthly climatology of rainfall at Conakry during the year. The variability of rainfall during the year shows an annual cycle with a peak recorded on August (>1000 mm). It is clear that the rainfall becomes significant in May, that corresponds to the beginning of summer (ICTZ northward migration), and it is followed by an abrupt upward jump in June (WAM onset) before reaching the peak in August (ICTZ at 10° N). During that period, the temperature decreases gradually from April to reach the minimum in August (Figure 4a). By the beginning of September, the rainfall is characterized by abrupt downward jump when the temperature starts increasing (ICTZ downward migration and weakening of WAM), and then the latest rains in the year are recorded in November. The beginning and end of rainy season are characterized by high frequency of strong storms in Guinea [9].

3.1.3. Temperature and Rainfall Anomalies

Temperature/rainfall anomaly from normal calculated for the period from 1961 to 1990 refers to the difference in degrees Celsius/in millimeter between the average annual temperature/annual rainfall observed from 1960 to 2016 in comparison with the average annual temperature/annual rainfall observed during the period from 1961 to 1990.

In this study, annual averaged temperatures were standardized by using the average of the period 1961–1990 (26.5 °C). In Figure 5a, the blue (red) bars indicate the negative (positive) anomalies and the fit line shows upward trend of temperature. Temperature anomalies could be classified in three classes:

- (a) the cold class: it corresponds to periods with negative anomalies (1960–1962; 1964–1965; 1967–1968; 1971; 1974–1976; 1986);
- (b) the quiet or normal class with temperature anomalies close to zero (1963; 1966; 1977–1978; 1985–1986; 1988–1992, 1994) and;
- (c) the warm class with positive anomalies (1969–1970; 1972–1973; 1979–1984; 1987; 1993; 1995–2006). The last period of the warm class is the warmest and longest one, it lasts about 12 years (1995–2006).

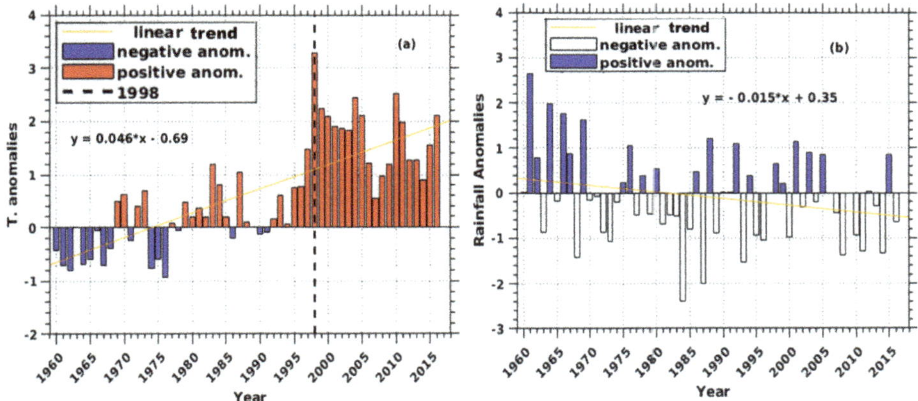

Figure 5. Temperature anomalies with dotted vertical line showing the year 1998, the blue (red) bars are negative (positive) anomalies and the increasing linear trend (**a**). Rainfall anomalies with decreasing linear trend, blue (white) bars indicate positive (negative) anomalies (**b**) of Conakry airport station: 1960–2016.

These results are consistent with those reported by Loua et al. [9]. On the whole, the inter-annual evolution of temperature shows a predominance of the warm class (positive anomalies) since 1992, and then all the following years are classified as warm, with a maximum in temperature anomalies obtained in 1998 (higher than + 3 °C). Among the most intense El Niño episodes of the last forty years, the one of 97–98 was the one that triggered the earliest and most severe. The countries most affected in terms of their infrastructure were USA, Indonesia and Brazil, but the highest human losses remain for Africa [58]. To confirm our result by the global analysis of surface temperature, Simmons et al. [59] highlighted that surface warming from 1998 to 2012 is larger than indicated by earlier versions of the conventional datasets used to characterize what the fifth assessment Report of the Intergovernmental Panel on Climate Change (IPCC) termed a hiatus in global warming.

Figure 5b illustrates the rainfall anomalies corresponding to the period from the year 1960 to 2016, using 1961–199 standard normal (3806.8 mm). It shows that there are both positive (blue) and negative (magenta) anomalies of rainfall during the study period. The positive (negative) anomalies correspond to wet (dry) years, and consecutive years (1970–1974) and (1981–1985) define two driest periods which correspond to two drought events in west Africa (1970s and 1980s). Peel et al. [60] highlighted that the consecutive dry years are associated with drought, which is a significant physical and economic phenomenon that imposes great stress on ecosystems and societies. However, drought is a part of natural climatic variability on the African continent, which is high at intra-annual, inter-annual, decadal and century timescales [61]. Where considering both the temperature and rainfall anomalies (Figure 5), we may notice that during the years 1970 and 1980 there were severe drought episodes in the study area. This was also reported by previous studies such as [62,63]. The West African Sahel is well known for the severe droughts that ravaged the region in the 1970s and 1980s [17].

This section allowed us to identify periods of hot consecutive years (70, 84 and 1992–2016) as well as periods of consecutive dry years (70–74 and 81–85). These periods served us as important references to take into account for the rest of the analysis on the variability of the trend and with the forcings used as well.

3.2. Trend Analysis of Temperature and Rainfall

In this study, the Mann–Kendall (MK) trend test was used. In regards to the temperature time series, a significant positive Z-scores value (9.3067) which is far greater than 1.96 was found, suggesting that the temperature trend is increasing. However, for rainfall, the MK trend test shows no-significant negative Z score (−0.17143) which above −1.96, suggesting weak decrease in rainfall variability.

Figure 6a shows the sequential statistic values of forward/progressive (Prog) u(t) (solid red line) and retrograde (Retr) u′(t) (black solid line) obtained by SQ-MK test for Conakry yearly mean temperature. In general, SQ-MK indicates and upwards trends of temperature in Conakry which is noticeable in both Prog. and Retr. SQ-MK statistic. The possibility is that the upwards trend started before the beginning of the time series (1960) because the change detection point, a point where Prog. and Retr. cross each other did not occur in the graph. What is noticeable in this figure is that, it is only from 1984 that this progressive SQ-MK statistic becomes positive and significant. At the same pace, it gradually increases until 1989 and then stands until 1998, the year from which the trend has increased significantly far above the confidence level (+3.866541) up to 2016 (+9.306717). There is a significant upward trend which seems to coincide with the 1970s 1980s droughts episodes and strongly the 1998 and 2014–2016 strong El Niño event. In a study that uses the similar non-parametric test method, Suhaila et al. [64] reported that the detection points captured by Pettitt and SQ–MK tests in Peninsular Malaysia temperature series during the years 1996, 1997 and 1998 are possibly related to climatic factors, such as El Niño and La Niña events. The retrograde statistic values are significant and negatives during the period from January 1960 to 1992 before it continues to be within the 95% confidence level limits (±1.96) except the year 1998 which the retrograde statistic value is significantly positive.

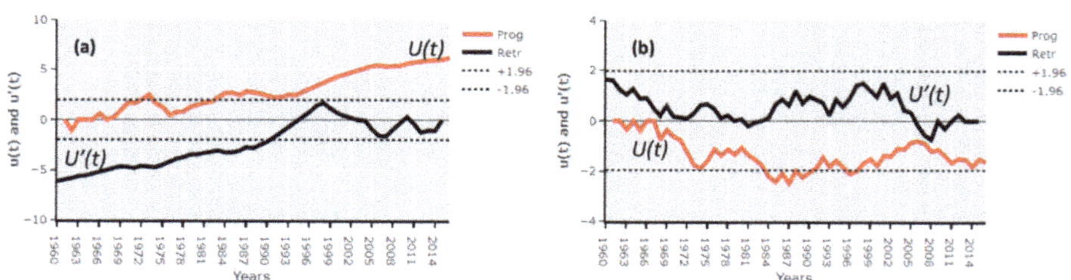

Figure 6. Sequential Mann-Kendal statistic values of progressive u(t) (solid redline) and retrograde u′(t) (black solid line), obtained by Sequential Mann-Kendall test for temperature (**a**) and rainfall (**b**) of Conakry airport: 1960–2016.

Figure 6b depicts the sequential statistic values of forward/progressive (Prog) u(t) (solid red line) and retrograde (Retr) u′(t) (black solid line) obtained by SQ-MK test for Conakry annual rainfall data for the period from 1960 to 2016. A strong significant upward trend was observed in late 1961, with the significant trend turning point observed in June 1962, which means that 1961 is the only year that is characterized by a positive and significant trend over the entire study period. But a careful analysis of the trend in progressive and retrograde which are non-significant (between ±1.96) and sometimes negative or positive shows two distinct periods that correspond to that found by the analysis of precipitation anomalies. For the first period (1970–1974) and the second period (1981–1985), the Retrograde curve is below the progressive curve in the negative band, which corresponds to periods of deficit rainfall. For the rest of the study period, the two curves intersect each other or the retrograde curve is above the progressive curve, that corresponds to periods with variable or normal rainfall. The response of the West African drought of 1970s and 1980s is clearly identified by the reduction in the rainfall at Conakry. Statistically there is a no-significant downward trend in rainfall since the end of 1960s.

In summary, the SQ-MK test and MK model for Conakry yearly data shows that the temperature and rainfall are subject to a significant increasing trend and a no-significant decreasing trend, respectively, during the period from 1960 to 2016. Thus, these methods seem to be useful for explaining the variability and trends of both temperature and rainfall.

In order to investigate physical relationships between climate forcing, precipitation and streamflow in the Mid-Atlantic region, Schulte et al. [4] selected eight climate indices. In the present study, four climate indices (Niño3.4, AMM, TNA and AN) were used as explanatory variables for this model because of their well-known possible influence on temperature and rainfall variability over West African region. The relevant time series of these climate indices are shown in Figure 7. Zebiak [27] specified that the dominant signature of ENSO is clearly focused on the Equator and its temporal variability is strongly focused at 3–5-year time scales.

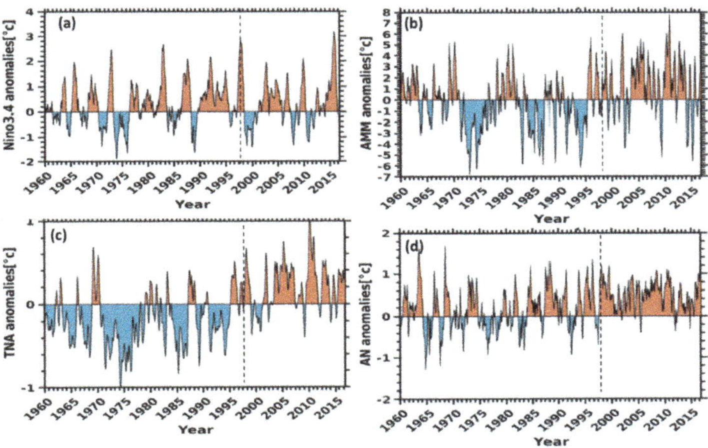

Figure 7. The standardized monthly Niño3.4 (**a**), AMM (**b**), TNA (**c**) and AN (**d**) time series for the period from 1960 to 2016. The vertical dashed lines indicate the year 1998.

There are two main forms of coupled ocean–atmosphere variability that exist in the tropical Atlantic Ocean, namely: the first one Atlantic Meridional Mode (AMM) [65] which is also called the interhemispheric mode [66]. It was originally identified by Servain [67]. This mode of variability is characterized by an interhemispheric gradient in sea surface temperatures and by oscillations in the strength of surface winds that cross the Equator, thereby reinforcing sea surface temperature anomalies [68]. The pronounced coupled ocean-atmosphere variability in the Tropical Atlantic is generated by fluctuations in the Atlantic Meridional Mode (AMM) [68]. The AMM is characterized by an anomalous meridional shift in the Intertropical Convergence Zone (ITCZ) that is caused by a warming (cooling) of SSTs and a weakening (strengthening) of the easterly trade winds in the northern (southern) tropical Atlantic [69]. And, the second one is the zonal mode, also called the Atlantic Niño [70]. Its seasonal evolution is due to surface wind variations associated with the northward migration of the ITCZ [71].

The tropical northern Atlantic (TNA) SST anomaly pattern is an important component of the tropical Atlantic SST variability, which is characterized by warm (or cold) SST anomalies in the TNA [72]. Sea surface temperatures in the tropical North Atlantic (TNA) affect the meridional movement of the

ITCZ and its band of heavy rainfall and cloud cover [73]. The Atlantic Niño (AN) is often regarded as something like the little brother of El Niño. During Pacific El Niño events, sea-surface temperatures (SSTs) in the central and eastern equatorial Pacific become warmer than average. Prevailing theories on the equatorial Atlantic Niño are based on the dynamical interaction between atmosphere and ocean [74]. In very much the same manner, SSTs in the central and eastern equatorial Atlantic become warmer than average (or anomalously warm) during Atlantic Niño events. The Atlantic Niño index used in this study is obtained by calculating the area average of SST in the cold tongue region, defined as 20° W to 0 and 3° S to 3° N [26]. While El Niño usually peaks in northern hemisphere winter, the Atlantic Niño peaks in summer [75]. Therefore, understanding of the Atlantic Niño (or lack thereof) has important implications for climate prediction in those regions. Although the Atlantic Niño is an intrinsic mode to the equatorial Atlantic [27].

There may be a tenuous causal relationship between climate parameters and the Atlantic Niño in some circumstances. Therefore, MLR and Wavelet analysis are used to identify the dependency and coherence between temperature, rainfall and climate forcings.

The correlations between the four indices used in our study are shown in Figure 8. There is strong correlation coefficient between AMM and TNA (0.76). As the two explanatory variables are strongly correlated, the MLR analysis may have difficulties to separate the contributions. For that purpose, the wavelet analysis was used by calculating the coherence between explanatory variables and the dependent variable separately.

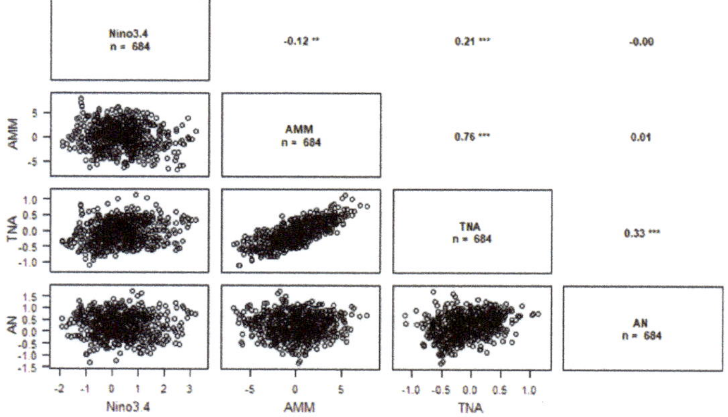

Figure 8. Correlation between the standardized monthly Niño3.4, AMM, TNA and AN time series for the period from 1960 to 2016, "n" is data number. The values are the correlation coefficient between the 4 parameters and we found that TNA and AMM are significantly correlated. The "***"; "**"; "*"; show that the correlation is significant to the 0.001; 0.01; 0.1 level and " " mean that there is no correlation.

For the MLR the explanatory variables (AMM + Niño3.4+ TNA + AN) were used. The output of the MLR statistical analysis of temperature and the independent variables is shown in Table 1. Statistically the results in Table 1 reveal a significant relationship between temperature and Niño3.4 AMM, and TNA, with p-values of 0.0138, 1.99×10^{7}, and less than 2×10^{16}, respectively. The p-value for AN indicates a statistically insignificant association with the temperature because of p-value which is far greater than 0.05.

Table 1. The output of Multiple Linear Regression (MLR) model in which temperature is a dependent variable and AMM, Niño3.4, TNA and AN are independent variables.

Variables	Estimate	Std. Error	t-Value	p-Value	Significance
Niño 3.4	−0.11653	0.04722	−2.468	0.0138	*
AMM	−0.14055	0.02675	−5.254	1.99×10^7	***
TNA	1.79296	0.20497	8.747	2×10^{16}	***
AN	0.14372	0.08790	1.635	0.1025	

Significant codes: 0 '***' 0.001 '**' 0.01 '*' 0.05 '.' 0.1 ' ' 1.

A comprehensive summary of the MLR analysis statistics encompassing rainfall, Niño3.4, AMM, TNA and AN is shown in Table 2. The results in Table 2 reveal a statistically significant relationship between rainfall and Niño3.4, AMM, and TNA with p-values of 0.04374, 0.00441, 0.00301. The p-value for AN indicates a statistically insignificant association with the rainfall because of p-value (0.73691) which is far greater than 0.05. A strong dependence between the two meteorological parameters, AMM and TNA were found. And then, Niño3.4 has a moderate influence on the temperature and rainfall of Conakry. The low dependency between the AN and these two meteorological parameters would be due to the distance between the Conakry site and the NA focus ($3° S–3° N$), and this could be verified for another station closer to the equator. To analyse the two-component dependence of which one in the temporal environment and the other in the frequency environment between the temperature, the rainfall and the forcings used in the study, unlike the MLR, the wavelet model has been evaluated and model outputs are explained in the next section.

Table 2. The output of Multiple Linear Regression (MLR) model in which rainfall is a dependent variable and AMM, Niño3.4, TNA and AN are independent variables.

Variables	Estimate	Std. Error	t-Value	p-Value	Significance
Niño 3.4	43.55	21.55	2.020	0.04374	*
AMM	34.89	12.21	2.857	0.00441	**
TNA	−278.57	93.56	−2.977	0.00301	**
AN	−13.49	40.12	−0.336	0.73691	

Significant codes: 0 '***' 0.001 '**' 0.01 '*' 0.05 '.' 0.1 ' ' 1.

An influential variable for most African rainfall areas is the zonal wind over the tropical Atlantic, the north-south SST gradient in the tropical Atlantic modulates rainfall in West Africa as expected [76].

Figure 9a depicts the time evolution of mean monthly temperatures, with the warming trend line superimposed. From Theil-Sen function, in this study, 684 points were used for trend estimation. The trend estimate is: $p < 0.001 = ***, p < 0.01 = **, p < 0.05 = *$ and $p < 0.1 = +$. The temperature increases at 0.02 °C per year (0.2 °C/decade) at Conakry. The superimposed red line indicates the obtained linear trend. And the dashed red lines indicate the 95% confidence interval. The annual evolution of rainfall exhibits a negative slope, which corresponds to decreasing trend (Figure 9b) at −8.14 mm per year (−81,4 mm/decade). Compared to other sites in West Africa, our results are similar to that found by [32], for stations in downstream Kaduna River Basin during 1975–2014, in Nigeria. The fifth Intergovernmental Panel on Climate Change assessment stated Africa surface temperature already increased by 0.5 °C–2 °C over the past hundred years and an observed drop in average annual rainfall of approximate 25–50 mm each decade from 1951–2010 in some parts of West Africa [77]. Globally, according to the IPCC Special Report [78], it has been reported that the warming of anthropogenic origin has already exceeded the environment.

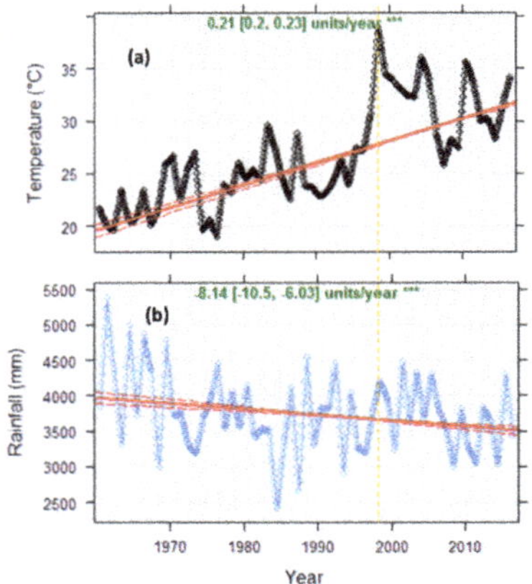

Figure 9. Shows the long-term trend of monthly temperature (**a**) and rainfall (**b**). The solid red line shows the trend estimate and the dashed red lines show the 95% confidence intervals for the trend based on resampling methods. The overall trend is shown at the top-left as 0.21 °C per year (**a**) and −8.14 mm per year (**b**), and the 95% confidence intervals in the slope from 0.2–0.23 °C/year (a) and −10.5–6.03 mm/year (b). On the figures, the sign "***" shows that the trend is significant to the 0.001 level.

The seasonal distribution of temperature is shown in Figure 10a. the increase in temperature is more significant in winter (December–January–February) of 0.03 °C per year than spring (March–April–May), summer (June–July–August) and autumn (September–October–November) of 0.02 °C per year. The Figure 10b depicts the seasonal distribution of rainfall; no significant trend is observed in the winter months and rainfall values seem to be stables. Negative linear trend was found in spring (−0.34 mm/year), autumn (−1.23 mm/year) and for summer, the trend is positive (0.1 mm/year).

3.3. Wavelet Analysis

The Figure 11 shows the normalised wavelet power spectrums calculated for the time series of temperature (a) and rainfall (b) for the period from year 1960 to 2016. In this figure, the "u" shaped solid lines represent the cone of influence (COI) which define the region of the spectrum which should be considered in the analyses. The COI actually indicates areas where edge effects occurs in the time series [44]. The thick black contours are the 95% significant regions of confidence level [42].

The main purpose of using the wavelet transform technique in our study is to identify any dominant variability mode that may be present within the two meteorological parameters (temperature and rainfall). In general, the wavelet transform of temperature shows two distinctive peaks (Figure 11a) corresponding at 6 and 12-month periods. The 12-month period shows strong power spectrums during the distinctive periods of 1961–1965, 1970, 1984 and 1992–2016. Its intensity increases with no-interruption from 1992 to 2016. These dominant wavelet peaks seem to be consistent with the results presented in Figure 3c. The wavelet power spectra for rainfall indicate a strong power spectral peak of 12-month cycle which starts from year 1960 to 2016 (Figure 11b). Moreover, there is a weak power that seems to appear at 6-month period for a few years and within the 95% significant regions of confidence level for distinctive years 1961, 1965, 1967, 1970, 1986–1992, 1998–2000 and 2002–2007.

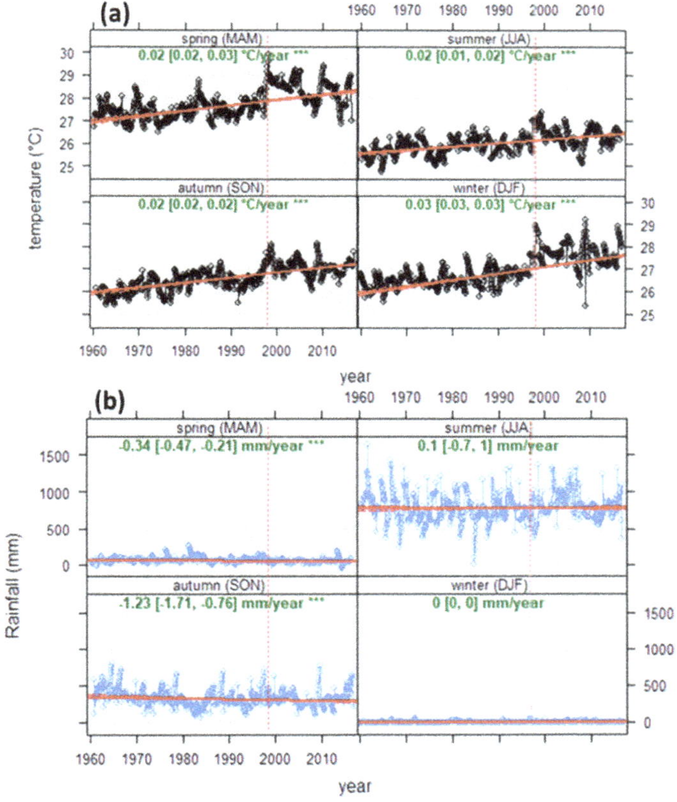

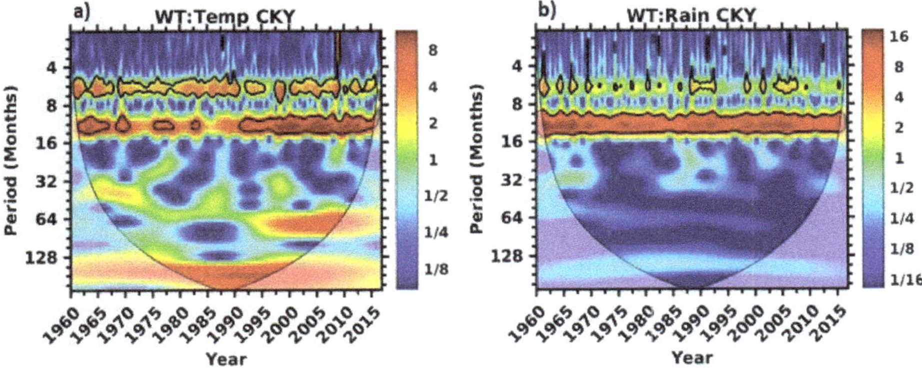

Figure 10. The plot shows the Seasonal trend distribution of monthly mean temperature (**a**) and rainfall (**b**) obtained with Theil-Sen's estimate. The solid red line shows the trend estimate and the dashed red lines show the 95% confidence intervals for the trend based on resampling methods. The vertical red line indicate year 1998. There is no trend in rainfall during winter (0 mm per year).

Figure 11. Wavelet transform of temperature (**a**) and rainfall (**b**) variability from 1960 to 2016 at Conakry. The black solid line contour delimits the region (red) where the power is strong and significant and the cone of influence indicates the 95% confidence level.

It is important to note that the distinctive power periodicities found in temperature may be associated with the annual and semi-annual cycles, which are controlled by the alternation between dry and wet seasons. The strong and continuous power spectrums shown by rainfall confirms the annual cycle variability of rainfall year-round. Sylla et al. [7] pointed out that depending on a given year, the onset of WAM may be strong or quiet in the second half of June and the West African rainfall is highly variable on intra-seasonal, interannual, and interdecadal time scales. The wavelet coherence analysis between temperature, rainfall and the four climate indices used in this study is shown below.

Wavelet coherence is a method for analysing the coherence and phase lag between two time series as both a function of time and frequency [43]. This method has been shown to be the best possible method to indicate teleconnection between two independent time series. Thus, this section focuses on investigating the teleconnection between both Conakry temperature and rainfall, and selected climate indexes. Figure 12 shows the cross-wavelet power spectra for (a) Temp–Niño, (b) Temp-AMM, (c) Temp–NA, and (d) Temp–AN, respectively. The phase relationship is represented by arrows. The regions where two cross-wavelet parameters are in phase is shown by arrows point to the right, anti-phase if the arrows point to the left, and temperature or rainfall leading (or lagging) if the arrows point upwards (or downwards), respectively. The vectors were only plotted for areas where the squared coherence is greater or equal to 0.5. More details about wavelet coherence calculations can be find in studies such as Grinsted et al. [41] and Schulte et al. [4]. The solid black line indicates the cone of influence (COI) where the edge effects become significant at different frequencies (scales), and the solid black line delimit the 95% significant regions of confidence level.

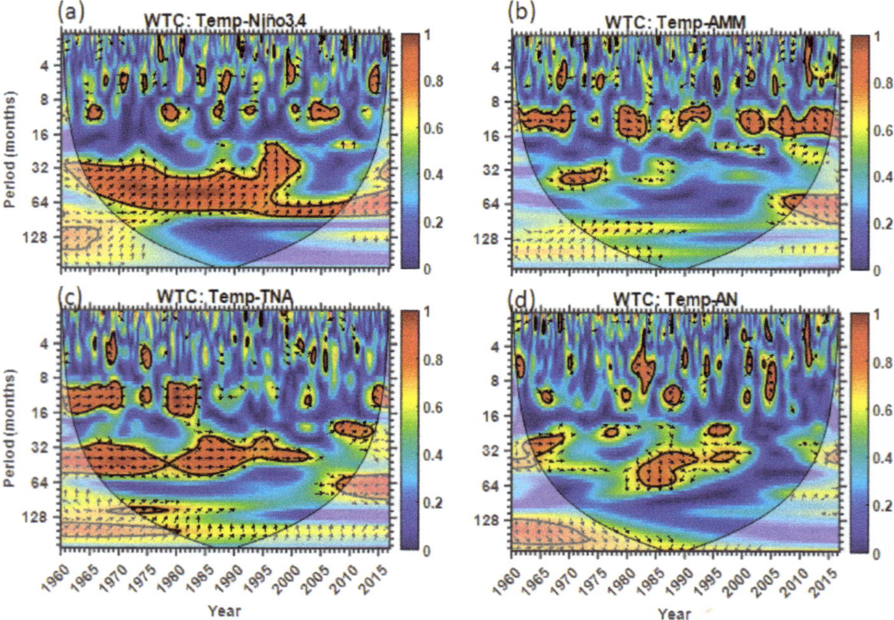

Figure 12. Wavelet coherence between temperature and Niño 3.4 (**a**), AMM (**b**), TNA (**c**) and AN (**d**) 1960–2016, the phase relationship is represented by arrows. The black solid line contour delimits the region (red) where the power is strong and significant and the cone of influence (solid black line) indicates the 95% confidence level.

Having found that the wavelet transform shows strong forcings with 6 and 12–months periods in temperature variability, we have proceeded to identify the wavelet coherence signature between temperature and the four climate modes (Niño3.4, AMM, TNA and AN). Figure 12a shows the coherence calculated for the monthly mean temperature and Niño 3.4 in both time and frequency domain. At the period band of 32–64 months, significant relationship between the temperature and Niño 3.4 is clearly visible from 1960 to 2016. According to the arrows (phase) which are pointing upward and then turning to the right at the period band 32–64, the temperature seems to lead the Niño 3.4. Also, there seems to be an in-phase relationship which may indicate a strong teleconnection of the Conakry temperature to Niño variability. It is also important to note that there is a distinctive appearance of periods which are less than 13 months in the time series, with varying phase relationships between the parameters.

The wavelet coherence between temperature and AMM (Figure 12b) is observed to delineate some areas that have high significant power at periods between 4–12 months with significant peaks of distinctive periods 1960–1970, 1984–1990 and 2005–2016. A study by Foltz et al. [79] also reported that there was cooling of SSTs in the equatorial North Atlantic (ENA; 2°–12° N) in 2009 in response to a strong Atlantic meridional mode event. It is also important to mention that there are significant peaks appearing in the period band of 32–48 months during 1969–1976 and 2005–2010, respectively. A key component of the AMM is a positive feedback between the ocean and the atmosphere. Surface air pressure responds to the SST anomalies, becoming higher than normal over the anomalously cold SSTs and lower than normal over anomalously warm SSTs.

In Figure 12c, there is a significant in-phase relationship between the temperature and TNA during the period band of 8–12 months with strong power during 1962–1970, 1975 and 1973–1984. In addition, it is noted that a significant in-phase relationship is also found during 1965–2004 and 2006 at the period band of 30–64 months. The Figure 12d shows an in-phase coherence between temperature and AN corresponding at the period band of 8–12 months during 1965–1970 and 1972–1984. At the period band of 32–64 months, the AN lead the temperature, so the significant power appears at the period from 1963 to 1968, 1982 to 1992 and 1994. From these results, it can be suggested that the four climate indices contribute to drier conditions across the Conakry region. Then, the wavelet coherence spectra show that the Niño 3.4, AMM, TNA and AMM are coherent with temperature at different time scales. We can summarise that temperature is subjected to climate indices forcing at Conakry. Anomalous surface wind flow from the cold to the warm hemisphere, strengthening the mean south-easterly trade winds in the South Atlantic and weakening the north-easterly trade winds in the North Atlantic. The surface wind anomalies thus provide a positive feedback onto the initial SST anomalies by forcing changes in wind-induced evaporative cooling of the ocean.

The wavelet coherences between four climate modes and rainfall are shown in Figure 13. Significant in-phase coherence was found with the Niño 3.4 at a band period of 8–12 months (Figure 13a). This coherence suggests that the negative phase of Niño 3.4 is in agreement with dry years and its positive phase with wet years at Conakry site. The secondary peak of significant coherence appears at the band period of 16–32 months from 1962 to 1967. The observed significant coherence at a period of 16–32 months seem to be partially linked to the response of wet condition of 1960s. An in-phase relationship between rainfall and Niño 3.4 is found too at the band period of 128–256 months from 1970 to 2005. Since ENSO events can have substantial influence on African rainfall [80], the equatorial region exhibits more rainfall during El-Niño years than La-Nina years [81]. In the Indian Ocean basin, Narasimha and Bhattacharyya [82] suggested that the stronger coherence between Homogeneous Indian Monsoon and Niño 3.4 is found in the 2–7-year band and that both rainfall and the Niño 3.4 index appear irregular and random. During 1950–99, there were seven most significant El Niño events (1957–58, 1965–66, 1972–73, 1982–83, 1986–87, 1991–92, and 1997–98) for which the SST anomalies in the Niño 3 region (5° S–5° N, 150° W–90° W) exceeded 1 °C [83].

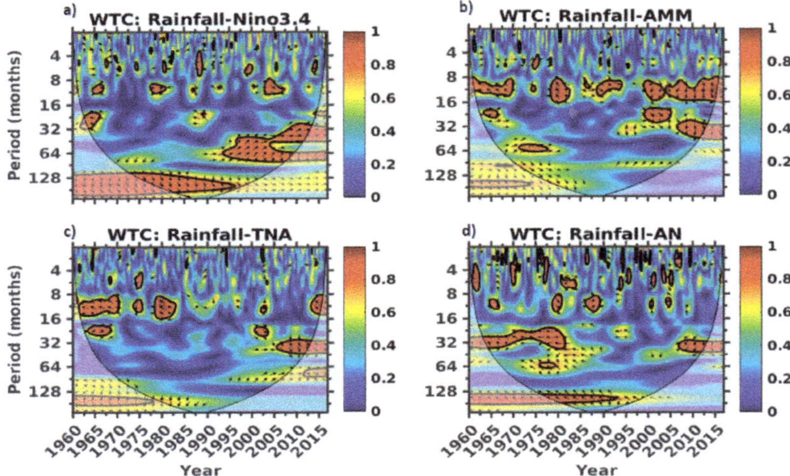

Figure 13. Coherence between rainfall and Niño3.4 (**a**), rainfall and AMM (**b**), rainfall and TNA (**c**) and rainfall and AN (**d**) 1960–2016, the phase relationship is represented by arrows. The black solid line contour delimits the region (red) where the power is strong and significant and the cone of influence (solid black line) indicates the 95% confidence level.

The Figure 13b depicts the wavelet coherence between rainfall and AMM and indicates that the AMM response to the rainfall variability shows an in-phase relationship at the band period of 8–12 months during the periods 1960–1970, 1978–1984, 1985–1990 and 1995–2016. The wavelet coherence analysis detected at the band period of 16–32 months lagged (i.e., AMM leading) relationship with the wet conditions during 1965, 1998–2004 and 2005–2016. Another peak is shown around 64 months from 1970 to 1984, which seems to be in relationship with the 1970s and 1980s droughts. The AMM is the dominant source of coupled ocean-atmosphere variability in the Atlantic and it affects rainfall in tropical cyclone development in the North Atlantic. During a positive phase of the AMM, the ITCZ is displaced northward. Warmer than normal SSTs and weaker than normal vertical wind shear during positive phases of the AMM tend to enhance tropical cyclone development in the Atlantic. The conditions are opposite for the negative phase of the AMM. The AMM exhibits strong variability on interannual to decadal timescales.

Figure 13c shows the coherence analysis between rainfall and TNA. A significant coherence and out-of-phase between rainfall and TNA is found at the band period of 8–12 months during 1961–1970 and 1980. A second significant coherence of in-phase relationship is shown at the band period of 16–32 months from 1965 to 1970 and 2005 to 2012. Comparing to the results of the case study for the northern part of Brazil, Uvo et al. [84] reported that the variations of April–May averaged precipitation are closely connected to the changes in the TNA SST. And sea surface temperatures in the tropical North Atlantic affect the meridional movement of the ITCZ and its band of heavy rainfall and cloud cover [73]. Furthermore, the results found by Sun et al. [85] clearly demonstrate that the climate indices have the influential consistent correlation relationship with the precipitation variation in Korea.

The wavelet coherence between AN and Conakry rainfall data was also computed (Figure 13d). The wavelet analysis detected a statistically significant coherence and in-phase relationship at the band period of 4–12 months during the years 1965, 1970, 1980, 1986 and 2005. Significant out-of-phase coherence was found at a band period of 32–48 months during the periods 1965–1984 and 2005–2013, suggesting that the positive phase of the AN contributes to drier and cooler conditions in Conakry. A period of significant coherence between the AN and rainfall extending from 1976 to 1994 was also identified at the band period of 128–190 months (~11-year), which may be due to solar cycle. Using wavelet techniques to examine the association between Indian monsoon and solar activity,

Bhattacharyya and Narasimha [86] found the power in the 8–16 y band during the period of higher solar activity at confidence levels exceeding 99.88%. The teleconnection between AN and both the temperature and rainfall measured at Conakry seems to be in agreement with previous studies (e.g., Hastenrath and Polzin. [87]; Rodriguez-Fonseca et al. [88] and others). In their study on the role of the SST anomalies in the West Africa droughts, Rodriguez-Fonseca et al., 2015 reported that the tropical Atlantic SST variability influence the West Africa rainfall in different time scales: the variability in areas closer to the equator and those at the south.

To compare our result to those of other areas, a study by Mbata et al. [89] reported for the sector of Democratic Republic of Congo that the wavelet analysis of the rainfall time series indicates an important fluctuation between practically 1960 and 1970. And then, Giannini et al. [90] suggested that the atmospheric convection and circulation changes due to the Atlantic Niño can cause increased precipitation across the equatorial Atlantic and decreases over the Sahel. These climate modes appear to have contributed substantially to the 1970s and 1980s drought in a different way and scales. The widespread influence of El Niño Southern Oscillation (ENSO) events on regional climate can have considerable socio-economic impact Climatic effects of ENSO, which vary substantially with region and season [91].

4. Discussion

Compared with previous studies on the variability of temperature and precipitation in West Africa [55,56], this study gives a more comprehensive investigation on variability of both warm and cool conditions that could be induced by temperature and of both wet and dry conditions by precipitation at Conakry. It provides more detailed information about their association with large-scale ocean–atmosphere oscillations as well as the trend analysis.

The findings revealed that the interannual evolution of temperature is characterized by a strong increase over the study period. As for the annual change, a semi-annual cycle and an annual cycle are found. The rainfall trend exhibits a slight decrease. Contrarily to Conakry site, Bose et al. [92] have found a significant positive increase in rainfall in the entire northern Nigeria within the period of 1970 to 2012.

Trends along the Guinea Coast are weak and non-significant except for extreme rainfall related indices, this missing significance is partly related to the hiatus in rainfall increase in the 1990s, but also to the larger interannual rainfall variability [93]. Temperature anomalies with an upward trend, remain positive since 1992 for all subsequent years, which corresponds to global continuous warming. The precipitation anomalies show a downward trend and the analysis has clearly shown the 1970s and 1980s drought periods', which caused significant material damage [94,95] and enormous loss of human life. Drought over all of West Africa is associated with the growth of positive SST anomalies in the eastern Pacific and in the Indian Ocean, and negative SST anomalies in the northern Atlantic and in the Gulf of Guinea [96]. The decrease of precipitation found by our study is in agreement with study by Aguilar et al. [97], who clearly specified that the measures of overall total precipitation are decreasing in Guinea. For the region that extend from 20° W–10° E and from 11° N–18° N, Panthou et al. [98] revealed the higher frequency of heavy rainfall and the return to wetter annual rainfall conditions since the beginning of the 2000s—succeeding the 1970–2000 drought. Furthermore, from our results, the 1970s and 1980s drought periods' have been exhibited, which were confirmed in Niger River Basin, by Djigbo F Badou et al. [55] who highlighted that the wetness of the decades, 1990s and 2000s and the manifold floods records of the first half of 2010s over West Africa are evidences that the droughts of 1970s and 1980s have stopped. An overview of the mechanisms that have been proposed to explain the influence of the AN with other climate modes within and outside the tropical Atlantic is given by Lübeckke et al. [99]. Most of part of the mechanism involve fluctuations in the wind field over the equatorial Atlantic. Part of these wind stress anomalies are excited by SST changes in the equatorial Atlantic itself [100]. They can also be due to a response to ENSO or variations in the South Atlantic subtropical high. Nnamchi et al., [101] suggested that thermodynamic feedbacks excited by stochastic atmospheric perturbations (driving surface heat fluxes), can explain a large part of the SST variability in the eastern equatorial Atlantic. The impact of AN on rainfall over Gulf of Guinea is direct because the warm SST reduce low level wind flow inland, leading to positive precipitation anomalies over the Gulf of Guinea and adjacent coastal region [102]. After confirming the drought of 1970s and 1980s, Masih et al. [56] reported that African continent is likely to face extreme and widespread droughts in future and this evident challenge is likely to aggravate due to slow progress in drought risk management, increased population and demand for water and degradation of land and environment.

To compare our result to global temperature, from 1979 to 2016 using ERA surface air temperature, Simmons et al. [59] clarified that, early in 2016 the global temperature appears to have first touched or briefly breached a level 1.5 °C above that during the industrial area, having touched the 1.0 °C level in 1998 during a previous El Niño. Thermodynamic feedbacks constitute the main source of Atlantic Niño variability [74]. Precipitation exhibits a clear and distinct pattern during different phases of ENSO. Dynamical parameters, specific humidity and horizontal wind also exhibit clear differences for both ENSO phases [81].

The upward trend in temperature and the downward trend in rainfall was verified by the Mann-Kendall test. To understand the influence of climatic forcings on both meteorological parameters, the linear regression model was evaluated and it has been found that TNA and AMM have a more significant dependence than other indices. For extremes analysis, Aguilar et al. [97] used long term daily temperature and precipitation data set of Guinea and other countries in Africa. For Conakry station, they found inhomogeneous data before 1950 and after around 1995. Then they used RClimDex to processes them in order to get homogeneous data. But after checking the archives of database of Conakry, we found that the exceptional shift of temperature in 1998 is not linked to instrument replacement nor any error of digitalization. The wavelet analysis of both signals showed the semi-annual and annual cycles in temperature and the annual cycle in rainfall. A study conducted by Adejuwon et al. [102], for 16 stations in west Africa (Nigeria) highlighted that for all the series analyzed, there is the general tendency towards increasing aridity and spectral analysis indicates prominent periods of between 2-and 8-year cycles.

Several previous studies have shown that there is existence of significant simultaneous covariability of ENSO with West African rainfall [80,103,104]. One of the possible teleconnection mechanisms that could explain the ENSO influence on West Africa rainfall is the eastward shift of the Walker circulation and subsequent decent over the Afro-Asia during the El Niño events [104]. The process of this Walker-type circulation is associated with reduced rainfall in the West Africa during the El Niño events. Also, El Niño events also increase transport of heat flux from ocean to the atmosphere which results in tropical warming.

The observed strong relationship between AMM and both temperature (Table 1) and rainfall (Table 2) is in agreement with a study by [71]. In their study, Doi et al., [71] found a significant link between the AMM and the interannual modulation in the seasonal variation of the Guinea Dome region. This study showed that during the preconditioning phase of the positive (negative) ANN, the Guinea Dome is anomalously weak (strong) and the mixed layer is anomalously deep (shallow), there is a late fall. This means that the AMM has a strong influence in both rainfall and temperature of the Conakry, Guinea.

The variability of the SST in the tropical North Atlantic region which can produce stronger or weaker trade winds is expected to have an influent in the rainfall and temperature of Conakry. Therefore, the strong correlation that is observed between TNA and rainfall is understandable because weaker trade winds are expected to bring moisture in the Guinea coast, while stronger trade winds are expected to bring dryer conditions.

5. Conclusions

The semi-annual and annual cycles of temperature and the annual cycle of rainfall could be generated indeed by the ITCZ oscillation and modulated by the WAM. The models used in this study highlighted the variability of temperature and rainfall that are characterized by a significant upward trend in temperature and a low downward trend in rainfall. The IPCC Fourth Assessment Report review of climate model projections of temperature shows a consistent warming in all subregions, but less consistent patterns for rainfall [105]. Temperature and rainfall variability at Conakry site were analysed and linked to dominant modes of climate variability at annual to multiannual timescales.

There is a strong teleconnection between the SST of both pacific and Atlantic in the variability of rainfall and temperature. So, the annual variability of temperature and rainfall at Conakry are largely influenced by climate forcings AMM, TNA and Niño3.4. However, there is no significant influence of AN on these meteorological parameters. Furthermore, the warming of 1998 seems to be a response of the 1997–1998 strong El Niño event. Physically, the influence of climate modes on temperature and rainfall were found to vary at different time scales. In Guinea, more important mining projects are under way. And, in this regard, it would also be important to take into account the anthropogenic impacts in order to deepen our knowledge. This would allow us to better understand their variability which is useful in planning sustainable development projects. It is hoped that the results from this study would help to better understand climate variability in order to get sufficient operational decision support, and resource management for a sustainable development of developing countries.

Author Contributions: Conceptualization, R.T.L.; Data curation, R.T.L.; Formal analysis, N.M. and N.B.; Resources, H.B.; Software, N.M.; Supervision, H.B., N.M., A.H., Z.B. and V.S.; Validation, A.H.; Writing—Original draft, R.T.L.; Writing—Review & editing, R.T.L., H.B., N.M. and V.S.

Funding: This research received no external funding.

Acknowledgments: This work is undertaken in the framework of the French South-African International Research Group LIA-ARSAIO (International Associated Laboratories—Atmospheric Research in Southern Africa and Indian Ocean) supported by the NRF and CNRS and by the Protea program. Authors are grateful to the Guinean Meteorological Service (GMS) for providing the temperature and rainfall data series used in this study, and to NOAA and KNMI web-teams for providing climate indices. Authors are thankful to Paulene Govender for proofreading. We are also thankful to the anonymous reviewers for their insightful comments.

Conflicts of Interest: The authors declare no conflict of interest.

References

1. Serra, C.; Burgueño, A.; Lana, X. Analysis of maximum and minimum daily temperatures recorded at Fabra Observatory (Barcelona, NE Spain) in the period 1917–1998. *Int. J. Climatol.* **2001**, *21*, 617–636. [CrossRef]

2. Conway, D. Adapting climate research for development in Africa. *Wiley Interdiscip. Rev. Clim. Chang.* **2011**, *2*, 428–450. [CrossRef]

3. Zerbo, J.-L.; Ouattara, F.; Nanéma, E. Solar Activity and Meteorological Fluctuations in West Africa: Temperatures and Pluviometry in Burkina Faso, 1970–2012. *Int. J. Astron. Astrophys.* **2013**, *3*, 408–411. [CrossRef]

4. Schulte, J.A.; Najjar, R.G.; Li, M. The influence of climate modes on streamflow in the Mid-Atlantic region of the United States. *J. Hydrol. Reg. Stud.* **2016**, *5*, 80–99. [CrossRef]

5. Gbobaniyi, E.; Sarr, A.; Sylla, M.B.; Diallo, I.; Lennard, C.; Dosio, A.; Dhiédiou, A.; Kamga, A.; Klutse, N.A.B.; Hewitson, B.; et al. Climatology, annual cycle and interannual variability of precipitation and temperature in CORDEX simulations over West Africa. *Int. J. Climatol.* **2014**, *34*, 2241–2257. [CrossRef]

6. Rodrigues, L.R.L.; García-Serrano, J.; Doblas-Reyes, F. Seasonal forecast quality of the West African monsoon rainfall regimes by multiple forecast systems. *J. Geophys. Res. Atmos.* **2014**, *119*, 7908–7930. [CrossRef]

7. Sylla, M.B.; Dell'Aquila, A.; Ruti, P.M.; Giorgi, F. Simulation of the intraseasonal and the interannual variability of rainfall over West Africa with RegCM3 during the monsoon period. *Int. J. Climatol.* **2010**, *30*, 1865–1883. [CrossRef]

8. Vizy, E.K.; Cook, K.H. Development and application of a mesoscale climate model for the tropics: Influence of sea surface temperature anomalies on the West African monsoon. *J. Geophys. Res. Atmos.* **2002**, *107*. [CrossRef]

9. Loua, R.T.; Beavogui, M.; Bencherif, H.; Barry, A.B.; BAMBA, Z.; Amory-Mazaudier, C. Climatology of Guinea: Study of Climate Variability in N'zerekore. *J. Agric. Sci. Technol. A* **2017**, *7*, 115–233.

10. Paeth, H.; Hense, A. On the linear response of tropical African climate to SST changes deduced from regional climate model simulations. *Theor. Appl. Climatol.* **2006**, *83*, 1–19. [CrossRef]

11. Okumura, Y.; Xie, S.-P. Interaction of the Atlantic equatorial cold tongue and the African monsoon. *J. Clim.* **2004**, *17*, 3589–3602. [CrossRef]

12. Joly, M.; Voldoire, A. Role of the Gulf of Guinea in the inter-annual variability of the West African monsoon: What do we learn from CMIP3 coupled simulations? *Int. J. Climatol.* **2010**, *30*, 1843–1856. [CrossRef]

13. Fitzpatrick, R.G.J.; Bain, C.L.; Knippertz, P.; Marsham, J.H.; Parker, D.J. The West African Monsoon Onset: A Concise Comparison of Definitions. *J. Clim.* **2015**, *28*, 8673–8694. [CrossRef]

14. Sultan, B.; Janicot, S. The West African Monsoon Dynamics. Part II: The "Preonset" and "Onset" of the Summer Monsoon. *J. Clim.* **2003**, *16*, 3407–3427. [CrossRef]

15. Sultan, B.; Janicot, S.; Diedhiou, A. The West African Monsoon Dynamics. Part I: Documentation of Intraseasonal Variability. *J. Clim.* **2003**, *16*, 3389–3406. [CrossRef]

16. Nicholson, S.E. Rainfall and Atmospheric Circulation during Drought Periods and Wetter Years in West Africa. *Mon. Weather Rev.* **1981**, *109*, 2191–2208. [CrossRef]

17. Nicholson, S.E. The West African Sahel: A Review of Recent Studies on the Rainfall Regime and Its Interannual Variability. Available online: https://www.hindawi.com/journals/isrn/2013/453521/ (accessed on 15 December 2017).

18. Citeau, J.; Guillot, B.; Lahuec, J.-P.; ORSTOM (Eds.) *Champs Thermiques de Surface de L'océan Atlantique: Convection et Fronts Thermiques Sur le Continent Africain: Essai de Mise en Relation*; Initiation-Documentations Techniques; Télédetection; ORSTOM: Lannion, France, 1985; ISBN 978-2-7099-0770-5.

19. Okoola, R.E.; Ambenje, P.G. Transition from the Southern to the Northern Hemisphere summer of zones of active convection over the Congo Basin. *Meteorol. Atmos. Phys.* **2003**, *84*, 255–265. [CrossRef]

20. Espinoza Villar, J.C.; Ronchail, J.; Guyot, J.L.; Cochonneau, G.; Naziano, F.; Lavado, W.; De Oliveira, E.; Pombosa, R.; Vauchel, P. Spatio-temporal rainfall variability in the Amazon basin countries (Brazil, Peru, Bolivia, Colombia, and Ecuador). *Int. J. Climatol.* **2009**, *29*, 1574–1594. [CrossRef]

21. Kouassi, A.; Assamoi, P.; Bigot, S.; Diawara, A.; Schayes, G.; Yoroba, F.; Kouassi, B. Étude du climat ouest-africain à l'aide du modèle atmosphérique régional MAR. *Climatologie* **2010**, *7*, 39–55.

22. Carter, R.C.; Parker, A. Climate change, population trends and groundwater in Africa. *Hydrol. Sci. J.* **2009**, *54*, 676–689. [CrossRef]

23. Tu, X.; Du, Y.; Singh, V.P.; Chen, X. Joint distribution of design precipitation and tide and impact of sampling in a coastal area. *Int. J. Climatol.* **2018**, *38*, e290–e302. [CrossRef]

24. Chen, S.-F.; Wu, R. An enhanced influence of sea surface temperature in the tropical northern Atlantic on the following winter ENSO since the early 1980s. *Atmos. Ocean. Sci. Lett.* **2017**, *10*, 175–182. [CrossRef]

25. Wang, C. ENSO, Atlantic Climate Variability, and the Walker and Hadley Circulations. In *The Hadley Circulation: Present, Past and Future*; Diaz, H.F., Bradley, R.S., Eds.; Springer: Dordrecht, The Netherlands, 2004; Volume 21, pp. 173–202. ISBN 978-90-481-6752-4.

26. Rodríguez-Fonseca, B.; Polo, I.; García-Serrano, J.; Losada, T.; Mohino, E.; Mechoso, C.R.; Kucharski, F. Are Atlantic Niños enhancing Pacific ENSO events in recent decades? *Geophys. Res. Lett.* **2009**, *36*. [CrossRef]

27. Zebiak, S.E. Air-Sea Interaction in the Equatorial Atlantic Region. *J. Clim.* **1993**, *6*, 1567–1586. [CrossRef]

28. WMO. *Directives de L'OMM Pour le Calcul Des Normales Climatiques*; OMM, Édition 2017; OMM: Geneva, Switzerland, 2017; ISBN 978-92-63-21203-0.

29. WMO. *Calculation of Monthly and Annual 30-Year Standat Normals*; WMO: Washington, DC, USA, 1989; Volume 10.

30. Mann, H.B. Nonparametric Tests Against Trend. *Econometrica* **1945**, *13*, 245–259. [CrossRef]

31. Yue, S.; Pilon, P.; Cavadias, G. Power of the Mann-Kendall and Spearman's rho tests for detecting monotonic trends in hydrological series. *J. Hydrol.* **2002**, *259*, 254–271. [CrossRef]

32. Okafor, G.C.; Jimoh, O.D.; Larbi, K.I. Detecting Changes in Hydro-Climatic Variables during the Last Four Decades (1975–2014) on Downstream Kaduna River Catchment, Nigeria. *Atmos. Clim. Sci.* **2017**, *7*, 161–175. [CrossRef]

33. Sneyers, R. *On the Statistical Analysis of Series of Observations*; World Meteorological Organization: Geneva, Switzerland, 1990; pp. 143–192.

34. Rahman, M.A.; Yunsheng, L.; Sultana, N. Analysis and prediction of rainfall trends over Bangladesh using Mann-Kendall, Spearman's rho tests and ARIMA model. *Meteorol. Atmos. Phys.* **2017**, *129*, 409–424. [CrossRef]

35. Chen, Y.; Guan, Y.; Shao, G.; Zhang, D. Investigating Trends in Streamflow and Precipitation in Huangfuchuan Basin with Wavelet Analysis and the Mann-Kendall Test. *Water* **2016**, *8*, 77. [CrossRef]

36. Mbatha, N.; Xulu, S. Time Series Analysis of MODIS-Derived NDVI for the Hluhluwe-Imfolozi Park, South Africa: Impact of Recent Intense Drought. *Climate* **2018**, *6*, 95. [CrossRef]

37. Khan, A.; Chatterjee, S.; Bisai, D. On the long-term variability of temperature trends and changes in surface air temperature in Kolkata Weather Observatory, West Bengal, India. *Meteorol. Hydrol. Water Manag.* **2015**, *3*, 9–16. [CrossRef]

38. Dang, X.; Peng, H.; Wang, X.; Zhang, H. Theil-Sen Estimators in a Multiple Linear Regression Model. Olemiss Edu, 2008. Available online: http://home.olemiss.edu/~{}xdang/papers/MTSE.pdf (accessed on 1 January 2019).

39. Theil, H. A rank-invariant method of linear and polynomial regression analysis, 3; confidence regions for the parameters of polynomial regression equations. *Inproceding.* **1950**, *53*, 386–392.

40. Sen, P.K. Estimates of the Regression Coefficient Based on Kendall's Tau. *J. Am. Stat. Assoc.* **1968**, *63*, 1379–1389. [CrossRef]

41. Grinsted, A.; Moore, J.C.; Jevrejeva, S. Application of the cross wavelet transform and wavelet coherence to geophysical time series. *Nonlinear Process. Geophys.* **2004**, *11*, 561–566. [CrossRef]

42. Torrence, C.; Compo, G.P. A practical guide to wavelet analysis. *Bull. Am. Meteorol. Soc.* **1998**, *79*, 61–78. [CrossRef]

43. Chang, C.; Glover, G.H. Time-frequency dynamics of resting-state brain connectivity measured with fMRI. *NeuroImage* **2010**, *50*, 81–98. [CrossRef]

44. Bilbao, J.; Román, R.; Yousif, C.; Mateos, D.; de Miguel, A. Total ozone column, water vapour and aerosol effects on erythemal and global solar irradiance in Marsaxlokk, Malta. *Atmos. Environ.* **2014**, *99*, 508–518. [CrossRef]

45. FAO Update on FAO's Activities in Relation to the 1997/98 El Niño and La Niña: October 1998. Available online: http://www.fao.org/english/newsroom/global/elnin3-e.htm (accessed on 1 January 2019).

46. Nicholson, S.E.; Some, B.; Kone, B. An Analysis of Recent Rainfall Conditions in West Africa, Including the Rainy Seasons of the 1997 El Niño and the 1998 La Niña Years. *J. Clim.* **2000**, *13*, 2628–2640. [CrossRef]

47. Angell, J.K. Comparison of surface and tropospheric temperature trends estimated from a 63-Station Radiosonde Network, 1958–1998. *Geophys. Res. Lett.* **1999**, *26*, 2761–2764. [CrossRef]

48. Maidment, R.I.; Allan, R.P.; Black, E. Recent observed and simulated changes in precipitation over Africa. *Geophys. Res. Lett.* **2015**, *42*, 8155–8164. [CrossRef]

49. Khodri, M.; Izumo, T.; Vialard, J.; Janicot, S.; Cassou, C.; Lengaigne, M.; Mignot, J.; Gastineau, G.; Guilyardi, E.; Lebas, N.; et al. Tropical explosive volcanic eruptions can trigger El Niño by cooling tropical Africa. *Nat. Commun.* **2017**, *8*, 778. [CrossRef]

50. Liu, F.; Chai, J.; Wang, B.; Liu, J.; Zhang, X.; Wang, Z. Global monsoon precipitation responses to large volcanic eruptions. *Sci. Rep.* **2016**, *6*, 24331. [CrossRef]

51. Wang, S.; Gong, D. Enhancement of the warming trend in China. *Geophys. Res. Lett.* **2000**, *27*, 2581–2584. [CrossRef]

52. Foster, G.; Rahmstorf, S. Global temperature evolution 1979–2010. *Environ. Res. Lett.* **2011**, *6*, 044022. [CrossRef]

53. Trenberth, K.E.; Caron, J.M.; Stepaniak, D.P.; Worley, S. Evolution of El Niño-Southern Oscillation and global atmospheric surface temperatures. *J. Geophys. Res. Atmos.* **2002**, *107*. [CrossRef]

54. Trenberth, K.E.; Fasullo, J.T. An apparent hiatus in global warming? *Earth's Future* **2013**, *1*, 19–32. [CrossRef]

55. Djigbo, F.B.; Afouda, A.; Diekkrüger, B.; Kapangaziwiri, E. Investigation on the 1970s and 1980s Droughts in Four Tributaries of the Niger River Basin (West Africa). In Proceedings, The Hague. 2015. Available online: http://rgdoi.net/10.13140/RG.2.1.1559.0485 (accessed on 1 January 2019).

56. Masih, I.; Maskey, S.; Mussá, F.E.F.; Trambauer, P. A review of droughts on the African continent: A geospatial and long-term perspective. *Hydrol. Earth Syst. Sci.* **2014**, *18*, 3635–3649. [CrossRef]

57. Sylla, M.B.; Diallo, I.; Pal, J.S. West African Monsoon in State-of-the-Science Regional Climate Models, In Tech. 2013. Available online: https://www.intechopen.com/books/climate-variability-regional-and-thematic-patterns/west-african-monsoon-in-state-of-the-science-regional-climate-models (accessed on 1 January 2019).

58. El Niño en 97-98 El Niño en 97-98|Espace des Sciences. Available online: https://www.espace-sciences.org/multimedia/blogs/57931/billet/el-nino-en-97-98 (accessed on 19 February 2019).

59. Simmons, A.J.; Berrisford, P.; Dee, D.P.; Hersbach, H.; Hirahara, S.; Thépaut, J.-N. A reassessment of temperature variations and trends from global reanalyses and monthly surface climatological datasets. *Q. J. R. Meteorol. Soc.* **2016**, *143*, 101–119. [CrossRef]

60. Peel, M.C.; Pegram, G.G.S.; Mcmahon, T.A. Global analysis of runs of annual precipitation and runoff equal to or below the median: Run length. *Int. J. Climatol.* **2004**, *24*, 807–822. [CrossRef]

61. Nicholson, S.E. The nature of rainfall variability over Africa on time scales of decades to millenia. *Glob. Planet. Chang.* **2000**, *26*, 137–158. [CrossRef]

62. Bonnecase, V. Retour sur la famine au Sahel du début des années 1970: La construction d'un savoir de crise, Building knowledge on a crisis. Famine in Sahelian Africa in the early 1970s. *Politique Afr.* **2010**, *3*, 23–42. [CrossRef]

63. Ouedraogo, M.; Paturel, J.-E.; Mahe, G. *Conséquences de la Sécheresse Observée Depuis le Début des Années 1970 en Afrique de L'Ouest et Centrale: Normes Météorologiques et Hydrologiques*; IAHS Press: Cape Town, South Africa, 2002; Volume 274, pp. 149–155.

64. Suhaila, J.; Yusop, Z. Trend analysis and change point detection of annual and seasonal temperature series in Peninsular Malaysia. *Meteorol. Atmos. Phys.* **2018**, *130*, 565–581. [CrossRef]

65. Chiang, J.C.H.; Vimont, D.J. Analogous Pacific and Atlantic Meridional Modes of Tropical Atmosphere-Ocean Variability. *J. Clim.* **2004**, *17*, 4143–4158. [CrossRef]

66. Ruiz-Barradas, A.; Carton, J.A.; Nigam, S. Role of the Atmosphere in Climate Variability of the Tropical Atlantic. *J. Clim.* **2003**, *16*, 2052–2065. [CrossRef]

67. Servain, J. Simple climatic indices for the tropical Atlantic Ocean and some applications. *J. Geophys. Res.* **1991**, *96*, 15137–15146. [CrossRef]

68. Xie, S.-P.; Carton, J.A. Tropical Atlantic Variability: Patterns, Mechanisms, and Impacts. In *Geophysical Monograph Series*; Wang, C., Xie, S.P., Carton, J.A., Eds.; American Geophysical Union: Washington, DC, USA, 2004; pp. 121–142. ISBN 978-1-118-66594-7.

69. Smirnov, D.; Vimont, D.J. Variability of the Atlantic Meridional Mode during the Atlantic Hurricane Season. *J. Clim.* **2011**, *24*, 1409–1424. [CrossRef]

70. Merle, J.; Fieux, M.; Hisard, P. Annual signal and interannual anomalies of sea surface temperature in the eastern equatorial Atlantic Ocean. In *Oceanography and Surface Layer Meteorology in the B/C Scale*; Elsevier: Amsterdam, The Netherlands, 1980; pp. 77–101. ISBN 978-1-4832-8366-1.

71. Doi, T.; Tozuka, T.; Yamagata, T. The Atlantic Meridional Mode and Its Coupled Variability with the Guinea Dome. *J. Clim.* **2010**, *23*, 455–475. [CrossRef]

72. Chen, S.; Wu, R.; Chen, W. The Changing Relationship between Interannual Variations of the North Atlantic Oscillation and Northern Tropical Atlantic SST. *J. Clim.* **2015**, *28*, 485–504. [CrossRef]

73. Rugg, A.; Foltz, G.R.; Perez, R.C. Role of Mixed Layer Dynamics in Tropical North Atlantic Interannual Sea Surface Temperature Variability. *J. Clim.* **2016**, *29*, 8083–8101. [CrossRef]

74. Nnamchi, H.C.; Li, J.; Kucharski, F.; Kang, I.-S.; Keenlyside, N.S.; Chang, P.; Farneti, R. Thermodynamic controls of the Atlantic Niño. *Nat. Commun.* **2015**, *6*, 8895. [CrossRef]

75. Philander, S.G. *El Niño, La Niña, and the Southern Oscillation*; International Geophysics Series; Academy Press: San Diego, CA, USA, 1990; ISBN 978-0-12-553235-8.

76. Jury, M.R.; Enfield, D.B.; Mélice, J.-L. Tropical monsoons around Africa: Stability of El Niño–Southern Oscillation associations and links with continental climate. *J. Geophys. Res.* **2002**. [CrossRef]

77. Intergovernmental Panel on Climate Change (Ed.) *Climate Change 2013—The Physical Science Basis: Working Group I Contribution to the Fifth Assessment Report of the Intergovernmental Panel on Climate Change*; Cambridge University Press: Cambridge, UK, 2014; ISBN 978-1-107-41532-4.

78. IPCC. Special Report on the impacts of global warming of 1.5 °C above pre-industrial levels and related global greenhouse gas emission pathways, in the context of strengthening the global response to the threat of climate change, sustainable development, and efforts to eradicate poverty. 2018. In Press. Available online: https://www.ipcc.ch/sr15/ (accessed on 1 January 2019).

79. Foltz, G.R.; McPhaden, M.J.; Lumpkin, R. A Strong Atlantic Meridional Mode Event in 2009: The Role of Mixed Layer Dynamics. *J. Clim.* **2012**, *25*, 363–380. [CrossRef]

80. Preethi, B.; Sabin, T.P.; Adedoyin, J.A.; Ashok, K. Impacts of the ENSO Modoki and Other Tropical Indo-Pacific Climate-Drivers on African Rainfall. *Sci. Rep.* **2015**, *5*, 16653. [CrossRef]

81. Ruchith, R.D.; Sivakumar, V. Influence of aerosol-cloud interaction on austral summer precipitation over Southern Africa during ENSO events. *Atmos. Res.* **2018**, *202*, 1–9. [CrossRef]

82. Narasimha, R.; Bhattacharyya, S. A wavelet cross-spectral analysis of solar–ENSO–rainfall connections in the Indian monsoons. *Appl. Comput. Harmon. Anal.* **2010**, *28*, 285–295. [CrossRef]

83. Wang, C. Atmospheric Circulation Cells Associated with the El Niño-Southern Oscillation. *J. Clim.* **2002**, *15*, 399–419. [CrossRef]

84. Uvo, C.B.; Repelli, C.A.; Zebiak, S.E.; Kushnir, Y. The Relationships between Tropical Pacific and Atlantic SST and Northeast Brazil Monthly Precipitation. *J. Clim.* **1998**, *11*, 551–562. [CrossRef]

85. Sun, M.; Li, X.; Kim, G. Correlation Analysis of Climate Indices and Precipitation Using Wavelet Image Processing Approach. In *Lecture Notes in Electrical Engineering, Proceedings of the International Conference on Frontier Computing*; Springer: Singapore, 2017; pp. 231–240.

86. Bhattacharyya, S. Possible association between Indian monsoon rainfall and solar activity. *Geophys. Res. Lett.* **2005**, *32*. [CrossRef]

87. Hastenrath, S.; Polzin, D. Long-term variations of circulation in the tropical Atlantic sector and Sahel rainfall. *Int. J. Climatol.* **2011**, *31*, 649–655. [CrossRef]

88. Rodríguez-Fonseca, B.; Mohino, E.; Mechoso, C.R.; Caminade, C.; Biasutti, M.; Gaetani, M.; Garcia-Serrano, J.; Vizy, E.K.; Cook, K.; Xue, Y.; et al. Variability and Predictability of West African Droughts: A Review on the Role of Sea Surface Temperature Anomalies. *J. Clim.* **2015**, *28*, 4034–4060. [CrossRef]

89. Mbata, A.; Jean Marie, T.; Phuku Phuati, E.; Kyamakya, K.; Keto, F.; Bopili Mbotia, R.; Ndotoni, Z. Continuous wavelet analysis of rainfall fluctuations at interannual and decennial scales on the south-eastern part in the Democratic Republic of the Congo between 1940 and 1997. *Int. J. Innov. Sci. Res.* **2016**, *27*, 2351–8014.

90. Giannini, A.; Saravanan, R.; Chang, P. Oceanic Forcing of Sahel Rainfall on Interannual to Interdecadal Time Scales. *Science* **2003**, *302*, 1027–1030. [CrossRef]

91. Davey, M.K.; Brookshaw, A.; Ineson, S. The probability of the impact of ENSO on precipitation and near-surface temperature. *Clim. Risk Manag.* **2014**, *1*, 5–24. [CrossRef]

92. Bose, M.; Kasim, I.; Mande, H.; Abdullah, A. Rainfall Trend Detection in Northern Nigeria over the Period of 1970–2012. *J. Environ. Earth Sci.* **2015**, *5*, 94–99.

93. Sanogo, S.; Fink, A.H.; Omotosho, J.A.; Ba, A.; Redl, R.; Ermert, V. Spatio-temporal characteristics of the recent rainfall recovery in West Africa. *Int. J. Climatol.* **2015**, *35*, 4589–4605. [CrossRef]

94. Andam-Akorful, S.A.; Ferreira, V.G.; Ndehedehe, C.E.; Quaye-Ballard, J.A. An investigation into the freshwater variability in West Africa during 1979–2010. *Int. J. Climatol.* **2017**, *37*, 333–349. [CrossRef]

95. Dennett, M.D.; Elston, J.; Rodgers, J.A. A reappraisal of rainfall trends in the sahel. *J. Climatol.* **1985**, *5*, 353–361. [CrossRef]

96. Fontaine, B.; Janicot, S. Sea Surface Temperature Fields Associated with West African Rainfall Anomaly Types. *J. Clim.* **1996**, *9*, 2935–2940. [CrossRef]

97. Aguilar, E.; Barry, A.A.; Brunet, M.; Ekang, L.; Fernandes, A.; Massoukina, M.; Mbah, J.; Mhanda, A.; do Nascimento, D.J.; Peterson, T.C.; et al. Changes in temperature and precipitation extremes in western central Africa, Guinea Conakry, and Zimbabwe, 1955–2006. *J. Geophys. Res. Atmos.* **2009**, *114*, 2039–2055. [CrossRef]

98. Panthou, G.; Lebel, T.; Vischel, T.; Quantin, G.; Sane, Y.; Ba, A.; Ndiaye, O.; Diongue-Niang, A.; Diopkane, M. Rainfall intensification in tropical semi-arid regions: The Sahelian case. *Environ. Res. Lett.* **2018**, *13*, 064013. [CrossRef]

99. Lübbecke, J.F.; Rodríguez-Fonseca, B.; Richter, I.; Martín-Rey, M.; Losada, T.; Polo, I.; Keenlyside, N.S. Equatorial Atlantic variability-Modes, mechanisms, and global teleconnections. *Wiley Interdiscip. Rev. Clim. Chang.* **2018**, *9*, e527. [CrossRef]

100. Bjerknes, J. Atmospheric teleconnections from the equatorial pacific. *Mon. Weather Rev.* **1969**, *97*, 163–172. [CrossRef]
101. Nnamchi, H.C.; Li, J.; Kucharski, F.; Kang, I.-S.; Keenlyside, N.S.; Chang, P.; Farneti, R. An Equatorial-Extratropical Dipole Structure of the Atlantic Niño. *J. Clim.* **2016**, *29*, 7295–7311. [CrossRef]
102. Adejuwon, J.O.; Balogun, E.E.; Adejuwon, S.A. On the annual and seasonal patterns of rainfall fluctuations in sub-saharan West Africa. *Int. J. Climatol.* **1990**, *10*, 839–848. [CrossRef]
103. Janicot, S.; Moron, V.; Fontaine, B. Sahel droughts and ENSO dynamics. *Geophys. Res. Lett.* **1996**, *23*, 515–518. [CrossRef]
104. Srivastava, G.; Chakraborty, A.; Nanjundiah, R.S. Multidecadal see-saw of the impact of ENSO on Indian and West African summer monsoon rainfall. *Clim. Dyn.* **2018**, 1–17. [CrossRef]
105. Christensen, J.H.; Hewitson, B.; Busuioc, A.; Chen, A.; Gao, X.; Held, R.; Jones, R.; Kolli, R.K.; Kwon, W.K.; Laprise, R.; et al. Regional Climate Projections. Available online: http://epic.awi.de/17617/ (accessed on 15 January 2019).

Article

Ushering in the New Era of Radiometric Intercomparison of Multispectral Sensors with Precision SNO Analysis

Mike Chu [1,*] and Jennifer Dodd [2]

[1] Cooperative Institute for Research in the Atmosphere, Colorado State University, Fort Collins, CO 80523, USA
[2] NASA Goddard Space Flight Center, Greenbelt, MD 20771, USA; dodd0032@gmail.com
* Correspondence: mchu061@yahoo.com; Tel.: +1-301-683-3369

Received: 15 May 2019; Accepted: 3 June 2019; Published: 10 June 2019

Abstract: A "nadir-only" framework of the radiometric intercomparison of multispectral sensors using simultaneous nadir overpasses (SNOs) is examined at the 1-km regimes and below using four polar-orbiting multispectral sensors: the twin MODerate-resolution Imaging Spectroradiometer (MODIS) in the Terra and Aqua satellites, the Visible Imaging Infrared Radiometer Suite (VIIRS) in the Suomi National Polar-orbiting Partnership (SNPP) satellite, and the Ocean and Land Colour Instrument (OLCI) in the Sentinel-3A satellite. The study is carried out in the context of isolating the on-orbit calibration of the reflective solar bands (RSBs) under the "nadir-only" restriction. With a homogeneity-ranked, sample size constrained procedure designed to minimize scene-based variability and noise, the overall approach successfully stabilizes the radiometric ratio and tightens the precision of each SNO-generated comparison event. Improvements to the multiyear comparison time series are demonstrated for different conditions of area size, sample size, and other refinements. The time series demonstrate the capability at 1% precision or better under general conditions but can attain as low as 0.2% in best cases. Solar zenith angle is examined not to be important in the "nadir-only" framework, but the spectral mismatch between two bands can give rise to significant yearly modulation in the comparison time series. A broad-scaled scene-based variability of ~2%, the "scaling phenomenon", is shown to have pervasive presence in both northern and the southern polar regions to impact inter-RSB comparison. Finally, this paper highlights the multi-instrument cross-comparisons that are certain to take on a more important role in the coming era of high-performing multispectral instruments.

Keywords: VIIRS; MODIS; OLCI; RSB; SNPP; Terra; Aqua; Sentinel-3A; reflective solar bands; intersensor comparison; intercalibration; SNO

---

1. Introduction

Earth science and climate studies have made significant progress in the recent decades along with continual advances in remote sensing technologies. Progressing along improving imaging capability is the intersensor comparison methodology, a method of evaluating the performance of sensor data or associated retrievals by comparing against a reference sensor, which is also certain to be utilized in greater capacity in the coming era. For multispectral sensors, it can be argued that the two units of the MODerate-resolution Imaging Spectradiometer (MODIS) [1,2], in the Terra and Aqua satellites launched on 18 December 1999 and 4 May 2002, respectively, are the forerunners leading the era of the high-performance instruments and big data. The twin MODIS, with 36 bands covering the spectral range of 0.45 to 12.4 μm are now closing in on two prolific decades of data acquisition. However, it is not until the launch of the next major multispectral sensor—the Visible Imaging Infrared Radiometer Suite (VIIRS) aboard the Suomi National Polar-orbiting Partnership (SNPP) satellite on 28 October

2011 [3,4]—that the twin MODIS finally has a comparable counterpart to generate high precision intersensor comparison result. Numerous radiometric intercomparisons of the reflective solar bands (RSBs) of Aqua MODIS and SNPP VIIRS ensued [5–7] utilizing the simultaneous nadir overpasses (SNOs) approach [8–11]. These studies demonstrated the capability of the radiometric intercomparison at the 1-km spatial resolution regime to be typically few percent. A main goal of this paper is to show that the capability, under an improved analysis procedure, is at the level of 1% precision or better.

The coming era is certain to make intersensor comparison a tool of increasing importance as more high-performance multispectral sensors are continually being launched into operation. For example, the Ocean and Land Colour (OLCI) Instrument and its companion Sea and Land Surface Temperature Radiometer (SLSTR) housed in the Sentinel-3A satellite [12] are the recently launched polar-orbiting multispectral sensors, with approximately 300-m spatial resolution. The first follow-on of VIIRS built is one on the Joint Polar Satellite System-1 (JPSS-1) satellite [13], or J1 satellite (officially designated as NOAA-20 post launch), was launched on 18 November 2017. A total of four follow-on VIIRS—J1 to J4 VIIRS—for which the SNPP VIIRS serves as the precursor, are scheduled to span the next 20 years of operation. The technological advancement also extends to geostationary sensors, with Himawari-8 Advanced Himawari Imager (AHI) [14,15], GOES-R Advanced Baseline Imager (ABI) [16–18], and GOES-S ABI [18], all with 1-km spatial resolution, recently launched. More follow-on geostationary sensors are also in the plan of succession. The demand for assessing sensor data quality and to monitor radiometric performance will certainly increase.

The overall accuracy of radiometric data depends on many inputs, but the regularly carried out on-orbit radiometric calibration operation to characterize the changing performance of detectors is one of central importance. One main commonality among the four instruments considered here—Terra and Aqua MODIS, SNPP VIIRS, and Sentinel-3A OLCI—is a fully equipped onboard calibration suite for carrying out the regularly scheduled on-orbit calibration. They following a similar strategy, including using a specially made solar diffuser (SD) panel for RSB performance characterization. This built-in calibration capability is a mark of the new generation high-performance multispectral sensors and makes them valuable radiometric references for other sensors or any climate studies to inter- or cross-calibrate. Thus establishing radiometric consistency between any pair of such independently calibrated sensor will be beneficial, and intersensor comparison is a most proper tool for this purpose. In addition, intercalibration using any of these sensors as a radiometric reference requires also a reliable intersensor comparison methodology. While there are numerous approaches for post-launch radiometric evaluation, the current knowledge points to intersensor comparison as potentially the most precise approach.

However, there are many other factors impacting the overall accuracy of the sensor data and also intersensor comparison result beyond just band or detector performance. One of such, outside the capability of the standard on-orbit calibration, is the response-versus-scan angle (RVS) effect of the scan mirror that can add in a systematic angle-dependent bias along-scan. For example, in the MODIS Collection 6 release [19,20], the addition of the time-dependent RVS characterization is a key upgrade to the RSB calibration methodology to mitigate the RVS effect that is beyond the capability of the on-orbit calibration methodology. For SNPP VIIRS, the time-dependent RVS effect is not yet an issue, but its potential emergence remains a possibility. Various common issues also create challenges for intersensor comparison analysis. Angle- or scene-dependent effects associated with the scenes, including the biredirectional reflectance distribution factor (BRDF) of the SNO scenes, also introduce additional complications into intersensor comparison result. For removing various confounding effects associated with larger angles or viewing areas, Chu et al. [7] utilized a "nadir-only" framework of SNO analysis in an Aqua MODIS versus SNPP VIIRS study, that limits the viewing angle to the Earth scenes to near 0° degree by using a small-sized comparison area. This study adopts the same "nadir-only" approach specifically in the context of examining the capability of intercomparison in evaluating on-orbit RSB calibration performance, and furthermore uses multiyear comparison time series that incorporate all high-quality SNO events as a tool of evaluation.

This study further distinguishes between statistical and physical constraints. For example, statistical analyses subject all physical conditions such as cloudy scenes or those of various geolocational conditions under the same criteria. This analysis carefully avoids any premature applications of physical constraints, such as the removal of cloudy scenes that can unnecessarily remove legitimate and usable data. Because statistical and physical attributes do not necessarily correlate, physical constraints should be applied only for targeted purposes. Also, for keeping data and results clean for achieving unambiguous and precise analysis, this work does not adjust or correct of data. It is often customary to adjust result such as using the spectral band adjustment factors (SBAFs) to account for spectral differences, but this study does not presume these practices to be reliable at the 1% precision level—the aim here is to first establish a clean groundwork before these other issues can be further examined.

In summary, this work examines the capability of intercomparison in a "nadir-only" refinement of SNO analysis that isolates the on-orbit RSB calibration from other large area-associated issues for a multiyear evaluation using comparison time series. In particular, Aqua MODIS versus SNPP VIIRS is used as the main case study because of their longer history with many studies already established numerous important findings. One more relevant point is that an intercomparison is a relative evaluation that is conclusive only when the reference sensor has already been established as reliable. Additional information, such using product retrievals derived from sensor data or another sensor for cross-examination, is often required to draw stronger conclusions. In other words, a stable radiometric comparison result can be deceptive due to both sensors containing coincidentally similar error in calibration. Nevertheless, intercomparison is most valuable when result shows deviating features that signals problems such as incorrect implementation, inadequate calibration or instrument anomalies. The current high-performance multispectral sensors with good imaging capability already has the 1% interscomparison capability that can ascertain various radiometric deviations of few percent; however, other types of sensors such as hyperspectral or microwave are either with insufficient spatial resolution or have not been shown with applicable precise intercomparison.

The organization of this paper is as follows. Section 2 briefly describes the four instruments and some general issues of radiometric intercomparison. Section 3 shows the results of the examination of the intercomparison methodology, emphasizing SNPP VIIRS versus Aqua MODIS in the 1-km regime. Section 4 shows the result of the examination for four different regimes of intercomparison, from 375 m to 1 km. Section 5 demonstrates cross-comparisons using Aqua MODIS, Sentinel-3A OLCI, and SNPP VIIRS. Section 6 provides a general discussion of relevant issues. Section 7 provides a summary and conclusion.

2. The Comparison Conditions

2.1. The Instruments

The flight and operational parameters of the satellites and instruments preset numerous limiting conditions for intercomparison. Table 1 lists some key parameters for Sentinel-3A, SNPP, Terra, and Aqua satellites and the key specifications for OLCI, VIIRS, and twin MODIS. Notable differences are the two different repeat cycles, at 27-day or 16-day, and the two nodes of flight, either ascending at 1:30 pm local time or descending at 10:00 am local time, that can influence SNO occurrences. The native spatial resolution at subsatellite point (SSP), determining the number of pixels per unit area, is a key parameter affecting the capability and the statistics of intercomparison. For example, at 1-km regime, a small area of 40 × 40 km-square contains 1600 pixels per SNO event, which is sufficient to attain robust statistics.

Table 1. Selected information and parameters for the four satellites and the multispectral instruments.

	Sentinel-3A: OLCI	SNPP: VIIRS	Terra: MODIS	Aqua: MODIS
Satellite Repeat Cycles (Days)	27	16	16	16
Satellite Local Crossing Time	Descending: 10:00 am	Ascending: 1:30 pm	Descending: 10:30 am	Ascending: 1:30 pm
Satellite Altitude (km)	814	834	705	705
Satellite Orbit Inclination	98.6	98.7	98.2	98.2
Sensor: Swath (km)	1270	3040	2330	2330
Sensor: Resolution at SSP (m)	300 m	750 m, 350 m	1 km, 500 m, 250 m	1 km, 500 m, 250 m
Sensor: Number of Bands	21	22	36	36
Sensor: RSB/TEB/Other	21/0	14/7/1	20/26	20/26

Figure 1 illustrates the spectral coverage, represented by the range of the relative spectral response (RSRs) or spectral response functions (SRFs), of selected RSBs considered in this study. Every SNPP VIIRS band up to M7 has a spectral counterpart in MODIS or OLCI, although the two latter sensors contain more bands not spectrally matched by SNPP VIIRS.

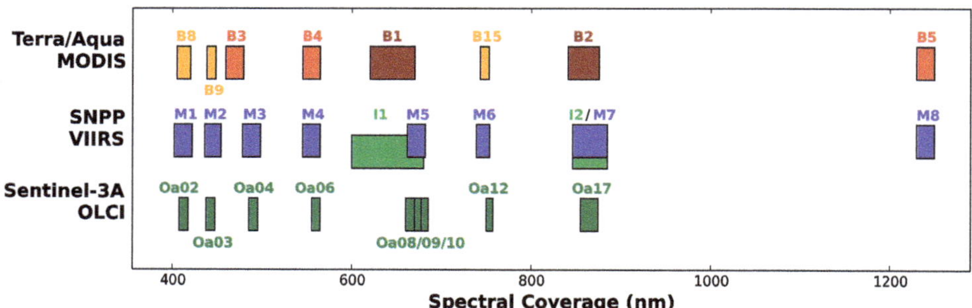

Figure 1. The spectral coverage of the selected reflective solar bands (RSBs) of MODIS, SNPP VIIRS, and Sentinel-3A OLCI considered in this study. The 250-m, 500-m, and 1-km spatial resolution MODIS bands are shown in orange, red, and purple, respectively. The 750-m and 375-m spatial resolution bands of SNPP VIIRS are shown in blue and light green.

Table 2 lists the specifications of the RSBs corresponding to those in Figure 1. The spatial resolutions of SNPP VIIRS moderate- and imagery-RSBs are at 750-m and 375-m. MODIS bands B1 and B2, B3–B7, and B8–B16 operate at 250-m, 500-m, and 1-km spatial resolution, respectively. MODIS bands B1–B7 are also aggregated at 1-km spatial resolution. All Sentinel-3A OLCI bands have approximately 300-m spatial resolution. The band-to-band comparisons between these sensors can be made in four spatial resolution regimes: 1-km, 750-m, 500-m, and 375-m. The maximum at-sensor radiance (L_{MAX}), with units watt/m^2/sr/μm, represents the maximum end of the band dynamic range. The spectral categories are visible (VIS), near infrared (NIR), and shortwave infrared (SWIR), in increasing order of wavelength.

Table 2. The specifications of the matching RSBs of Sentinel-3A OLCI, SNPP VIIRS, and MODIS.

| | | Sentinel-3A OLCI | | | | | SNPP VIIRS | | | | | Terra/Aqua MODIS | | | |
Type	Band	Spectral Range (nm)	Center λ (nm)	Spatial Resolution (m)	Lmax	Band	Spectral Range (nm)	Center λ (nm)	Spatial Resolution (m)	Lmax	Band	Spectral Range (nm)	Center λ (nm)	Spatial Resolution (m)	Lmax
VIS	Oa02	407–417	412.5	300	501.3	M1	402–422	410	750	135/615	B8	405–420	412	1000	175
	Oa03	438–448	442.5	300	466.1	M2	436–454	443	750	127/687	B9	438–448	443	1000	133
	Oa04	485–495	490	300	483.3	M3	478–498	486	750	107/702	B3	459–479	469	500/1000	593
	Oa06	555–565	560	300	524.5	M4	545–565	551	750	78/667	B4	545–565	555	500/1000	518
	Oa08/ (Oa09)/ (Oa10)	660–670/ (670–677.5)/ (677.5–685)	665/ (673.75)/ (681.25)	300	364.9/ (443.1)/ (350.3)	M5	662–682	671	750	59/651	B1	620–670	645	250/500/100	685
						I1	600–680	640	375	718					
NIR	Oa12	750–757.5	753.75	300	377.7	M6	739–754	745	750	41	B15	743–753	748	1000	26
	Oa17	856–876	865	300	229.5	M7	846–885	862	750	29/349	B2	841–876	859	250/500/1000	285
						I2	846–885	862	375	349					
SWIR		(N/A)				M8	1230–1250	1238	750	165	B5	1230–1250	1240	500/1000	110

The study uses SNPP VIIRS as the common reference against other three sensors. In particular, Aqua MODIS versus SNPP VIIRS has been well studied [5–7] and its result establishes many key baselines. The band pair of Aqua MODIS B5 versus SNPP VIIRS M8 is an ideal case study of the comparison capability given the near identical spectral coverage and the very stable time series [6,7]. The four spectrally well-matched bands of Aqua MODIS B2, SNPP VIIRS I2/M7, and OLCI Oa17, centered at ~860 nm, represent a uniquely interesting set to make a consistent study on the impact of spatial resolution at all four possible regimes at 375 m, 500 m, 750 m, and 1 km. Both Terra MODIS and Sentinel-3A OLCI are not as reliably established as Aqua MODIS but still provide additional useful result. The datasets used in this study are the MODIS Collection 6.0 release [21], the operational SDR version generated by the Interface Data Processing Segment (IDPS) for SNPP VIIRS [22], and the current mission release for Sentinel-3A OLCI [23].

2.2. General Issues

The general conditions of SNOs and radiometric intercomparison have been well described [9–11,24] and the details are not repeated here. The focus here is on issues pertaining to the capability of the radiometric intercomparison methodology.

2.2.1. SNO Geolocation and Scenes

The occurrences of SNOs are determined by the flight trajectories of the satellites. Figure 2 displays the SNO locations for three pairs of satellites—Sentinel-3A versus SNPP for the year 2017 (red triangles), Terra (green stars), and Aqua (blue squares) versus SNPP (green stars) up to end of 2017—showing the northern polar region in Figure 2a and the southern polar region in Figure 2b. The 2017 SNO subsets are highlighted for Terra versus SNPP (magenta triangles) and Aqua versus SNPP (cyan squares); this is to illustrate that SNO locations do not repeat yearly. The SNOs of Sentinel-3A versus SNPP, with descending node for the former and ascending node for the latter, are concentrated within a tight circular band at around N71° latitude just inside the Artic Zone, with no occurrences in the southern region. On the other hand, the SNOs of Terra versus SNPP, also in opposing descending and ascending node, occur over both northern and southern regions, tracing out near the N68 and S68 circulars. The SNOs of Aqua versus SNPP, both ascending node, occur over both northern and southern polar regions in an interesting three-arm spiral pattern. This illustrates that different orbits and flight parameters map out different SNO locations, and therefore the reflectance property and the common atmospheric conditions of these SNO scenes are important factors. For example, Aqua versus SNPP commonly crosses over icy scenes of Antarctica, which have scene radiance commonly above 50 watt/m^2/sr/μm, thus easily saturate SNPP VIIRS M6 and many MODIS bands of low dynamic range.

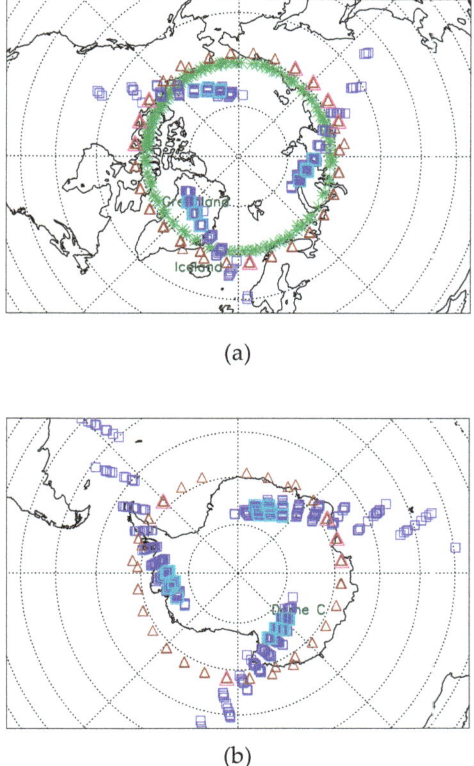

(a)

(b)

Figure 2. The precise SNOs of Sentinel-3A versus SNPP satellite (green stars) in 2017, Aqua versus SNPP satellite for the entire SNPP mission (blue squares) and in 2017 (cyan squares), and Terra versus SNPP satellite for the entire SNPP mission (red diamonds) and in 2017 (magenta diamonds), over (**a**) northern polar region and (**b**) southern polar region.

Figure 3 shows the daily frequency of precise SNO events for the year 2017. An interesting finding is the extended four-month periods of missing SNOs events for Sentinel-3A versus SNPP satellites (green bars) that run from October to February. Although not shown, the late 2016 and early 2018 periods are also without SNO occurrences for Sentinel-3A versus SNPP. A quick check confirms that Sentinel-3A OLCI observational coverage changes throughout the year and does not extend beyond 71° latitude during those four-month gaps, and therefore, despite any actual SNO events of the two satellites, there is no OLCI data available. Another interesting result of Sentinel-3A versus SNPP is that the SNOs cluster in distinctive days, 45 days of multiple SNO occurrences that further group into 13 clusters, thus showing that mismatching flight parameters, such as 16-day versus 27-day repeat cycle for this case, can generate interesting occurrences.

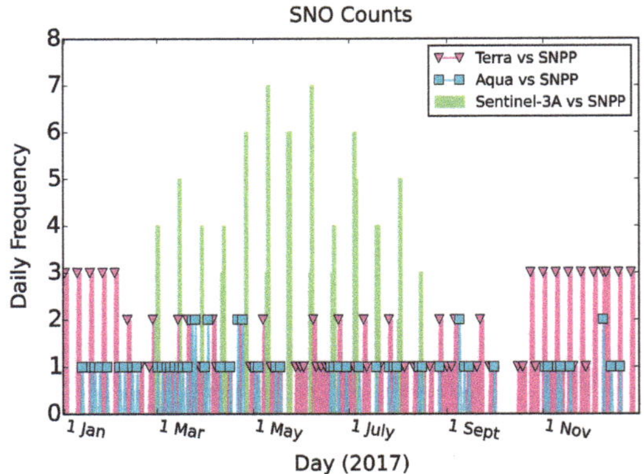

Figure 3. SNO occurrences in 2017 for Sentinel-3A versus SNPP (green bars), Terra versus SNPP (magenta triangles), and Aqua versus SNPP (cyan squares).

2.2.2. Spectral Match

The matching of two bands for radiometric comparison is customarily made according to their spectral proximity to ensure comparable radiometric responses over SNO scenes. In reality, most band pairs have RSR differences that induce yearly variability into the comparison time series. On the other hand, some band pairs not showing good spectral match, such as Aqua MODIS B3 versus SNPP VIIRS M3 with limited RSR overlap as shown in Figure 1, can still generate usable comparison time series [6,7]. A more extreme example is Aqua MODIS B7 (2130 nm) versus SNPP VIIRS M11 (2257 nm) [25] for which the two RSRs do not overlap but a marginally usable time series can still be generated. The impact of RSR mismatch and the full range of possibilities beyond the standard spectral-matching practice are not fully understood and should be pursued in future studies.

2.2.3. Dynamic Range

The limitation due the dynamic range is briefly presented here only for clarifications. The narrow dynamic range of a band can set a limitation impossible to overcome. For example, Aqua MODIS B15 (748 nm) versus SNPP VIIRS M6 (746 nm), with L_{MAX} of 3.5 and 41 watt/m^2/sr/μm, respectively, hardly generates any successful outcomes as both bands saturate over the higher-latitude icy polar scenes where their SNOs commonly occur. However, future sensors with progressively wider dynamic range are less likely to encounter saturation. For instance, Sentinel-3A OLCI bands already have sufficient dynamic range and show no saturation issue for this study. But for band M6 of all VIIRS builds at only 41 watt/m^2/sr/μm L_{MAX}, the success of SNOs involving VIIRS M6 is limited.

2.2.4. Spatial Resolution

The central goal of this study is to assess the capability of radiometric intercomparison and the achievable statistics in different regimes of spatial resolution. That is, how well can intersensor comparison assess radiometric performance of a sensor at different pixel sizes? As the regime reaches the 1-km resolution or so, the number of pixels in a small but realistic sized area selected for comparison becomes sufficient to allow standard statistical sampling. For example, at 1-km regime, a small area of 32 × 32 km-square contains 1024 pixels, which is sufficient for robust statistics under favorable scene conditions. The current result, such as shown in Chu et al. [7], suggests that the precision result of the time series at 1-km spatial regime is ~1%. Below the 1-km regime, the greater pixel density then give more samples per unit area as well as greater flexibility to enable more powerful sampling analysis—it may be possible to reach precision result much tighter than 1%. At coarser spatial resolution, for example at 5-km pixel size, to have 1000 pixels require an area size of 160 × 160 km-square, and that extent is too large to realize "nadir-only" condition. Therefore, the result using large coarser pixel size is likely to have large-area effects to render the result unreliable.

Most SNPP VIIRS and Aqua MODIS bands are moderate bands, at 750-m and 1-km spatial resolution respectively, and their intercomparison at the 1-km regime have demonstrated precision result at ~1% [7]. But Aqua MODIS and SNPP VIIRS also contain imagery bands with resolutions as fine as 250 m, and furthermore, the OLCI spatial resolution is ~300 m. The examination at regimes finer than 1-km can therefore assess the capability at higher imaging capability. The coming era will have more such higher spatial resolution imaging sensors, such as OLCI already in operation.

3. The Examination of Radiometric Intersensor Comparison

This study generalizes the methodology used in the Aqua MODIS versus SNPP VIIRS inter-RSB comparison by Chu et al. [7] to focus on three key criteria—area size, pixel homogeneity, and pixel sample size. The small area is a way to approximate the "nadir-only" condition, while homogeneity and sample size constraints are containment strategies to minimize a generally persistent scene-based variability of ~2% significantly impacting the comparison result. This persistent broad-scale variability—the "scaling phenomenon"—renders the use of larger area and sample size to improve statistics useless and is a key motivator of this study. The earlier assessment [7] suggests that the scaling phenomenon arises out of some mid- to large-scale scene conditions in the southern polar region, including Antarctica, where SNOs commonly occur. The application of the constraints to each SNO event successfully circumvented the variability to achieve a better precision at about 1%. This study clarifies how scene-based variability can impact intercomparison and why improvements can be made. The northern polar region is also shown to have the same 2% scene-based variability.

3.1. Procedure and Setup

Given an SNO event precisely determined within a single pixel of nadir crossing, a small area centered on the nadir crossing is used for pixel-based radiometric intercomparison. The radiance pixels of the two sensors within the small area are matched pair-by-pair via geolocation information. Each pair of collocated pixels is used to compute a pixel-based radiometric ratio of radiance. For this analysis, SNPP VIIRS radiance is taken to be the common radiometric reference against that of MODIS or OLCI. A fixed number of pixels of the best homogeneity quality, to be explained below, is selected for the computation of population statistics. The population average and the relative standard deviation (STD) of all qualified pixel-based ratios represent the ratio and the precision, or error bar, of the SNO event. The low radiance bias of MODIS and the impact of the solar zenith angle (SZA) are two issues briefly discussed here for clarification but are not used for analysis.

First, specific only to the inter-RSB comparison of MODIS versus SNPP VIIRS, a radiance cut of the 20% of the lowest radiance is imposed to remove biased cases occurring at low radiance, as was first done by Chu et al. [7] for Aqua MODIS-based result. The low radiance values from the two sensors are actually in good agreement on absolute terms, but nevertheless can result in large relative bias primarily as numerical artifact due to the low radiance value in the denominator. This low-radiance bias is also quickly confirmed to be true for Terra MODIS versus SNPP VIIRS. High radiance cases also possesses a few outliers, possibly associated with band response near saturation, thus the highest 10% of the radiance are removed as a safety measure. On the other hand, the OLCI-based comparison result does not exhibit bias at either low or high radiance. This points to MODIS possibly having some calibration issues, such as incorrect characterization of nonlinearity at low radiance, but is in any case a calibration issue not examined here.

The second issue concerns the impact of the SZA dependence, which imparts to radiance a distinctive seasonal pattern. However, the "nadir-only" restriction effectively cancels out the SZA effect in the radiometric comparison because the SZAs of the two sensors are effectively identical across the small area. Figure 4 shows the SZA correction ratio of Terra versus SNPP (magenta triangle), Aqua versus SNPP (cyan squares), and Sentinel-3A versus SNPP (green stars) for the year 2017. The error bars are mostly ~0.1% and smaller. Given the yearly cycle of the SZA, the demonstrated stability in the year 2017 is sufficient to show that the SZA correction factor will not impart to comparison time series any seasonal modulation or multiyear drift. The small random variability can be attributed to the time difference that is also random from one SNO event to next; furthermore, the accuracy of geolocation data can also be questioned at the level of 0.1%. Thus it is not necessary to include the SZA correction factor in the "nadir-only" framework.

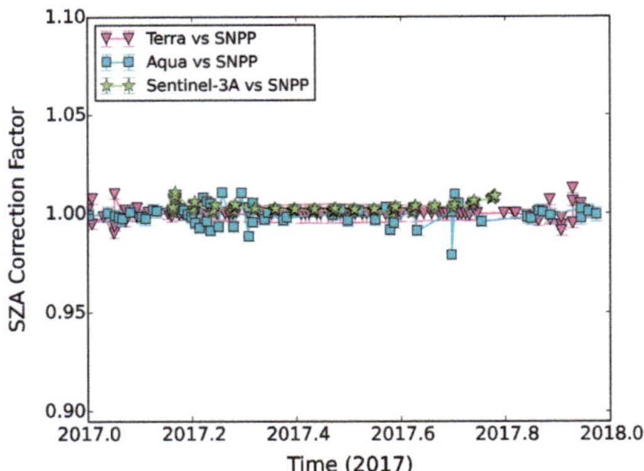

Figure 4. Solar zenith angle (SZA) correction factors in the year 2017 for Terra versus SNPP satellite (magenta triangles), Aqua versus SNPP satellites (cyan squares), and Sentinel-3A versus SNPP satellite (green stars) demonstrating stable trends.

3.2. Homogeneity

Homogeneity, or spatial uniformity, quantifies the variability of a pixel. It is most straightforwardly represented by the percentage STD calculated using the pixel itself and its eight neighbors—that is, the STD of the value of the nine pixels in the 3 × 3 square divided by the value of the center pixel. A few options exist for its application—at pixel-based radiance of each sensor, at pixel-based ratios computed from collocated radiance pixels, or both; it is tested for this study that for as long as homogeneity is applied, the result differs very modestly only in rare events. For simplicity, this analysis applies homogeneity to radiance data.

The primary importance of homogeneity lies with it being a proxy to statistical quality of pixel-based data to be used in tandem with a sample size constraint condition, to be described below. Homogeneity in this analysis is not simply an imposed threshold, but is used to generate a sorting of pixel quality to allow a selection procedure under a sample size constraint. Using only a simple homogeneity threshold will include all pixels satisfying the threshold, and different SNO events will have different sample size. On the other hand, using size-constrained selection forces all qualified SNO events to contain the same number of pixels, and this has the advantage of allowing more straightforward interpretations and comparison among events. A threshold of homogeneity, such as 4.5% as a reasonable level can always be imposed, but its importance to contain noise or variability becomes secondary when sample size constraint, itself a mechanism of containment, is used.

3.3. Area Size and Sample Size Constraint

The impact of area size, sample size constraint, the scaling phenomenon, and other associated issues of the comparison sampling analysis are examined here under expanded scope. The band pair of Aqua MODIS B5 (1240 nm) and SNPP VIIRS M8 (1238 nm) is used as the representative case study because their comparison result has shown to be the most stable [6,7]—this is primarily due to their well-matched spectral coverage and partly to the radiometric stability of the SWIR bands. For each SNO event, an examination of the impact of area and sample size is carried out at each spatial scale from the 20-km to 160-km scale. That is, 20×20 km-square area centered on nadir crossing is analyzed, then on to 32×32 km-square and so on until up to 160×160 km-square. At each scale or area size, statistics are computed for two separate cases. For the sample-unconstrained case, all pixels within the 4.5% homogeneity are used to compute the population statistics. In the sample-constrained case, only a fixed number of pixels of the best homogeneity quality, also necessarily below 4.5%, are used. The sample size of 500 samples and 1000 samples are used for the size-constrained cases.

The usable SNO events range from those of clearest scene conditions to those of variable conditions. A best-scenario SNO event is shown in Figure 5, that of 15 December 2016, for the sample-unconstrained case (red stars) and two sample-constrained cases, at 500 (green diamonds) and 1000 samples (blue squares). The top panel shows the average ratio of the qualified pixel-based ratios at each scale or area size, and the bottom panel shows the corresponding error bar, or the relative STD. The ratio result shows near-perfect broad-scale agreement among three cases that is remarkably stable at 0.991. The error bars are also very tight for all three cases, and are practically identical for the two sample-constrained cases at 0.2%. The overall result indicates a very clean scene condition that can generate very robust result at all scales shown up to 160 km. The occurrences of SNO events with this level of pristine clarity are only of several per year, but they remarkably reveal the capability of intercomparison to be fundamentally at the 0.2% level. The broad-scale agreement also reflects the sampling procedure to be meaningfully constructed and correct. All similarly clear-scene cases at other times have been checked to generate stable ratios and tight error bars across all scales as well. When clear SNO conditions exist, such as with low cloud or aerosol content, then using any small area size within the SNO scene will generate a robust and the correct result. It is here pointed out that Chu et al. [7,25] have examined one such high-precision event to confirm its clear-scene condition.

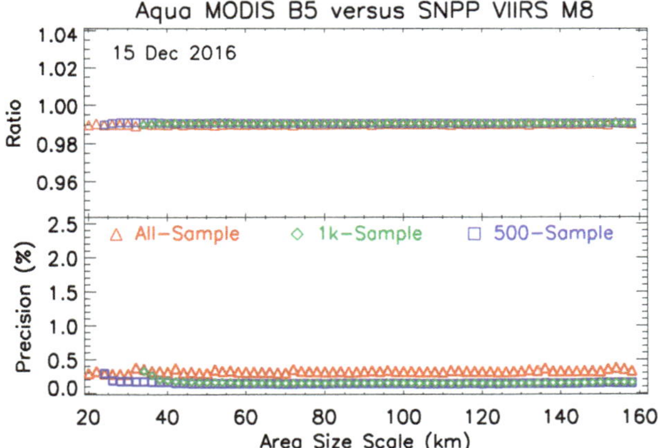

Figure 5. The ratio (top) and the relative precision (bottom) versus area size for the three cases of unconstrained sample size (red stars), constrained size at 1000 samples (blue squares), and constrained size at 500 samples (green diamonds) for the clear-scene SNO event on 15 December 2016 for Aqua MODIS M8 versus SNPP VIIRS M8.

The primarily important SNO cases are those of marginal statistical quality with broad-scale error bar of few percent, approximately 2% to 4%, that can be improved to be below 2% to be added to the comparison time series. Thus the number of these marginal cases can determine the success or failure of a time series. Figure 6 illustrates two representative cases with ~2% broad-scale error bar. The labels are the same as those of Figure 5. The most outstanding feature to note is that, consistent over the entire range of scale or area size shown, the sample-constrained ratios are stable with tighter error bars, while the sample-unconstrained ratios are unstable at the level of 1.5% or worse. In particular, the ratio-versus-scale result of each sample-unconstrained case demonstrates worsening scatter toward larger scales—this decisively demonstrates the use of larger areas on its own does not improve comparison result and can in fact make it worse. Thus a strategy such as using larger areas or all available pixels, even when many noisy pixels have been removed via a homogeneity filter, is not a reliable procedure. On the other hand, the two sample-constrained cases—at 500 and 1000 samples—show stable ratio with broad-scale agreement. This finding shows that robust results are not achieved by having more samples but on the contrary by limiting them, specifically by using only the best-quality pixels. The error bar results of the constrained case are also tighter and continue to smoothly tighten further with increasing scale. The overall strong conclusion is that the application of sample size constraint, in conjunction with a homogeneity-ranked selection, stabilizes the ratio and tightens the error bar at each scale. Because of this stabilization, area size actually becomes statistically conforming—that is, by increasing the area size, more samples become available for selection and the error bar tightens as expected. The caveat is that fixing the number of best quality pixels is a necessary middle step to facilitate this conforming behavior.

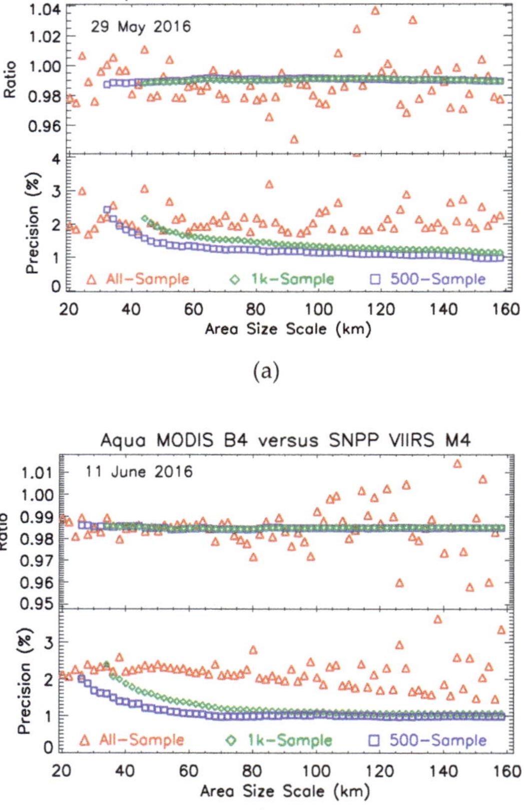

Figure 6. The scale-dependent result of radiometric comparison of (**a**) Aqua MODIS B5 versus SNPP VIIRS M8 on 29 May 2016 and (**b**) Aqua MODIS B4 versus SNPP VIIR M4 on 11 June 2016, for ratio (top panel) and the error bar (bottom panel) versus area size for the three cases of unconstrained sample size (red stars), constrained size at 1000 samples (blue squares), and constrained size at 500 samples (green diamonds).

The precision-versus-scale result (bottom panel) also illustrates distinctively different behavior between the sample-constrained and the sample-unconstrained cases. While the sample-unconstrained case exhibits unstable large scatter, the two sample-constrained cases instead show a smooth exponentially decreasing patterning of error-bar tightening that begins to agree at the 60-km scale and finally settling at ~1%. It is clear that the unconstrained case uses all pixels necessarily including all those of worse statistical quality, thus the inclusion of all pixels does not help to tightening the error bar but in fact worsens the result. An examination of the pixel quality (shown later) illuminates this point. The precision of the unconstrained cases also shows consistent clustering at around the 2% level throughout all scales, thus demonstrating instances of "scaling phenomenon" within individual SNO events. However the phenomenon is herein explicitly revealed to be only loosely scale-invariant, and that the error bar can vary with scale or area size to some degree. This is a common feature for a majority of SNO events.

The exponential shape of the error bar results also indicates some well-behaved property. For the 1000-sample case, the 32-km scale is where the area size is minimally large enough to have more than 1000 pixels, specifically at 1024, for the analysis to be applied. At this scale, both the constrained and the unconstrained results contain almost the same set of pixels, thus the two precisions necessarily are closely matched, as shown in both dates at ~2.3%. As the area size increases to include more pixels, the constrained case will have more available pixels from which to select those of best homogeneity quality to further tighten the error bar. The error bar stabilizes at larger scales when most pixels of best homogeneity quality have been found, and that finding more pixels of better homogeneity from larger area becomes both less probable and less leveraging. This finding suggests that the selection of area size and sample size should not be too tightly matched, and instead, given any sample size constraint, the area size should be made larger to allow more samples. For example, for comparison at the 1-km regime using 1000 samples, an area of 50×50 km-square with 2500 available pixels will be better than a 32×32 km-square area with only 1024 pixels. The precision result in Figure 6, showing tightening precision at larger area, proves this point.

The relative left-right shift in the error bar versus scale result demonstrates another aspect of the sample-size condition. As explained above, the 1000-sample case starts its first point at the 32-km scale; for the 500-sample case, 23-km is the starting point with 529 available pixels. In any given spatial resolution regime, sample-constraint size determines the minimal scale. Therefore future sensors with finer imaging capability will push the minimum area even lower, allowing for more refined studies and improved capability.

3.4. Examination of Pixel Quality

A closer examination into the homogeneity of pixels reveals some insights into their statistical quality. Figure 7 shows the homogeneity of 2000 pixels from the 50×50 km-square area in the 11 June 2016 event of Aqua MODIS B4 versus SNPP VIIRS M4, corresponding to Figure 6b, ranked from best to worst. The two vertical lines mark 500 and 1000 samples. The first 500 samples have homogeneity better than 3.5%, but the next set of 500 samples, from number 501 to 1000, ranges from 3.5% to 4.7%. It is clear that the first 500 ranked samples will generate smaller variability then the next 500 samples and so on. This is consistent with Figure 6 which shows the 500-sample case is actually more precise then the 1000-sample case. The ranking of homogeneity as in Figure 7 exposes that includes more pixels can bring in those pixels with greater variability and make statistics worse. While obvious as presented, this runs counter to the common expectation that a larger sample size would generate better, not worse, statistics. The continually rising pattern of homogeneity of ranked-pixels indicates different variability pixel-wise, thus a sampling analysis over SNO scenes does not conform to standard sampling where each data point conforms to the same variability. This is neither an obvious nor trivial property that is anticipated, but nevertheless is consistent with physical reality in hindsight. Therefore cleaning processes based on physical conditions, such as cloud removal, that focuses on a subset of pixels with specific physical attributes does nothing for this pixel-based variability and will not stabilize the statistics. The issue is not if the pixels have been cleansed of certain physical attributes but whether or not if too many pixels of higher variability have been sampled. The real physical conditions of Earth scene data can vary, and cannot be expressed by a single well-defined distribution. Inclusion of more samples to improve statistics is inherently erroneous and can end up broadening the distribution and worsen the error bar. Therefore the containment of the worsening statistics, such as limiting the sample size and using only the lowest variability pixels, is the necessary remedy.

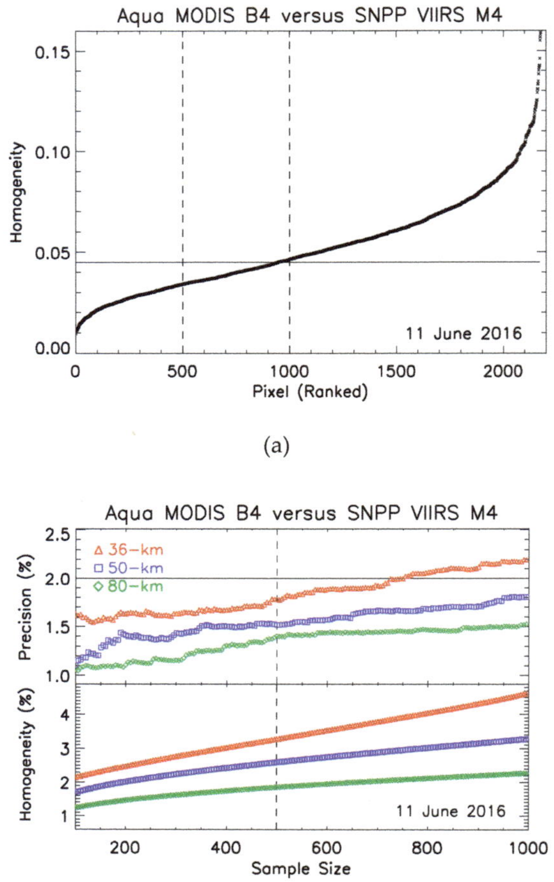

Figure 7. (**a**) Homogeneity versus ranked pixels for Aqua MODIS B4 versus SNPP VIIRS M4 for the 11 June 2016 SNO event and (**b**) precision (top panel) and homogeneity (bottom panel) with respect to sample size constraint for the 11 June 2016 event for the 36-km (red triangles), 50-km (blue squares), and 80-km (green diamonds) scales.

Figure 7b demonstrates how average precision and average homogeneity increases with respect to the number of sample at three area sizes—36-km (red triangles), 50-km (blue squares), and 80-km scale (green diamonds). For each of three testing area sizes, average precision and average homogeneity are computed for each given number of the homogeneity-ranked pixels. For example, for the 36 × 36 km-square area case which has 1296 pixels, the best 100 pixels in terms of homogeneity are used for computation of statistics for the first point, and then 101 pixels of the best quality are used and so on. Expectedly, the average homogeneity and precision worsens with inclusion of more pixels. The three cases also show that statistics improve with larger area under sample constraint. The 11 June 2016 event is a marginal case, and its 36-km, 1000-sample precision result at 2.2% would have been excluded by a 2% precision requirement for the time series; but its 50-km, 500-sample result shows that a different set of criteria can improve pixel selection leading to significant improvement to 1.7% precision. A quick summary of the sample size constraint is that, large constraint size can worsen statistics but using larger area can improve them.

It is natural to want to find the optimal scale and sample size choice, but the answer does not require another thorough study, but rather simply on the caution of keeping the area size small enough to avoid potential hidden bias. While Figure 7 may shows that the 80-km scale (green diamonds) generates the best statistics, the overall finding including that of the unconstrained cases in Figure 6 also suggests the presence of some underlying bias over larger area. For the 1-km regime, the 50-km scale is an acceptable balance between having an area small enough to minimize the large area bias and one large enough for good, but not necessarily the best, statistics. The result also shows that sample size range of 400 to 600 to be reasonable.

The distributions of qualified pixel-based ratios per each SNO event are examined for the three different sample size conditions. Figure 8 shows the three distributions of the 29 May 2016 event at the 70-km scale to be normal-like, indicating that the samples as a set are well behaved in each case. The key point is that the 500-sample case has the tightest distribution, followed by the 1000-sample case and finally the unconstrained case. This is consistent with the result of Figure 6 showing 500-sample cases having lower error bars. Other scales are checked to have the same behavior. The broadening of the distribution from the 500-sample to the unconstrained case is the most direct demonstration of the lack of an underlying stable distribution, showing that the sampling in intercomparison involves physical data of different variability. By including more samples in the homogeneity-ranked scheme into the distribution, the result increasingly contains worse statistics to broaden the distribution.

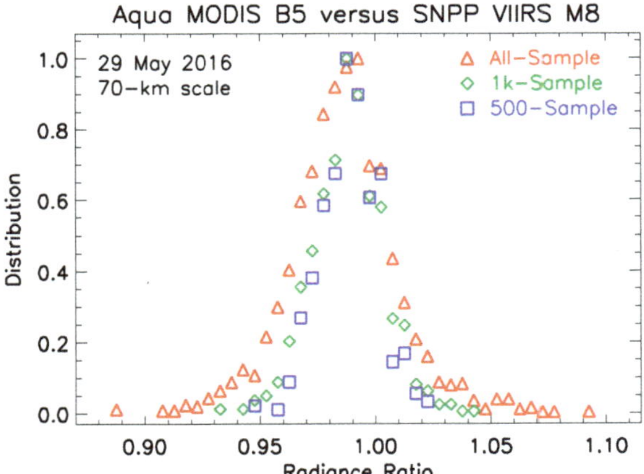

Figure 8. The three ratio distributions of Aqua MODIS B5 versus SNPP VIIRS M8 of the 29 May 2016 event, taken at the 70-km scale, for the sample-unconstrained condition (red stars), the constrained size 1000 samples (green diamonds), and the constrained size at 500 samples (blue squares).

It is worthy to clarify that the impact of homogeneity on error bar is neither direct nor absolute. Homogeneity as applied in this study has been shown to be a beneficial metric to help stabilize statistics, but pursuing into greater details is not necessary at the 1% precision level. It has been examined that slight variation at ~4.5% leads only to the slightest difference in a few SNO events. The sample size limitation and the selection procedure as described thus far are the main factors impacting the error bar result.

3.5. Application and Result

The long-term stability of the Aqua MODIS B5 and SNPP VIIRS M8 time series makes it an ideal case to illustrate the impact of various selection criteria. Figure 9 shows the three cases of 36-km with 1000 samples (red crosses), 36-km with 500 samples (green diamonds), and 50-km with 500 samples (blue squares). The solid line is the series mean set at 0.988 and the two dash lines mark the 2% boundaries above and below the mean. A precision threshold of 3% is applied to all three time series. The mean and the precision results are computed using the best 100 events in each time series for consistent comparison purpose.

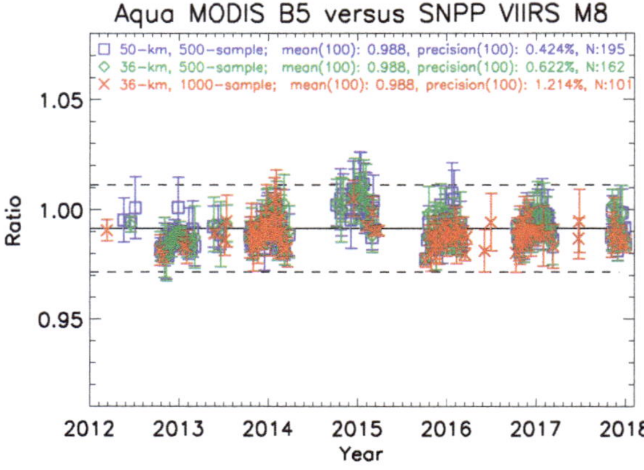

Figure 9. The time series of Aqua MODIS B5 versus SNPP VIIRS M8 under three combinations of area scale and sample size. A 3% precision threshold is imposed on each SNO event.

It is seen that lowering constraint size has impact. Lowering the sample size constraint from 1000 (red crosses) to 500 (green diamonds) increases successful comparison outcomes from 101 to 162 and tightens the 100-event error bar average from 1.214% to 0.622%. The 50-km, 500-sample (blue squares) time series, contrasting against the 36-km, 500-sample case, increases the number of successful SNO events to 195 and tightens the error bar to 0.424%.

Nevertheless, all three time series generate statistically indistinguishable series means at 0.988, thus it may appears at first that different conditions do not matter. However, for other purposes such as generating a fuller time series with fewer data gaps, i.e., better regularity, a larger area size and a less stringent sample size constraint may be better. For example, the 50-km, 500-sample time series (blue squares) contains more outcomes in the year 2012 and 2013 than the other two cases. What is demonstrated is that the area size and the sample size constraint can be tuned to improve some characteristics of the time series such as regularity that can be helpful to evaluate the radiometric performance at certain period. Larger area sizes beyond 80-km scale and lower sample size down to 250 samples have been examined to result in no improvement, thus supporting the 50-km with 500-sample condition to be sufficiently optimal for Aqua MODIS B8 versus SNPP VIIRS M8.

Yet the same result reveals a limitation—the existence of data gaps, such as the 5-month Austral winter period years 2014, 2015, and 2016. While many SNO events do exist in these periods and the refined analysis here has improved the situation somewhat, the challenging conditions of low radiance and noisy scenes are difficult to overcome. This is definitely one area for continual improvement.

Figure 10 further illustrates Aqua MODIS B5 versus SNPP VIIRS M8 result for three scenarios at the 50-km scale—the constrained case with 500 samples of best homogeneity (blue square), the unconstrained case with all samples without homogeneity condition (red stars) and clear-scene subset of the 500-sample constrained case (green diamonds). The 500-sample constrained case is the same time series in Figure 9, also in blue squares, repeated here for comparison. The same 3% precision threshold is applied for the constrained and the constrained cases, and a 100-event choice is similarly made for consistent comparison. For the clear-scene time series, a 0.35% precision is further imposed on the constrained case to extract this subset. The first notable improvement is that the constrained case is significantly better at 195 outcomes and a 100-event precision of 0.424%, comparing with the unconstrained case at 166 outcomes and 1.059% precision. This is consistent with the result in Figure 9. Also, the clean scene time series with 38 best outcomes, those with error bars below 0.35%, trace out a long-term baseline at 0.989 that is very consistent with other cases. As already revealed by the event of 15 December 2016 shown in Figure 5, the clear-scene result should be closest to the "truth" of comparison result. The clean scene time series with ~0.3% average precision suggests that 0.989 reflects the true comparison baseline, and that other times series are highly consistent with this result—this finding can be very helpful in pinpointing the radiometric baseline and helping to ascertain other features. It is worthy to note again that "nadir-only" condition by using a small area, here at 50-km scale, is already itself a sufficiently constraining condition, and therefore even the unconstrained case can appear to have comparably acceptable result.

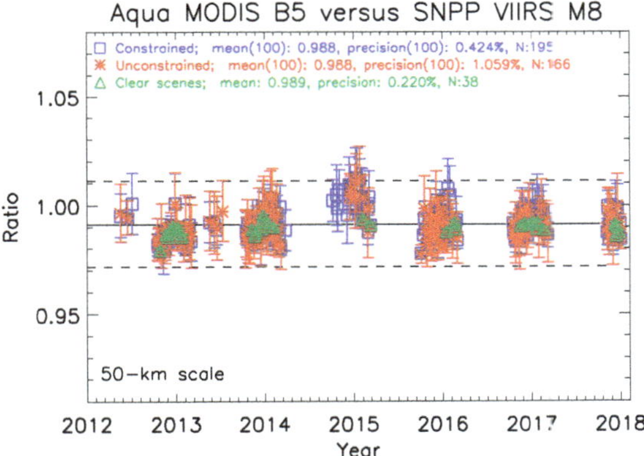

Figure 10. The three time series of Aqua MODIS B5 versus SNPP VIIRS M8 correspond to the constrained analysis using homogeneity and sample size constraint, the unconstrained case, and the clear-scene result.

Intercomparison can expose a variety of different outcomes and features. Figure 11 shows the corresponding comparison result of Aqua MODIS B4 (555 nm) versus SNPP VIIRS M4 (551 nm) for the constrained (blue squares), unconstrained (red stars), and clear-scene (green triangles) scenarios. The same 3% precision threshold is applied for both the constrained and the unconstrained cases, and the best 100 events are used to compute the time series statistics. In comparison with Figure 10, it is clear that different band pairs have clear qualitative difference; for Aqua MODIS B4 and SNPP VIIRS M4, which center near the 550 nm spectral range, the stronger scene radiance leads to more successful events.

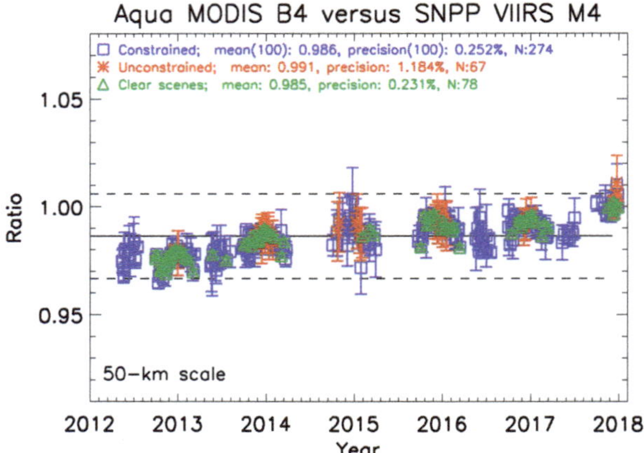

Figure 11. The three time series of Aqua MODIS B4 versus SNPP VIIRS M4 correspond to the constrained analysis using homogeneity and sample size constraint, the unconstrained case and the clear-scene result.

Although the examination into the physical cause of any deviation is not a purpose of this study, one Aqua MODIS B4 versus SNPP VIIRS M4 result reveals an important feature: in the four-year period from 2013 to 2017, an upward drift of ~2% can be seen, thus exposing some worsening on-orbit calibration error in either the IDPS-generated SNPP VIIRS M4 or the Aqua MODIS B4 of Collection 6 release. The clear-scene time series (green diamonds) is particularly lucid in tracing out both the multiyear drift and the yearly oscillation. The worsening calibration error comes from within the IDPS-generated radiance due to some nontrivial angular dependence in the reflectance property of the SD degradation [26–29] that has not been correctly captured by the standard on-orbit calibration methodology. This calibration error is neither trivial nor negligible, and can severely compromise product retrievals and climate studies. Thus establishing a meaningful and reliable time series, along with robust ratios and tight error bars, is a fundamentally important aspect of intercomparison methodology to enable correct assessments of the sensor data. Additionally, the seasonal modulation exhibited in the time series is typical of inter-RSB comparison of Aqua MODIS versus SNPP VIIRS [5–7]. Figure 4 has shown that the SZA correction is not a contributor to this modulation; the RSR mismatch is necessarily one of the contributing causes.

Also, in Figure 11, the unconstrained case shows significantly worse statistics than the sample-constrained case, again demonstrating the utility of these constraints despite using fewer samples.

3.6. Impact of Precision Threshold on the Time Series

The current finding so far suggests a 0.2% stability of the overall ratio mean of time series under different scenarios, but additional examination of the dependence on the threshold over a larger threshold range yields some confirmation. Figure 12 shows the time series mean versus precision threshold of Aqua MODIS B5 versus SNPP VIIRS M8 for the four different constraint conditions over a 0.6% range. For each precision threshold, all SNO events under the threshold are included in the computation of the time series mean. As the precision threshold is relaxed, more SNO events with larger error bar are included, and the series mean changes accordingly.

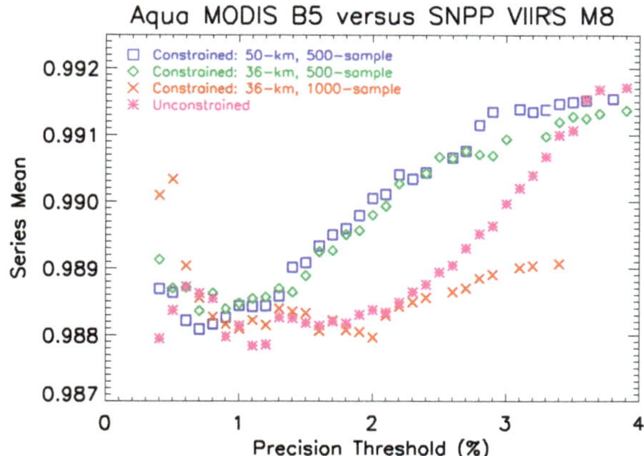

Figure 12. The mean of the radiometric comparison time series at each level of precision threshold, for the constrained and the unconstrained cases in Aqua MODIS B5 versus SNPP VIIRS M8.

The most important result is that the time series mean varies, primarily upward for this particular comparison case, over a 0.4% range with respect to precision threshold. The overall pattern is consistent with those events of tightest precision being more likely representative of the true radiometric comparison result, and those events of worse precision contain more radiometric bias. Therefore keeping a tight precision threshold is recommended to reduce any nontrivial variability or bias in the time series mean. The 2% precision threshold appears to be a reasonable choice with variability of the mean on the level of 0.2% variability in the mean for this context of the constrained procedure; while a more generous choice to achieve fuller time series must be cautious about making the time series mean less reliable.

The long-term stability of the Aqua MODIS B5 versus SNPP VIIRS M8 time series is what makes clearer the existence of any deviation or variability. In contrast, cases such as Aqua MODIS B4 versus SNPP VIIRS M4 with significant drift, as shown in Figure 11, are more difficult for interpreting the dependence on the precision threshold since the 2% drift complicates the result. For these cases, the mitigating the on-orbit calibration error should take top priority over any intercomparison issue. As emphasized already, intercomparison analysis is most valuable when it reveals some deviating that requires correction.

3.7. Scaling Phenomenon in MODIS versus SNPP VIIRS

The "scaling phenomenon" [7] is a broad-scaled and persistent variability pervading into the SNO results as illustrated in Figure 6 in selected events. Figure 13 illustrates the phenomenon for the Aqua MODIS B8 versus SNPP VIIRS M1 time series as a whole and includes the new result under the constrained analysis. Each point represents the error bar versus sample size outcome of an SNO event in the time series. Time series results of three different area sizes are shown: 36-km scale (red triangles), 50-km (blue squares), and 80-km (green diamonds). The result demonstrates how the scene-based scaling phenomenon blocks the use of the larger area size to improve statistics and how the constrained procedure overcomes this limitation.

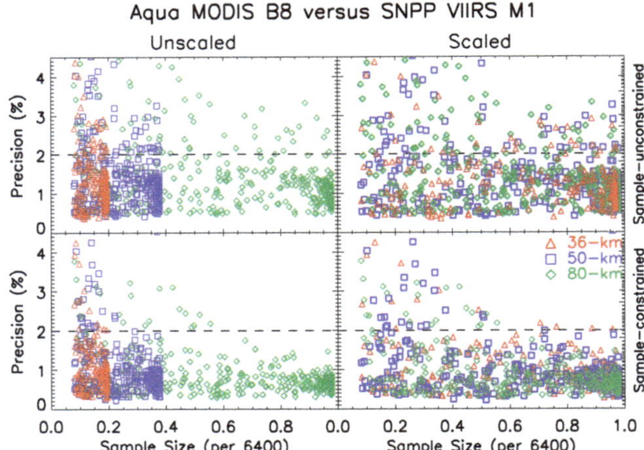

Figure 13. Scaling phenomenon in Aqua MODIS B8 versus SNPP VIIRS M1 for both sample-unconstrained and sample-constrained cases.

The top left panel of Figure 13 displays the time-series result of error bar versus sample size (for all events) without size constraint. The maximum sample size for the 36-km scale is 1296, or close to 0.2 per 6400 on the plot, and similarly for the 50-km scale at 2500, at near 0.39, and 80-km at 6400, at 1.0. It can be seen that the pattern of scatter of error bar values, apart from different sample size ranges, appear similar for all three scales. The top right panel of Figure 13, the scaled version, explicitly demonstrates the scaling phenomenon by linearly scaling the sample size of 36-km and the 50-km result to match 6400, i.e., stretching the result in the horizontal direction rightward until 6400. The scaled result shows that scatter pattern of three cases are effectively indistinguishable. The clear implication is that enlarging the area size to increase the number of pixels ends up generating same statistics and does not improve the quality of the time series. In contrast, the time series results in Figure 9 under homogeneity-ranked sample constraint, demonstrate clear improvement with lager areas. More detailed examination into each SNO event reveals that the scaling phenomenon is only an approximate effect of some common scene-based effect. As shown in Figure 6, the error bar result in the sample-unconstrained case (red stars) in each single NO event can slightly change with increasing scale.

The sample size constraint, originally applied to stabilize the error bar [7], necessarily impacts any scene-based effects including the "scaling phenomenon". The bottom two panels of Figure 13 demonstrate the impact of the constraints, for sample size of 500, on error bar versus sample size result. The label of "Sample Size" on the horizontal axis refers to the original available number of pixels for each event before the constraint is applied—thus it corresponds to the sample size for the corresponding unconstrained case. However all actual outcomes have the same final sample size of 500. In the bottom-right panel, the error bar scatter pattern of the 80-km result (green diamonds) is seen to become tighter than those of the 36-km and the 50-km scales, thus showing that scaling phenomenon is no longer true in the constrained analysis. In the same plot, the range of the error bar shows more obvious and faster tightening with increasing size for all three cases, reaching below 2% at higher sample size, showing that the constrained procedure is effective.

For completion and illustration, the corresponding ratio versus sample size of the Aqua MODIS B8 versus SNPP VIIRS M1 time series is shown in Figure 14. The 4 to 6% range of spread makes it less obvious to discern any 0.1 to 0.5% effect, but many resulting points can be seen to have shifted from the unconstrained case (top panels) toward the center of the range in the constrained case (bottom panels). The 4 to 6% spread of Aqua MODIS B8 versus SNPP VIIRS M1 ratio result is among the worst comparison results, whereas cases such as Aqua MODIS B5 versus SNPP VIIRS M8, as in Figure 8, spreads over a smaller 2% range. In general, ratio result is not an effective discriminator of statistical quality among SNO events given its large spread, and the final selection of the times series events should not rely on using ratio. On the other hand, error bar result, as shown by the bottom panels of Figure 13 as well as in earlier figures, has demonstrated to be stronger discriminator of statistical quality of SNO events that can be utilized as a selection filter.

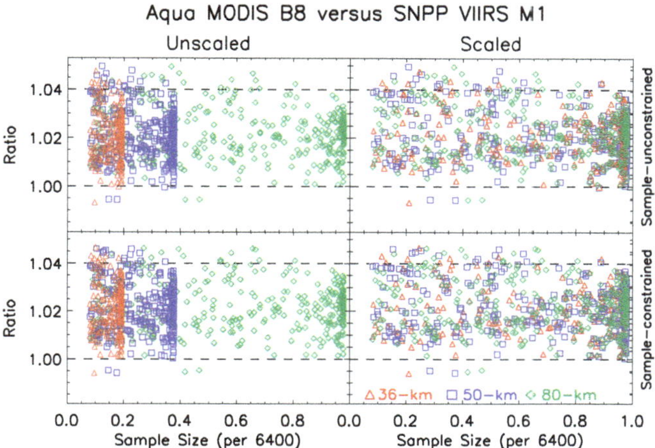

Figure 14. Ratio versus sample size for Aqua MODIS B8 versus SNPP VIIRS M1 for unscaled (top) and the scaled (top right) ratio result under the sample-unconstrained condition, and for the corresponding unscaled (bottom left) and scaled (bottom right) ratio results under the sample-constrained condition.

The scaling phenomenon exists in effectively identical fashion for all inter-RSB comparisons of Aqua MODIS versus SNPP VIIRS. Figure 15 shows the scaling phenomenon for six inter-RSB comparisons of Aqua MODIS versus SNPP VIIRS. Be it thin clouds, aerosol, or any combination of scene conditions, it appears that some atmospheric conditions in the polar scenes impact all RSBs in nearly identical way. A general implication is that any inter-RSB comparison between any two polar-orbiting multispectral sensors that generate SNO scenes over the polar regions necessarily needs to take this scene-based effect into account.

Figure 16 demonstrates that the scaling phenomenon also exists for Terra MODIS versus SNPP VIIRS, exemplified by Terra MODIS B8 versus SNPP VIIRS M1. As Terra versus SNPP SNO events trace out completely different locations (see Figure 2), this result generalizes this scene-based variability over both northern and southern polar scenes.

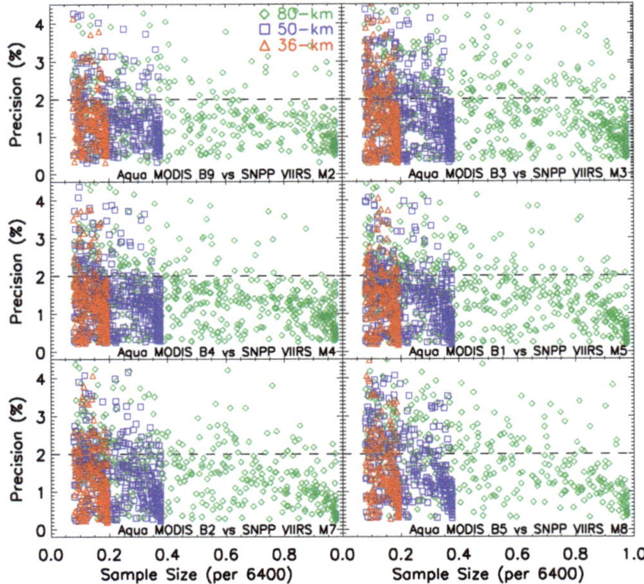

Figure 15. Precision versus sample size for six inter-RSB comparisons of Aqua MODIS versus SNPP VIIRS under no sample constraint demonstrating scaling phenomenon.

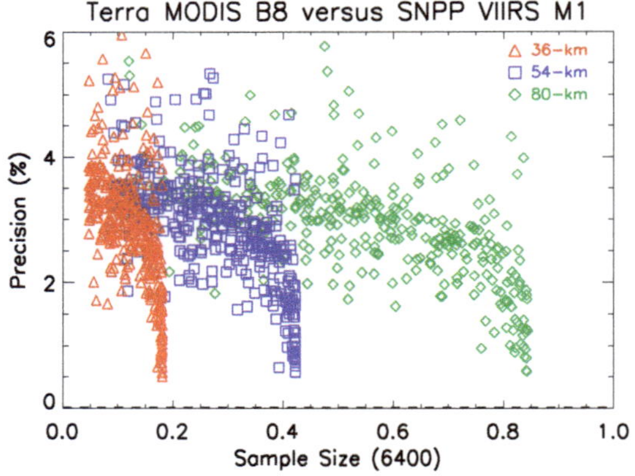

Figure 16. Precision versus sample size for Terra MODIS B8 versus SNPP VIIRS M1 under no sample constraint demonstrating scaling phenomenon.

3.8. Scale-Dependence in Sentinel-3A OLCI versus SNPP VIIRS

Sentinel-3A OLCI is yet without enough SNO data to demonstrate the scaling phenomenon in full a time series result, but the scale-dependence can be examined within individual SNO events as done in Figure 6. Figure 17 shows the dependence of ratio (top) and error bar (bottom) on area scale for Sentinel-3A OLCI Oa02 (412.5 nm) versus SNPP VIIRS M1 (410 nm), for a 13 April 2017 event for the three cases of unconstrained sample size (red triangles), constrained size at 1000 samples (blue squares), and constrained size at 500 samples (green diamonds).

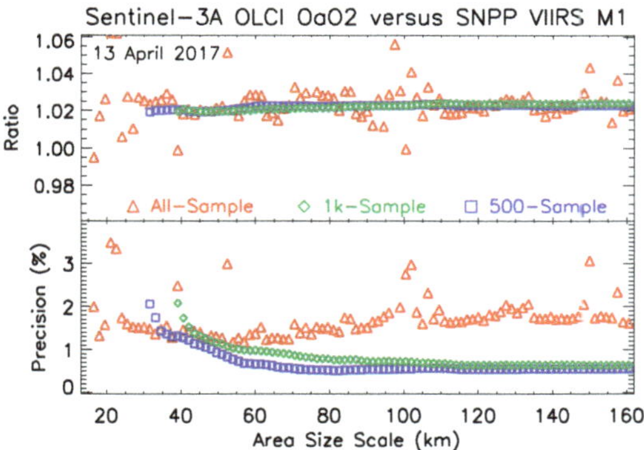

Figure 17. The scale-dependent result of radiometric comparison of Sentinel-3A OLCI Oa02 versus SNPP VIIRS M1 for the 13 April 2017 event for ratio (top panel) and the error bar (bottom panel) shown for the three cases of unconstrained sample size (red triangles), constrained size at 1000 samples (blue squares), and constrained size at 500 samples (green diamonds).

All features of the OLCI-based result effectively repeat identically. This result reinforces the recommendation to confine the SNO analysis to a "nadir-only" condition using small area and that scaling phenomenon is a general effect impacting any inter-RSB comparison of two polar-orbiting instruments.

3.9. Discussion and Summary

The key finding is that a homogeneity-ranked, sample size constrained sampling procedure under a small-area restriction stabilizes the ratio against some broad-scale variability to generate result that is reliable and robust. A smaller area size such as under the 50-km scale contains enough pixels for the refined sampling procedure but simultaneously avoids the pitfall of large-area or large-angle bias. As the ratio result has been stabilized, the application of various criteria, such as scale or homogeneity threshold, is further shown to have impact on the comparison time series.

Since Aqua MODIS B5 versus SNPP VIIRS M8 is one the most stable inter-RSB comparisons due to good spectral match and long-term radiometric stability, the average precision of the time series at ~1.0% very well represents the general statistical capability of inter-RSB comparison at the 1-km regime. While the clear-scene result such as in Figure 5 is remarkably stable and precise at 0.2% or so, its number is not sufficient for full evaluation. In general, radiometric comparison time series are best used as a tool of discovery of deviating features such as the multiyear drift.

Also important is the generality of the scene-based variability over both polar regions as shown in the inter-RSB comparison results of MODIS and OLCI versus SNPP VIIRS. Therefore, any inter-RSB comparisons of polar-orbiting multispectral sensors necessarily need to treat this polar scene variability with some care.

4. Capability at Different Regimes of Spatial Resolution

An assessment of the intercomparison at finer regimes of spatial resolution provides an understanding of what capability can be achieved in the coming era. For this purpose, the inter-RSB comparisons using SNPP VIIRS M7 (862 nm; 750 m) and I2 (862 nm; 375 m) against Aqua MODIS B2 (859 nm; 250 m), and Sentinel-3A OLCI Oa17 (865 nm; 300 m) are used to test four regimes of spatial resolution. Because the RSRs of SNPP VIIRS M7 and I2 are effectively identical, intercomparisons against them directly shows the impact of different spatial resolutions. In addition, Aqua MODIS B2, at 250-m native spatial resolution, also comes with aggregated data at 500-m and 1-km resolutions (Table 2) and provides direct testing of different spatial resolutions

The four regimes of spatial resolutions to be tested are described as follows. First, SNPP VIIRS M7, at 750-m, can be matched with the aggregate 1-km and 500-m data of Aqua MODIS B2, generating comparisons at the 1-km and 750-m regimes. Second, SNPP VIIRS I2, at 375-m, can be matched with the aggregate 500-m and the native 250-m data of Aqua MODIS B2, generating comparisons at the 500-m and 375-m regimes. For each pairing, the regime of intercomparison is defined by the lower spatial resolution. For Sentinel-3A OLCI at 300 m, the match with SNPP VIIRS M7 will be at the 750-m regime, and the match with SNPP VIIRS I2 will be at the 375-m regime.

Figure 18a shows the time series of Aqua MODIS B2 versus SNPP VIIRS M7 at the 1-km (blue squares) and the 750-m (green stars) regime for the first six years of SNPP VIIRS mission; Figure 18b shows the time series of Aqua MODIS B2 versus SNPP VIIRS I2 at the 500-m (red triangles) regime and the 375-m (cyan crosses). The precision threshold for each SNO event is 3%. The two times series in each plot have been carefully selected and matched to allow unambiguous event-to-event comparison. The time series cleansed and used here for illustration are otherwise slightly different from result strictly from the prescribed constrained procedure. The key and unexpected finding is that the three finer regimes appear only fractionally better than the 1-km regime—this hints at a lower limit of the statistical capability of the inter-RSB comparison methodology, or perhaps an additional physical effect at the level of 750-m scale. This can be an issue worthy of future pursuit.

For a more explicit demonstration, Figure 18c shows the precision result ranked from the tightest to the worst, but using the SNO events of the 375-m regime time series shown in Figure 18b as the reference of ranked events—the purpose is to reveal the statistical quality of individual events at different regimes. The SNO events of the 375-m regime (cyan crosses) are first sorted according to their precision from best to worst, and then results of other three regimes following the same SNO event sequence are plotted accordingly. That is, the 1-km, 750-m and the 500-m regime result are not separately sorted, but follow the same sorting event-by-event as that of the sorted 375-m regime result for comparison.

First, all of the most precise SNO events converge toward the beginning of the plot at about 0.15% to 0.2% precision, and this is because of the excellent homogeneity of clear-scene events. This indicates the comparison analysis has the inherent capability to reach 0.15% level. Second, the ranked result shows different intervals of slightly different pattern—a smooth pattern up to event 200 under 0.6% precision, followed by a stronger increasing pattern with more noise from event 200 to 350 and up to 2% precision, and finally the sharply rising and noisy pattern after event 350 and 2% precision. This even-by-event showing of the precision quality reveals the how precision threshold may be decided for a time series. For these cases, a 1.0% precision threshold seems a good balanced choice between having tight error bars and number of events. Third and most importantly as a focus of this examination, the result of the 750-m (green stars) and 500-m (red triangles) regimes can be seen to evenly straddle around the 375-m (cyan crosses) regime result, showing consistent agreement among the three finer regimes. On the other hand, the 1-km precision result (blue squares) is on the average higher than the result of three other regimes, as already revealed in Figure 18a,b. The capability of the radiometric intercomparison methodology, at least in the context of the constrained procedure, may have reached optimal result at the 750-m regime.

The time series also reveal some deviating features indicative of some basic on-orbit calibration issues. Although it may deceptively appear that the time series exhibits long-term drift, the result is more consistent with a series of radiometric jumps, suggesting numerous calibration adjustments for Aqua MODIS B2 or SNPP VIIRS M7/I2.

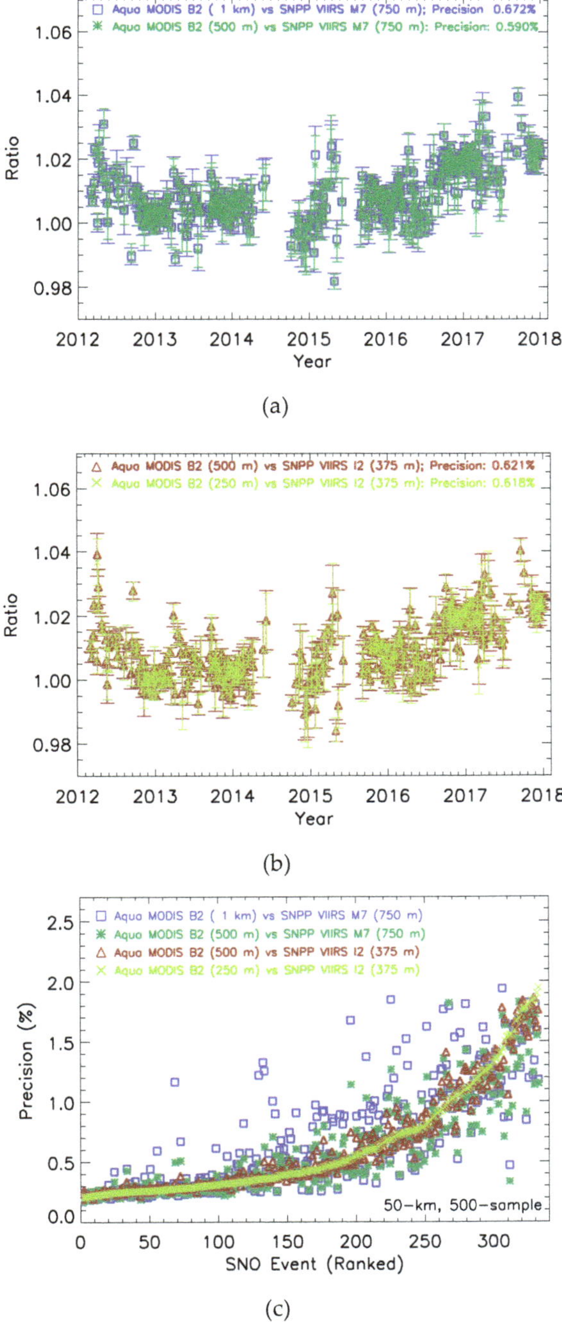

Figure 18. Results of four different regimes of intersensor comparison demonstrated by (**a**) Aqua MODIS B2 versus SNPP VIIRS M7 time series at the 1-km and 750-m regimes, (**b**) Aqua MODIS B2 versus SNPP VIIRS I2 time series at the 500-m and 375-m regimes, and (**c**) the precision versus ranked SNO events for all four cases.

The Aqua MODIS versus SNPP VIIRS results shown in Figure 18 are statistically dominated by events over the southern polar scenes, easily noticeable for the clustering of events during the Austral summer period from October to March. To demonstrate events over the northern region, Figure 19 shows the two inter-RSB comparisons of Sentinel-3A OLCI Oa17 (865 nm) versus SNPP VIIRS M7 at the 750-m regime (green diamonds), and versus I2 (862 nm) at the 350-m regime (orange crosses). The two OLCI-based times series are also time-matched to ensure event-by-event correspondence. The subset of the Aqua MODIS B2 versus SNPP VIIRS I2 comparison in Figure 18b occurring in the northern polar region is also shown (cyan diagonal crosses) in Figure 19 for comparison. The three precision results illustrate similar statistical performance at the 750-m and the 375-m regimes, at ~1%, with no clear advantage of the 375-m regime over the 750-m regime. The two OLCI-based times series also demonstrate an overall event-by-event consistency of precision between the two regimes, as also shown by the Aqua MODIS-based result in Figure 18. The combined findings of Aqua MODIS-based and OLCI-based results show that precision result for comparison under 1-km regimes in either polar regions is consistently at ~1% and slightly less.

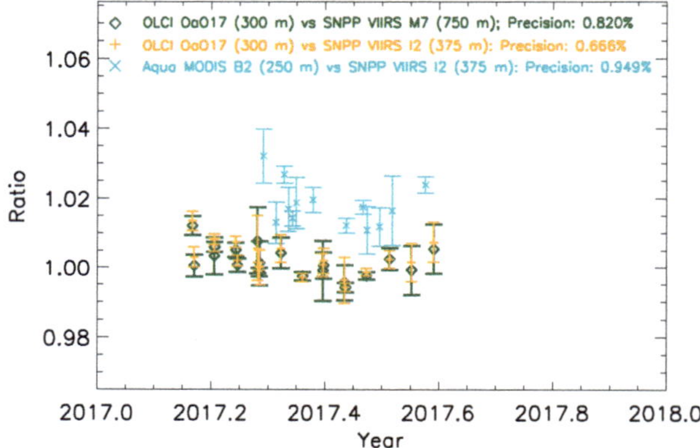

Figure 19. Inter-RSB comparisons of OLCI Oa17 versus SNPP VIIRS M7 and I2, occurring exclusively over the northern polar region, demonstrate the 750-m (green diamonds) and 375-m (orange crosses) regimes. The subset of Aqua MODIS B2 versus SNPP VIIRS I2 comparison occurring over the northern polar region (cyan diagonal crosses) is shown for comparison.

5. Multi-instrument Cross-Comparison

Intercomparison becomes even more useful when three or more sensors of comparable performance capability can be cross-checked. The next few figures exemplify the cross-comparisons of Aqua MODIS, Terra MODIS, and Sentinel-3A OLCI against SNPP VIIRS for the year 2017. The MODIS versus SNPP VIIRS comparison is carried out at the 1-km regime while that of OLCI versus SNPP VIIRS is at the 750-m regime. The time series are plotted over a 20% range centering on the time series means of the OLCI versus SNPP VIIRS, with two dashed lines marking the 2% level above and below the series mean. The applied precision threshold is 3%. The final figure shows the comparison result of three OLCI bands overlapping with SNPP VIIRS M5, explicitly demonstrating the impact of various level of mismatching RSRs. The impact of the spectral mismatch on time series remains one fundamental issue not yet adequately explored by the intersensor community.

5.1. Aqua MODIS, Sentinel-3A OLCI and SNPP VIIRS Comparisons

Figure 20a shows the three comparison time series using Aqua MODIS B8 (blue squares), Terra MODIS B8 (green triangles), and Sentinel-3A OLCI Oa02 (red crosses) against SNPP VIIRS M1. It is seen that OLCI Oa02-based result is stable within 1% without strong seasonal modulation and short-term drift. On the other hand, the Aqua MODIS B8-based result reveals a 3% peak-to-trough seasonal variation beyond the 1% error bar, while the Terra MODIS B8-based result is of 2% seasonal variability along with a 2% difference with that of Aqua MODIS B8. Certainly, spectral mismatch between MODIS B8 and VIIRS M1 can induce seasonal pattern in both time series, but it remains possible that some physical or optical effect is impacting both Terra and Aqua MODIS B8. In addition, the discrepancy between the two MODIS-based results points to some inconsistency in the on-orbit calibration of MODIS B8.

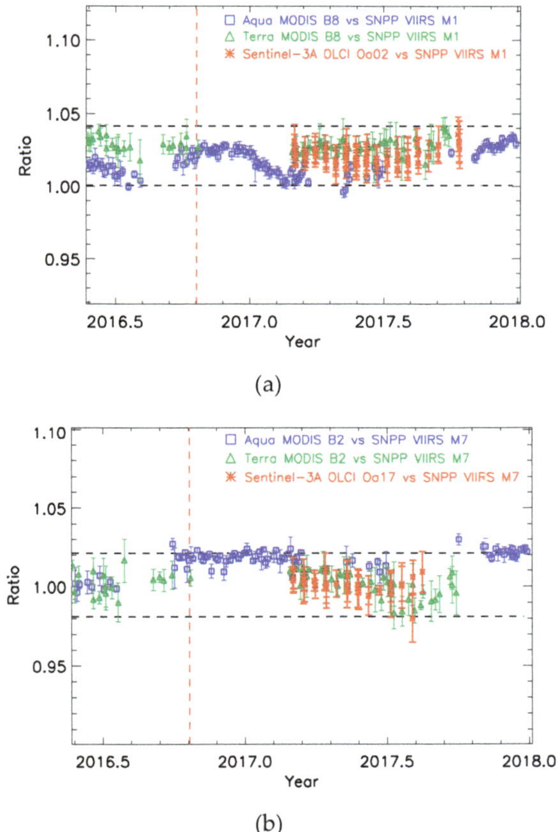

Figure 20. The radiometric comparison time series, from May 2017 to September 2017, of (**a**) MODIS B8 and Sentinel-3A OLCI Oa02 versus SNPP VIIRS M1 and (**b**) MODIS B2 and Sentinel-3A OLCI Oa17 versus SNPP VIIRS M7.

Chu et al. [7] have previously concluded that IDPS-generated radiance for SNPP VIIRS M1 contains a long-term drift of approximately 0.4% over the four-year period from February 2012 to February 2016. For this recent 16-month period, the drift in IDPS-generated SNPP VIIRS M1 radiance is estimated to only ~0.15%, which is too small to be seen in these time series. The result suggests that OLCI Oa02 is not likely to have any significant short-term drift over the 16-month period.

Figure 20b shows the three comparison time series against SNPP VIIRS M7 using Aqua MODIS B2 (blue squares), Terra MODIS B2 (green triangles), and Sentinel-3A OLCI Oa17 (red stars). This set of bands is a clean case study due to well-matched RSRs, thus providing a good example of multisensor cross-check that can identify radiometric deviations. The precision is 1.15% for Aqua MODIS B2 result, 1.55% for Terra MODIS result, and 1.74% for OLCI Oa17 result. It is seen that both Terra MODIS and OLCI agree well with SNPP VIIRS M7, with time series consistent at ~1.0, as expected. However, Aqua

MODIS B2 shows a clear upward shift of ~2% against SNPP VIIRS M7 starting sometime between May and October 2017. This discontinuity had been documented in a preliminary study [30] but the times series is extended here to make clearer the upward discontinuity. Based on the overall result, it is concluded that Aqua MODIS B2 went through the 2% radiometric jump just before October 2016.

5.2. Impact of RSR Mismatch: Sentinel-3A Oa08-Oa10 versus SNPP VIIRS M5

The impact of the spectral coverage mismatch between two bands is nontrivial to quantify for intercomparison and is so far not well addressed or even well understood. The two primary effects of the mismatch are the offset from 1.0 in radiometric ratio and the emergence of yearly modulation. Here, only a demonstration of the effect is intended through an illustrative example using SNPP VIIRS M5 as a fixed reference and a set of three adjacent bands in Sentinel-3A OLCI. As shown in Table 2, the three OLCI RSBs, Oa08 (660–670 nm), Oa09 (670–677.5 nm), and Oa010 (677.5–685 nm), cover the 660 to 685 nm spectral region in sequence, with each having some spectral overlap with SNPP VIIRS M5 (662–682 nm).

The three inter-RSB comparison time series are shown in Figure 21 for the year 2017, for Oa08 (red stars), Oa09 (blue diagonal crosses), and Oa10 (green triangles). The three time series have nearly identical precision at 1.76%, yet differences among them are clearly shown. First, different radiometric offsets away from 1.0 expectedly show the dependence on the level of RSR mismatch. Second, while Oa08-based time series seems stable, both Oa09- and Oa10-based time series exhibit greater seasonal modulation, in particular, the Oa10-based result has the largest deviation at ~3%, not accounting for the three outliers below 0.95. This is the definitive demonstration of the different responses to the same set of SNO scenes arising only because the effect of mismatching RSRs. As Oa10 result indicates of it having the largest impact of the spectral mismatch with SNPP VIIRS M5, it shows both the largest downward offset and the most variable seasonal modulation in a consistent manner that is expected. However, what is not clear is how certain mismatch has less impact on the time series than others, such as OLCI Oa08-based result having weaker modulation. Nevertheless, the connection between the offset and the seasonal modulation is direct, that both being the manifestation of the spectral mismatch. Specifically, this connection may be useful for quantifying the impact of spectral mismatch.

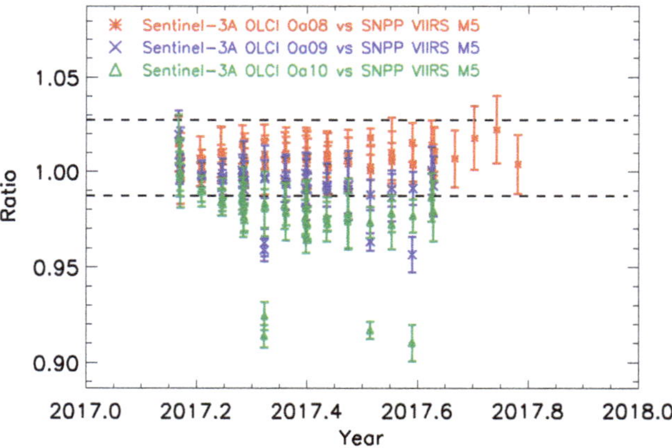

Figure 21. The inter-RSB comparisons of Sentinel Oa08–Oa10 with SNPP VIIRS M5 demonstrating the impact of different level of spectral mismatch. The precision threshold is 3%.

The three outliers are briefly discussed here as an instance of multimodality. The outliers of each time series correspond to the same SNO events of the other two but at different ratios. These cases can arise from some scenes of less stable condition, such as cloudy or ocean scenes, which on occasions can still be stable enough to pass selection criteria. These cases are technically the result of a different mode, arising from the nontrivial effect of mismatching RSRs responding to different scene conditions. As has been pointed out previously, the presence of outliers or additional modes is one scenario where a targeted removal of certain scene condition, such as cloud, can be applied.

6. General Discussions

Intercomparison of radiometric data, as in any statistical sampling, is not entirely useful without a reliable estimate of error bars. The procedure described herein establishes the reliability of the error bars, or precision, of the comparison events, further making error bar a usable discriminator for selecting best-quality SNO events. The overall result shows that an overall 1% precision is reachable at the 1-km resolution. The constructed multiyear time series, "as is" without adjustment, are capable of capturing various features illustrative of some underlying radiometric or calibration issues as listed below.

1. Long-term drift reveals a systematically worsening error in the on-orbit calibration of the sensor data, as exemplified by Aqua MODIS B4 versus SNPP VIIRS B4 in Figure 11.

2. Sudden radiometric shift reveals a likely one-time calibration adjustment or instrument change as exemplified by the jumps in Aqua MODIS B2 versus SNPP VIIRS M7 before Oct 2016 shown in Figure 20b.

3. Noise and variability reveals scene-based effects as exemplified by almost all inter-RSB comparisons of Aqua MODIS and SNPP VIIRS in Figures 11–17.

4. Seasonal modulation reveals impact of RSR or other physical effects as exemplified by Aqua MODIS B8 versus SNPP VIIRS M1 in Figure 20a and most demonstratively the three OLCI-based time series in Figure 21. Multimodality can also manifest from RSR mismatch. Definitively the seasonal modulation is not an issue related to the SZA.

5. Non-seasonal and sporadic shifts reveal possible calibration instability as exemplified by Aqua MODIS B2 versus SNPP VIIRS M7/I2 in Figure 18a,b.

6. Discrepancy between the different intercomparisons, in addition to possibilities listed above, can reveal additional biases and calibration inconsistencies, as exemplified in the cross-comparisons of Figures 19–21.

So far, this study focuses on the on-orbit performance of the multispectral sensor data in the context of standard operational on-orbit RSB calibration. But the complete evaluation must include sensor data over all extent beyond nadir. It is therefore important to continue to distinguish between the issues of on-orbit RSB characterization from those of other additional calibration adjustments. One such important associated issue is the time-dependent RVS effect of the scan mirror that is known for MODIS [20], although not known in VIIRS and not yet addressed in OLCI. The full calibration of the sensor data for MODIS Collection 6 [19,20] involves additional correction necessary to mitigate this angle-dependent effect throughout the entire spatial extent that cannot be analyzed by the standard on-orbit calibration analysis. While the "nadir-only" framework of intercomparison can expose issues of standard operational on-orbit calibration, it is not designed to address any large-extent issues such as RVS. Nevertheless, this study puts forth a spatial scale-dependent analysis possibly extendable to examine off-nadir issues. The result of this study supports a strategy to first isolate and examine on-orbit calibration before studying other effects.

Nevertheless, some built-in limitations are difficult to overcome, including narrow-band dynamic range, lack of spectral counterparts, or simply missing data. Approaches entirely different from intercomparison, such as using stable Earth scenes, even if less reliable, must necessarily be included to build a full-evaluation strategy. This study also does not isolate the impact of geolocational error, but the overall result highly suggests geolocational issue not to be significant. Regardless, the increasing number of high-performing multispectral sensors in and to be in operation definitively expands the overall usability of intercomparison. OLCI is a prime example—given its dense spectral coverage from 400 nm to 900 nm by 21 bands, 300-m spatial resolution and the built-in on-orbit RSB calibration capability—of a new a powerful radiometric reference in the VIS/NIR range.

Lastly, a recent study by Chu and Dodd [31] demonstrates that the radiometric intercomparison of MODIS versus SNPP VIIRS thermal emissive bands (TEBs) can be analyzed under the "nadir-only" framework, along with homogeneity-ranked and sample size constrained procedure. Although an in-depth study of the capability of the radiometric intercomparison has not yet been carried out for TEBs, the general applicability of the prescribed procedure to RSBs and TEBs is expected.

7. Conclusions

The capability of the radiometric intersensor comparison of multispectral sensors using four major sensors has been examined to attain robust 1% precision and better in multiyear time series. The "nadir-only" restriction of SNO-based comparison analysis provides a framework within which the operational on-orbit RSB calibration performance can be evaluated in isolation from other issues arising from larger area size or viewing angles, such as the RVS effect or scene-BRDF. With the use of pixel-based homogeneity and sample size constraint, the procedure successfully stabilizes ratio, tightens error bars, and makes fuller time series. The procedure makes error bar a meaningful discriminator of SNO events of varying level of statistical quality. A well-behaved time series can attain even better precision making it possible to detect a persistent multiyear drift as small as 0.3%. This study also clarifies that the application of targeted removal algorithm, such as cloud removal, not to be effective in overcoming variability at least not on the level of reaching 1% result. Various issues are also discussed and presented, such as the SZA impact not being important under the "nadir-only" framework, the impact of RSR mismatch to be radiometric ratio offset and seasonal modulation, and that the 2% scene-based effect, loosely called the "scaling phenomenon", is pervasively present in both the northern and southern polar scenes to affect all polar-orbiting RSBs. However, arguably the most important aspect is the multisensor cross-comparisons made even more useful by the 1% precision capability. Limitations in intercomparison certainly exist, and the lack of spectrally matching bands between sensors is arguably the most basic one making full intercomparison impossible, thus requiring a more comprehensive strategy. Nevertheless, this study strengthens intersensor comparison as a powerful tool of monitoring and discovery for multispectral sensors in the coming era.

Author Contributions: M.C. is responsible for initiating and leading this effort. J.D. has provided significant contribution to operational information, data, processing, and plotting using python.

Funding: This research received no external funding.

Acknowledgments: The authors thank Junqiang Sun for continual support.

Conflicts of Interest: The authors declare no conflicts of interest.

References

1. Barnes, W.L.; Salomonson, V.V. MODIS: A global imaging spectroradiometer for the Earth Observing System. *Crit. Rev. Opt. Sci. Technol.* **1993**, *CR47*, 285–307.
2. Guenther, B.; Barnes, W.; Knight, E.; Barker, J.; Harnden, J.; Weber, R.; Roberto, M.; Godden, G.; Montgomery, H.; Abel, P. MODIS Calibration: A brief review of the strategy for the at-launch calibration approach. *J. Atmos. Ocean. Technol.* **1996**, *12*, 274–285. [CrossRef]
3. Suomi NPP Home Page. Available online: https://www.nasa.gov/mission_pages/NPP/main/index.html (accessed on 1 February 2018).
4. Cao, C.; Deluccia, F.; Xiong, X.; Wolfe, R.; Weng, F. Early on-orbit performance of the Visible Infrared Imaging Radiometer Suite onboard the Suomi National Polar-orbiting Partnership (S-NPP) satellite. *IEEE Trans. Geosci. Remote Sens.* **2014**, *52*, 1142–1156. [CrossRef]
5. Wu, A.; Xiong, X.; Cao, C.; Chiang, K. Assessment of SNPP VIIRS VIS/NIR radiometric calibration stability using Aqua MODIS and invariant surface targets. *IEEE Trans. Geosci. Remote Sens.* **2016**, *54*, 2918–2924. [CrossRef]

6. Uprety, S.; Blonski, S.; Cao, C. On-orbit radiometric performance characterization of S-NPP VIIRS reflective solar bands. In Proceedings of the Earth Observing Missions and Sensors: Development, Implementation and Characterization IV, New Delhi, India, 4–7 April 2016; Volume 9881, p 98811H.

7. Chu, M.; Sun, J.; Wang, M. Performance evaluation of on-orbit calibration of SNPP reflective solar bands via intersensor comparison with Aqua MODIS. *J. Atmos. Ocean. Technol.* **2018**, *35*, 385–403. [CrossRef]

8. Cao, C.; Heidinger, A.K. Inter-comparison of the long-wave infrared channels of MODIS and AVHRR/NOAA-16 using simultaneous nadir observations at orbit intersections. In *Earth Observing Systems VII*; Barnes, W.L., Ed.; International Society for Optical Engineering: Bellingham, WA, USA, 2002; Volume 4814, pp. 306–316. [CrossRef]

9. Heidinger, A.K.; Cao, C.; Sullivan, J.T. Using Moderate Resolution Imaging Spectroradiometer (MODIS) to calibrate Advanced Very High Resolution Radiometer reflectance channels. *J. Geophys. Res.* **2002**, *107*, 4702. [CrossRef]

10. Cao, C.; Weinreb, M.; Xu, H. Predicting simultaneous nadir overpasses among polar-orbiting meteorological satellites for the intersatellite calibration of radiometers. *J. Atmos. Ocean. Technol.* **2004**, *21*, 21537–21542. [CrossRef]

11. Chander, G.; Hewison, T.J.; Fox, N.; Wu, X.; Xiong, X.; Blackwell, W.J. Overview of Intercalibration of Satellite Instruments. *IEEE Trans. Geosci. Remote Sens.* **2013**, *51*, 1056–1080. [CrossRef]

12. Donlon, C.; Berruti, B.; Buongiorno, A.; Ferreira, M.-H.; Féménias, P.; Frerick, J.; Goryl, P.; Klein, U.; Laur, H.; Mavrocordatos, C.; et al. The global monitoring for environment and security (GMES) sentinel-3 mission. *Remote Sens. Environ.* **2012**, *120*, 37–57. [CrossRef]

13. JPSS Series Satellites: NOAA-20 Home Page. Available online: https://www.nesdis.noaa.gov/jpss-1 (accessed on 1 February 2018).

14. Tabata, T.; Andou, A.; Bessho, K.; Date, K.; Dojo, R.; Hosaka, K.; Mori, N.; Murata, H.; Nakayama, R.; Okuyama, A.; et al. Himawari-8/AHI latest performance of navigation and calibration. In Proceedings of the Earth Observing Missions and Sensors: Development, Implementation, and Characterization IV, New Delhi, India, 4–7 April 2016; Volume 9881, p. 98811H.

15. Meteorological Satellite Center (MSC) of JMA, Himawari-8 Imager (AHI) Home Page. Available online: https://www.data.jma.go.jp/mscweb/en/himawari89/space_segment/spsg_ahi.html.

16. Schmit, T.J.; Gunshor, M.M.; Menzel, W.P.; Gurka, J.J.; Li, J.; Bachmeier, A.S. Introducing the next-generation Advanced Baseline Imager on GOES-R. *Bull. Am. Meteorol. Soc.* **2005**, *86*, 1079–1096. [CrossRef]

17. Schmit, T.J.; Griffith, P.; Gunshor, M.M.; Daniels, J.M.; Goodman, S.J.; Lebair, W.J. A Closer Look at the ABI on the GOES-R Series. *Bull. Am. Meteorol. Soc.* **2017**, *98*, 681–698. [CrossRef]

18. GOES-R Series Home Page. Available online: https://www.goes-r.gov (accessed on 10 June 2019).

19. Sun, J.; Angal, A.; Xiong, X.; Chen, H.; Geng, X.; Wu, A.; Choi, T.; Chu, M. MODIS reflective solar bands calibration improvements in Collection 6. In Proceedings of the Earth Observing Missions and Sensors: Development, Implementation, and Characterization II, Kyoto, Japan, 29 October–1 November 2012; Volume 8528, p. 85280N.

20. Sun, J.; Xiong, X.; Angal, A.; Chen, H.; Wu, A.; Geng, X. Time-dependent response versus scan angle for MODIS reflective solar bands. *IEEE Trans. Geosci. Remote Sens.* **2014**, *52*, 3159–3174. [CrossRef]

21. NASA EARTHDATA: LAADS DAAC Home Page. Available online: https://ladweb.modaps.eosdis.nasa.gov (accessed on 1 February 2018).

22. NOAA CLASS Home Page. Available online: https://www.bou.class.noaa.gov (accessed on 1 February 2018).

23. ESA Copernicus Open Access Hub Home Page. Available online: https://schub.copernicus.eu (accessed on 1 February 2018).

24. Tansock, J.; Bancroft, D.; Butler, J.; Cao, C.; Datla, R.; Hansen, S.; Helder, D.; Kacker, R.; Latvakoski, H.; Mylnczak, M.; et al. Guidelines for Radiometric Calibration of Electro-Optical Instruments for Remote Sensing, *Space Dynamics Lab Publications 2015*. Paper 163. Available online: https://digitalcommons.usu.edu/sdl_pubs/163 (accessed on 10 June 2019). [CrossRef]

25. Chu, M.; Sun, J.; Wang, M. Radiometric Evaluation of SNPP VIIRS Band M11 via Sub-Kilometer Intercomparison with Aqua MODIS Band 7 over Snowy Scenes. *Remote Sens.* **2018**, *10*, 413. [CrossRef]

26. Sun, J.; Wang, M. Visible Infrared Imaging Radiometer Suite solar diffuser calibration and its challenges using solar diffuser stability monitor. *Appl. Opt.* **2014**, *53*, 8571–8584. [CrossRef] [PubMed]

27. Sun, J.; Wang, M. On-orbit calibration of the Visible Infrared Imaging Radiometer Suite reflective solar bands and its challenges using a solar diffuser. *Appl. Opt.* **2015**, *54*, 7210–7223. [CrossRef] [PubMed]
28. Sun, J.; Wang, M. Radiometric calibration of the Visible Infrared Imaging Radiometer Suite reflective solar bands with robust characterizations and hybrid calibration coefficients. *Appl. Opt.* **2015**, *54*, 9331–9342. [CrossRef] [PubMed]
29. Sun, J.; Chu, M.; Wang, M. Degradation nonuniformity in the solar diffuser bidirectional reflectance distribution function. *Appl. Opt.* **2016**, *55*, 6001–6016. [CrossRef] [PubMed]
30. Chu, M.; Sun, J.; Wang, M. The inter-sensor radiometric comparison of SNPP VIIRS reflective solar bands with Aqua MODIS updated through June 2017. In Proceedings of the Earth Observing Systems XXII, San Diego, CA, USA, 6–10 August 2017; Volume 10402, p. 1040222.
31. Chu, M.; Dodd, J. Examination of Radiometric Deviations in Bands 29, 31 and 31 of MODIS. *IEEE Trans. Geosci. Remote Sens.* **2019**. [CrossRef]

Article

The 10-Year Return Levels of Maximum Wind Speeds under Frozen and Unfrozen Soil Forest Conditions in Finland

Mikko Laapas [1],*, Ilari Lehtonen [1], Ari Venäläinen [1] and Heli M. Peltola [2]

[1] Weather and Climate Change Impact Research, Meteorological and Marine Research Programme, Finnish Meteorological Institute, P.O. Box 503, FI-00101 Helsinki, Finland: ilari.lehtonen@fmi.fi (I.L.); ari.venalainen@fmi.fi (A.V.)

[2] School of Forest Sciences, Faculty of Science and Forestry, University of Eastern Finland, Yliopistonkatu 7, FI-80101 Joensuu, Finland; heli.peltola@uef.fi

* Correspondence: mikko.laapas@fmi.fi

Received: 8 March 2019; Accepted: 29 April 2019; Published: 30 April 2019

Abstract: Reliable high spatial resolution information on the variation of extreme wind speeds under frozen and unfrozen soil conditions can enhance wind damage risk management in forestry. In this study, we aimed to produce spatially detailed estimates for the 10-year return level of maximum wind speeds for frozen (>20 cm frost depth) and unfrozen soil conditions for dense Norway spruce stands on clay or silt soil, Scots pine stands on sandy soil and Scots pine stands on drained peatland throughout Finland. For this purpose, the coarse resolution estimates of the 10-year return levels of maximum wind speeds based on 1979–2014 ERA-Interim reanalysis were downscaled to 20 m grid by using the wind multiplier approach, taking into account the effect of topography and surface roughness. The soil frost depth was estimated using a soil frost model. Results showed that due to a large variability in the timing of annual maximum wind speed, differences in the 10-year return levels of maximum wind speeds between the frozen and unfrozen soil seasons are generally rather small. Larger differences in this study are mostly found in peatlands, where soil frost seasons are notably shorter than in mineral soils. Also, the high resolution of wind multiplier downscaling and consideration of wind direction revealed some larger local scale differences around topographic features like hills and ridgelines.

Keywords: boreal region; extreme wind speed; wind climate; soil frost; wind damage risk management; wind multiplier; downscaling; topography; surface roughness

---

1. Introduction

In the last few decades, wind storms have caused the most damage and economic losses in European forests, compared to all abiotic and biotic damage agents [1–4]. So far, winter storms have caused the most destructive damage in Western and Central Europe [3,5,6], e.g., storms like Vivian in 1990 (over 100 million m^3 of timber), Lothar and Martin in 1999 (over 175 million m^3), Kyrill in 2007 (54 million m^3) and Klaus in 2009 (50 million m^3), respectively. Damages have increased in recent years also in northern Europe [4,6,7], where in 2015 Gudrun damaged 70 million m^3 and in 2007 Per damaged 12 million m^3 of timber, mainly in Sweden. In Finland, over 25 million m^3 of timber has been damaged during storms since 2000, the most in autumn storms in 2001 (Pyry and Janika, 7.3 million m^3) and in summer storm in 2010 (Asta, Veera, Lahja and Sylvi, 8 million m^3), respectively. The increasing amount of damages in European forests may at least partially be explained by increasing volume of growing stock and changes in forest structure (e.g., age, tree species) related to changes in forest management practices [1,5,8,9]. Forest disturbances may also amplify or even cancel out the expected increase in productivity of forests under changing climate [4,10].

Some recent studies indicate increased storminess for some regions in Europe (see e.g., review by [11]). However, the majority of studies point towards decadal variation in storminess without any clear trend for a direction or another [12–15]. In Finland, slight weakening of annual mean (-0.09 ms^{-1} decade^{-1}) and maximum (-0.32 ms^{-1} decade^{-1}) wind speeds across 33 weather stations have been observed in the period of 1959–2015 [16], which is in accordance with widespread weakening of terrestrial near-surface wind speeds [17,18]. For future projections, the change in the extreme wind speed during the coming decades is still a somewhat unsolved issue and the outcome is largely dependent on the climate model used for the simulation [11,19,20].

However, the risk of wind damage to forests may still increase in Northern Europe under climate change even if the frequency and severity of wind storms do not increase. This is due to the shortening of the frozen soil period, which improves tree anchorage during the windiest season of the year from late autumn to early spring [21–24]. Moreover, storms may be accompanied by heavier rainfall, leading to more saturated soils and increased risk of wind damage [5]. When estimating the forest wind damage risk it is thus essential to know whether the extreme wind speeds occur during the frozen or unfrozen soil conditions. Typically, the windiest season in Finland is from October to March [25] and soil frost season starts in October–November and ends in April–May [26]. However, there is large year-to-year and regional variation in soil frost duration.

Even a 20 cm thick frozen soil increases the anchorage of trees and reduces substantially the risk of uprooting [27,28]. According to tree-pulling experiments in Finland, under frozen soil the type of failure was stem breakage, whereas under unfrozen soil conditions, about 80% of trees uprooted, respectively [28]. From the three economically and ecologically most important boreal tree species in Finland, Norway spruce (*Picea abies*) with the shallow rooting is the most vulnerable to uprooting, followed by Silver and Downy birches (*Betula pendula* and *Betula pubescens*), and Scots pine (*Pinus sylvestris*), respectively [27,28]. However, from late autumn to early spring, birches (without leaves) are not vulnerable to wind damage and therefore excluded from this study.

For snow free surfaces, the soil frost modelling can be done using cold season frost sum and soil characteristics alone [29]. The presence of snow and vegetation complicates the modelling, and requires a more sophisticated approach including the modelling of heat and water transfer [23]. An example of a relatively simple approach accounting for the main controlling factors was published by [30]. It was further developed and tested in the Finnish conditions by [31] for the calculation of soil temperatures in three common combinations of soil and forest types in Finland, i.e., dense Norway spruce stands on clay or silt soil, Scots pine stands on sandy soil, and Scots pine stands on drained peatlands. Soil frost conditions can vary a lot, even up to few months in mean duration, depending on soil type. Peat is effective insulator compared to mineral soils, therefore having shorter soil frost periods in similar climatic conditions [31].

The estimation of the return levels of maximum wind speed values (extreme winds) can be done using observational data representing conditions at the observing station location or using reanalyzed data like ERA-Interim [32], representing a larger area's averaged value, respectively. When studying the high-resolution spatial variation of extreme winds, the data has to be either downscaled from the reanalyzed coarse grid to a local value or upscaled from station point observations to areas located between the stations. Downscaling can be done by applying various spatial statistical tools, e.g., [33,34], or complex airflow models like e.g., WAsP [35], which are typically applied for wind power potential predictions. GIS-based methods for mapping the areas having highest wind damage risk have also been introduced, e.g., [36–38]. One computationally feasible approach for the estimation of the return levels of extreme wind speeds for large geographical areas with very high spatial resolution is the wind multiplier approach [39–41]. In this method, return levels obtained, e.g., from the reanalysed data, are downscaled to local wind speeds with help of land cover (roughness) and topography data. By applying GIS-tools such as ArcGIS, QGIS or R, it is rather straightforward to produce the required multipliers.

The reliable high-resolution information on the spatial variation of extreme wind speeds can enhance wind damage risk management in forest planning and forestry. In the above context, the objective of this study was to produce spatially detailed estimates (maps) of the 10-year return level maximum wind speed under current climate for unfrozen and frozen soil conditions in some of the most common combinations of forest and soil types in Finland. By utilizing soil frost calculations of [31] to determine the duration of soil frost seasons, the wind speed return level calculations were done for dense Norway spruce stands on clay or silt soil, Scots pine stands on sandy soil, and Scots pine stands on drained peatland. The coarse resolution estimates of the 10-year return level of maximum wind speed were based on 1979–2014 ERA-Interim dataset [32]. Downscaling to a 20 m grid was done by applying the wind multiplier approach [41].

2. Materials and Methods

2.1. Soil Frost Modelling

Soil frost conditions were modelled by using an extended version of the original soil temperature model [30]. It was derived from the law of conservation of energy and mass assuming constant water content in the soil. Model was further developed to take into account the heat flow below soil layer of consideration [42]. Following [42], soil temperature at depth Z_S (m) can be calculated as follows:

$$T_Z^{t+1} = T_Z^t + \frac{\Delta t * K_T}{(C_S + C_{ICE}) * (2 * Z_S)^2} * \left[T_{AIR}^t - T_Z^t \right] * \left[e^{-f_S * D_S} \right] + \frac{\Delta t * K_{T,LOW}}{(C_{S,LOW} + C_{ICE}) * 2 * (Z_l - Z_S)^2} * \left[T_{LOW} - T_Z^t \right], \qquad (1)$$

where T_Z^t (°C) is the soil temperature on a previous day, T_{AIR} (°C) is the air temperature, Δt is the length of a time step (s), K_T (W m^{-1} °C^{-1}) is the thermal conductivity of the soil above Z_S, C_S (J m^{-3} °C^{-1}) is the specific heat capacity of the soil above Z_S, C_{ICE} (J m^{-3} °C^{-1}) is the specific heat capacity due to freezing and thawing, f_S (m^{-1}) is an empirical damping parameter due to snow cover, D_S (m) is snow depth, $K_{T,LOW}$ (W m^{-1} °C^{-1}) is the thermal conductivity of the soil below Z_S, $C_{S,LOW}$ (J m^{-3} °C^{-1}) is specific heat capacity of the soil below Z_S, and T_{LOW} (°C) is soil temperature at the depth of Z_l. Following to [31], Z_l was set to 6.8 m.

By using soil temperature observations from several stations across Finland, the soil temperature model was parametrized for three different soil types: clay or silt soil, sandy soil, and peatlands [31]. Between the depths of 20 and 100 cm, the parametrized model explained approximately 90–99% of the observed variability in soil temperatures.

In a study by [31], also a snow depth model (based largely on the work of [43]) was used to simulate the snow depth, using daily temperature and precipitation observations [44], for different forest conditions in addition to open areas. In this study, we used the soil frost data calculated by [31] for different combinations of forest and soil types, based on combined use of soil temperature and snow depth model. The soil frost data in 0.1° × 0.2° grid has been calculated for dense spruce stands on clay or silt soil (hereafter CSS), pine stands on sandy soil (hereafter SP) and pine stands on peatlands (hereafter PP), respectively. Calculations for each of the forest and soil types were performed on every grid cell. The soil was assumed to be frozen and provide sufficient anchorage for trees when the modelled soil frost extended at least to a depth of 20 cm continuously from the surface and unfrozen otherwise. The expectation of the sufficient anchorage was based on the typical rooting depth of main boreal tree species, see e.g., [21,24,27,28].

2.2. Estimation of the 10-Year Return Levels of Wind Speed

The 10-year return levels, corresponding to an annual probability of exceeding the 90th percentile, of maximum wind speeds were calculated using the ERA-Interim dataset [32] covering years 1979–2014 and the generalized extreme value method (GEV) [45]. We used the block maxima approach, in this case for seasonal maximum wind speeds of both frozen and unfrozen soil season, with the maximum-likelihood fitting of GEV distribution [46,47]. We analysed 10-minute instantaneous wind speeds available at 6-hour intervals given as grid box averages, each covering an area of 0.75° × 0.75°.

The maximum wind speed dependence on wind direction was estimated by making the calculations wind direction wise, i.e., the 10-year return levels were estimated separately for cardinal and intercardinal wind direction sectors. For comparison and validation purposes, 10-year return levels were also calculated for 40 weather stations (Figure 1a) across Finland (on mainland) using wind speed observations covering the same period of 1979–2014 as in the ERA-Interim dataset. Observational data consisted of synoptic observations of 10-minute average wind speed with 3-hour measurement interval.

Compared to our data period of 1979–2014, the 10-year return level estimates can be expected to be quite robust. Estimated 10-year return level, i.e., wind speed equalled or exceeded on average once every 10 years, is relatively short period compared to our data period of 35 years. However, in coastal regions of southern and southwestern Finland, uncertainties related to the statistical estimation of return level is somewhat increased. This applies particularly for PP as, based on soil frost calculations by [31], mild winters with no soil frost, or at least not exceeding 20 cm in depth, are quite common (not shown). In these cases, the dataset from which soil frost season return levels are calculated is smaller, leading to a wider return level estimate confidence intervals. Also, regions with only short soil frost period, when the window for seasons maximum wind speed can often be only e.g., 1–2 months, have larger variability in the used dataset for return level estimation, therefore increasing uncertainty even if totally frost-free years are rare.

Considering normal approximation 95% confidence intervals of calculated 10-year return level estimates for weather stations, range is on average +/−1 m/s over all 1920 return level calculations consisting of 40 stations, eight wind directions, three forest soil types, and distinction between frozen and unfrozen soil season, respectively.

2.3. Downscaling of the 10-Year Return Levels of Wind Speed

The impact of local terrain features on maximum wind speed cannot fully be taken into account in the relatively coarse 0.75° × 0.75° grid of ERA-Interim, because for example, hills, lakes, and changes in land-use are not considered in detail. For this reason, in order to downscale the wind speed return level from the coarse grid to a high-resolution grid we used a wind multiplier approach tested recently by [41] for boreal forest conditions. In this study, only topographic and terrain (surface roughness) properties are taken into account when assessing local maximum wind speeds (and their return levels) separately for the eight cardinal and intercardinal wind directions. For the application in forested landscapes, the shielding factor, i.e., the effect of upwind buildings providing cover to the place of interest and only relevant in urban areas, was not considered. The wind multiplier method has been presented earlier in [40,48] for more details.

Following the study by [41], the return level of regional maximum wind speed (U_R) in an open terrain at a 10-m height is downscaled into site-specific return level (U_{site}) by applying two multipliers, i.e., terrain multiplier (M_z), and topographic (hill-shape) multiplier (M_h):

$$U_{site} = U_R \times M_Z \times M_h \tag{2}$$

We used a 20 × 20-m grid, which is in line with the CORINE Land Cover 2012 dataset [49] providing the information on land cover and land use that enabled the calculation of terrain multiplier. When defining the terrain multiplier (M_z), we used a 500 m fetch length and weighted the grid points close to place of interest more than the further upwind grid points.

The topographic (hill-shape) multiplier (M_h) was calculated by taking into account the variations and change of elevation 1000 m upwind from the place of interest. As well, the elevation of the place of interest was taken into account. Development of the M_z and M_h multipliers used in this study were described more elaborately in the Sections 2.3 and 2.4 of [41]. According to [41], for areas with no extreme variations of elevation, the wind multiplier approach was a feasible method to identify at a high spatial resolution locations having the highest forest wind damage risks.

2.4. Comparison of Return Levels Derived from Point Observations, Reanalysed Data, and Wind Multiplier Downscaled Data

The 10-year return levels of maximum wind speeds calculated from the observations (OBS) of 40 weather stations (Figure 1a) were compared to corresponding return levels calculated from the original ERA-Interim reanalysed data (ERA) and return levels downscaled with wind multiplier approach (WM). Weather station location coordinates were used to derive data from ERA and WM gridded datasets.

Besides general visual scatterplot comparison of ERA and WM values to OBS values, also the coefficient of determination R^2, D statistic of two-sample Kolmogorov–Smirnov test, and mean differences were used to analyse the performance of ERA and WM to produce return level values similar to OBS. These statistics were also analysed at the station level. Comparisons were considered more at a qualitative than quantitative level, i.e., are return level values produced with WM approach improvement to original ERA values when considering similarity to values derived from weather station observations.

The D statistic of the two-sample Kolmogorov–Smirnov test was used as a measure of similarity of ERA and WM to OBS. Smaller values of D are considered as a good result, i.e., EDF (empirical distribution function) of WM is more similar than EDF of ERA compared to EDF of OBS. Mean difference statistic used was simply the mean of differences between ERA and WM to OBS at station level, over all the combinations of eight wind directions, three soil types, and distinction between frozen and unfrozen soil. Again, smaller values were considered as a good result as a difference between WM and OBS is smaller than a difference between ERA and OBS. R^2 was used as a goodness of fit of a simple linear regression between OBS and ERA or WM. Here, increasing R^2 was considered as an improvement when comparing a regression of OBS and WM to a regression of OBS and ERA.

2.5. Structure and Restrictions of Data Analyses

For deeper understanding of results for calculated return levels of maximum wind speeds and their differences between frozen and unfrozen soil, we first considered independently underlying soil frost conditions (e.g., number of soil frost days and duration of soil frost) and wind conditions (e.g., timing of maximum wind speeds throughout year and between frozen and unfrozen seasons), respectively.

We also restricted our analysis to mainland Finland (see Figure 1a). The reasoning for this is the lack of years with soil frost in the archipelago, leading to increased uncertainty in the calculation of wind speed return levels. Also, the insufficient performance of wind multiplier method for the small Baltic Sea islands found by [41] supports our decision.

The territory of Finland was moreover divided into three sub-regions in the analysis of results. The three sub-regions were based roughly on the mean annual growing degree day sum (GDD) calculated using the threshold of 5 °C. The limits are GDD > 1200 °C days for southern, 1000 °C days < GDD ≤ 1200 °C days for central and GDD ≤ 1000 °C days for northern sub-regions, following also roughly the borders of boreal subzones.

Also, one smaller area (30 × 30 km) from northern Finland (Figure 1a) with a more complex topography (Figure 1b) was used to examine and present the more local scale behavior and influences of wind multiplier downscaling to 10-year return levels of wind speeds and differences between frozen and unfrozen soil seasons.

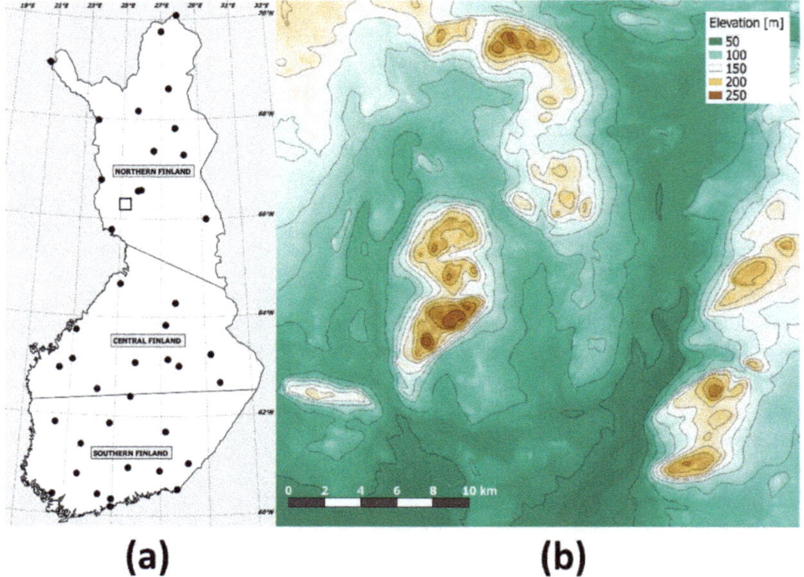

Figure 1. (**a**) Locations of 40 weather stations (black dots), division of the Finland into three parts, and location of detailed study area (square with black borders). (**b**) Topography and elevation (meters above sea level) of detailed study area.

3. Results

3.1. Soil Frost Conditions

Figure 2 presents the modelled annual mean, minimum and maximum number of soil frost days for three different forest and soil-type combinations in the period 1979–2014. In general, upland forests on sandy soil had the most soil frost days and forests on drained peatlands least. Duration of soil frost season in northern Finland was on average approximately 5–7 months for dense spruce stands on clay or silt soil (CSS) and pine stands on sandy soil (SP). For pine stands on peatlands (PP) soil frost season was considerably shorter, 3–5 months, with large spatial variability. In the central parts of the country, the season lasts on average about 3–4 months for PP and roughly 4–6 months for SP and CSS. Length of the average soil frost season in southern Finland range from less than two months in the coastal areas for PP up to five months for SP in the northern part of southern Finland.

However, especially for PP, frost-free seasons are possible almost everywhere in Finland. For SP, soil frost season can also be as short as about a month in southern and southwestern coastal areas. Conversely for SP, soil frost season is at least 5 months long in the whole northern part of Finland. The maximum length of soil frost season differs less from the average than the minimum. Here CSS and SP are quite similar, the maximum length of frost season ranging from about 5 months even at the coast to over 8 months in the most northwestern part of Finland. For PP, the longest soil frost periods are roughly a month shorter.

Years with zero soil frost days are virtually nonexistent in CSS and SP, but in PP there are rather large areas with roughly one-third of the 36-year study period with no soil frost (not shown). These areas are mainly in the southern part of Finland, but also in the northern parts, respectively. Years with less than 60 soil frost days are rare in CSS and SP apart from the southwestern part of Finland, where especially in coastal areas about every third year is this kind. Again, PP is substantially different with some areas having a majority of years with soil frost season less than 60 days.

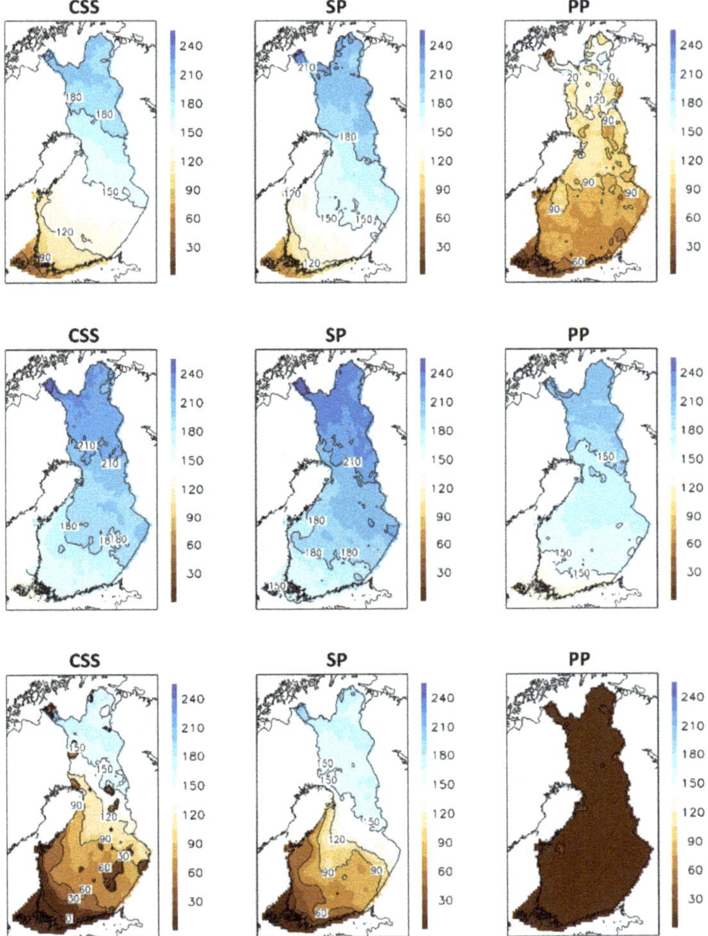

Figure 2. Annual mean (top row), maximum (middle row), and minimum (bottom row) number of modelled soil frost days over the period 1979–2014 in three forest and soil types. CSS (spruce on clay/silt), SP (pine on sand), PP (pine on peat).

3.2. Wind Conditions

We found large year-to-year variability in the timing of the annual maximum wind speed in the period 1979–2014. Among all 40 stations, the most common month for annual maximum wind speed was December (Figure 3a). However, also months from October to May are rather common for annual maximum wind speeds. Direction wise the annual maximum wind speeds from N, NE, and E were most commonly observed during spring and early summer (April–June), whereas for rest of the directions it was usually observed from October to March (not shown).

It was rather common that the annual maximum wind speed was observed multiple times during a year, partly driven by wind observations having no digits in the first part of the study period. About 25% of all the years among 40 stations had annual maximum wind speeds observed on more than one month of that year. 49% for PP, 59% for CSS, and 62% for SP of these years were ones with similar maximums during frozen and unfrozen seasons (not shown).

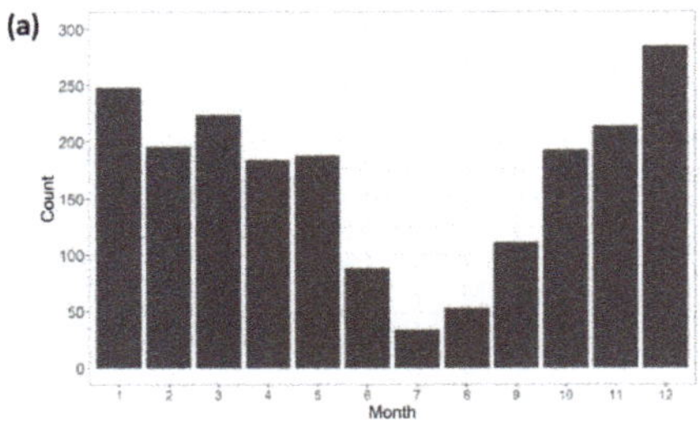

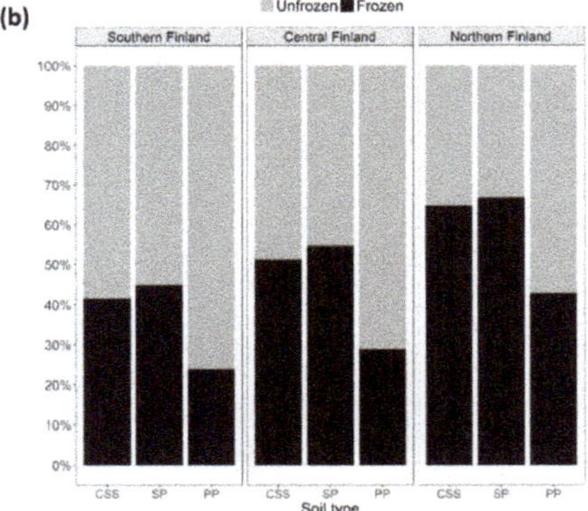

Figure 3. (**a**) Distribution of annual maximum wind speed observation months over 40 weather stations and period 1979–2014. (**b**) Proportion of annual maximum wind speed (1979–2014) observed either during frozen (black) or unfrozen (grey) soil frost season at weather stations located in southern, central, and northern Finland. CSS (spruce on clay/silt), SP (pine on sand), PP (pine on peat).

Figure 3b presents how observed annual maximum wind speed was split up between frozen and unfrozen seasons on different forest and soil type combinations in southern (14 stations), central (12 stations), and northern (14 stations) Finland. The annual maximum wind speed was observed rather evenly in both seasons in the case of CSS and SP, occurring slightly more often during unfrozen season in southern Finland and frozen season in northern Finland, respectively. For PP the difference was more pronounced, especially in southern and central Finland where annual maximum wind speed was clearly more often observed during unfrozen season.

We also further compared the underlying distributions of seasonal maximum wind speeds, from where return level estimates for 40 stations were calculated, using two-sample Kolmogorov–Smirnov tests to determine if there was statistically significant ($p < 0.05$) difference between frozen and unfrozen soil seasons. For CSS, difference was statistically significant in southern and northern, and non-significant in central Finland. For SP, significant in central and northern, and non-significant in southern Finland. And for PP, significant in southern and central, and non-significant in northern Finland, respectively.

3.3. The 10-Year Return Levels of Maximum Wind Speed for Frozen vs. Unfrozen Soil

Generally differences in maximum 10-year return level of wind speed between seasons of frozen and unfrozen soil are rather small (difference +/−1 m/s) in large part of Finland. Small differences were observed especially for CSS (Figure 4a maps) and SP (Figure 4b maps), of which results as a whole resemble each other closely with similar spatial patterns and direction of differences. On large scale, larger differences were observed for CSS and SP only in parts of northernmost Finland and in the coastal are of southwestern Finland, i.e., maximum 10-year return level of wind speed was about 1–2 m/s larger in soil frost season. Differences larger than +/−1 m/s were a bit more common for PP

(Figure 4c maps). Areas with stronger winds during the unfrozen season were found across coastal areas, parts of eastern Finland, and northernmost Finland, respectively. Noteworthy, compared to CSS and SP, sign of the difference is opposite in the coastal areas and in the most northwestern part of Finland. Notable positive differences in PP restricts to western and central parts of Lapland.

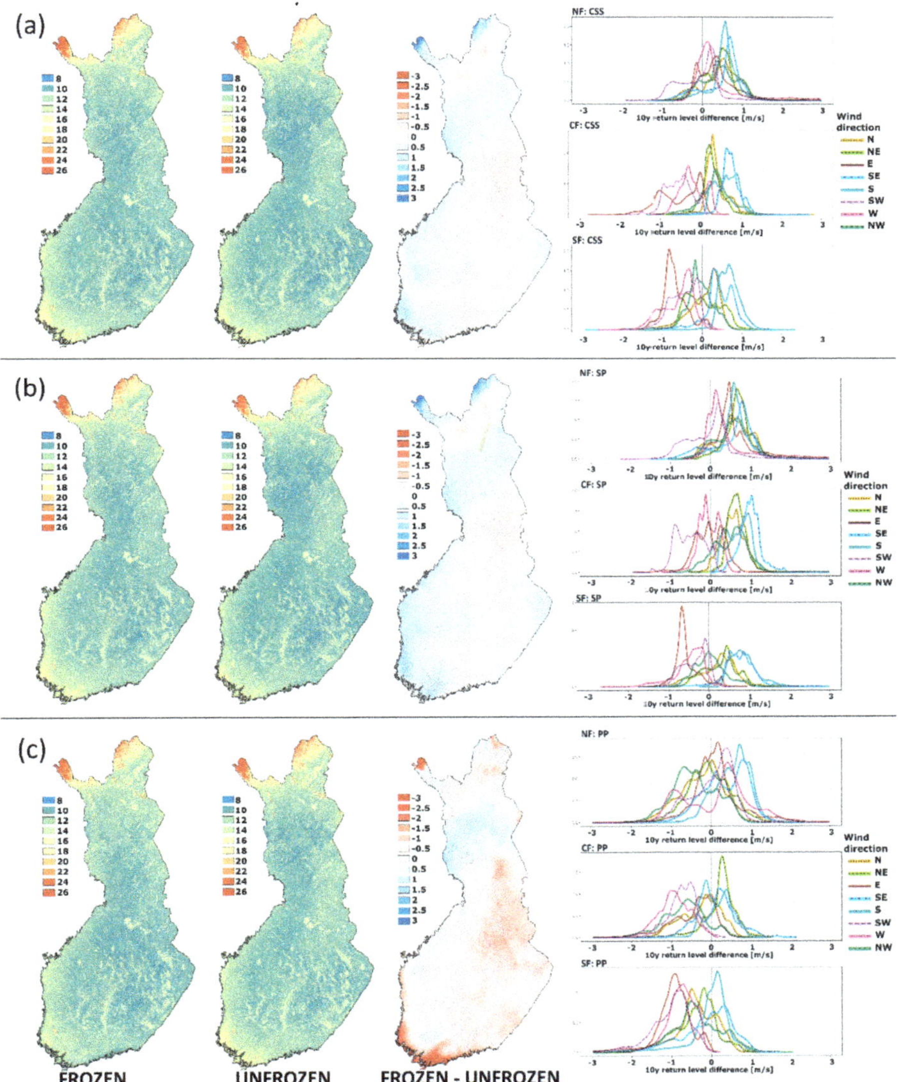

Figure 4. Aggregated maps presenting the maximum of 10-year return level of maximum wind speed [m/s] from cardinal and intercardinal wind directions during season of frozen (left) and unfrozen (middle) soil on: (**a**) soil type CSS (spruce on clay/silt), (**b**) SP (pine on sand), and (**c**) PP (pine on peat). Map on right presents the difference between two seasons (m/s). Distributions present the differences in 10-year return levels between frozen and unfrozen soil seasons wind direction wise, divided into northern (NF, top), central (CF, middle), and southern (SF, bottom) Finland.

There were also some quite notable differences direction wise whether the bulk/peak of the distribution of differences was over or below zero (Figure 4a–c distributions). The effect of wind direction on distribution of differences was smallest in northern Finland, regardless of combination of forest and soil type. In southern Finland, for S and SE wind directions return level of wind speeds were stronger during soil frost season. Conversely, winds from W, SW, and E were characterized by stronger winds during unfrozen soil season, whereas for the rest of the directions, differences were more or less evenly distributed around zero. In central Finland, dependence on wind direction was similar for CSS and SP, with N, NE, SE, S, and NW directions having dominantly stronger winds during frozen soil season and W, SW, and E directions were characterized by stronger winds during unfrozen soil season or the differences were distributed rather evenly. For PP, more directions were characterized by stronger winds during unfrozen soil season, namely N, E, SW, W, and NW. Only NE, SE, and S directions had stronger winds mainly during frozen soil season.

All in all, large-scale differences were in general quite subtle and/or restricted to few areas. On the other hand, small-scale features were visible in the maps over the whole Finland (Figure 4). In detail, these local scale nuances in the behavior of wind multiplier downscaled return levels of 10-year maximum wind speeds are demonstrated, only for PP in this study, on 30 × 30 km area from northern Finland with more complex topography (Figure 1b) including multiple hills/fells with elevation changing between 40 and 270 m above sea level.

In the example area (Figure 1) used for more detailed analysis about the effects of wind multiplier downscaling, the strongest winds were from the south (Figure 5). This dictates the general large-scale characteristics of aggregated differences (Figure 6, middle), i.e., difference was positive on the majority of the study area. However, generally weaker winds from NW (Figure 5) also had a significant role when the effect of topography was taken into account via wind multipliers. As winds from NW were conversely stronger during unfrozen soil season, this together with relatively strong topographical forcing created isolated areas on hillsides where wind speed return level characteristics are deviating quite a lot from the general conditions of the area (Figure 6, middle). In this example case, stronger winds occurred during soil frost season, but there were also areas, mainly northwestern hillsides, where the situation was opposite.

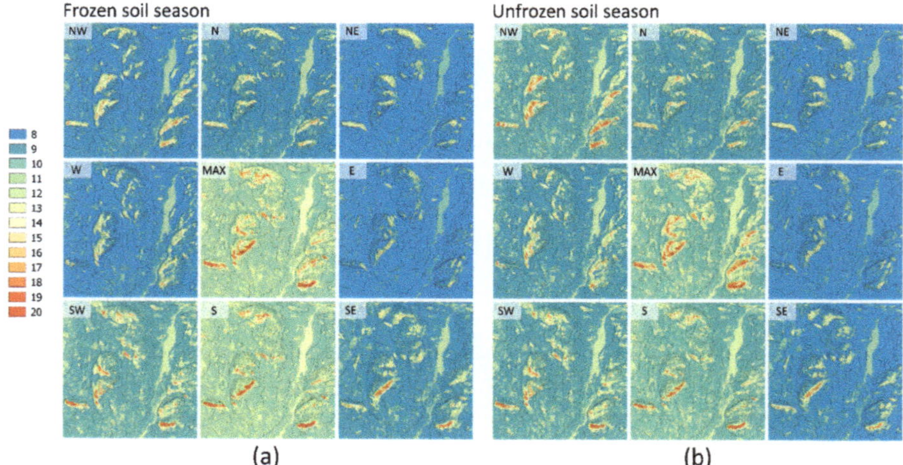

Figure 5. Ten-year return levels of wind speed [m/s] for PP (pine on peat) (**a**) frozen soil season and (**b**) unfrozen soil season in the detailed study area of northern Finland (location and topography, see Figure 1). The aggregated maximum values are in the middle, surrounded by the return levels of wind speeds from each of the cardinal and intercardinal directions.

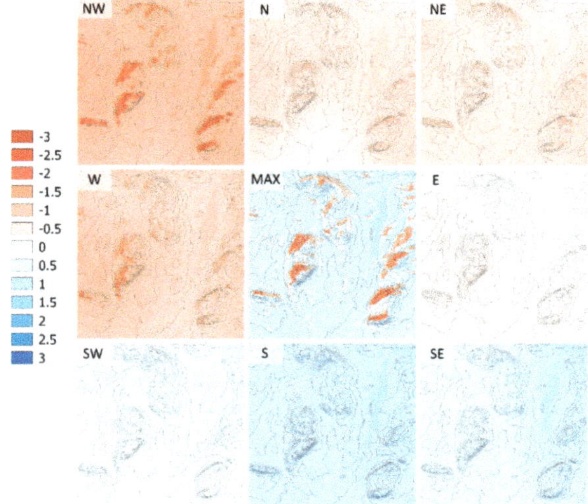

Figure 6. Differences of values presented in the Figure 5a,b. Positive values correspond to stronger winds during frozen soil season.

3.4. Comparison of the 10-Year Return levels of Maximum Wind Speed between OBS vs. ERA and WM

Figure 7 show that even if there is still large variability and some systematic biases, the majority of the OBS and WM comparisons are closer to 1:1 line than the corresponding OBS and ERA comparisons. Also in Figure 8, all three statistics show improvement when comparing differences between OBS and WM to differences between OBS and ERA. Taken over all the comparisons, the D statistic of two-sample Kolmogorov–Smirnov test is decreasing from 0.543 to 0.195, R^2 of the linear regression increasing from 0.228 (95% confidence interval 0.196–0.262) to 0.320 (95% confidence interval 0.286–0.355), and the mean difference decreasing from −1.96 to −0.77 (Mann–Whitney U-test $p < 2.2$e-16).

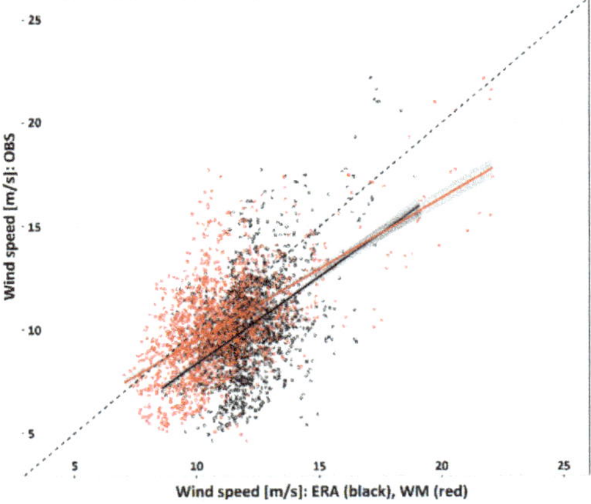

Figure 7. Scatterplot comparisons of 10-year return levels of wind speed between OBS and ERA (black) and between OBS and WM (red) at the station locations of 40 weather stations, including eight wind directions, three soil types, and two states of soil frost. OBS stands for return levels derived from observed weather station data, ERA for return levels derived straight from ERA-Interim reanalysis, and WM for ERA-Interim derived return levels downscaled with wind multiplier approach.

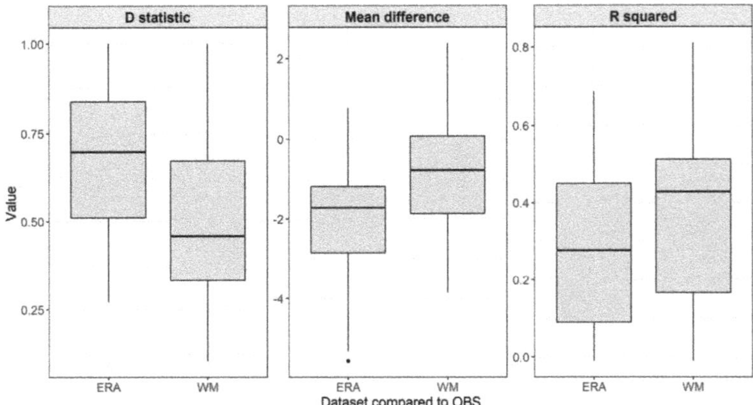

Figure 8. Boxplots of 40 station level comparisons of D statistic of two-sample Kolmogorov–Smirnov test (left), mean differences (middle), and coefficient of determination R^2 (right) between return levels derived from observations and ERA-Interim reanalysis (ERA) and between observations and wind multiplier downscaling (WM).

At some station locations, the application of wind multipliers led to more biased return levels compared to ones derived straight from the ERA-Interim. When considering all the 40 stations and each of three comparison statistic, 23% of the cases showed deterioration of results. However, for only three of the 40 stations, all three comparisons statistics were worsening (not shown).

A majority of the most pronounced overestimations of WM compared to OBS above return levels of 15 m/s (Figure 7) were from a single station at relatively high altitude combined with a large lake in the direction of the largest overestimations. There was also some similarity in the stations characterized by restricted openness, for which wind multipliers overestimated return levels around 10 m/s. For example, the station producing largest overestimations around return levels of 12–13 m/s was located in the residential area on the top of a hill.

4. Discussion

4.1. Main Findings

In this study, we produced high-resolution results (dataset) of 10-year return levels of maximum wind speed, separately for seasons of frozen and unfrozen soil. This was done by utilizing wind speed data from ERA-Interim reanalysis, modelled soil frost data, and surface roughness and topography-based wind multiplier downscaling. Differences in wind speed return levels between seasons of frozen and unfrozen soil are of interest for practical forestry as frozen soil reduces wind damage risk in terms of uprooting of trees due to stronger tree anchorage, in opposite to unfrozen soil during strong winds. Mapping of the most exposed areas to wind damage risk could also provide support for risk management in forest planning and forestry.

Relatively small differences found in this study between the 10-year return levels of maximum wind speeds during the frozen and unfrozen soil conditions can mainly be explained by the large year-to-year variability in the part of year when the annual maximum wind speeds are occurring. In Finland, maximum annual wind speed is only rarely observed during the warmest season from June to September, and those cases are typically connected to more isolated convective weather events. The period from October to December is characterized by frequent large-scale wind storms, however, in most of Finland, the soil can still be unfrozen. During the coldest season from January onwards, the possibility to have soil frost is largest and the occurrence of high wind speeds is almost as frequent as in October to December period. As a result, datasets from where wind speed return levels were derived for both frozen and unfrozen soil seasons were rather similar in the end, especially for spruce stands on clay or silt soil (CSS) and pine stands on sandy soil (SP). Peat is an efficient insulator and it takes longer cold periods before soil frost penetrates deeper into the soil, explaining the higher return levels of maximum wind speed for the unfrozen period in the case of pine stands on peatlands (PP).

The high resolution of wind multiplier downscaling produces also some fine local scale spatial variation in differences between seasons of soil frost. These differences can be quite large and have different sign in the opposing sides of topographic features like hills, ridgelines etc. Our results revealed that taking into account the impact of terrain variability and wind direction, the occurrence of maximum wind speed can change from frozen to unfrozen soil season.

4.2. Strengths and Limitations of Study Approaches

There are multiple sources of uncertainty in our work due to the combination of results from several modelled datasets, use of statistical estimation approaches and aim to produce high-resolution information for the whole of Finland apart from the archipelago. Uncertainties related to return level estimation are somewhat increased in coastal areas. This applies especially for PP as some of the years have no frost at all, and therefore smaller dataset for return level estimation of wind speed in soil frost season. Generally, uncertainty related to 10-year wind speed return level estimates can be considered to be small (+/−1 m/s).

Also, the use of a single threshold of 20 cm for needed soil frost depth to provide sufficient support for anchorage of trees is a rather large simplification in our approach. In reality, needed soil frost depth is affected by various factors, like e.g., soil type, soil wetness, and tree species and other tree and stand characteristics. Furthermore, in the used soil frost model soil water content is assumed to be constant, therefore not representing the extreme cases of very wet or dry conditions. One rather unnecessary source of uncertainty is our use of modelled snow depth, especially as we restricted our work to the observed climate, with the possibility to use gridded observational datasets presumably having smaller uncertainties. However, our approach is justified as we are planning to take projected climate change into account in future work. In the end, [31] concluded that despite many uncertainties in soil frost modelling; their results were, in general, reasonable.

Unfortunately, point observations from operational weather stations do not provide the best possible basis for the validation of wind multiplier downscaled return levels of maximum wind speeds. Operational weather stations are usually founded on locations representing regional climatic conditions or for practicality reasons on specific locations. E.g., 17 out of 40 stations used here are located within flat and open airports and airfields for purposes of aviation. This, compared to the objective of wind multiplier downscaling, i.e., to take into account the influence of small-scale topographic and surface roughness derived variations in wind speed, does not provide an ideal starting point for validation of our results. This applies especially to our application in forestry, where the interest is focused more on areas where wind damage risk is increased. In fact, considering all 40 stations and each of eight wind directions, wind multiplier values are over 1.0, i.e., increasing the wind speed return level estimate derived from ERA-Interim reanalysis, in only 15% of the cases. More detailed and accurate validation would need a more specified measurement campaign in a more complex terrain, which would also make possible the fine tuning of the wind multiplier method. Still, our validation results indicated that wind multipliers improve the wind speed return levels derived straight from ERA-Interim reanalysis by producing return levels less biased as a whole.

The high-resolution wind speed return levels, taking into account the upwind surface roughness and topography, produced in this study may provide a valuable support for wind damage risk assessment. In wind damage risk assessment it is first calculated the critical wind speed (CWS) needed for wind damage of trees to occur based on different tree and stand properties and forest configurations [27,50–53], and further on estimated the probability of wind damage and the amount of damage, respectively, based on the probability of CWS see e.g., [52]. Our results could be used e.g., when probabilities of exceeding CWSs are assessed and how these probabilities are affected by the soil frost season.

Despite differences in wind speed return levels between soil frost seasons are small, even statistically non-significant for some forest-soil type combinations and parts of Finland when station level estimates are considered, it should be kept in mind that strong winds occurring during soil frost season are excluded from unfrozen soil season. As there was such a large year-to-year variability in the timing of annual maximum wind speeds, divided surprisingly evenly between soil frost seasons found in this study, return level estimates for both seasons are therefore smaller than ones estimated from annual maximum wind speeds. In this context, differences are expected to increase in the future climate of Finland, as according to [30], the ongoing climate change is expected to reduce frozen soil conditions by several weeks until the end of this century in Finland. This might also lead to more pronounced differences of wind damage risk between different parts of Finland.

5. Conclusions

In this study, we found in general small differences in the 10-year return levels of maximum wind speeds under frozen and unfrozen soil, associated with large variability in the timing of annual maximum wind speeds. On the other hand, larger differences can be expected in the warmer climate. When the soil frost period gets even shorter, there is also a shorter window, and thus smaller probability, for strongest winds to occur in the time of frozen soil.

Further validation of used wind multiplier method could benefit from wind observations measured in a more variable topography compared to observations at operational weather stations used in this study. However, the wind multiplier approach is a pragmatic and computationally feasible way to produce extensive high-resolution dataset to identify local scale areas with elevated wind damage risk compared to regional characteristics. Data produced here is made openly available to promote its further use as a part of a more comprehensive wind damage risk assessment in forest planning and forestry.

Author Contributions: Conceptualization, A.V. and M.L.; methodology, M.L., I.L. and A.V.; software, M.L. and I.L.; formal analysis, M.L. and I.L.; writing—original draft preparation, M.L.; writing—review and editing, M.L., A.V., H.M.P. and I.L.; visualization, M.L. and I.L.; supervision, A.V.; project administration, A.V.; funding acquisition, H.M.P. and A.V.

Funding: This research was supported by the Strategic Research Council of the Academy of Finland (FORBIO project, grant number 314224).

Acknowledgments: We thank Pentti Pirinen for technical support with wind multipliers. We also thank anonymous reviewers for their constructive comments on the manuscript.

Conflicts of Interest: The authors declare no conflict of interest. The funders had no role in the design of the study; in the collection, analyses, or interpretation of data; in the writing of the manuscript, or in the decision to publish the results.

References

1. Schelhaas, M.-J.; Nabuurs, G.-J.; Schuck, A. Natural disturbances in the European forests in the 19th and 20th centuries. *Glob. Chang. Biol.* **2003**, *9*, 1620–1633. [CrossRef]
2. Schelhaas, M.-J. Impacts of Natural Disturbances on the Development of European Forest Resources: Application of Model Approaches from Tree and Stand Levels to Large-Scale Scenarios. 2008. Available online: http://edepot.wur.nl/146827 (accessed on 29 April 2019).
3. Seidl, R.; Schelhaas, M.-J.; Rammer, W.; Verkerk, P.J. Increasing forest disturbances in Europe and their impact on carbon storage. *Nat. Clim. Chang.* **2014**, *4*, 806–810. [CrossRef]
4. Reyer, C.P.O.; Bathgate, S.; Blennow, K.; Borges, J.G.; Bugmann, H.; Delzon, S.; Faias, S.P.; Garcia-Gonzalo, J.; Gardiner, B.; Gonzalez-Olabarria, J.R.; et al. Are forest disturbances amplifying or canceling out climate change-induced productivity changes in European forests? *Environ. Res. Lett.* **2017**, *12*, 034027. [CrossRef] [PubMed]

5.   Gardiner, B.; Blennow, K.; Carnus, J.M.; Fleischner, P.; Ingemarson, F.; Landmann, G.; Lindner, M.; Marzano, M.; Nicoll, B.; Orazio, C.; et al. *Destructive Storms in European Forests: Past and Forthcoming Impacts. Final Report to European Commission-DG Environment*; European Forestry Institute: Joensuu, Finland, 2010.

6.   Schuck, A.; Schelhaas, M.-J. Storm damage in Europe—An overview. In *Living with Storm Damage to Forests*; What Science Can Tell Us 3; Gardiner, B., Schuck, A., Schelhaas, M.-J., Orazio, C., Blennow, K., Nicoll, B., Eds.; European Forestry Institute: Joensuu, Finland, 2013; pp. 15–23.

7.   Gregow, H.; Laaksonen, A.; Alper, M.E. Increasing large scale windstorm damage in Western, Central and Northern European forests, 1951–2010. *Sci. Rep.* **2017**, *7*, 46397. [CrossRef]

8.   Schelhaas, M.-J.; Hengeveld, G.; Moriondo, M.; Reinds, G.J.; Kundzewicz, Z.W.; ter Maat, H.; Bindi, M. Assessing risk and adaptation options to fires and windstorms in European forestry. *Mitig. Adapt. Strateg. Glob. Chang.* **2010**, *15*, 681–701. [CrossRef]

9.   Thom, D.; Seidl, R. Natural disturbance impacts on ecosystem services and biodiversity in temperate and boreal forests. *Biol. Rev.* **2016**, *91*, 760–781. [CrossRef]

10.   Kellomäki, S.; Peltola, H.; Nuutinen, T.; Korhonen, K.T.; Strandman, H. Sensitivity of managed boreal forests in Finland to climate change, with implications for adaptive management. *Philos. Trans. R. Soc. B* **2008**, *363*, 2341–2351. [CrossRef] [PubMed]

11.   Feser, F.; Barcikowska, M.; Krueger, O.; Schenk, F.; Weisse, R.; Xia, L. Storminess over the North Atlantic and northwestern Europe—A review. *Q. J. R. Meteorol. Soc.* **2015**, *141*, 350–382. [CrossRef]

12.   Bärring, L.; von Storch, H. Scandinavian storminess since about 1800. *Geophys. Res. Lett.* **2004**, *31*, L20202. [CrossRef]

13.   Bärring, L.; Fortuniak, K. Multi-indices analysis of southern Scandinavian storminess 1780-2005 and links to interdecadal variations in the NW Europe-North Sea region. *Int. J. Climatol.* **2009**, *29*, 373–384. [CrossRef]

14.   Brönnimann, S.; Martius, O.; von Waldow, H.; Welker, C.; Luterbacher, J.; Compo, G.P.; Sardeshmukh, P.D.; Usbeck, T. Extreme winds at northern mid-latitudes since 1871. *Meteorol. Z.* **2012**, *21*, 13–27. [CrossRef]

15.   Dawkins, L.C.; Stephenson, D.B.; Lockwood, J.F.; Maisey, P.E. The 21st century decline in damaging European windstorms. *Nat. Hazards Earth Syst. Sci.* **2016**, *16*, 1999–2007. [CrossRef]

16.   Laapas, M.; Venäläinen, A. Homogenization and trend analysis of monthly mean and maximum wind speed time series in Finland, 1959–2015. *Int. J. Climatol.* **2017**, *37*, 4803–4813. [CrossRef]

17.   Vautard, R.; Cattiaux, J.; Yiou, P.; Thépaut, J.-N.; Ciais, P. Northern Hemisphere atmospheric stilling partly attributed to an increase in surface roughness. *Nat. Geosci.* **2010**, *3*, 756–761. [CrossRef]

18.   McVicar, T.R.; Roderick, M.L.; Donohue, R.J.; Li, L.T.; Van Niel, T.G.; Thomas, A.; Grieser, J.; Jhajharia, D.; Himri, Y.; Mahowald, N.M.; et al. Global review and synthesis of trends in observed terrestrial near-surface wind speeds: Implications for evaporation. *J. Hydrol.* **2012**, *416–417*, 182–205. [CrossRef]

19.   Nikulin, G.; Kjellström, E.; Hansson, U.; Strandberg, G.; Ullerstig, A. Evaluation and future projections of temperature, precipitation and wind extremes over Europe in an ensemble of regional climate simulations. *Tellus A* **2011**, *63*, 41–55. [CrossRef]

20.   Pryor, S.C.; Barthelmie, R.J.; Clausen, N.E.; Drews, M.; MacKellar, N.; Kjellström, E. Analyses of possible changes in intense and extreme wind speeds over northern Europe under climate change scenarios. *Clim. Dyn.* **2012**, *38*, 189–208. [CrossRef]

21.   Peltola, H.; Kellomäki, S.; Väisänen, H. Model computations of the impact of climatic change on the windthrow risk of trees. *Clim. Chang.* **1999**, *41*, 17–36. [CrossRef]

22.   Venäläinen, A.; Tuomenvirta, H.; Heikinheimo, M.; Kellomäki, S.; Peltola, H.; Strandman, H.; Väisänen, H. Impact of climate change on soil frost under snow cover in a forested landscape. *Clim. Res.* **2001**, *17*, 63–72. [CrossRef]

23.   Kellomäki, S.; Maajärvi, M.; Strandman, H.; Kilpeläinen, A.; Peltola, H. Change Effects on Snow Cover, Soil Moisture and Soil Frost in the Boreal Conditions over Finland. *Silva Fenn.* **2010**, *44*, 213–234. [CrossRef]

24.   Gregow, H.; Peltola, H.; Laapas, M.; Saku, S.; Venäläinen, A. Combined occurrence of wind, snow loading and soil frost with implications for risks to forestry in Finland under the current and changing climatic conditions. *Silva Fenn.* **2011**, *45*, 35–54. [CrossRef]

25.   Pirinen, P.; Simola, H.; Aalto, J.; Kaukoranta, J.-P.; Karlsson, P.; Ruuhela, R. *Tilastoja suomen Ilmastosta 1981–2010 (Climatological Statistics of Finland 1981–2010)*; Reports 2012:1; Finnish Meteorological Institute: Helsinki, Finland, 2012. Available online: https://helda.helsinki.fi/handle/10138/35880 (accessed on 28 February 2019).

26. Korhonen, J.; Haavanlammi, E. *Hydrological Yearbook 2006–2010*; Finnish Environment Institute: Helsinki, Finland, 2012. Available online: https://helda.helsinki.fi/handle/10138/38812 (accessed on 28 February 2019).

27. Peltola, H.; Kellomäki, S.; Väisänen, H.; Ikonen, V. A mechanistic model for assessing the risk of wind and snow damage to single trees and stands of Scots pine, Norway spruce, and birch. *Can. J. For. Res.* **1999**, *29*, 647–661. [CrossRef]

28. Peltola, H.; Kellomäki, S.; Hassinen, A.; Granander, M. Mechanical stability of Scots pine, Norway spruce and birch: An analysis of tree-pulling experiments in Finland. *For. Ecol. Manag.* **2000**, *135*, 143–153. [CrossRef]

29. Venäläinen, A.; Tuomenvirta, H.; Lahtinen, R.; Heikinheimo, M. The influence of climate warming on soil frost on snow-free surfaces in Finland. *Clim. Chang.* **2001**, *50*, 111–128. [CrossRef]

30. Rankinen, K.; Karvonen, T.; Butterfield, D. A simple model for predicting soil temperature in snow-covered and seasonally frozen soil: Model description and testing. *Hydrol. Earth Syst. Sci.* **2004**, *8*, 706–716. [CrossRef]

31. Lehtonen, I.; Venäläinen, A.; Kämäräinen, M.; Asikainen, A.; Laitila, J.; Anttila, P.; Peltola, H. Projected decrease in wintertime bearing capacity on different forest and soil types in Finland under a warming climate. *Hydrol. Earth Syst. Sci.* **2019**, *23*, 1611–1631. [CrossRef]

32. Dee, D.P.; Uppala, S.M.; Simmons, A.J.; Berrisford, P.; Poli, P.; Kobayashi, S.; Andrae, U.; Balmaseda, M.A.; Balsamo, G.; Bauer, P.; et al. The ERA-Interim reanalysis: Configuration and performance of the data assimilation system. *Q. J. R. Meteorol. Soc.* **2011**, *137*, 553–597. [CrossRef]

33. Etienne, C.; Lehmann, A.; Goyette, S.; Lopez-Moreno, J.-I.; Beniston, M. Spatial Predictions of Extreme Wind Speeds over Switzerland Using Generalized Additive Models. *J. Appl. Meteor. Climatol.* **2010**, *49*, 1956–1970. [CrossRef]

34. Jung, C.; Schindler, D. Statistical Modeling of Near-surface Wind Speed: A Case Study from Baden-Wuerttemberg (Southwest Germany). *Austin J. Earth Sci.* **2015**, *2*, 1–11.

35. Mortensen, N.G. *Wind Resource Assessment Using the WAsP Software (DTU Wind Energy E-0135)*; DTU Wind Energy: Roskilde, Denmark, 2016. Available online: http://orbit.dtu.dk/en/publications/wind-resource-assessment-using-the-wasp-software-dtu-wind-energy-e0135(259e26f3-1828-4e3f-9c37-17de375cd057).html (accessed on 28 February 2019).

36. Talkkari, A.; Peltola, H.; Kellomäki, S.; Strandman, H. Integration of component models from the tree, stand and regional levels to assess the risk of wind damage at forest margins. *For. Ecol. Manag.* **2000**, *135*, 303–313. [CrossRef]

37. Zeng, H.; Talkkari, A.; Peltola, H.; Kellomäki, S. A GIS-based decision support system for risk assessment of wind damage in forest management. *Environ. Model. Softw.* **2007**, *22*, 1240–1249. [CrossRef]

38. Schindler, D.; Grebhan, K.; Albrecht, A.; Schönborn, J.; Kohnle, U. GIS-based estimation of the winter storm damage probability in forests: A case study from Baden-Wuerttemberg (Southwest Germany). *Int. J. Biometeorol.* **2012**, *56*, 57–69. [CrossRef] [PubMed]

39. Cechet, R.; Sanabria, L.A.; Divi, C.B.; Thomas, C.; Yang, T.; Arthur, W.C.; Dunford, M.; Nadimpalli, K.; Power, L.; White, C.J.; et al. *Climate Futures for Tasmania: Severe wind Hazard and Risk Technical Report*; Record 2012/43; Geoscience Australia: Canberra, Australia, 2012.

40. Yang, T.; Nadimpalli, K.; Cechet, B. *Local Wind Assessment in Australia: Computation Methodology for Wind Multipliers*; Record 2014/033; Geoscience Australia: Canberra, Australia, 2014. [CrossRef]

41. Venäläinen, A.; Laapas, M.; Pirinen, P.; Horttanainen, M.; Hyvönen, R.; Lehtonen, I.; Junila, P.; Hou, M.; Peltola, H.M. Estimation of the high-spatial-resolution variability in extreme wind speeds for forestry applications. *Earth Syst. Dyn.* **2017**, *8*, 529–545. [CrossRef]

42. Jungqvist, G.; Oni, S.K.; Teutschbein, C.; Futter, M.N. Effect of climate change on soil temperature in Swedish boreal forests. *PLoS ONE* **2014**, *9*, e93957. [CrossRef] [PubMed]

43. Vehviläinen, B. *Snow Cover Models in Operational Watershed Forecasting*; Publication of the Water and Environment Research Institute 11; National Board of Waters and the Environment: Helsinki, Finland, 1992. Available online: https://helda.helsinki.fi/handle/10138/25706 (accessed on 28 February 2019).

44. Aalto, J.; Pirinen, P.; Jylhä, K. New gridded daily climatology of Finland: Permutation-based uncertainty estimates and temporal trends in climate. *J. Geophys. Res. Atmos.* **2016**, *121*, 3807–3823. [CrossRef]

45. Coles, S. *An Introduction to Statistical Modeling of Extreme Values*; Springer Series in Statistics; Springer: London, UK, 2001. [CrossRef]

46. Stephenson, A.G. EVD: Extreme Value Distributions. *R News* **2002**, *2*, 31–32.

47. Gilleland, E.; Katz, R.W. extRemes 2.0: An Extreme Value Analysis Package in R. *J. Stat. Softw.* **2016**, *72*, 1–39. [CrossRef]

48. *AS/NZS 1170.2: Structural Design Actions, Part 2: Wind Actions*, 2nd ed.; Standards Australia International/Standards New Zealand: Sydney, Australia; Wellington, New Zealand, 2011.

49. Büttner, G.; Soukup, T.; Kosztra, B. CLC2012: Addendum to CLC2006 Technical Guidelines. 2014. Available online: https://land.copernicus.eu/user-corner/technical-library (accessed on 28 February 2019).

50. Peltola, H.; Ikonen, V.; Gregow, H.; Strandman, H.; Kilpeläinen, A.; Venäläinen, A.; Kellomäki, S. Impacts of climate change on timber production and regional risks of wind-induced damage to forests in Finland. *For. Ecol. Manag.* **2010**, *260*, 833–845. [CrossRef]

51. Dupont, S.; Ikonen, V.-P.; Väisänen, H.; Peltola, H. Predicting tree damage in fragmented landscapes using a wind risk model coupled with an airflow model. *Can. J. For. Res.* **2015**, *45*, 1065–1076. [CrossRef]

52. Ikonen, V.-P.; Kilpeläinen, A.; Zubizarreta-Gerendiain, A.; Strandman, H.; Asikainen, A.; Venäläinen, A.; Kaurola, J.; Kangas, J.; Peltola, H. Regional risks of wind damage in Boreal forests under changing management and climate projections. *Can. J. For. Res.* **2017**, *47*, 1–45. [CrossRef]

53. Suvanto, S.; Henttonen, H.M.; Nöjd, P.; Mäkinen, H. High-resolution topographical information improves tree-level storm damage models. *Can. J. For. Res.* **2018**, *48*, 721–728. [CrossRef]

Article

Characterization of Meteorological Droughts Occurrences in Côte d'Ivoire: Case of the Sassandra Watershed

Natacha Santé [1], Yao Alexis N'Go [1], Gneneyougo Emile Soro [1], N'Diaye Hermann Meledje [2] and Bi Tié Albert Goula [1,*]

[1] Unit Training and Research in Science and Environment Management, University Nangui Abrogoua, 02 BP 801 Abidjan 02, Côte d'Ivoire; natachasante@gmail.com (N.S.); nyaoalexis@yahoo.fr (Y.A.N.G.); ge_soro@yahoo.fr (G.E.S.)
[2] Ecology Research Center, Marine Geology Laboratory, Sedimentology and Environment, 08 BP 109 Abidjan 08, Côte d'Ivoire; meledjendiay@yahoo.fr
* Correspondence: goulaba2002@yahoo.fr; Tel.: +225-0710-4569

Received: 5 March 2019; Accepted: 11 April 2019; Published: 20 April 2019

Abstract: The Sassandra Basin, like most regions of Côte d'Ivoire, is increasingly affected by droughts that involve many environmental, social and economic impacts. This basin is full of several amenities such as hydroelectric dams, hydraulic and agricultural dams. There is also a strong agricultural activity. In the context of climate change, it is essential to analyze the occurrence of droughts in order to propose mitigation or adaptation measures for water management. The methodological approach consisted initially in characterizing the dry sequences by the use of the SPI (Standardized Precipitation Index) and secondly in determining the probabilities of occurrence of successive dry years using by Markov chains 1 and 2. The results indicate that most remarkable droughts in terms of intensity and duration occurred after the 1970s. A comparison of Markov matrices 1 and 2 between the period considered 1953–2015 with the periods 1953–1970 and 1971–2015 shows a profound change in the distribution of droughts at the different station. Thus, the probability of having two successive dry years is greater over the period 1970–2015 and is accentuated to the Southern and Northern regions (probabilities ranging from 71% to 80%) of the basin. Over the 1970–2015 period, the probability of obtaining three successive dry years is significantly high in this watershed (between 20% and 70%).

Keywords: drought; sassandra watershed; Côte d'Ivoire

1. Introduction

Drought is one of the greatest natural hazards with effects on water resources, natural ecosystems and agriculture. Frequent and severe droughts limit the development of vegetation cover and make the soil more susceptible to erosion by leaching due to heavy rainfall [1]. They are responsible for famine, epidemics and land degradation in developing countries and cause major economic losses in developed regions [2]. From a meteorological point of view, drought can be defined as an abnormal but recurrent behavior of the climate essentially linked to the absence of rainfall received by a region within a certain period of time [3,4].

West and Southern Africa are experiencing severe drought, disrupting agricultural and livestock production systems in nearly 14 countries. Agriculture is nearly 95% rainfed in the region. It therefore remains highly vulnerable to rainfall fluctuations [5]. Work on climate fluctuations in this part of the world has made it possible to identify periods of drought since the 1970s. In Côte d'Ivoire, this deficit situation has resulted in major climatic disruptions, including a significant drop in rainfall [6–10]. There has also been an abnormal extension of the dry season [11], irregular and uneven rainfall distribution and a significant decline in hydroelectric production [12]. These disruptions have had serious consequences such as forest and plantation fires accompanied by a sharp drop in agricultural production and power outages. For example, in December 1983, fires destroyed 60,000 ha of forests and 108,000 ha of plantations and crops [13]). Given the magnitude of the environmental impacts of droughts, public authorities should attach greater importance to the development of an early warning and adaptation strategy that would announce the beginning, end and future intensity of drought. The example of the Sassandra Watershed chosen for this study is interesting because this region undergoes more and more dry season sequences. These deficit periods caused a disruption of cropping seasons in rural areas [14,15] and the decrease in in stream flows [16,17]. This basin is mainly marked by strong anthropogenic pressures. Indeed, this basin, which is also part of Côte d'Ivoire's cocoa and coffee economy, is experiencing a reduction in plant cover linked to systematic large-scale deforestation of the forest heritage for the creation of plantations [14]. There are socio-economic infrastructures (hydroelectric dams, agricultural dams, etc.) and this basin are subject to many water-related projects.

Given the impact and occurrence of droughts that are likely to increase in the coming years under certain scenarios of global change [18], it is essential to better understand how irregularity and rainfall distribution is manifested and to adopt preventive measures [19]. It is in this context that the present study was initiated on the Sassandra watershed. This study aims to highlight the occurrence of meteorological droughts in this basin using the Markov chains method based on annual rainfall over the period 1953–2015.

2. Study Area

The Sassandra basin is located between longitude 5°75 and 8°16 West and latitude 5° and 9°75′ North (Figure 1). It covers Odienné, Touba, Séguéla, Daloa, Man, Guiglo, Soubré, Sassandra, Gagnoa cities. It has a total area of about 75,000 km^2, of which the Ivorian part occupies an area of about 67,000 km^2. The relief of the study area consists of plains and uplands at varying altitude from 1100 to 1180 m. There are some rock chains that have resisted to erosion. The zenith sun movement controls the migration of the ITCZ (Intertropical Convergence Zone) in Côte d'Ivoire, which explains the introduction of different seasonal regimes. Thus the basin of Sassandra is subdivided into four climatic units according to rainfall patterns [20]. The equatorial transitional climate with four seasons (a large rainy season from April to June, a small rainy season from September to November, a large dry season from December to March and a small dry season from July to August). The interannual rainfall average is 1441.5 mm; the equatorial climate of attenuated transition is marked by two seasons (a major rainy season covering the months of August to October and a major dry season from November to March). The interannual average is 1305.2 mm; the tropical transitional climate has a unimodal pattern. It is characterized by a rainy season that occurs from June to October. The dry season covers the months of November to March. The interannual rainfall recorded at the Odienné station is 1473 mm; the mountain climate is characterized by an azonal type pattern. The highest rainfall peak is recorded in September (279 mm). The dry season covers the month of November to March. The average interannual rainfall is 1578.5 mm. The average monthly temperatures range from 23°C to 28°C and are generally uniform from one region to another. The average monthly relative humidity varies from 77 to 96% in Guinea environment and from 44 to 83% in the North [21].

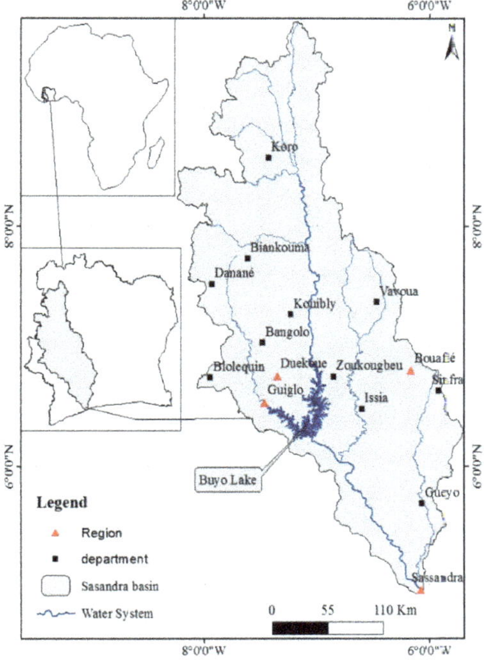

Figure 1. Geographic localization of the study area.

3. Data and Methods

3.1. Historical Time Series Data

Historical rainfall data used in this work cover the entire study area at the yearly scale. Climate data over the period 1953–2015 was provided by the National Meteorology Directorate (DMN) of the Society of Development and Exploitation, Aeoroportuary, Aeronautic and Meteorology (SODEXAM) of Côte d'Ivoire. These data consist of daily rainfall readings from ten rainfall stations (Table 1) selected to provide the most homogeneous coverage of the different climatic areas across in the Sassandra basin. The study variable is the annual rainfall for the period 1953–2015.

Table 1. Rainfall stations selected for the study.

Climatic Area	Name	Latitude North	Longitude West	Code
Attiean climate	Sassandra	4°57′	6°50′	1090017200
	Gagnoa	6°07′	5°56′	1090010300
Baoulean climate	Soubré	5°47′	6°36′	1090018100
	Guiglo	6°32′	7°28′	1090011200
	Vavoua	7°22′	6°28′	1090021400
	Seguéla	7°57′	6°40′	1090017500
	Daloa	6°53′	6°27′	1090008200
	Touba	8°17′	7°41′	1090020500
Mountain climate	Man	7°24′	7°3_′	1090014200
Sudanese climate	Odienné	9°30′	7°34′	1090016000

3.2. Methodology

3.2.1. Characterization of Meteorological Drought Sequences

❖   *The choice of the statistical index*

The Standardized Precipitation Index (SPI), developed by [22], is used in this study to characterize meteorological droughts. It has advantages in terms of statistical consistency and the ability to describe both short-term and long-term drought impacts through different time scales [22]. The development of this index is based solely on the use of rainfall as a baseline data to determine wet and dry periods, and to specify their duration and intensity. The probabilistic nature of the SPI index allows it to be comparable between different sites [23].

❖   *Standardized Precipitation Index*

The standardized precipitation index (SPI) [22,24] was developed to quantify the rainfall deficit for multiple time scales that will reflect the impact of drought on the availability of different types of water resources over a given period. It is expressed as follows (Equation (1)):

$$SPI = (Pi - Pm)/S \tag{1}$$

Pi: Total rainfall over year i (mm); Pm: Average precipitation over the period 1953–2015 (mm); S: Standard deviation of precipitation over the period 1953–2015 (mm).

According to [22], a drought occurs when the SPI is consecutively negative and its value reaches an intensity of −1 or less and ends when the SPI becomes positive. A drought classification is performed according to the SPI values (Table 2).

Table 2. Classification of drought sequences according to Standardized Precipitation Index (SPI) [22].

SPI Value	Drought Sequence
−0.99 to 0.99	Near the Normal
−1.00 to −1.49	Moderately dry
−1.50 to −1.99	Severely dry
−2.00 and under	Extremely dry

❖   *Descriptive parameters of drought sequence*

-   *Maximum duration of drought sequences*

Duration is an important characteristic of drought. In fact, if a drought starts quickly under some weather conditions, it usually takes at least two to three months before it can spread to other regions. It can then persist for months or even years. The calculation of the duration is as follows [25]. (Equation (2)).

$$D = (A_{end} - A_{initial}) \tag{2}$$

$A_{initial}$: Year of the initial dry period; A_{end}: Year of end of the dry period

-   *Intensity of drought sequences*

Intensity of drought can be defined as the magnitude and severity of the consequences for the rainfall deficit. It can be evaluated using the SPI values. In this study, the extreme value of the SPI was considered as a reference value for drought intensity.

3.2.2. Time Series Change Detection

Rupture is defined as a sudden change in the properties of a random process [26]. Rupture tests are complementary to standard indices because the existence of sudden change in time series is a possible cause of the rupture of the homogeneity of these series [27]. In this study, the Cumulative Gap (CG) test was used to detect possible sudden changes in rainfall series. This non-parametric procedure, based on rank, analyzes whether the means of the two parts of the series are different for an unknown break date [28]. The statistic of this test is calculated from the cumulative sum of the "sign" function of the difference between the observed values and the median. This statistical processing is performed with Hydrospect 2.0 software. The test statistic is defined as follows (Equation (3)):

$$|TS| = (2\backslash n) \max |S_k| \text{ with } S_k = \sum_{i=1}^{k} \text{sign} (x_i - X_m) \text{ and } (k = 1, \ldots, n) \tag{3}$$

where x_i is the extreme hydrometric observation of rank i ($i = 1 \ldots n$); X_m is the median of the extreme hydrometric series; S_k is the statistic test; n is the number of value for the rank i.

3.2.3. Characterization of Meteorological Droughts Occurrence by Markov Chains

Several statistical techniques for analyzing precipitation data have been published in the literature. The most used technique is still the one based on the Markov chains. This method is widely used for rainfall analysis and modelling [29–36]. A Markov string is a series of random variables ($Xn, n \in N$) that allows to model the dynamic evolution of a random system: Xn represents the state of the system at time n. The fundamental property of Markov chains, known as "Markov property", is that its future evolution depends on the past only through its current value. In other words, conditionally to Xn ($X0$, $\ldots$, Xn) and ($Xn+k, k \in N$) are independent [37].

o   *Markov chain with two states of order 1*

For a first order Markov chain, the state of the variable $E(t)$ at time t depends only on its state at time ($t - 1$). Thus, we have four situations: [31]

$$\begin{aligned}
P_{00} &= pr(E(t + 1) = 0| (E(t) = 0)) \\
P_{01} &= pr(E(t + 1) = 1 | (E(t) = 0)) \\
P_{10} &= pr(E(t + 1) = 0 | (E(t) = 1)) \\
P_{11} &= pr(E(t + 1) = 1| (E(t) = 1))
\end{aligned} \tag{4}$$

P_{ij} is the probability of going to state j knowing that you are in state i. These probabilities were calculated using the following relationship:

$$P_{ij} = N_{ij}/N_i \text{ with: i and j} = 0 \text{ or } 1 \tag{5}$$

N_{ij} is the transition number from state i to state j and N_i is the number of transitions from state i to any other state. The pairs of years N_{ij} are determined [35] (Equation (6)):

$$\begin{cases}
N_0 = N_{00} + N_{01} \\
N_1 = N_{10} + N_{11} \\
N = N_0 + N_1
\end{cases} \tag{6}$$

N_0; N_1 and N are the number of dry, wet years and the total number of years of observation, respectively. N_{01} and N_{10} respectively represent the number of years of state change from a dry year to a wet year and from a wet year to a dry year. The transition matrix P of the conditional probabilities P_{ij}, is presented so that each line is equal to 1 [35]. Resulting in a set of possible P_{ij} values (Equation (7)):

$$P = \begin{bmatrix}
P_{00} & P_{01} & \cdots \\
P_{10} & P_{11} & \cdots \\
\cdots & \cdots & \cdots \\
P_{i0} & P_{i1} & \cdots
\end{bmatrix} \tag{7}$$

○   *Markov chain with two states of order 2*

For a Markov string of order 2, the state of the variable E(t) at time t depends on its state E(t − 1) at time (t − 1) as well as its state E(t − 2). The probability of having this state can be written:

$$P_{ijk} = pr\ (E(t) = k\ |(E(t-1) = j, E(t-2) = i)) \tag{8}$$

P_{ijk} represents the conditional probability of having a state doublet (*j, k*) following the state doublet (*i, j*) and *i, j, k* = 0 or 1, calculated using the following relationship [31]:

$$P_{ijk} = N_{ijk}/N_{ij} \tag{9}$$

where N_{ijk} is the number of transitions from the state doublet (i, j) to the state doublet (j, k).

The process of transition of conditional probabilities with the Markov 2 chain is as follows (Table 3):

Table 3. Markov process of order 2 [35].

State at Day k−1 and k−2	State at Day k−1 and k			
	00	01	10	11
00	P000	P001	0	0
01	0	0	P010	P011
10	P100	P101	0	0
11	0	0	P110	P111

4. Results

4.1. Analysis of Meteorological Drought Sequences

The application of the cumulative gap test identified change point detection in the series. The null hypothesis of no rupture was rejected at the 99% and 95% confidence levels. These ruptures were identified mainly after 1970 except in the Man, Guiglo and Gagnoa regions, which had their ruptures between 1967 and 1969. Temporal analysis shows a slight downward trend in SPIs after ruptures, confirming a decrease in rainfall. SPI values over the 1953–2015 period show very few dry sequences before the rupture years (Figure 2).

For the equatorial transition regime (attiean climate), change point detection were detected in 1966 and 1983 at the Gagnoa and Sassandra stations respectively. Before these change point detection, the index counted four dry sequences in Gagnoa and three dry sequences in Sassandra. Thus, the most remarkable sequences have a duration of nine successive years in Gagnoa and 18 in Sassandra.

For the equatorial regime of attenuated transition (Baoulean climate): before the change point detection years, the SPI index recorded four dry sequences in Soubré, three in Guiglo, seven in Daloa, four in Vavoua, 12 in Seguéla and seven in Touba. In this period, the Daloa and Touba stations recorded two sequences of two successive dry years. For Daloa, the index shows the periods 1964–1965 and 1969–1970 and for Touba, the periods 1960–1961 and 1970–1971. As for the Seguela station, the index detected a sequence of five successive dry years. The most remarkable dry episodes after the rupture have lasted 13 successive years in Soubré, six years in Guiglo, seven years in Daloa, four years in Vavoua, eight years in Seguéla and 10 years in Touba.

For the mountain regime, the Man station recorded five dry sequences before 1968. This number increased to 27 after the change point detection. The index has also recorded sequences of successive dry years, the most remarkable of which is five years (2011–2015).

For the Tropical Attenuated Transition Regime (Sudanese climate), Odienné station recorded 8 dry sequences, including a sequence of two and three successive dry years in 1980–1981 and 1973–1975. After the change point detection, the SPIs show 24 dry sequences, the most remarkable in terms of duration being 11 successive dry years (1983–1993).

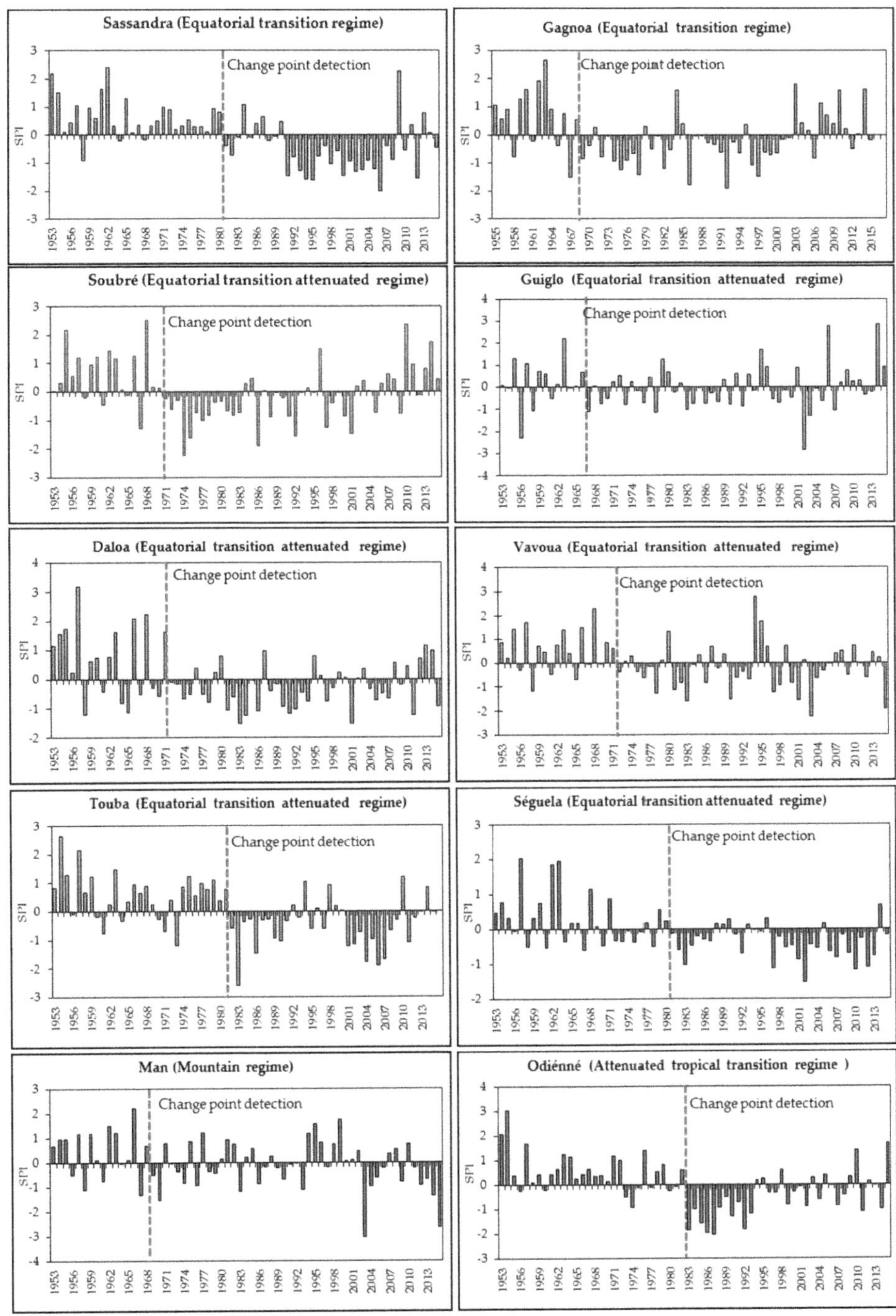

Figure 2. SPI index evolution over the period 1953–2015.

4.2. Intensity and Duration Parameter Analysis

The most remarkable drought years in terms of intensity and duration during the 62 years analyzed differ according to the climatic zones of the watershed (Table 4). For the attean climate, the driest years were those of 1992 and 1998. These dry events were classified as "very dry" in terms of intensity. The Gagnoa station recorded the highest intensity. As for the Sassandra station, it experienced the longest drought period with 18 years of consecutive dry sequences.

At the level of the Attenuated Equatorial Transition Regime (Baoulean climate), the stations of Soubré, Guiglo, Vavoua and Touba recorded the highest intensities in terms of drought. These events characterized as "extremely dry" were detected in 1974, 2002, 2003 and 1983, respectively. In this region of the basin, the Guiglo station was the most affected in terms of intensity while the Touba station recorded the longest dry period (10 successive dry years).

In the Tropical transitional Attenuated Regime (Sudanese climate), the most remarkable droughts were those of 1987. These dry episodes were described as extremely severe in terms of intensity. Odienné station was marked by a longer dry period with (11 successive years).

As for the Mountain Regime, it recorded a deficit of an intensity described as extremely severe in 2003. This area was marked by a long drought period of seven years.

The application of the SPI to rainfall data for the period 1953–2015 shows the western part of the basin, which is the most affected by drought in terms of intensity (−3.06) while the southern part of the basin has the longest drought period (18 years).

Table 4. Intensity and duration of meteorological drought sequences during the period 1953–2015.

Stations	Intensity (SPI)	Type	Maximum Duration	Date of Occurrence
Daloa	−1.53	Severely dry	7	2001
Vavoua	−2.26	Extremely dry	4	2003
Man	−3.06	Extremely dry	5	2003
Guiglo	−2.86	Extremely dry	6	2002
Seguéla	−1.53	Severely dry	8	2002
Touba	−2.61	Extremely dry	10	1983
Odienné	−2.05	Extremely dry	11	1987
Gagnoa	−1.90	Severely dry	7	1992
Soubré	−2.25	Extremely dry	13	1974
Sassandra	−1.87	Severely dry	18	1998

4.3. Analysis of the Meteorological Droughts Occurrence

4.3.1. Transition States Probability of Markov Chains 1

The probability of obtaining two successive dry years and two successive wet years is high in the Gagnoa and Sassandra areas (reaching 70%). In this climatic zone in the southern part of the basin, when a dry year is followed by a non-dry year or a non-dry year is followed by a dry year, the probability is low (Table 5). The Soubré, Seguéla, Daloa and Touba regions recorded high probabilities (over 50%) of obtaining a doublet of successive dry years. As for the Guiglo, Vavoua and Man regions, these probabilities were average. The probabilities of having a wet year after a dry year and a wet year after a wet year are low in these regions of the basin. The Man region recorded average probabilities (55%) for two successive dry years and two successive wet years. As for the other probabilities, they are very low in this climate regime.The chances of obtaining successively dry and successively wet episode doublets are very high in Odienné. When a year is dry and is followed by a humid year or a humid year is followed by a dry year, the probabilities are very low. The analysis of the occurrences of two successive dry years over the 1953–2015 series shows very high probabilities over the entire basin (62% on average). However, the regions of Sassandra, Gagnoa, Seguéla, Touba and Odienné recorded the highest probability (up to 70%) of dry spells over this period.

Table 5. Occurrence of meteorological droughts using Markov Chains 1 over the period 1953–2015.

Climate Regimes	Stations	Probability (%)			
		W-W	D-W	W-D	D-D
Attiean climate	Sassandra	68	30	30	70
	Gagnoa	54	32	46	70
Baoulean climate	Soubré	57	36	40	64
	Guiglo	43	55	57	45
	Vavoua	40	52	60	50
	Seguéla	46	31	54	70
	Daloa	44	40	56	60
	Touba	62	30	38	70
Mountain climate	Man	55	45	45	55
Soudanese climate	Odienné	60	38	40	63

Note: D: Dry year; W: Humid year.

The results of the occurrence analysis for two successive dry years before and after 1970 are presented in Table 6.

Table 6. Occurrence of meteorological droughts using Markov Chains 1 over the periods 1953–1970 and 1971–2015.

Period	Climate Regimes	Stations	Probability (%)			
			W-W	D-W	W-D	D-D
1953–1970	Attiean climate	Sassandra	28	36	71	60
		Gagnoa	38	50	63	50
	Baoulean climate	Soubré	25	70	75	30
		Guiglo	38	60	63	40
		Vavoua	11	78	67	22
		Seguéla	29	40	71	55
		Daloa	40	63	60	23
		Touba	44	45	60	45
	Mountain climate	Man	40	56	55	44
	Soudanese climate	Odienné	50	15	50	85
1971–2015	Attiean climate	Sassandra	77	20	23	80
		Gagnoa	55	39	45	61
	Baoulean climate	Soubré	61	36	40	64
		Guiglo	35	54	65	46
		Vavoua	41	53	59	45
		Seguéla	61	22	39	78
		Daloa	48	45	52	55
		Touba	70	30	30	71
	Mountain climate	Man	58	43	42	55
	Soudanese climate	Odienné	55	39	41	61

Note: D: Dry year; W: Humid year.

During the period 1953–1970, the probability of obtaining two successive humid years (W-W) and two consecutive dry years (D-D) is relatively low in the basin (less than 50%) except at the Odienné, Sassandra and Seguéla stations which have a high probability (85%, 60% and 55%, respectively) of obtaining two successive dry years (D-D). When a year is dry, the probability of having the following year not dry (D-W) is high at the Daloa (70%), Soubré (70%) and Vavcua (78%) stations. In the event that a year is not dry, the probability that the following year will be dry is high throughout the basin (greater than 50%) and varies from 55% to 75%.

Over the period 1971–2015, a trend contrary to the previous period is observed. Indeed, if a year is dry at the beginning, the probability of having a dry year (D-D) is high for most regions (above 50%) and reaches its maximum (80%) at the Sassandra station. In the case where the dry year is followed by a non-dry year (D-W), the probability is relatively low and varies from 25% to 48% except in the Guiglo area (54%). When a humid year is followed by a dry year (W-D), the probability is high at Guiglo (62%) and Vavoua (59%). For the other parts of the basin, this probability is moderate. When two years are non-dry successively (W-W), the probability is greater than 50% over most of the basin except in the Daloa, Vavoua and Guiglo regions where the values are low (less than 50%).

The analysis shows the increase in drought episodes over the period 1971–2015. This is reflected in the high probabilities, the most notable of which were recorded in the North (Touba (71%) and Seguéla (78%)) and the South (Sassandra (80%)) of the basin.

4.3.2. Transition States Probability of Markov Chains 2

The results of the Markovian approach on the 1953–2015 series show that the probability of obtaining a dry year after two successive dry years (D-D-D) is high in the areas of Sassandra (67%) and Gagnoa (59%). As for the stations of Seguéla (51%), Soubré (55%) and Touba (50%) located in an equatorial regime of attenuated transition (Baoulean climate), they recorded average values. At the other stations, this probability is low (<50%) (Table 7). The probability of having a dry year between two wet years (W-D-W), a wet year followed by two successive dry years (W-D-D) and a wet year after two dry years (D-D-W) is very low over the entire basin. The averages recorded are 24%, 20.4% and 16.9%, respectively. Overall, the probability of having three consecutive dry years (D-D-D) is relatively low in the basin (44.8% on average). The most affected regions are those in the southern part of the basin.

Table 7. Occurrence of meteorological droughts using Markov Chains 2 over the period 1953–2015.

Climate Regimes	Stations	Probability (%)			
		W-D-W	W-D-D	D-D-W	D-D-D
Attiean climate	Sassandra	15	6	7	67
	Gagnoa	23	20	11	59
Baoulean climate	Soubré	30	10	10	55
	Guiglo	40	17	20	30
	Vavoua	35	25	21	25
	Seguéla	25	25	15	51
	Daloa	22	32	22	40
	Touba	14	21	18	50
Mountain climate	Man	22	25	23	30
Soudanese climate	Odienné	16	23	22	41

Note: D: Dry year; W: Humid year.

The results of the occurrence of three successive dry years over the periods 1953–1970 and 1971–2015 are shown in Table 8.

During the period 1953–1970, the probability of having three successive dry years is relatively low at all stations except the Odienné station where this probability is 60%. In the regions of Man, Soubré, Vavoua and Daloa, the probability of having three successive dry years is zero.

During the period 1971–2015, the probability to observe (D-D-D) is high and higher than 50% at Seguéla (60%), Touba (63%) and the maximum is reached at Sassandra (70%). As for the rest of the stations, the probability is low and less than 50%.

Table 8. Occurrence of meteorological droughts using Markov Chains 2 over the periods 1953–1970 and 1971–2015.

Period	Climate Regimes	Stations	Probability (%)			
			W-D-W	W-D-D	D-D-W	D-D-D
1953–1970	Attiean climate	Sassandra	29	43	20	30
		Gagnoa	38	25	10	20
	Baoulean climate	Soubré	50	25	20	0
		Guiglo	50	13	20	20
		Vavoua	60	22	25	0
		Seguéla	14	57	30	20
		Daloa	40	20	20	0
		Touba	22	33	30	15
	Mountain climate	Man	33	33	30	0
	Soudanese climate	Odienné	0	60	20	60
1971–2015	Attiean climate	Sassandra	10	10	10	70
		Gagnoa	25	20	17	43
	Baoulean climate	Soubré	20	20	20	50
		Guiglo	40	30	21	25
		Vavoua	30	30	30	20
		Seguéla	11	25	15	60
		Daloa	20	30	30	30
		Touba	20	10	8	63
	Mountain climate	Man	17	25	0	25
	Soudanese climate	Odienné	20	23	22	40

Note: D: Dry year; W: Humid year.

4.4. Analysis of the Spatial Variability of Drought Occurrence

4.4.1. Spatial Variability of the Probability for Two Successive Dry Years

Figure 3 shows the probability spatial distribution for two successive dry years (D-D) over the Sassandra basin for the periods 1953–2015, 1953–1970 and 1971–2015. The period 1953–2015 is marked by an increase in meteorological droughts over almost the entire basin with the probability of having a very high doublet of dry years that reaches 70%. The most affected areas are those in the North and South of the catchment area.

From 1953 to 1970, the drought was particularly severe in the extreme north of the basin (around the Odienné region) and in the south (precisely around Sassandra) with probabilities of more than 60%. These drought episodes spread over almost the entire basin during the period 1971–2015 except in the central eastern areas (Daloa and Vavoua) and the western areas (Guiglo and Man) of the basin. Over this time period, the probability of obtaining two consecutive dry years is high, averaging 61.6%.

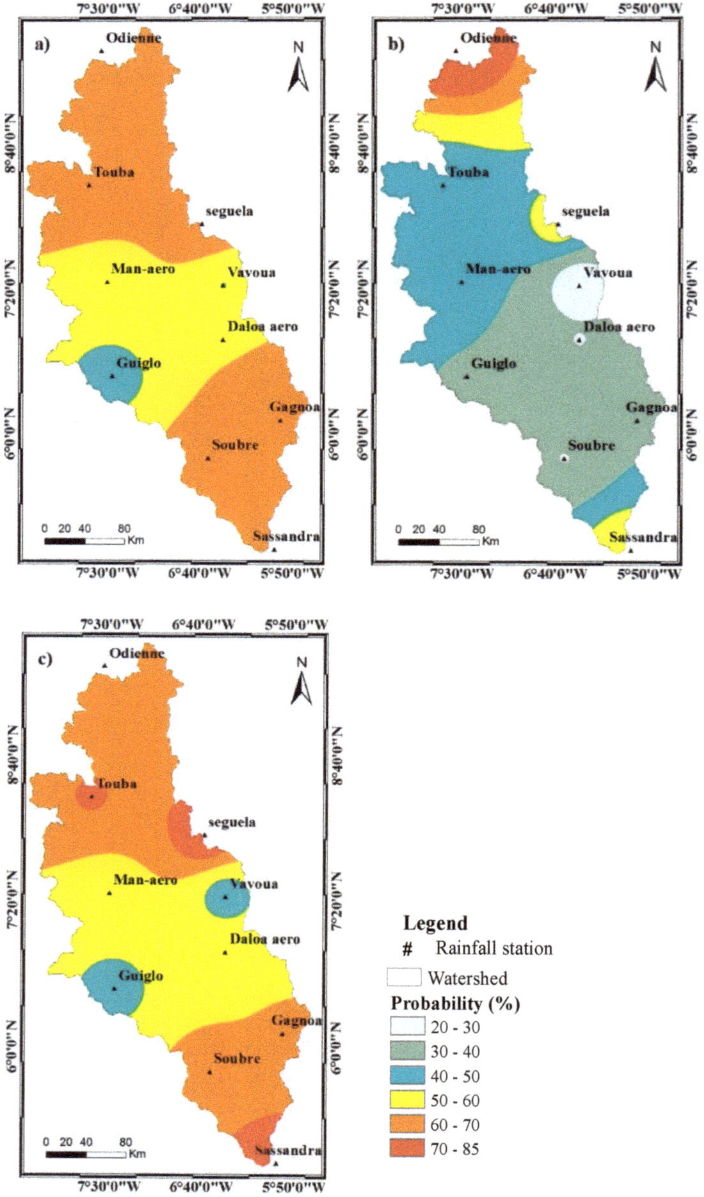

Figure 3. Spatial distribution of the occurrence for two successive dry years (D-D) over the Sassandra watershed (**a**) D-D probability over the period 1953–2015; (**b**) D-D probability over the period 1953–1970; (**c**) D-D probability over the period 1971–2015.

4.4.2. Spatial Variability of Probabilities for Three Consecutive Dry Years

The spatial variability of the probabilities of obtaining three successive dry years over the periods 1953–2015, 1953–1970 (before the rupture) and 1971–2015 (after the rupture) is presented in Figure 4. An intensification of droughts in the South (around the Sassandra, Gagnoa and Soubré regions) and in the North-East (Seguéla) of the basin, with probabilities higher than 50%, is observed over the period 1953–2015.

Droughts intensify over time in the watershed. The dry areas which over the period 1953–1970 were located in the extreme North of the basin (covering the Odienné region), increased over the period 1971–2015 and spread with high probabilities of occurrence (up to 70%) over the Touba, Seguéla and South (Sassandra) regions. As for the northern tip of the basin, the probabilities were low.

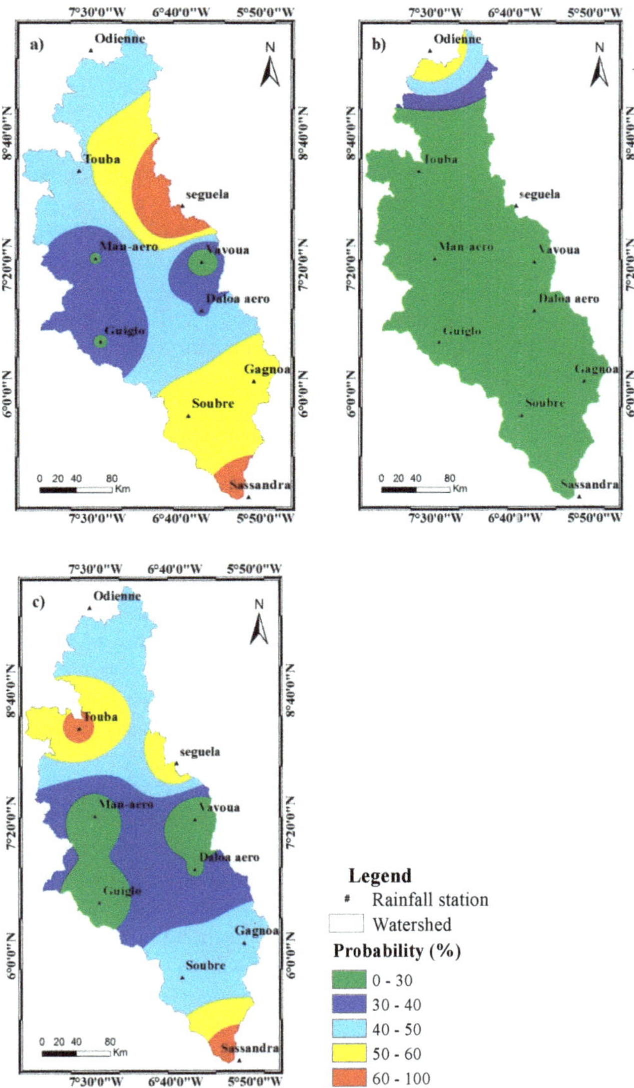

Figure 4. Spatial distribution of the occurrence for three successive dry years (D-D-D) over the Sassandra watershed (**a**) D-D-D probability over the period 1953–2015; (**b**) D-D-D probability over the period 1953–1970; (**c**) D-D-D over the period 1971–2015).

5. Discussion

The statistical test used to determine the breaks in the SPI series made it possible to detect breaks, for the most part, after the 1970s, which is the pivotal date observed almost throughout West Africa. The stations of Man, Guiglo and Gagnoa experienced early ruptures between 1967 and 1969. These years of disruption coincide with previous studies on rainfall in West Africa and Côte d'Ivoire [6,27,38–41]. This work reveals the appearance of a rainfall deficit from 1970 onwards and its continuation during the decades 1970–1979, 1980–1989 and 1990–1999. In the Sassandra basin, studies by [16] on the Lobo basin also mentioned a decrease in rainfall over the period 1970–2009 with a deficit of between 10% and 20% compared to the rupture years.

The analysis of the SPI index over the period 1953–2015 shows an increase of 54.35% to 82.35% in the number of dry years. This phenomenon was accentuated around the 1980s, with a persistence of dry sequences over the period 1970–2015.The peak of the most remarkable droughts occurred in 1974, 1983, 1987, 1987, 2002 and 2003. These peaks are characterized by "extremely severe" droughts. Moreover, changes in SPI values at the various stations indicate that, in terms of intensity, the regions of Man (−3.06), Guiglo (−2.86), Touba (−2.61), Vavoua (−2.26), Odienné (−2.05) and Soubré (−2.25), were the most affected by the droughts. As for duration, the SPIs indicate an increase in dry years after ruptures. Thus, Sassandra Station records the longest dry period (18 successive years). Our results are in line with studies conducted in West Africa. According to [42], the drought was more severe in the second half of the period 1900–2013, i.e., from the 1970s onwards. This trend is confirmed by several studies at the continental level and in West Africa [43,44]. These studies locate the most remarkable drought events in 1961, 1970, 1983, 1984, 1984, 1992 and 2001.

However, very few studies have focused on investigating the causes of droughts in West Africa. However, the work of [45,46] has shown that recent droughts remain linked to the emanation of ocean warming (southward warming gradient of the Atlantic Ocean and steady warming of the Indian Ocean) and fluctuations in the inter-tropical convergence zone (ITCZ). The ITCZ is defined as a convergence zone of northeastern Harmattan winds from the Sahara and the southwestern monsoon flow from the Atlantic [47]. The zenithal movement of the sun will command the displacement of the inter-tropical convergence zone towards the South to reach its southern position around 5° N in West Africa. This explains the introduction of the dry seasons. In addition to these factors, there is the effect of the land-atmosphere feedback through natural vegetation and land cover change. Indeed, deforestation can cause significant reduction of the rainfall and effect on the monsoon circulation [42]. The application of Markov chains over the periods 1953–2015 and the after change point detection year made it possible to highlight the areas that were most affected by droughts in terms of occurrence probability. The analysis shows that during the period 1953–2015 the probability of obtaining two successive dry years is high in the Northern and South of this basin. As for the probabilities of obtaining three dry years, they are high over part of the South and the northeast of the watershed. Over the periods 1953–1970 and 1971–2015, a large variation in probabilities was observed. Indeed, during 1953–1971, the dry areas that were observed in the extreme North of the basin according to Markov 1, spread over the entire basin over the period 1971–2015, with high values recorded in the North and South (ranging from 71% to 80%). As for the second order chains, they show the probabilities of obtaining high D-D-D in the far North over the period 1953–1970. During 1971–2015, drylands spread over part of the North (Touba), the North-East (Seguéla) and the South (Sassandra). These results indicate that the succession of dry conditions increased during the period 1971–2015 compared to the previous period (1953–1970) with the southern and northern areas of the basin most affected. This situation could be related to the effects of climate change observed on rainfall in West Africa. These results complement the work done of [35] on the transboundary watershed of the Bia River in eastern Côte d'Ivoire. The conclusions of this study are that the succession of two to three dry years is more marked in this basin after 1970.

The fact that the southern region located in the forest zone has the highest probability of two or three successive dry years is possibly due to the effect of droughts on the spatial and temporal variation of the 1200 isohyet in the southwestern part of the basin. According to [17], these variations over the decades 1981–1990 and 1991–2000, created a dry corridor focused on the city of Sassandra, thus promoting a microclimate surrounded by watered areas that coincide with forest reserves and forest areas. In addition, there may be a significant reduction in forest cover in favor of an increase in cultivation areas, fallows and habitat.

6. Conclusions

Application of the standardized rainfall index (SPI) has made it possible to characterize drought situations in the Sassandra catchment area. During the 1971–2015 period, the driest sequences in terms of intensity (values between −1.53 and −3.06) and duration occurred. These dry events reached their peak in 1974, 1983, 1987, 1987, 2002 and 2003, with extremely severe droughts. Among the 10 stations that have been studied, those of Man, Guiglo, Touba, Odienné, Soubré, Seguéla and Vavoua seem to be the most affected by droughts. As for the Sassandra station, it seems to be more affected by a long dry period. As for the study of drought persistence using Markov chains 1 and 2, it made possible to determine the probability occurrence of droughts as well as to analyze their behavior in the basin. The results indicated that the succession of dry conditions increased during 1971–2015. The greatest probabilities for obtaining a doublet of successive dry years (D-D) and three consecutive dry years (D-D-D) were recorded in the southern and northern regions of the watershed. Markov chains 1 and 2 applied to a representative sample of 10 stations are found to provide a good regional drought indicator. Thus, the probability of a dry year in this basin will depend on the previous year's situation and even more on the condition of the year before. This study will enable populations, decision-makers, etc., to develop new water resource management strategies for the proper functioning of existing socio-economic infrastructures (hydroelectric and agricultural dams, etc.), water projects and the development of new farming systems to cope with the effects of climate change.

Author Contributions: N.S. and Y.A.N.G. developed the ideas; G.E.S. and N D.H.M. contributed to the data processing. N.D.H.M. contributed to creation of the maps. N.S. analyzed the data and wrote the article with contributions from Y.A.N.G., N.E.S., N.D.H.M., and B.T.A.G.. Supervision and validation were provided by Y.A.N.G. and B.T.A.G.

Funding: This research received no external fundi.

Acknowledgments: The authors thank the National Meteorology Directorate (DMN) of SODEXAM (Company operating and Developing airports, Aeronautics and Meteorological) of Côte d'Ivoire, for data acquisition.

Conflicts of Interest: The authors declare no conflict of interest.

References

1. Lécuyer, C. *Evolution de la Désertification en Afrique de L'ouest*; Rapport de Stage de Master 1 à l'Institut de Recherche pour le Développement (IRD): Bondy, France-Nord, 2012; 101p.
2. Nicholson, S.E. Climatic and environmental change in Africa during the last two centuries. *Clim. Res.* **2001**, *17*, 123–144. [CrossRef]
3. Kouassi, A. Caractérisation d'une Modification Éventuelle de la Relation Pluie-débit et ses Impacts sur les Ressources en eau en Afrique de l'Ouest: Cas du Bassin Versant du N'Zi (Bandama) en Côte d'Ivoire. Ph.D. These, Université de Cocody, Abidjan, Côte d'Ivoire, 2008; 217p.
4. Bedoum, A.; Bouka, B.C.; Alladoum, M.; Adoumi, I.; Baohoutoul, L. Impact de la variabilité pluviométrique et de la sécheresse au Sud du Tchad: Effet du changement climatique. *Rev. Ivoir. Sci. Technol.* **2014**, *23*, 13–20.
5. BAD. Banque Africaine de Développement. Lutte Contre la Sécheresse en Afrique de l'Ouest et Australe, 2016. La BAD Octroie une Enveloppe de 549 Millions. Available online: https://www.afdb.org/fr/news-and-southern-africa-15547 (accessed on 20 March 2019).
6. Goula, B.T.A.; Savané, I.; Konan, B.; Fadika, V.; Gnamien, K.B. Impact de la variabilité climatique sur les ressources hydriques des bassins de N'Zo et N'Zi en Côte d'Ivoire (Afrique tropicale humide). *Vertigo* **2006**, *7*, 1–12. [CrossRef]
7. Noufé, D.; Lidon, B.; Mahé, G.; Servat, E.; Brou, T.; Koli Bi, Z. Variabilité climatique et production de maïs en culture pluviale dans l'Est ivoirien. *J. Sci. Hydrol.* **2011**, *56*, 152–167. [CrossRef]

8. Otchoumou, K.F.; Saley, M.B.; Aké, G.E.; Savane, I.; Djê, K.B. Variabilité climatique et production de café et cacao dans la zone tropicale humide: Cas de la région de Daoukro (Centre-Est de la Côte d'Ivoire). *Int. J. Innov. Appl. Stud.* **2012**, *1*, 194–215.

9. Sorokoby, V.M.; Saley, M.B.; Kouamé, K.F.; Djagoua, E.M.V.; Affian, K.; Biemi, J. Variabilité spatio-temporelle des paramètres climatiques et son incidence sur le tarissement dans les bassins versants de Bô et Debo (département de Soubré au Sud-Ouest de la Côte d'Ivoire). *Int. J. Innov. Appl. Stud.* **2013**, *2*, 287–299.

10. Yao, F.Z.; Reynard, E.; Ouattara, I.; N'Go, Y.A.; Fallet, J.M.; Savané, I. A new statistical approach to assess climate variability in the white Bandama watershed, Northern Cote d'Ivoire. *Atmos. Clim. Sci.* **2018**, *8*, 410–430.

11. Savané, I.; Coulibaly, K.; Gion, P. Variabilité climatique et ressources en eaux souterraines dans la région semi-montagneuse de Man. *Sécheresse* **2001**, *12*, 231–237.

12. Kouadio, B.H.; Kouamé, K.F.; Saley, M.B.; Biémi, J.; Ibrahima, T. Insécurité climatique et géorisques en Côte d'Ivoire: Étude du risque d'érosion hydrique des sols dans la région semi-montagneuse de Man (Ouest de la Côte d'Ivoire). *Sécheresse* **2007**, *18*, 29–37.

13. FAO. *Situation des Forêts Dans le Monde*; FAO: Rome, Italy, 1999; 154p.

14. Brou, Y.T. Variabilité climatique, déforestation et dynamique agro-démographique en Côte d'Ivoire. *Sécheresse* **2010**, *21*, 1–6.

15. N'go, Y.A.; Ama-Abina, J.T.; Kouadio, A.Z.; Kouassi, H.K.; Savané, I. Environmental Change in Agricultural Land in Southwest Cote d'Ivoire: Driving Forces and Impacts. *J. Environ. Prot.* **2013**, *4*, 1373–1382. [CrossRef]

16. Yao, A.B.; Goula, B.T.A.; Kouadio, Z.A.; Kouakou, K.E.; Kanté, A.; Sambou, S. Analyse de la variabilité climatique et quantification des ressources en eau en zone tropicale humide: Cas du bassin versant de la Lobo au Centre-Ouest de la Cote d'Ivoire. *Rev. Ivoir. Sci. Technol.* **2012**, *19*, 136–157.

17. N'go, Y.A. Hydrologie et Dynamique de L'état de Surface des Terres Dans le sud-Ouest de la Côte d'Ivoire: Impacts et Moteurs de Dégradation. Ph.D. These, Université Nangui Abrogoua, Abidjan, Côte d'Ivoire, 2015.

18. Watson, W.E.; Kumar, K.; Michaelsen, L.K. Cultural diversity's impact on interaction process and performance: Comparing homogeneous and diverse task groups. *Acad. Manag. J.* **1993**, *36*, 590–602.

19. Doudja, S.G.; Nour, E.D.; Abd, E.M.B. Simulation de la pluviométrie journalière en zone semi-aride par l'analyse en composantes principales. *Sécheresse* **2007**, *18*, 97–105.

20. Eldin, M. Le climat de la Côte d'Ivoire. In *Le Milieu Naturel de la Côte d'Ivoire*; Mémoires de l'ORSTOM: Paris, France, 1971; Volume 50, pp. 73–108.

21. Savané, I. Contribution à L'étude Géologique et Hydrogéologique des Aquifères Discontinues du socle Cristallin d'Odienné (Nord- Ouest de la Côte D'ivoire). Apport de la Télédétection et d'un Système D'information Hydrogéologique à Référence Spatiale. Ph.D. These, Université de Cocody, Abidjan, Côte d'Ivoire, 1997.

22. Mckee, T.B.; Doesken, N.J.; Kliest, J. The relationship of drought frequency and duration of time scale. In Proceedings of the Eight Conference on Applied Climatology, Anaheim, CA, USA, 7–22 January 1993; American Meteorological Society: Boston, MA, USA, 1993; pp. 179–184.

23. Mirabbasia, R.; Anagnostoub, E.N.; Fakheri-Farda, A.; DInpashoha, Y.; Eslamianc, S. Analysis of meteorological drought in northwest Iran using the Joint Deficit Index. *J. Hydrol.* **2013**, *492*, 35–48. [CrossRef]

24. Hayes, M.J.; Svoboda, M.D.; Wilhite, D.A.; Vanyarkho, O.V. Monitoring the 1996 drought using the standardized precipitation index. *Bull. Am. Meteorol. Soc.* **1999**, *80*, 429–438. [CrossRef]

25. Soro, G.E.; Anouman, D.G.L.; Goula Bi, T.A.; Srohorou, B.; Savané, I. Caractérisation des séquences de sécheresse météorologique à diverses échelles de temps en climat de type Soudanais: Cas de l'extrême nord-ouest de la Côte d'Ivoire. *LARHYSS J.* **2014**, *18*, 107–124.

26. Renard, C.; Garreta, V.; Lang, M. An application of Bayesian analysis and Markov chain Monte Carlo methods to the estimation of a regional trend in annual maxima. *Water Resour. Res.* **2006**, *42*, W12422. [CrossRef]

27. Kouakou, K.E.; Goula, B.T.A.; Savané, I. Impacts de la variabilité climatique sur les ressources en eau de surface en zone tropicale humide: Cas du bassin versant transfrontalier de la Comoé (Côte d'Ivoire-Burkina Faso). *Eur. J. Sci. Res.* **2007**, *16*, 31–43.

28. Chiew, F.H.S.; Mc Mahon, T.A. Assessing the adequacy of catchment stream flow yield estimates. *Aust. J. Soil Res.* **1993**, *31*, 665–680. [CrossRef]

29. Javier, M.V.; Lidia, G. Regionalization of peninsular Spain based on the length of dry spells. *Int. J. Climatol.* **1999**, *19*, 537–555.

30. Mark, T.; George, K.A. Hidden Markov model for modelling long-term persistence in multi-site rainfall time series 1. Model calibration using a Bayesian approach. *J. Hydrol.* **2003**, *275*, 12–26.

31. Lazri, M.; Ameur, S.; Haddad, B. Analyse des données de précipitations par approche Markovienne. *LARHYSS J.* **2007**, *6*, 7–20.

32. Jan, L.; Anastassia, B.; Deliang, C. Modelling precipitation in Sweden using multiple step Markov chains and a composite model. *J. Hydrol.* **2008**, *363*, 42–59.

33. Justin, G.M. Markov chain modeling of sequences of lagged Numerical Weather Prediction ensemble probability forecasts: An exploration of model properties and decision support applications. *Mon. Weather Rev.* **2008**, *136*, 3655–3669.

34. Alam, J.A.T.M.; Sayedur, M.R.; Saadat, A.H.M. Monitoring meteorological and agricultural drought dynamics in Barind region Bangladesh using standard precipitation index and Markov chain model. *Int. J. Geomat. Geosci.* **2013**, *3*, 511–524.

35. Meledje, N.D.H.; Kouassi, K.L.; N'Go, Y.A.; Savane, I. Caractérisation des occurrences de sécheresse dans le bassin hydrologique de la Bia transfrontalier entre la Côte d'Ivoire et le Ghana: Contribution des chaînes de Markov. *Cah. Agric.* **2015**, *24*, 186–197.

36. Khadr, M. Forecasting of meteorological drought using Hidden Markov Model (case study: The upper Blue Nile river basin, Ethiopia). *Ain Shams Eng. J.* **2016**, *7*, 47–56. [CrossRef]

37. Radu, S.S. *Modélisation Probabiliste et Inférence Statistique pour L'analyse des Données Spatiales*; Habilitation à Diriger des Recherches (HDR) de l'Université de Lille: Lille, France, 2014; Volume 1, 232p.

38. Kanohin, F.; Saley, M.B.; Savané, I. Impacts de la variabilité climatique sur les ressources en eau et les activités humaines en zone tropicale humide: Cas de la région de Daoukro en Côte d'Ivoire. *Eur. J. Sci. Res.* **2009**, *26*, 209–222.

39. Kouassi, A.M.; Kouamé, K.F.; Koffi, Y.B.; Djé, K.B.; Paturel, J.E.; Oularé, S. Analyse de la variabilité climatique et de ses influences sur les régimes pluviométriques saisonniers en Afrique de l'Ouest: Cas du bassin versant du N'zi (Bandama) en Côte d'Ivoire. *Cybergeo Eur. J. Geogr.* **2010**, *513*, 29. [CrossRef]

40. Soro, D.T.; Soro, N.; OGA, M.S.; Lasm, T.; Soro, G.; Ahoussi, E.K. La variabilité climatique et son impact sur les ressources en eau dans le degré carré de Grand-Lahou (Sud-Ouest de la Côte d'Ivoire). *Physio-Géo* **2011**, *5*, 55–73. [CrossRef]

41. Kouakou, E.; Koné, B.; N'Go, A.; Cissé, G.; Ifejika, S.C.; Savané, I. Groundwater Sensitivity to Climate Variability in the White Bandama Basin, Ivory Coast. *SpringerPlus* **2014**, *3*. 226. [CrossRef]

42. Masih, I.; Maskey, S.; Mussá, F.E.F.; Trambauer, P.A. Review of droughts on the African continent: A geospatial and long-term perspective. *Hydrol. Earth Syst. Sci.* **2014**, *18*, 3635–3649. [CrossRef]

43. Ouassou, A.; Ameziane, T.; Ziyad, A.; Belghiti, M. Application of the drought management guidelines in Morocco. *Options Méditerr. Ser. B* **2007**, *58*, 343–372.

44. Kasei, R.; Diekkrüger, B.; Leemhuis, C. Drought frequency in the Volta Basin of West Africa. *Sustain. Sci.* **2010**, *5*, 89–97. [CrossRef]

45. Lebel, T.; Cappelaere, B.; Galle, S.; Hanan, N.; Kergoat, L.; Levis, S.; Vieux, B.; Descroix, L.; Gosset, M.; Mougin, E.; et al. Studies in the Sahelian region of West-Africa: An overview. *J. Hydrol.* **2009**, *375*, 3–13. [CrossRef]

46. Dai, A.; Lamb, P.J.; Trenberth, E.K.; Hulme, M.; Jones, D.P.; Xie, P. The Recent Sahel Drought Is Real. *Int. J. Clim.* **2004**, *24*, 1323–1331. [CrossRef]

47. Nicholson, S.E. The West African Sahel: A Review of Recent Studies on the Rainfall Regime and Its Interannual Variability. *ISRN Meteorol.* **2013**, *2013*, 453521. [CrossRef]

Article

Constraints to Vegetation Growth Reduced by Region-Specific Changes in Seasonal Climate

Hirofumi Hashimoto [1,*], Ramakrishna R. Nemani [2], Govindasamy Bala [3], Long Cao [4], Andrew R. Michaelis [1], Sangram Ganguly [5], Weile Wang [1], Cristina Milesi [6], Ryan Eastman [7], Tsengdar Lee [8] and Ranga Myneni [9]

[1] California State University Monterey Bay/NASA Ames Research Center, Moffett Field, CA 94035, USA; amichaelis@csumb.edu (A.R.M.); weile.wang@nasa.gov (W.W.)

[2] NASA Advanced Supercomputing Division, Ames Research Center, Moffett Field, CA 94035, USA; rama.nemani@nasa.gov

[3] Center for Atmospheric and Oceanic Sciences, Indian Institute of Science, Bangalore, Karnataka 560012, India; gbala@iisc.ac.in

[4] Department of Atmospheric Sciences, ZheJiang University, HangZhou 310007, China; longcao@zju.edu.cn

[5] Bay Area Environmental Research Institute/NASA Ames Research Center, Moffett Field, CA 94035, USA; sangram.ganguly@nasa.gov

[6] CropSnap LLC, Sunnyvale, CA 94087, USA; cmilesi@cropsnap.com

[7] Department of Atmospheric Sciences, University of Washington, Seattle, WA 98195, USA; rmeast@atmos.washington.edu

[8] NASA Headquarters, Washington, DC 20546, USA; tsengdar.j.lee@nasa.gov

[9] Department of Earth and Environment, Boston University, Boston, MA 02215, USA; rmyneni@bu.edu

* Correspondence: hirofumi.hashimoto@gmail.com; Tel.: +1-650-604-6446

Received: 31 October 2018; Accepted: 10 January 2019; Published: 1 February 2019

Abstract: We qualitatively and quantitatively assessed the factors related to vegetation growth using Earth system models and corroborated the results with historical climate observations. The Earth system models showed a systematic greening by the late 21st century, including increases of up to 100% in Gross Primary Production (GPP) and 60% in Leaf Area Index (LAI). A subset of models revealed that the radiative effects of CO_2 largely control changes in climate, but that the CO_2 fertilization effect dominates the greening. The ensemble of Earth system model experiments revealed that the feedback of surface temperature contributed to 17% of GPP increase in temperature-limited regions, and radiation increase accounted for a 7% increase of GPP in radiation-limited areas. These effects are corroborated by historical observations. For example, observations confirm that cloud cover has decreased over most land areas in the last three decades, consistent with a CO_2-induced reduction in transpiration. Our results suggest that vegetation may thrive in the starkly different climate expected over the coming decades, but only if plants harvest the sort of hypothesized physiological benefits of higher CO_2 depicted by current Earth system models.

Keywords: terrestrial ecosystems; GPP; LAI; CMIP5; CO_2 fertilization effect; feedback

---

1. Introduction

Climate change caused by increasing atmospheric carbon dioxide (CO_2) concentrations has been extensively studied in the context of global warming, and the land carbon cycle feedback is recognized as one of the biggest sources of uncertainty in climate projection [1]. Global warming is proceeding with a greening trend of the Earth, as shown by satellite and ground observations of increases in leaf area index [2,3], canopy cover [4], and biomass [5]. A greening Earth has significant consequences for the terrestrial carbon sink, the integrity of ecosystem, and climate [6,7]. Numerous mechanisms appear to underlie the observed greening, including changes in the climate system [8,9]. Among the mechanisms supporting the greening Earth, CO_2 fertilization is considered the dominant factor in enhancing vegetation, with evidence from free-air CO_2 enrich (FACE) experiments [10], satellite observations [2,4], and ground observations [11]. Climate change is also substantially contributing to the increase in global vegetation productivity because of the indirect effect of increasing CO_2 concentrations [9]. For instance, global warming is enhancing vegetation growth in high latitudes [12,13]. As such, each mechanism contributing to vegetation growth has been scrutinized independently, but, in contrast to climate change studies, the global factors affecting vegetation response have not been well studied and summarized. Therefore, the combination and interactions of multiple different climatological and biophysical mechanisms make it difficult to predict the future growth of vegetation at the global scale. As a result, the big discrepancy between modeled and observed sensitivity to CO_2 concentrations is always a source of controversy in the prediction of the future carbon cycle [14]. If climate constrains increase, climate change can cancel the positive effects of CO_2 or of other biogeochemical fertilization (e.g. nitrogen deposition) on vegetation, and possibly accelerate global warming. Therefore, understanding the relative strength of climate variables or increasing CO_2 concentrations, leading to greening or browning of the Earth, is imperative for future projections.

We firstly summarized and analyzed the trends in climate and vegetation responding to increasing CO_2 concentrations from the subset of the Coupled Model Intercomparison Project Phase 5 (CMIP5). The CMIP5 dataset includes present run and future projection data produced by Earth system models following several experimental scenarios. Future vegetation growth depends on the type, magnitude, and seasonal timing of climatic changes and their interactions with vegetation physiology. To simplify the understanding of these complex mechanisms, we divided the vegetated areas into three categories, namely temperature-limited, water-limited, and radiation-limited areas, following Nemani et al. [9].

Then, we decomposed the mechanisms enhancing vegetation growth into three factors: CO_2 concentration fertilization effect, radiative climate change, and local climate feedback by vegetation growth. By assessing these three factors quantitatively, we can answer the question as to whether increasing CO_2 concentrations will tighten or relax climate constrains on vegetation at the global scale. Friedlingstein et al. [1] evaluated the strengths of the effects of CO_2 fertilization and temperature increase on land vegetation carbon storage. Lemordant et al. [15] decomposed the climate effects on evapotranspiration into net radiation, precipitation, and vapor pressure deficit (VPD). However, none of them counted the local climate feedback effect at the global scale. Therefore, contradicting results of vegetation growth by different mechanisms made future predictions confusing. For instance, it is hard to discuss the regional trend in precipitation [16–18] and the changes in water use efficiency [11] at the same time without knowing which mechanism is relatively stronger. Finally, we discussed the validity of the findings through the analysis of historical observations.

2. Materials and Methods

2.1. CMIP5

CMIP5 is a set of model experiments for assessing past and future climate change in the Intergovernmental Panel for Climate Change Assessment Report number 5 (IPCC AR5) [19]. To objectively select datasets, the models of CMIP5 data used in this paper met the following criteria.

- The models had monthly data of near-surface air temperature (output variable name in the standard output is tas; the other variables are showed the same way hereafter), precipitation (pr), surface downwelling shortwave radiation (rsds), and Leaf Area Index (LAI) (lai) data for the specific years (1875–2005: historical; and 2006–2099: Representative Concentration Pathway (RCP) 8.5).
- The land sub-model had year-to-year changes in LAI.

We used three sets of outputs from 21 models from CMIP5 (Table 1).

Table 1. Coupled Model Intercomparison Project Phase 5 (CMIP5) models and the forcing characteristics. Models used in the sensitivity analysis (experiment IDs are esmFixClim1, esmFdbk1, and 1pctCO2) shown in the analysis are highlighted in red.

Model	Modeling Group	Land Component	N Cycle	Dynamic Vegetation
bcc-csm1-1	Beijing Climate Center, China Meteorological Administration, CHINA	AVIM1.0	N	N
bcc-csm1-1-m	Meteorological Administration, CHINA	AVIM1.0	N	N
BNU-ESM	Beijing Normal University, CHINA	CoLM3 & BNU DGVM (C/N)	-	-
CanESM2	Canadian Center for Climate Modelling and Analysis, CANADA	CLASS2.7 & CTEM1	N	N
CESM1-CAM5	Community Earth System Model Contributors, NSF-DOE-NCAR, USA	CLM4	Y	N
CESM1-WACCM	Community Earth System Model Contributors, NSF-DOE-NCAR, USA	CLM4	Y	N
CESM1-BGC	Community Earth System Model Contributors, NSF-DOE-NCAR, USA	CLM4	Y	N
GFDL-CM3	NOAA Geophysical Fluid Dynamics Laboratory, USA	LM3	N	Y
GFDL-ESM2G	NOAA Geophysical Fluid Dynamics Laboratory, USA	LM3	N	Y
GFDL-ESM2M	NOAA Geophysical Fluid Dynamics Laboratory, USA	LM3	N	Y
HadGEM2-ES	Met Office Hadley Centre, UNITED KINDOM	MOSES2 & TRIFFID	N	Y
INMCM4	Russia			
IPSL-CM5A-LR	Institut Pierre-Simon Laplace, FRANCE	ORCHIDEE	N	N
IPSL-CM5A-MR	Institut Pierre-Simon Laplace, FRANCE	ORCHIDEE	N	N
MIROC5	JAMSTEC, University of Tokyo, and NIES, JAPAN	MATSIRO & SEIB-DGVM	N	Y
MIROC-ESM-CHEM	JAMSTEC, University of Tokyo, and NIES, JAPAN	MATSIRO & SEIB-DGVM	N	Y
MIROC-ESM	JAMSTEC, University of Tokyo, and NIES, JAPAN	MATSIRO & SEIB-DGVM	N	Y
MPI-ESM-LR	Max Planck Institute for Meteorology, GERMANY	JSBACH	N	Y
MPI-ESM-MR	Max Planck Institute for Meteorology, GERMANY	JSBACH	N	Y
MPI-ESM1	Max Planck Institute for Meteorology, GERMANY	JSBACH	N	Y
NorESM1-ME	Norwegian Climate Centre, NORWAY	CLM4	Y	N
NorESM1-M	Norwegian Climate Centre, NORWAY	CLM4	Y	N

The experiments are:

1. Historical run: Runs covering the historical period 1850–2005. For this period, model forcings include: greenhouse gases (GHG), volcanoes, aerosols, and land cover.

2. RCP 8.5: Projections forced by pre-determined increasing CO_2 concentrations covering 2006 to 2100. For this analysis we chose the RCP 8.5 scenario, a pathway with the highest greenhouse gas emissions, leading to 8.5 W/m^2 radiative forcing at the end of the 21st century [20]. Although just a decade passed since 2006, it has been reported that the emission concentration in 2100 is projected to follow RCP 8.5 [21].

3. Sensitivity experiments: In order to assess the contribution of CO_2 fertilization and climate effects on vegetation separately, we used an eight-model (highlighted in Table 1) ensemble to compare three CMIP5 experiments, each of which was run for 140 years and experiences a constant CO_2 of preindustrial level and/or CO_2 increasing by 1%/year to 4xCO_2: (1) In the fertilization experiment CO_2 increases by 1%/year to 4xCO_2 for the land surface, but stays constant at preindustrial level for the atmosphere, and thus the climate effect is suppressed and the CO_2 fertilization effect is dominant on land (the official experiment ID is esmFixClim1); (2) in the climate experiment CO_2 increases for the atmosphere, but stays constant for the land surface, and hence the CO_2 fertilization effect is suppressed and the climate impact dominates (esmFdbk1); (3) in the combined experiment CO_2 concentration increases for the full Earth system (1pctCO2).

To calculate the ensemble mean, at first, we remapped all the CMIP5 data into quarter degree grid data using the bilinear interpolation method. Then, we calculated the ensemble mean for each quarter degree grid from the available modeled data.

In this paper, gross primary production (GPP) was chosen from available CMIP5 land variables as the representation of photosynthesis. GPP is the amount of photosynthesis by vegetation per unit area, from which respiration is not subtracted. In the budget analysis, net biome production (NBP), which accounts for respiration and disturbance, should be the key flux of vegetation response. However, these experiments are not CO_2 emission driven, but rather CO_2 concentration driven, and the results of the carbon budget of vegetation do not change the atmospheric CO_2 concentration. Thus, GPP, which is equivalent to photosynthesis, can represent vegetation growth better than NBP. We also used LAI as the representative of carbon storage because LAI can be compared to satellite estimates.

The vegetation response in the historical run, RCP 8.5, and combined experiments in sensitivity experiments can be simplified by the linear models as follows:

$$\Delta GPP = a\,\Delta CO_2 + b\,\Delta Clim + f\,\Delta Clim_{feedback}, \tag{1}$$

where ΔCO_2, $\Delta Clim$, and $\Delta Clim_{feedback}$ represent change in CO_2 concentration, change in one of the limiting climate factors but only caused by the radiative effect of the change in CO_2 concentration, and feedback of climate by changing GPP through the fertilization effect, respectively. The coefficients a, b, and f assume a simple linear system. The term $\Delta Clim_{feedback}$ represents the feedback of GPP through effects on climate, such as the effect of a change in cloud cover due to increasing evapotranspiration.

The fertilization experiment examines how higher CO_2 affects climate and vegetation via increases in leaves' internal CO_2 concentration, which should in turn reduce stomatal conductance transpiration. As a result, climate feedback occurs as decreasing cloud cover whilst increasing soil moisture, runoff, and solar radiation [11,22,23]. So, the fertilization experiment can be expressed as:

$$\Delta GPP = a\,\Delta CO_2 + f\,\Delta Clim_{feedback}, \tag{2}$$

It is noteworthy that ΔCO_2 includes the effect of changing water use efficiency because it is directly affected by increasing CO_2 concentrations, not through changing climate.

The climate experiment shows how higher CO_2 affects vegetation via the traditional greenhouse effect on climate. The fertilization effect of increasing CO_2 on vegetation was suppressed. The climate experiment can be expressed as:

$$\Delta GPP = b\,\Delta Clim, \tag{3}$$

For mapping purposes, the outputs were firstly re-gridded to 0.5 × 0.5 degree resolution, using the bilinear interpolation method, to be consistent with the limiting factor data [9].

2.2. GIMMS-LAI3G

The GIMMS-LAI3G data were derived from the Global Inventory Modeling and Mapping Studies (GIMMS) Normalized Differential Vegetation Index (NDVI) using the neural network algorithm [24]. We aggregated the 1/12 degree spatial resolution LAI data into the half degree data prior to monthly and annual analysis. To be consistent with historical CMIP5 runs that end in 2005, we used GIMMS-LAI3g data from overlapping the period 1982–2005.

2.3. CRU/CRUNCEP

We used 0.5 × 0.5 degree monthly temperature and precipitation from Climate Research Unit (CRU-TS3.23) and solar radiation data from CRU National Centers for Environmental Prediction version 4 (CRUNCEP-V4, a blend of CRU data and NCEP- National Center for Atmospheric Research (NCAR) reanalysis data) [25]. CRU data are interpolated gridded datasets from monthly observations. To be consistent with historical CMIP5 runs, we used CRU/CRUNCEP data from overlapping the period with Global Inventory Modeling and Mapping Studies (GIMMS)-LAI data during 1982–2005.

2.4. NDP026

We reanalyzed the ground-based total cloud cover observations in Numeric Data Package NDP06 [26], following methodology outlined in Warren et al. [27]. Briefly, the methodology consists of estimating seasonal trends at each of the stations, then averaging these trends over a 10 × 10 degree grid. We compared these estimated trends with those predicted by CMIP5 models for the period of 1971–2005 using a chi-squared test for independence. Several modeling studies showed the connection between low clouds over land and increased levels of CO_2 in the atmosphere [28,29]. As such, it would have been more appropriate to use changes in low cloud cover in our analysis. However, CMIP5 did not mandate modeling teams to submit low, medium, and high cloud simulations separately. Only total cloud cover was required from the modeling teams, hence we used total cloud cover data in our analysis.

2.5. Limiting Factor Analysis

We analyzed global changes in the three regions defined by the climate factor—temperature, precipitation, or radiation—that is most limiting to GPP. Following Nemani et al. [9], we calculated the strength of the limiting factor based on monthly climate data of minimum temperature, precipitation, and cloudy day. Then, we defined the three regions using the highest value among the three data (Figure 1a).

2.6. Estimation of Precipitation Equivalent Water

Increases in atmospheric CO_2 are known to increase the water use efficiency (WUE) of vegetation, estimated as amount of carbon gained per unit amount of water used [21,22,30]. While WUE increases globally with CO_2, its changes over precipitation-limited regions are of particular significance for plant growth. To assess the contribution of improved WUE in precipitation-limited regions, we used a slightly different metric called precipitation use efficiency (PUE). PUE is estimated as GPP per unit amount of precipitation. The changes in PUE between the first (2006–2015) and last decades (2090–2099) are then translated to changes in precipitation equivalent, PE_q (mm), as follows:

$$PE_q = \left(GPP_{last} - GPP_{first} \right) \times \frac{P_{first}}{GPP_{first}} - \left(P_{last} - P_{first} \right), \tag{4}$$

where GPP_{first} and GPP_{last} are GPP (kgC/m^2) for the first and last decades of the 21st century, and P_{first} and P_{last} are precipitation received at each of the grid cells during the first and last decades.

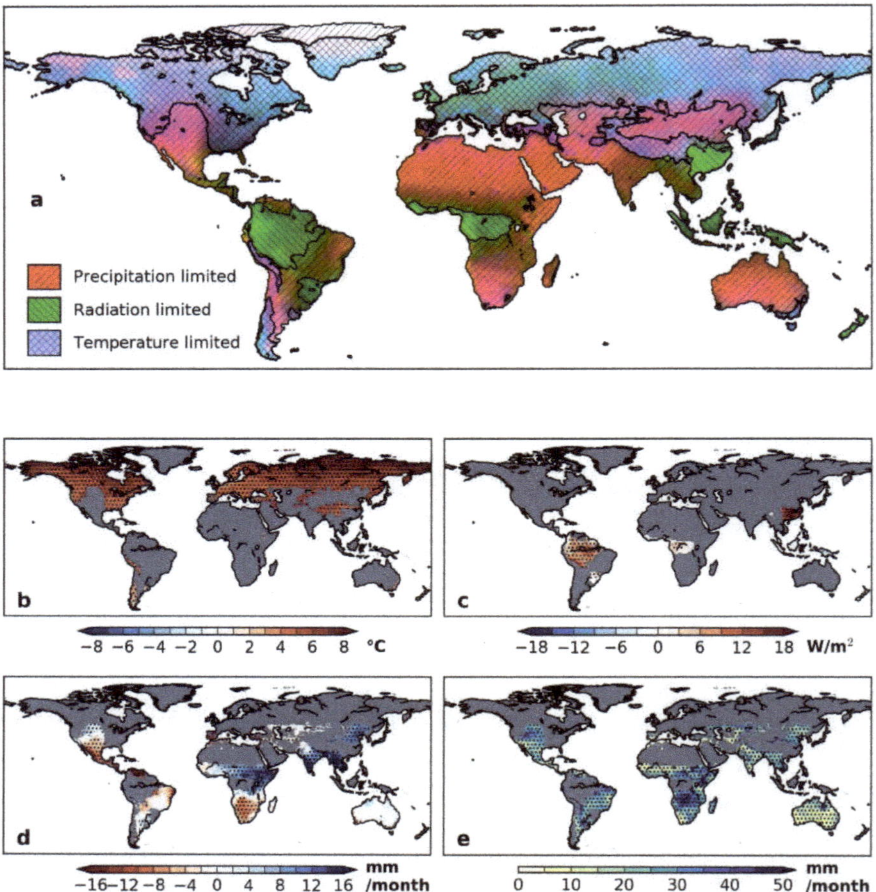

Figure 1. Earth system models project easing of temperature, precipitation, and radiation constraints to growth. A map of potential climate limiting factors to plant growth ((**a**) air temperature in blue, precipitation in red, solar radiation in green) was used to guide the spatial analysis of simulated changes in climate and how such changes could impact plant growth around the world. Using outputs from Earth system models of CMIP5 we estimated ensemble mean differences in 2090–2099 minus 2006–2015 monthly air temperature (**b**), solar radiation (**c**), precipitation (**d**), and precipitation use efficiency (expressed as precipitation equivalent; see Methods) (**e**). Changes in constraints by regional mean: temperature-limited, easing 94%, and no significant change 6%; precipitation-limited, easing 23%, tightening 16%, and no significant change 61%; radiation-limited, easing 45%, tightening 19%, and no significant change 36%.

2.7. Mann–Kendall Test

The Mann–Kendall test detects the presence of a monotonic trend in time-series data [31]. The test is broadly used because it does not require the assumption of normal distribution of the time-series data. We use the Mann–Kendall test as the significance test for the annual or seasonal data.

3. Results

3.1. Annual Mean Trend and Climate Feedback to Vegetation

The multi-model ensemble mean shows pervasive future changes in vegetation structure and function by the end of this century under the high-emission RCP 8.5 scenario: LAI increases by up to 60% and GPP increases by up to 100% (Figure 2). LAI significantly increases globally except for Amazon, Mexico, and Southern Africa (Figure 2a). GPP also significantly increases for nearly the entire vegetated planet (Figure 2b), though the magnitudes may be uncertain [32]. The percentage changes are higher in the high latitude regions for LAI, while changes in magnitude of LAI and GPP are higher in both the high latitude and tropical regions. A few tropical and semi-arid areas show decreases in LAI, but none of the changes are statistically significant. The results of increasing LAI are consistent with Mahowald et al. [33].

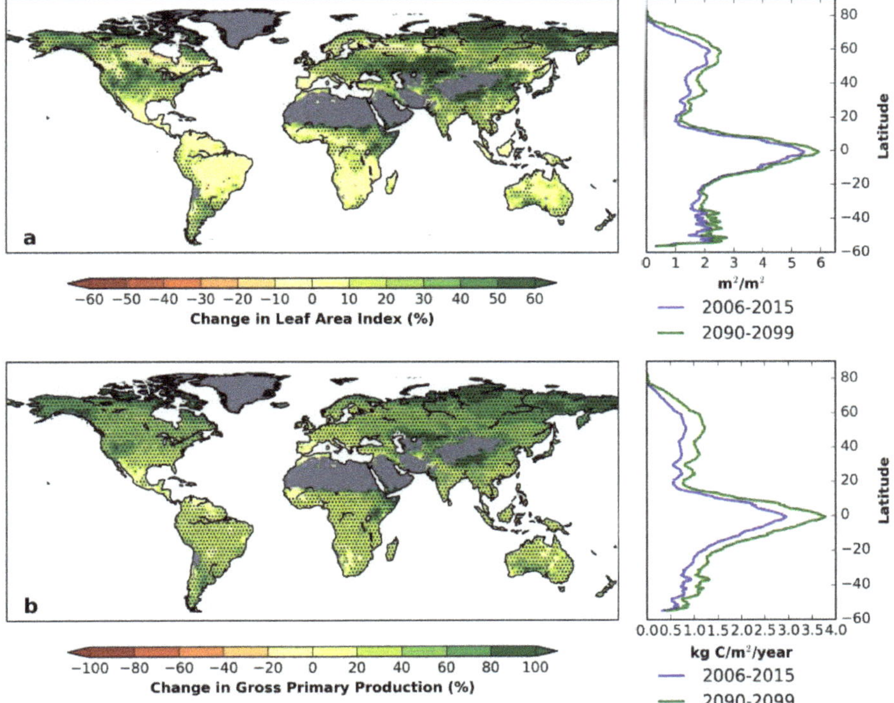

Figure 2. Greening of the Earth. Simulated changes in leaf area index (LAI) and gross primary production (GPP) from the Earth system models of CMIP5. Shown are differences in annual average mean LAI (**a**) and annual total GPP (**b**) at each grid cell for 2090–2099 minus 2006–2015. Stippling shows statistically significant differences among models from a Wilcoxon signed rank test at the 95% level. Right subplots are latitude average of LAI and GPP for 2006–2015 (blue) and 2090–2099 (green).

We divided the climate (Figure 3) and the vegetation (Figure 4) response in the RCP 8.5 scenario into the temperature-limited, precipitation-limited, and radiation-limited regions. Temperature and precipitation increase for all the three regions, while radiation increases only in the radiation-limited area (Figure 3). Thus, all the climate factors contribute to vegetation growth in addition to the CO_2 fertilization effect. As a result, all the three different climate-limited regions experience increases in ensemble GPP and LAI throughout the 21st century (Figure 4). LAI in the temperature-limited region shows large variability among the models compared to GPP, which implies the difficulty in modeling respiration and allocation ratios. LAI in the radiation-limited region shows a significant increase, but the magnitude of increase is small due to the saturation of the leaf increase.

By the last decade of the 21st century, the summary of the projections shows that annual climate constraints will ease for 51% of the Earth's vegetated land area (i.e., warmer in the temperature-limited region), tighten in 11% of the land area, with the remainder experiencing no change. The degree of easing varies, from 94% in the temperature-limited region and 23% in the precipitation-limited region, to 45% in the radiation-limited region (Figure 1b–d).

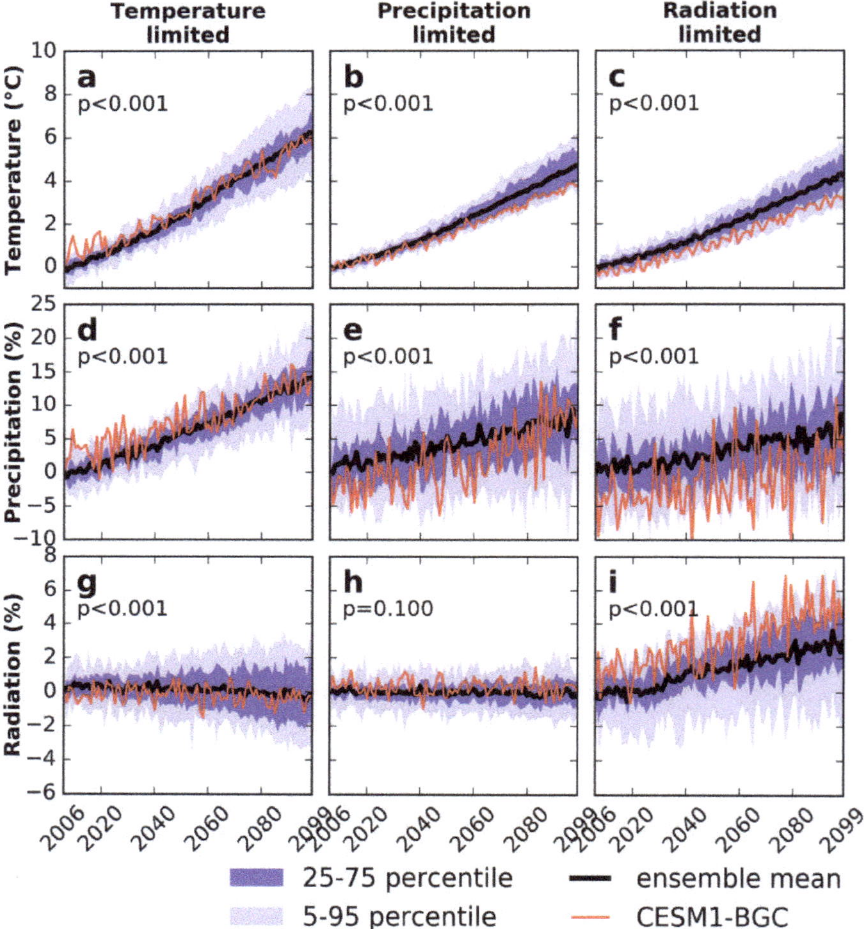

Figure 3. Climate responses from the Earth system models of CMIP5, summarized over the three climate-limiting regions. The simulations represent RCP 8.5, a pathway with the highest greenhouse gas emissions. Ensemble means and the percentiles show progressive relief of the main limiting factor for each region. Models diverge substantially towards the end of the simulation period, but almost all trends are statistically significant (see p-values in each panel). Changes in temperature (**a–c**), precipitation (**d–f**), and radiation (**g–i**) are expressed as percent of initial values in 2006. Outputs of Community Earth System Model, version 1–Biogeochemistry (CESM1-BGC), red line, are shown as examples of results from Earth system models incorporating nitrogen cycling. The p-value was calculated from Mann–Kendall trend test.

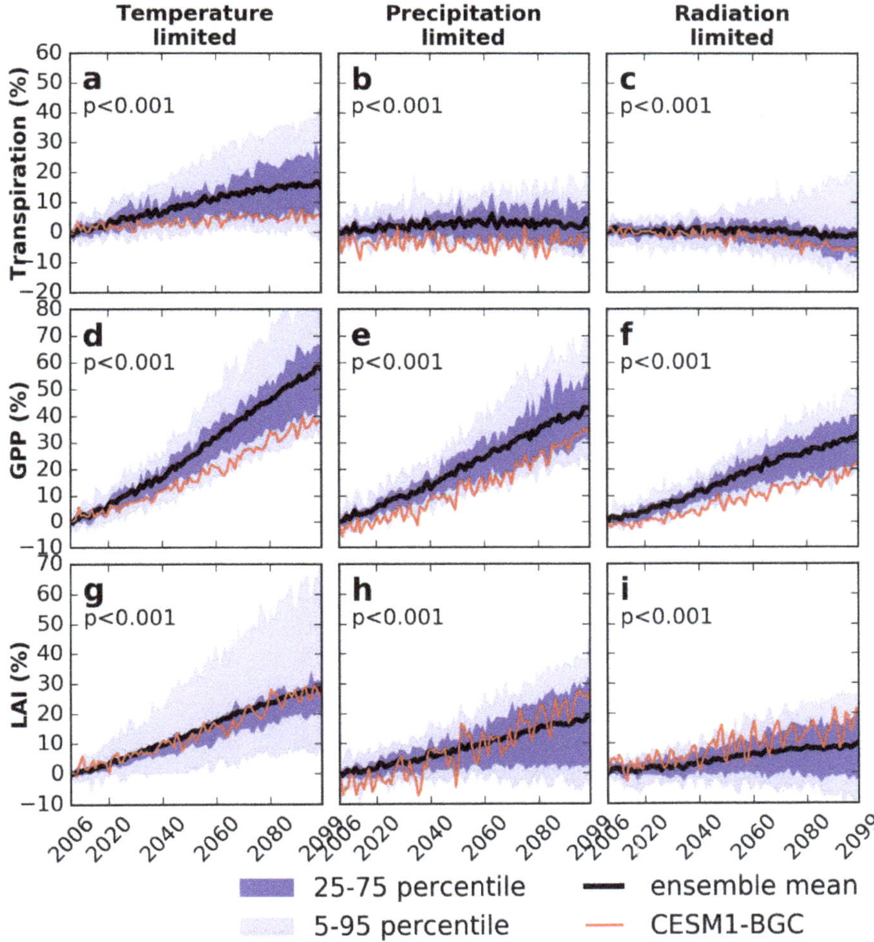

Figure 4. Vegetation responses from the Earth system models of CMIP5, summarized over the three climate-limiting regions. Same as Figure 3, except for changes in transpiration (**a–c**), GPP (**d–f**), and LAI (**g–i**).

3.2. Seasonal Trend in Climate and Vegetation

The changes in LAI and GPP are underlain not only by changes in annual climate but also by seasonal climate changes. Seasonal changes in climate (Figure 5) are associated with increased LAI and GPP (Figure 6), particularly for the temperature-limited region (note that we present seasonal analyses for the Northern Hemisphere only, to avoid contrasting seasonal patterns in the Southern Hemisphere and because the Northern Hemisphere accounts for 68% of total land area).

Boreal summers (June, July, and August: JJA) in temperature-limited regions become warmer, wetter, and brighter (Figure 5a,d,g); conditions that, at least in current Earth system models, lead to a summertime spike in LAI and GPP increases (Figure 6m,p). The precipitation increase is largest in the boreal fall (September, October, and November: SON); warming is at least 4 °C in all months, and increases to nearly 10 °C in winter. The JJA increase in solar radiation approaches 10 W/m^2, but is offset by winter decreases (Figure 5g), leading to no net change in the annual mean (Figure 3g).

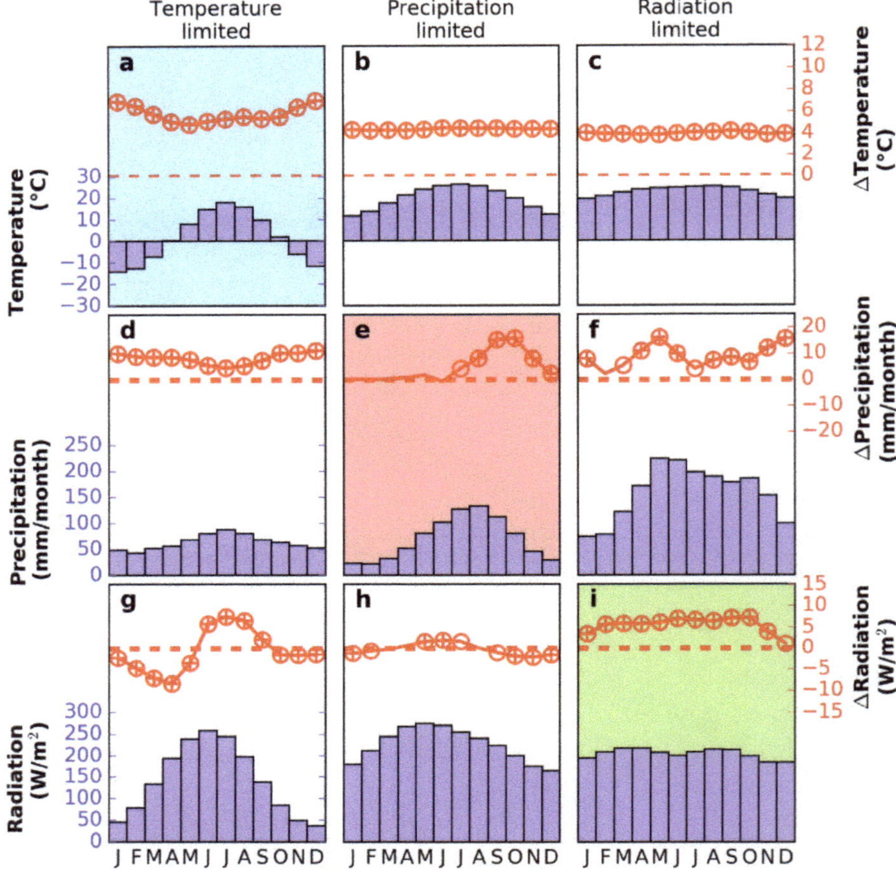

Figure 5. Easing of seasonal constraints to vegetation growth. Simulated changes in monthly climate and vegetation in the three climate-limiting regions: temperature, (**a–c**); precipitation, (**d–f**); and radiation, (**g–i**). Bars show climatological values (2006–2015), lines show ensemble mean monthly changes from Earth system models of CMIP5. Circles indicate that the trend in the 2006–2099 ensemble mean is significant at the 95% level from a Mann–Kendall trend test, while plus signs show that the Wilcoxon signed rank test is significant at the 95%. Shading is used to highlight changes in the limiting factor for each of the three regions (e.g., blue shading highlights temperature changes in the temperature-limited region).

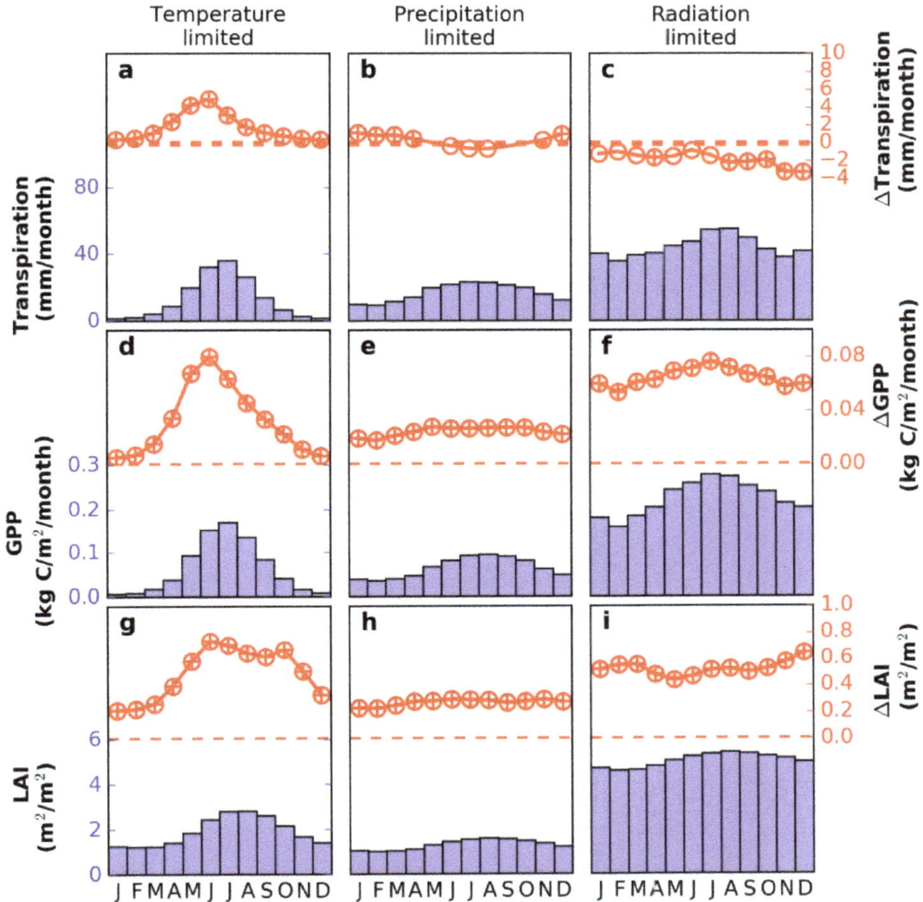

Figure 6. Easing of seasonal constraints to vegetation growth. Same was Figure 5, except for transpiration, **a–c**; GPP, **d–f**; and LAI, **g–i**.

3.3. Sensitivity Experiments for Climate and Vegetation

Increased CO_2 affects LAI and GPP through climate change, through both the greenhouse effect and CO_2 fertilization effect, but the CMIP5 ensemble for the RCP 8.5 scenario cannot separate the importance of the two processes [19]. To isolate their varying importance, we generate an eight-model ensemble (highlighted in Table 1), comparing results from the sensitivity experiments.

At first, we summarized the climate responses to increasing CO_2 concentration in these sensitivity experiments. The esmFdbk1 experiment shows that radiative effects clearly dominate projected warming, the signature feature of climate change (compare black and red lines in Figures 7a–c and 8a,d,g). Radiative effects also drive other key aspects of climate change, such as Arctic amplification and land–sea warming contrast (Figure 8a), equatorial and high-latitude increases in precipitation (Figure 8b), and high-latitude dimming from increases in fall/winter cloud cover (Figure 8c) [34]. Meanwhile, the esmFixClim experiment revealed that, except for increases in radiation for the temperature-limited region in JJA (Figure 7g), the vegetation physiological effect has little effect on region-averaged climate (e.g., near-zero changes in Figure 7a–f). Spatially, the physiological effects on climate are more consistent than the radiative effects, and include a slight drying (Figure 8e) and brightening because of reduced cloud cover (Figure 8f) over most land surfaces.

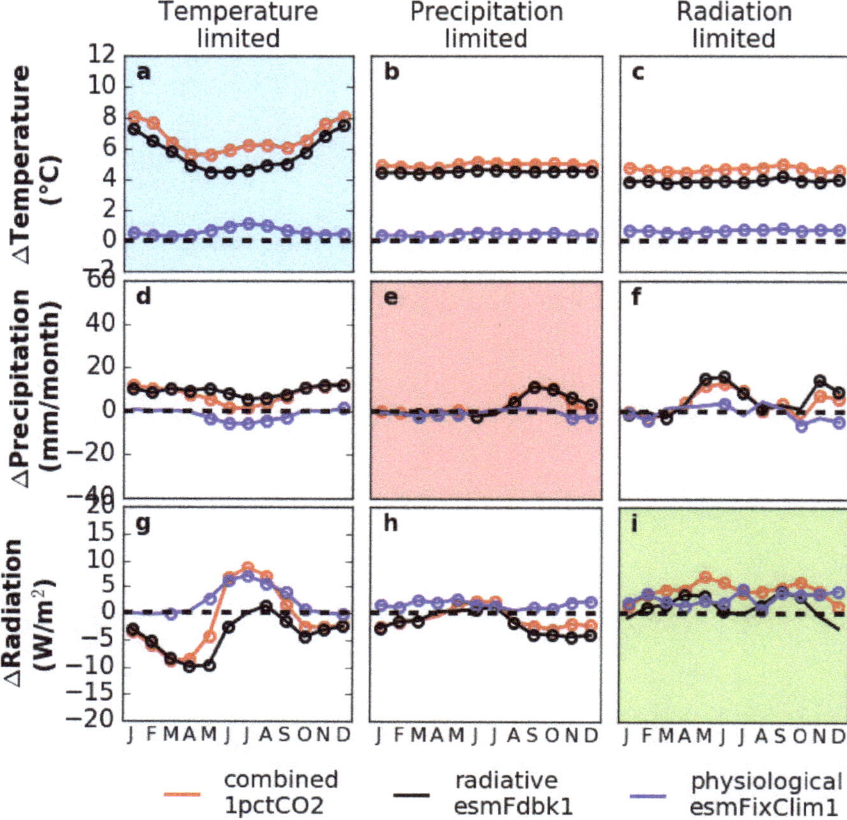

Figure 7. Additive and counteracting impacts of CO_2 on climate. Simulated changes in climate and vegetation from the CMIP5 eight-model ensemble in response to increasing CO_2 from 280 ppm by 1%/year for 140 years: temperature, **a–c**; precipitation, **d–f**; and radiation, **g–i**. The changes are the mean of the last ten years minus the mean of the first ten years (see Figures 8 and 10 for a spatial representation of the changes). In order to assess the contribution of radiative and vegetation physiological effects on climate and vegetation, three experiments were carried out: (1) CO_2 has a radiative forcing on climate but no direct effect on vegetation; (2) CO_2 has a vegetation physiological impact, primarily on internal CO_2 concentration and stomatal conductance, but does not directly alter radiative forcing; and (3) CO_2 has a combined effect on both radiative forcing and physiological impacts. Shading is used to highlight changes in the limiting factor for each of the three regions (e.g., blue shading highlights temperature changes in the temperature-limited region). Circles indicate that the trend in the 140-year ensemble mean is significant at the 95% level from a Mann–Kendall trend test.

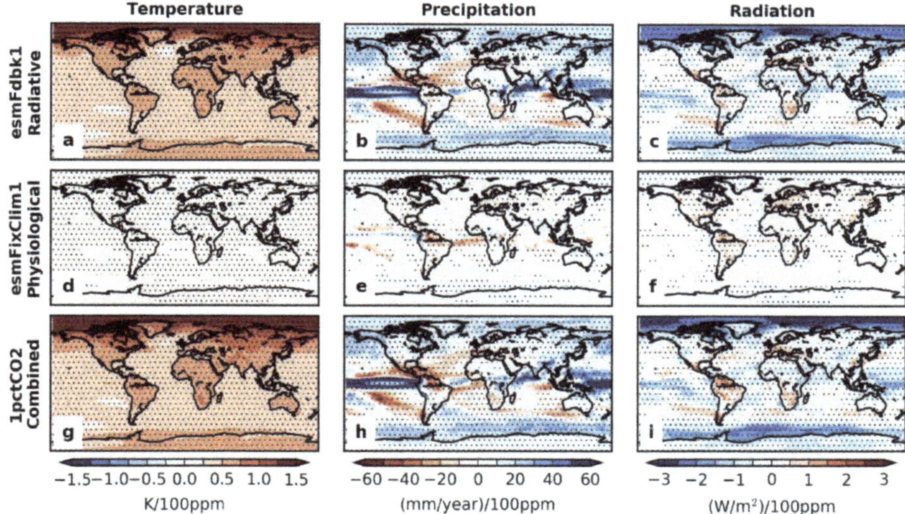

Figure 8. Spatial distribution of the additive and counteracting impacts of CO_2 on climate limiting factors. Temperature, (**a,d,g**); precipitation, **b,e,h**; and solar radiation, (**c,f,i**). In order to assess the contribution of radiative and vegetation physiological effects on climate and vegetation, we used an eight-model [highlighted in Table 1] ensemble to compare three CMIP5 experiments, each of which was run for 140 years and experiences a constant CO_2 at pre-industrial levels and/or CO_2 increasing by 1%/year to 4xCO_2: (1) In the radiative experiment (**a–c**), CO_2 increases for the atmosphere but stays constant for vegetation and the carbon cycle and hence the direct effects of CO_2 on plants are suppressed; (2) in the vegetation physiology experiment (**d–f**), CO_2 increases by 1%/year to 4xCO_2 for vegetation and the carbon cycle—thereby reducing stomatal conductance and providing CO_2 fertilization—but stays constant at 280 ppm for the atmosphere and thus the radiative effect is suppressed; (3) in the combined experiment (**g–i**), CO_2 concentration increases for the full Earth system. The changes are the mean of the last ten years minus the mean of the first ten years. Stippling indicates that the trend in the 140-year ensemble mean is significant at the 95% level from a Mann–Kendall trend test.

On the other hand, the summary of the vegetation response to increasing CO_2 concentration shows the opposite of the climate response in the sensitivity experiment. In spite of inducing climate changes to temperature and precipitation, the CO_2 fertilization effect alone can account for much, and in some cases, almost all, of the simulated changes in GPP and LAI in the combined experiment (compare blue and red lines in Figure 9d–i). Increases in LAI and GPP in the FixClim1 experiment appear to be driven by more radiation (because of reduced cloud cover) and reduced transpiration (therefore increased soil water), both of which are consistent with stomatal down-regulation following CO_2 increases [35]. As shown in the esmFdbk1 experiment, in contrast, the climate change induced by radiative forcing has near-zero effects on simulated GPP for all three regions (black line in Figure 9d–f), except for a small positive effect on GPP in the boreal spring of the temperature-limited region. Spatially, the ordinal impact on GPP and LAI often reverses between the esmFdbk1 and esmFixClim1 experiments, with much of the Southern Hemisphere switching from a reduction (Figure 10b,c) to an enhancement (Figure 10e,f).

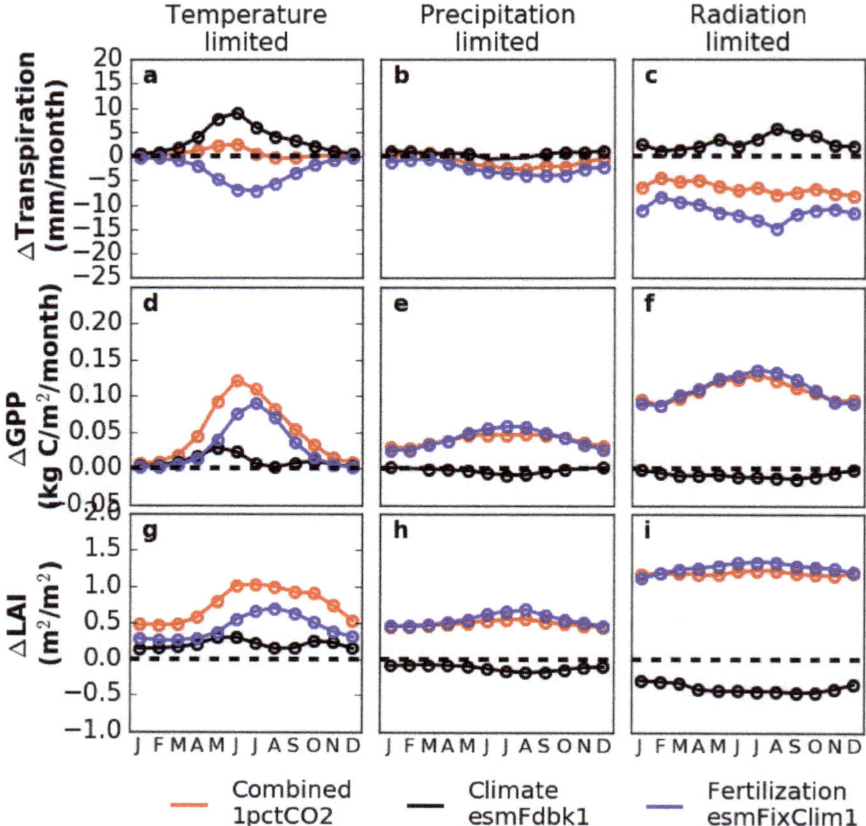

Figure 9. Additive and counteracting impacts of CO_2 on vegetation. Same as Figure 7 except for transpiration, **a–c**; GPP, **d–f**; and LAI, **g–i**.

For seven out of nine comparisons, the esmFdbk1 and esmFixClim1 experiments produce offsetting impacts on LAI and GPP (Figure 9). The largest differences are in the radiation-limited region, where the esmFdbk1 experiment slightly reduces GPP and the esmFixClim1 experiment increases GPP by as much as 0.15 kg C/m^2/month in JJA (compare blue and red lines in Figure 9f); LAI changes switch from about −0.5 to 1.0 (compare blue and red lines in Figure 9i). The temperature-limited region is an exception, where, for both LAI and GPP, the esmFdbk1 and esmFixClim1 experiments are additive (Figure 9m,p). In particular, high northern latitudes are the one clear location where both the esmFdbk1 and esmFixClim1 experiments increase LAI (Figure 10c,f) and GPP (Figure 10b,e), supporting recent conclusions of a strong climate imprint on the broad region of high-latitude during the observational era [36].

The critical role of vegetation physiology is clear in the precipitation-limited region, which experiences the least easing and an almost equal area with precipitation reductions (Figure 1d). But the region also sees a statistically significant increase in simulated equivalent precipitation water in 64% of its area (equivalent to up to 50 mm/month of water in some areas, Figure 1e) and increases in LAI and GPP (Figure 9e,h), in spite of near-zero changes in simulated annual (Figure 4b) and seasonal (Figure 6b) transpiration. While these patterns depict a more efficient use of available water resources and a progressive greening, the physiological response of plants to higher CO_2 in semi-arid regions appears to depend on local variations in simulated precipitation, which remain highly uncertain in CMIP5 simulations [37].

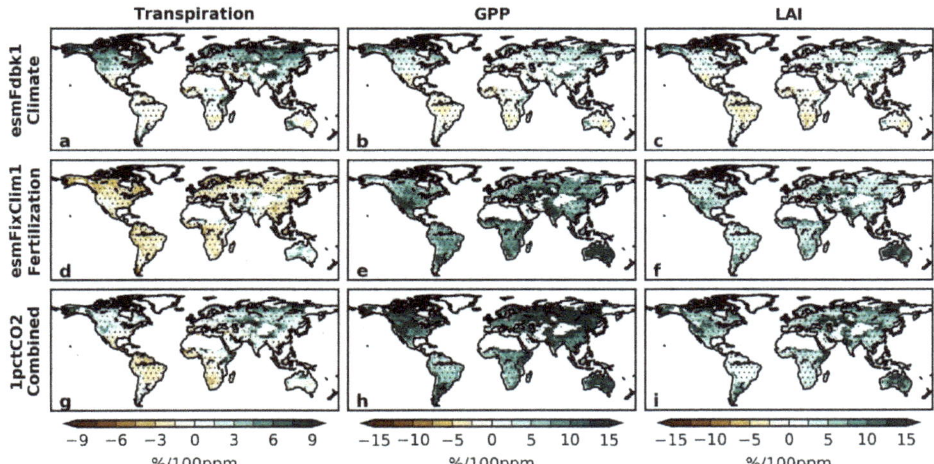

Figure 10. Spatial distribution of the additive and counteracting impacts of CO_2 on vegetation. Same as Figure 8 except for transpiration, **a, d, g**; GPP, **b, e, h**; and LAI, **c, f, i**.

3.4. Decomposing Vegetation Growth Into Three Factors

The results indicate that the climate feedback substantially contributed to the growth of vegetation by relaxing climate constraints. Although it is well known that the climate feedback can positively or negatively influence the growth of vegetation, it has not been quantitatively assessed at the global scale. The sensitivity experiments allow us to quantitatively evaluate the climate feedback to vegetation ($f \Delta Clim_{feedback}$ in Equation (1)). We assumed that the ratio of GPP change to change in climate variables is constant (i.e., $b = f = const.$). At first, we calculated $\Delta GPP / \Delta Clim$ (i.e., b) in the climate experiment using the linear regression of annual GPP on annual mean climate variables. Then, we calculated $a \Delta CO_2$ by subtracting $f \Delta Clim_{feedback}$ from GPP increase in Equation (2). Finally, $b \Delta Clim$ was derived from Equation (3) by subtracting $a \Delta CO_2$ and $f \Delta Clim_{feedback}$ by assuming the additive relationship of the fertilization effect and the climate change effect.

The percentage ratios of the contribution of each term (i.e., $a \Delta CO_2$, $b\Delta Clim$, and $f \Delta Clim_{feedback}$) were shown for each limited region in Figure 11. In the temperature-limited region, all the three terms substantially contributed to the increase in GPP, and the annual average of the climate feedback contribution was 17%. The snow-albedo feedback can account for the climate feedback [38,39]. The climate feedback added 37% more increase in GPP than the radiative warming effect alone. The total contribution of climate feedback and climate is 63%, which is the highest contribution among the three climate-limited regions. The contribution is much higher in winter than summer because the temperature did not limit GPP in the summer.

In the precipitation limited area, there was almost no contribution of the climate feedback effect. This result can be explained by the relatively low water-recycling ratio compared to the humid area [40]. The moisture from the other regions controls the precipitation trend in the water-limited region, so that the influence of changing water use efficiency on the region is negligible.

In the radiation-limited area, the contribution of the climate feedback is 7%, while the radiative climate change negatively affects 24% of the increase in GPP. The feedback was caused by the decreasing trend in cloud cover through change in water use efficiency [35]. The magnitude of the contribution of the climate feedback changes with model selection due to the difficulty in modeling clouds in GCMs. Thus, the feedback contribution can be underestimated, especially when a low resolution GCM cannot represent well the increase in regional convective clouds caused by the enhanced water cycle that results from added vegetation growth.

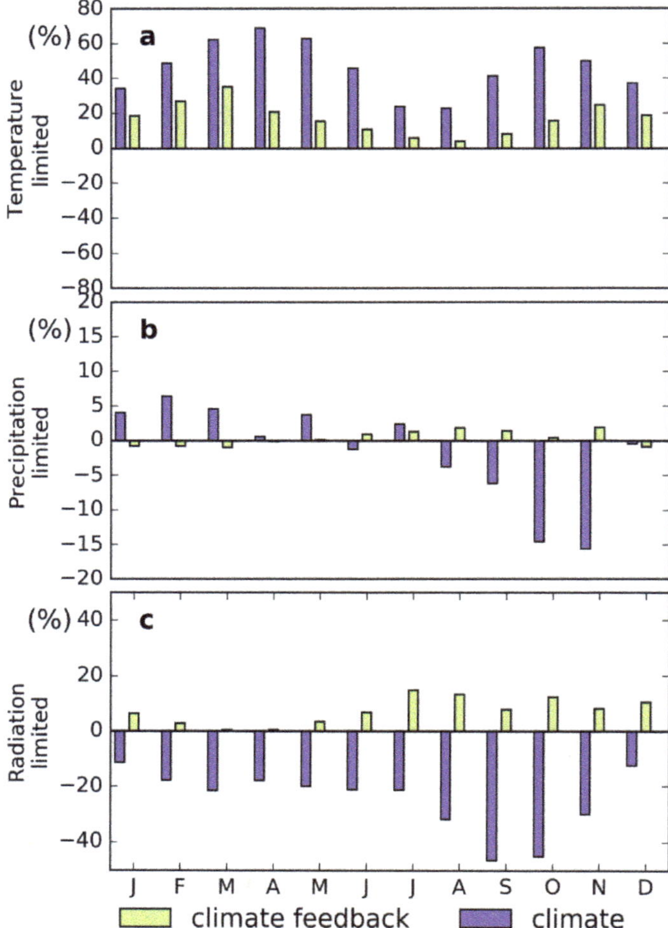

Figure 11. Monthly contribution of climate feedback and radiative climate change to vegetation growth in increases in CO_2 by 1%/year to 4xCO_2 experiment (1pctCO2) for each climate-limited area. The monthly contribution was calculated for each climate-limited region. The green and purple bars show the contribution of climate feedback and radiative climate change, respectively. In each region, the total of the contribution (CO_2 fertilization, climate feedback, and radiative climate change) was summed up to 100%.

4. Discussion

Our analysis suggests fundamental future increases in the amount of vegetation and photosynthesis (LAI and GPP in Figure 1), mainly arising from relaxing climate constraints on vegetation growth. This feedback effect can explain the discrepancy between the models and the observation in the β factor [14]. We argue that the results are consistent with three lines of observational evidence and a considerable body of paleoclimatic evidence of dramatically different vegetation composition during past high-CO_2 periods [41,42].

First, if our central claim that vegetation physiological process reduces transpiration, reduces cloud cover, and increases radiation is correct, then cloud cover, particularly low-level clouds, which strongly influence the planetary shortwave radiation budget, should decrease. Recent climate modeling studies indeed simulate a decrease in low-level cloudiness due to the vegetation physiological effect [23,29,35]. Further, several modeling studies indicate that the rapid adjustments of the troposphere for the combined radiative and physiological effects of increased CO_2 are associated with a decrease in low-level cloud cover over land, but increased boundary layer cloud cover over oceans [29,43,44]. We re-analyzed the NDP026 ground observation of cloud cover [26] and show that the modeled processes have indeed occurred over the period 1971 to 2005 using the Mann–Kendall trend test,. During this period, cloud cover significantly decreased by a few percent points per decade over much of the land surface, and increased over the ocean (Figure 12).

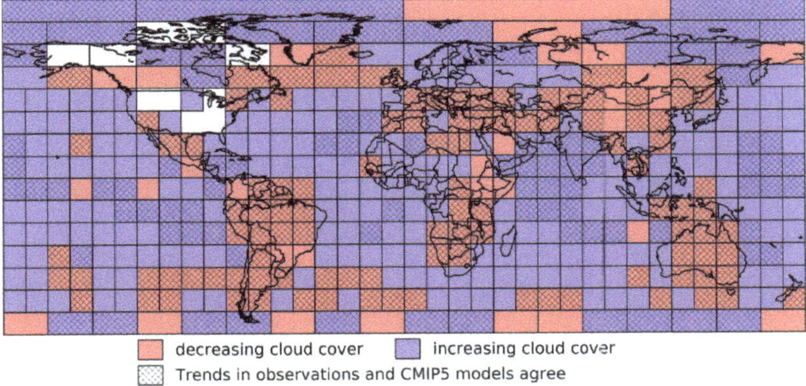

Figure 12. Changes in observed (NDP026) and CMIP5-simulated cloud cover are consistent with a CO_2-induced down-regulation of stomatal conductance, resulting in reduced transpiration. Trends from observations are shown as decreasing (red) or increasing (blue) annual average cloud cover for 1971–2005: Red would tend to support our hypothesis of reduced transpiration and cloud cover as a result of stomatal down-regulation. If the CMIP5 trend and the observed trend have the same sign, the corresponding box is hatched. Location and shape of boxes corresponds to the coverage in the observational dataset, the 1971–2005 comparison period represent the overlap between the observational dataset and the historical CMIP5 runs. A change during the mid-1990s in cloud cover observation methodology in the US precluded their use in trend evaluation in NDP026. A chi-squared test between the two data sets rejected the null-hypothesis that they are independent ($p = 0.05$). Chi-squared tests performed on the two data sets at seasonal scales yielded the following p-values: December January February (0.22), March April May (0.05), June July August (0.26), and September October November (0.01).

Second, the same models we used for future projections under RCP 8.5 produce simulations of the historical climate and vegetation that are broadly consistent with independent observations (Figure 13). Although historical skill does not guarantee future performance, region-level simulations of climatological temperature, and precipitation are statistically indistinguishable from the Climate Research Unit (CRU) product [25] in all months (Figure 13a,b). Simulated radiation, assessed for the radiation-limited regions of the Northern hemisphere against the CRUNCEP radiation dataset, has climatological differences of up to 15 W/m^2, but a similar seasonal cycle (Figure 13e). The Earth system models also capture seasonal satellite-observed variations in LAI in all three regions (Figure 13b,d,f).

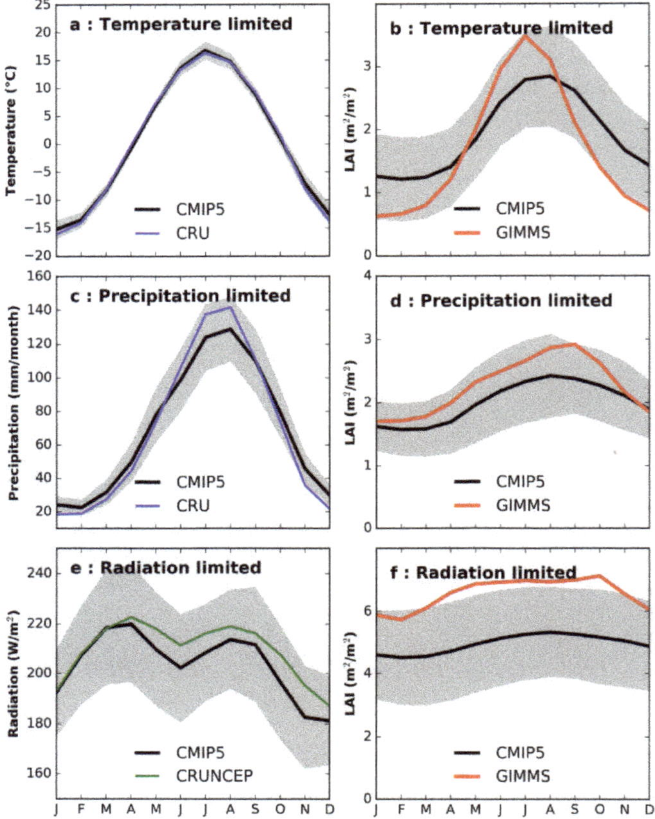

Figure 13. Performance of Earth system models used in CMIP5 against observations (1982–2005) in each of the climate-limiting regions. CMIP5 seasonality for climate and vegetation are ensemble means of each of the parameters. Shading represents standard deviation around the ensemble mean from the CMIP5 models. Observed seasonality in climate and vegetation over the same period is calculated from Climate Research Unit (CRU) (temperature, precipitation), CRU National Centers for Environmental Prediction version (CRUNCEP) (radiation), and Global Inventory Modeling and Mapping Studies Leaf Area Index (GIMMS) LAI.

Third, the projected changes in LAI and climate are already apparent in the observational era. Satellite data show that LAI has increased from 1982 to 2005 for all three regions (Figure 14b,d,f); the CRU product shows warming in the temperature-limited region (Figure 14a), increased precipitation in the precipitation-limited region (Figure 14c), and reduced cloud cover in the radiation-limited region (Figure 14e). Consistent with these changes in climate and the biosphere, terrestrial ecosystems have been shown as net sink for carbon in recent decades [45,46]. Thus, the 21st century changes to climate and greening do not appear anomalous or implausible, when viewed in the context of recent history.

Numerous processes, including extreme climatic events [47], could reduce the projected changes in LAI and GPP. But the paleoclimate record also shows that profound changes in vegetation have occurred in the past, particularly in high latitudes, where the temperature-limited region appears to benefit the most from physical climate changes, mediated through vegetation physiological mechanisms. The early Eocene greenhouse climate, for example, supported redwoods at 78° N paleo-latitude under CO_2 levels that are similar to the modern levels [42]. The deep-time perspective, albeit associated with different time scales and continental configurations, therefore, does not appear to rule out the sort of major changes to vegetation seen in the 21st century projections [48].

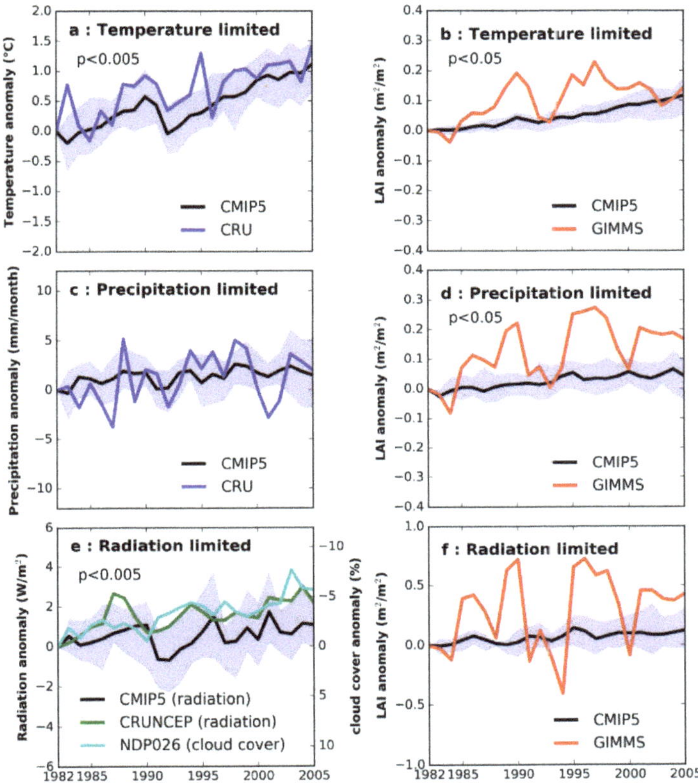

Figure 14. Performance of Earth system models in capturing the observed trends in climate and vegetation. Observed trends in annual mean climate and satellite-derived LAI from 1982 to 2005 are shown as average responses over the three climate-limiting regions. Ensemble mean of CMIP5 models over the same period do not capture the inter-annual variability, but appear to capture the overall trends in climate and vegetation. The first year of each series is set to zero to emphasize the magnitude of the trend and deviation between the trend lines. The *p*-values indicate the level of significance for the trends in observations only (CRU, CRUNCEP, GIMMS, and cloud cover data from NDP026). Shading around the CMIP5 ensemble mean indicates 25–75 percentile.

We focused on two elemental components of terrestrial ecosystems—the amount of leafy material and gross carbon fixation—but do not provide insights into respiratory and net carbon fluxes, carbon stocks, such as biomass and soil carbon, and vegetation dynamics. The CMIP5 models, especially low-resolution models, cannot count the extreme events, such as forest fires or hurricanes. The FACE experiments also suggest that non-climate limiting factors, such as nitrogen and phosphorous [49], might supersede climate limitations in the future (although the inclusion of a nitrogen cycle produces results that are within the uncertainties of the full ensemble; see red lines in Figures 3 and 4). The available state-of-the-art Earth system models; however, depict a late 21st century world in which vegetation physiology interacts with pervasive changes to annual and seasonal climate to create a greener land surface.

5. Conclusions

We analyzed the climate feedback on vegetation using CMIP5 model experiments for each climate-limited region. In contrast to the climate trend induced by the radiative effect, the positive trend in GPP and LAI can be attributed mainly to the CO_2 fertilization effect. While CO_2 fertilization was the main driver of the increasing trend in vegetation, the climate feedback on vegetation also contributed to 17% and 7% of vegetation growth in temperature-limited and radiation-limited regions, respectively. These feedbacks provide additional sensitivity to the CO_2 fertilization, and can explain the discrepancy of the β factor between models and observation. The observed trend corroborates the importance of the climate feedback in explaining the greening earth.

Author Contributions: Conceptualization, H.H., R.R.N., G.B., L.C., S.G., W.W., C.M., R.E., T.L. and R.M.; data curation, A.R.M.

Funding: This research was funded by the NASA Earth science program.

Acknowledgments: Computational resources from NASA Earth Exchange helped facilitate the analysis.

Conflicts of Interest: The authors declare no conflict of interest.

References

1. Friedlingstein, P.; Meinshausen, M.; Arora, V.K.; Jones, C.D.; Anav, A.; Liddicoat, S.K.; Knutti, R. Uncertainties in CMIP5 climate projections due to carbon cycle feedbacks. *J. Clim.* **2013**, *27*, 511–526. [CrossRef]
2. Zhu, Z.; Piao, S.; Myneni, R.B.; Huang, M.; Zeng, Z.; Canadell, J.G.; Ciais, P.; Sitch, S.; Friedlingstein, P.; Arneth, A.; et al. Greening of the Earth and its drivers. *Nat. Clim. Chang.* **2016**, *6*, 791–795. [CrossRef]
3. Mao, J.; Ribes, A.; Yan, B.; Shi, X.; Thornton, P.E.; Séférian, R.; Ciais, P.; Myneni, R.B.; Douville, H.; Piao, S.; et al. Human-induced greening of the northern extratropical land surface. *Nat. Clim. Chang.* **2016**, *6*, 959–963. [CrossRef]
4. Donohue, R.J.; Roderick, M.L.; McVicar, T.R.; Farquhar, G.D. Impact of CO_2 fertilization on maximum foliage cover across the globe's warm, arid environments. *Geophys. Res. Lett.* **2013**, *40*, 3031–3035. [CrossRef]
5. Pan, Y.; Birdsey, R.A.; Fang, J.; Houghton, R.; Kauppi, P.E.; Kurz, W.A.; Phillips, O.L.; Shvidenko, A.; Lewis, S.L.; Canadell, J.G.; et al. A large and persistent carbon sink in the world's forests. *Science* **2011**, *333*, 988–993. [CrossRef] [PubMed]
6. Ciais, P.; Sabine, C.; Bala, G.; Bopp, L.; Brovkin, V.; Canadell, J.; Chhabra, A.; DeFries, R.; Galloway, J.; Heimann, M.; et al. Carbon and other biogeochemical cycles. In *Climate Change 2013: The Physical Science Basis. Contribution of Working Group I to the Fifth Assessment Report of the Intergovernmental Panel on Climate Change*; Stocker, T.F., Qin, D., Plattner, G.K., Tignor, M., Allen, S.K., Boschung, J., Nauels, A., Xia, Y., Bex, V., Midgley, P.M., Eds.; Cambridge University Press: Cambridge, UK; New York, NY, USA, 2013; pp. 465–570. ISBN 9781107661820.
7. Schimel, D.; Stephens, B.B.; Fisher, J.B. Effect of increasing CO_2 on the terrestrial carbon cycle. *Proc. Natl. Acad. Sci. USA* **2015**, *112*, 436–441. [CrossRef] [PubMed]
8. Devaraju, N.; Bala, G.; Caldeira, K.; Nemani, R. A model based investigation of the relative importance of CO_2-fertilization, climate warming, nitrogen deposition and land use change on the global terrestrial carbon uptake in the historical period. *Clim. Dyn.* **2016**, *47*, 173–190. [CrossRef]
9. Nemani, R.R.; Keeling, C.D.; Hashimoto, H.; Jolly, W.M.; Piper, S.C.; Tucker, C.J.; Myneni, R.B.; Running, S.W. Climate-driven increases in global terrestrial net primary production from 1982 to 1999. *Science* **2003**, *300*, 1560–1563. [CrossRef] [PubMed]
10. Norby, R.J.; Zak, D.R. Ecological Lessons from Free-Air CO_2 Enrichment (FACE) Experiments. *Annu. Rev. Ecol. Evol. Syst.* **2011**, *42*, 181–203. [CrossRef]
11. Keenan, T.F.; Hollinger, D.Y.; Bohrer, G.; Dragoni, D.; Munger, J.W.; Schmid, H.P.; Richardson, A.D. Increase in forest water-use efficiency as atmospheric carbon dioxide concentrations rise. *Nature* **2013**, *499*, 324–327. [CrossRef] [PubMed]
12. Bonan, G.B.; Pollard, D.; Thompson, S.L. Effects of boreal forest vegetation on global climate. *Nature* **1992**, *359*, 716–718. [CrossRef]

13. Myneni, R.B.; Keeling, C.D.; Tucker, C.J.; Asrar, G.; Nemani, R.R. Increased plant growth in the northern high latitudes from 1981 to 1991. *Nature* **1997**, *386*, 698–702. [CrossRef]

14. Smith, W.K.; Reed, S.C.; Cleveland, C.C.; Ballantyne, A.P.; Anderegg, W.R.L.; Wieder, W.R.; Liu, Y.Y.; Running, S.W. Large divergence of satellite and Earth system model estimates of global terrestrial CO_2 fertilization. *Nat. Clim. Chang.* **2015**, *6*, 306–310. [CrossRef]

15. Lemordant, L.; Gentine, P.; Swann, A.S.; Cook, B.I.; Scheff, J. Critical impact of vegetation physiology on the continental hydrologic cycle in response to increasing CO_2. *Proc. Natl. Acad. Sci. USA* **2018**, *115*, 4093–4098. [CrossRef] [PubMed]

16. Chadwick, R.; Good, P.; Martin, G.; Rowell, D.P. Large rainfall changes consistently projected over substantial areas of tropical land. *Nat. Clim. Chang.* **2015**, *6*, 177. [CrossRef]

17. Kumar, S.; Allan, R.P.; Zwiers, F.; Lawrence, D.M.; Dirmeyer, P.A. Revisiting trends in wetness and dryness in the presence of internal climate variability and water limitations over land. *Geophys. Res. Lett.* **2015**, *42*, 10867–10875. [CrossRef]

18. Polson, D.; Hegerl, G.C. Strengthening contrast between precipitation in tropical wet and dry regions. *Geophys. Res. Lett.* **2017**, *44*, 365–373. [CrossRef]

19. Taylor, K.E.; Stouffer, R.J.; Meehl, G.A.; Taylor, K.E.; Stouffer, R.J.; Meehl, G.A. An overview of CMIP5 and the experiment design. *Bull. Am. Meteorol. Soc.* **2012**, *93*, 485–498. [CrossRef]

20. Riahi, K.; Rao, S.; Krey, V.; Cho, C.; Chirkov, V.; Fischer, G.; Kindermann, G.; Nakicenovic, N.; Rafaj, P. RCP 8.5—A scenario of comparatively high greenhouse gas emissions. *Clim. Chang.* **2011**, *109*, 33–57. [CrossRef]

21. Christensen, J.H.; Krishna, K.K.; Aldrian, E.; An, S.-I.; Cavalcanti, I.F.A.; de Castro, M.; Dong, W.; Goswami, P.; Hall, A.; Kanyanga, J.K.; et al. Climate Phenomena and their Relevance for Future Regional Climate Change. In *Climate Change 2013: The Physical Science Basis. Contribution of Working Group I to the Fifth Assessment Report of the Intergovernmental Panel on Climate Change*; Stocker, T.F., Qin, D., Plattner, G.K., Tignor, M., Allen, S.K., Boschung, J., Nauels, A., Xia, Y., Bex, V., Midgley, P.M., Eds.; Cambridge University Press: Cambridge, UK; New York, NY, USA, 2013; pp. 1217–1308. ISBN 978-1-107-66182-0.

22. Dekker, S.C.; Groenendijk, M.; Booth, B.B.B.; Huntingford, C.; Cox, P.M. Spatial and temporal variations in plant water-use efficiency inferred from tree-ring, eddy covariance and atmospheric observations. *Earth Syst. Dyn.* **2016**, *7*, 525–533. [CrossRef]

23. Boucher, O.; Randall, D.; Artaxo, P.; Bretherton, C.; Feingold, G.; Forster, P.; Kerminen, V.M.; Kondo, Y.; Liao, H.; Lohmann, U.; et al. Clouds and aerosols. In *Climate Change 2013: The Physical Science Basis. Contribution of Working Group I to the Fifth Assessment Report of the Intergovernmental Panel on Climate Change*; Stocker, T.F., Qin, D., Plattner, G.K., Tignor, M., Allen, S.K., Boschung, J., Nauels, A., Xia, Y., Bex, V., Midgley, P.M., Eds.; Cambridge University Press: Cambridge, UK; New York, NY, USA, 2013; pp. 571–658. ISBN 9781107661820.

24. Zhu, Z.; Bi, J.; Pan, Y.; Ganguly, S.; Anav, A.; Xu, L.; Samanta, A.; Piao, S.; Nemani, R.R.; Myneni, R.B. Global Data Sets of Vegetation Leaf Area Index (LAI)3g and Fraction of Photosynthetically Active Radiation (FPAR)3g Derived from Global Inventory Modeling and Mapping Studies (GIMMS) Normalized Difference Vegetation Index (NDVI)3g for the period 1981 to 2. *Remote Sens.* **2013**, *5*, 927–948. [CrossRef]

25. New, M.; Lister, D.; Hulme, M.; Makin, I. A high-resolution data set of surface climate over global land areas. *Clim. Res.* **2002**, *21*, 1–25. [CrossRef]

26. Eastman, R.; Warren, S.G. A 39-yr survey of cloud changes from land stations worldwide 1971–2009: Long-term trends, relation to aerosols, and expansion of the tropical belt. *J. Clim.* **2013**, *26*, 1286–1303. [CrossRef]

27. Warren, S.G.; Eastman, R.M.; Hahn, C.J.; Warren, S.G.; Eastman, R.M.; Hahn, C.J. A Survey of Changes in Cloud Cover and Cloud Types over Land from Surface Observations, 1971–96. *J. Clim.* **2007**, *20*, 717–738. [CrossRef]

28. Zelinka, M.D.; Klein, S.A.; Taylor, K.E.; Andrews, T.; Webb, M.J.; Gregory, J.M.; Forster, P.M.; Zelinka, M.D.; Klein, S.A.; Taylor, K.E.; et al. Contributions of Different Cloud Types to Feedbacks and Rapid Adjustments in CMIP5. *J. Clim.* **2013**, *26*, 5007–5027. [CrossRef]

29. Cao, L.; Bala, G.; Caldeira, K. Climate response to changes in atmospheric carbon dioxide and solar irradiance on the time scale of days to weeks. *Environ. Res. Lett.* **2012**, *7*, 034015. [CrossRef]

30. Drake, B.G.; Gonzàlez-Meler, M.A.; Long, S.P. More efficient plants: A consequence of rising atmospheric CO_2? *Annu. Rev. Plant Biol.* **1997**, *48*, 609–639. [CrossRef]

31. Kendall, M.G. *Rank Correlation Methods*; Griffin: London, UK, 1975; ISBN 9780852641996.

32. Mystakidis, S.; Davin, E.L.; Gruber, N.; Seneviratne, S.I. Constraining future terrestrial carbon cycle projections using observation-based water and carbon flux estimates. *Glob. Chang. Biol.* **2016**, *22*, 2198–2215. [CrossRef]

33. Mahowald, N.; Lo, F.; Zheng, Y.; Harrison, L.; Funk, C.; Lombardozzi, D.; Goodale, C. Projections of leaf area index in earth system models. *Earth Syst. Dyn.* **2016**, *7*, 211–229. [CrossRef]

34. Norris, J.R.; Allen, R.J.; Evan, A.T.; Zelinka, M.D.; O'Dell, C.W.; Klein, S.A. Evidence for climate change in the satellite cloud record. *Nature* **2016**, *536*, 72–75. [CrossRef]

35. Cao, L.; Bala, G.; Caldeira, K.; Nemani, R.; Ban-Weiss, G. Importance of carbon dioxide physiological forcing to future climate change. *Proc. Natl. Acad. Sci. USA* **2010**, *107*, 9513–9518. [CrossRef] [PubMed]

36. Forkel, M.; Carvalhais, N.; Rödenbeck, C.; Keeling, R.; Heimann, M.; Thonicke, K.; Zaehle, S.; Reichstein, M. Enhanced seasonal CO_2 exchange caused by amplified plant productivity in northern ecosystems. *Science* **2016**, *351*, 696–699. [CrossRef] [PubMed]

37. Bathiany, S.; Claussen, M.; Brovkin, V.; Bathiany, S.; Claussen, M.; Brovkin, V. CO_2-induced Sahel greening in three CMIP5 earth system models. *J. Clim.* **2014**, *27*, 7163–7184. [CrossRef]

38. Wang, L.; Cole, J.N.S.; Bartlett, P.; Verseghy, D.; Derksen, C.; Brown, R.; von Salzen, K. Investigating the spread in surface albedo for snow-covered forests in CMIP5 models. *J. Geophys. Res. Atmos.* **2016**, *121*, 1104–1119. [CrossRef]

39. Loranty, M.M.; Berner, L.T.; Goetz, S.J.; Jin, Y.; Randerson, J.T. Vegetation controls on northern high latitude snow-albedo feedback: Observations and CMIP5 model simulations. *Glob. Chang. Biol.* **2014**, *20*, 594–606. [CrossRef] [PubMed]

40. Dirmeyer, P.A.; Brubaker, K.L.; Dirmeyer, P.A.; Brubaker, K.L. Characterization of the Global Hydrologic Cycle from a Back-Trajectory Analysis of Atmospheric Water Vapor. *J. Hydrometeorol.* **2007**, *8*, 20–37. [CrossRef]

41. Franks, P.J.; Adams, M.A.; Amthor, J.S.; Barbour, M.M.; Berry, J.A.; Ellsworth, D.S.; Farquhar, G.D.; Ghannoum, O.; Lloyd, J.; McDowell, N.; et al. Sensitivity of plants to changing atmospheric CO_2 concentration: From the geological past to the next century. *New Phytol.* **2013**, *197*, 1077–1094. [CrossRef]

42. Maxbauer, D.P.; Royer, D.L.; LePage, B.A. High Arctic forests during the middle Eocene supported by moderate levels of atmospheric CO_2. *Geology* **2014**, *42*, 1027–1030. [CrossRef]

43. Kamae, Y.; Watanabe, M. Tropospheric adjustment to increasing CO_2: Its timescale and the role of land-sea contrast. *Clim. Dyn.* **2013**, *41*, 3007–3024. [CrossRef]

44. Modak, A.; Bala, G.; Cao, L.; Caldeira, K. Why must a solar forcing be larger than a CO_2 forcing to cause the same global mean surface temperature change? *Environ. Res. Lett.* **2016**, *11*, 044013. [CrossRef]

45. Le Quéré, C.; Peters, G.P.; Andres, R.J.; Andrew, R.M.; Boden, T.A.; Ciais, P.; Friedlingstein, P.; Houghton, R.A.; Marland, G.; Moriarty, R.; et al. Global carbon budget 2013. *Earth Syst. Sci. Data* **2014**, *6*, 235–263. [CrossRef]

46. Graven, H.D.; Keeling, R.F.; Piper, S.C.; Patra, P.K.; Stephens, B.B.; Wofsy, S.C.; Welp, L.R.; Sweeney, C.; Tans, P.P.; Kelley, J.J.; et al. Enhanced seasonal exchange of CO_2 by Northern ecosystems since 1960. *Science* **2013**, *341*, 1085–1089. [CrossRef] [PubMed]

47. Reichstein, M.; Bahn, M.; Ciais, P.; Frank, D.; Mahecha, M.D.; Seneviratne, S.I.; Zscheischler, J.; Beer, C.; Buchmann, N.; Frank, D.C.; et al. Climate extremes and the carbon cycle. *Nature* **2013**, *500*, 287–295. [CrossRef] [PubMed]

48. Salzmann, U.; Haywood, A.M.; Lunt, D.J. The past is a guide to the future? Comparing Middle Pliocene vegetation with predicted biome distributions for the twenty-first century. *Philos. Trans. R. Soc. A Math. Phys. Eng. Sci.* **2009**, *367*, 189–204. [CrossRef] [PubMed]

49. Zhang, Q.; Wang, Y.P.; Matear, R.J.; Pitman, A.J.; Dai, Y.J. Nitrogen and phosphorous limitations significantly reduce future allowable CO_2 emissions. *Geophys. Res. Lett.* **2014**, *41*, 632–637. [CrossRef]

Article

Influence of Bias Correction Methods on Simulated Köppen—Geiger Climate Zones in Europe

Beáta Szabó-Takács [1,*], Aleš Farda [1,2], Petr Skalák [1,2] and Jan Meitner [1]

[1] Global Change Research Institute CAS, 60300 Brno, Czech Republic; farda.a@czechglobe.cz (A.F.);
skalak.p@czechglobe.cz (P.S.); meitner.j@czechglobe.cz (J.M.)
[2] Czech Hydrometeorological Institute, 143006 Prague, Czech Republic
[*] Correspondence: szabo.b@czechglobe.cz

Received: 7 December 2018; Accepted: 19 January 2019; Published: 22 January 2019

Abstract: Our goal was to investigate the influence of bias correction methods on climate simulations over the European domain. We calculated the Köppen—Geiger climate classification using five individual regional climate models (RCM) of the ENSEMBLES project in the European domain during the period 1961—1990. The simulated precipitation and temperature data were corrected using the European daily high-resolution gridded dataset (E-OBS) observed data by five methods: (i) the empirical quantile mapping of precipitation and temperature, (ii) the quantile mapping of precipitation and temperature based on gamma and Generalized Pareto Distribution of precipitation, (iii) local intensity scaling, (iv) the power transformation of precipitation and (v) the variance scaling of temperature bias corrections. The individual bias correction methods had a significant effect on the climate classification, but the degree of this effect varied among the RCMs. Our results on the performance of bias correction differ from previous results described in the literature where these corrections were implemented over river catchments. We conclude that the effect of bias correction may depend on the region of model domain. These results suggest that distribution free bias correction approaches are the most suitable for large domain sizes such as the pan-European domain.

Keywords: Regional Climate Model; climate classification; bias correction methods; precipitation; temperature

---

1. Introduction

Climate classifications are frequently applied tools for evaluating the real climate system. One of the oldest and still widely accepted systems of climate types was introduced by Wladimir Köppen [1] and later modified by Geiger [2] and additionally by Trewartha [3–7]. Köppen divided eleven climate types based on annual and monthly changes in temperature and precipitation. Trewartha modified the Köppen classification so that the classifications based on the main quality differences and the vegetation characteristics were better taken into consideration. The so-called Köppen–Geiger (K-G) climate classification is derived directly from eco-biological vegetation characteristics within the individual regions of the Earth, which make it suitable for assessing climate change impacts on ecosystems. It is based on annual and monthly mean values of temperature and precipitation and distinguishes five main vegetation groups: the equatorial zone (A), the arid zone (B), the warm temperate zone (C), the snow zone (D) and the polar zone (E). The main groups are further divided into subtypes, reflecting the annual course of air temperature or precipitation and their monthly values compared to a defined threshold. For a detailed overview of all K-G classes and their spatial distribution around the world, we refer to [8]. K-G classification can be applied either to the real observed data of the Earth's climate or present or future conditions simulated by climate models [5,7,9,10]. Some studies, e.g., [11,12] have used the Köppen–Trewartha classification [13] to map the extent of climate change in Europe using an ensemble mean of regional climate models (RCMs) and simulations, considering the uncertainty related to driving global climate models (GCMs). However, the fact remains that all studies based on climate models should deal with model errors carefully before drawing conclusions.

According to [14], model errors can be caused by the initial and boundary conditions, parameterization, physical formulation, internal variability or model shortcomings [15–19]. Model errors can be divided into two categories: unsystematic errors (random) and systematic errors (bias). Random errors stem from the internal variability of climate models, which are a dominant source of uncertainty for shorter (decadal) timescales in model simulations [20]. Bias is defined as any systematic discrepancy of model simulation and observation. Systematic errors can originate either from inadequately constrained parameters or from model structures that are unable to describe the physical process of interest [21]. Model bias is the most prevalent source of uncertainty for longer (century) timescales [20]. Moreover, bias corrected climate model outputs may lead to a significant response in some impact models as decision support tools [22–24].

In our previous work [25] we applied the K-G classification as a diagnostic tool of climate change for six RCM experiments originally produced as a part of the EU FP6 project, ENSEMBLES [26]. Every experiment represented one specific RCM, driven by one of two GCMs. The simulations followed the A1B emission scenario of Intergovernmental Panel on Climate Change (IPCC) [27,28], and the results were evaluated for the near (2021−2050) and far (2071−2100) future periods. The model simulations were subjected to validation and bias correction using the empirical distribution mapping technique on E-OBS [29] observed data as a reference. We found that warmer climate type increased in each RCM for the future but the degree of their extension was different among them. These differences came from the different GCM applications as the driver, the different physical packages of RCMs and the different representations of natural variability in individual models.

Owing to the fact that any choice of bias correction method can be an additional source of uncertainty [23], in this study, we aim to quantify the impacts of different bias correction techniques on the simulated distribution of K-G zones over Europe. The influence of different bias correction methods has been studied over small geographical domains, usually select river basins in Scandinavia [30], North America [31], or North-estern China [32]. In these studies, the performance of bias correction methods was investigated by statistical indices. References [31] and [30] suggested distribution-based methods, while [32] found that the quantile mapping and power transformation of precipitation methods performed equally best in terms of the frequency-based indices, while the local intensity scaling (LOCI) method performed the best in terms of the time-series-based indices. We intend to test the performance of bias correction over a large pan-European domain, as the bias varies in regions of the domain. Moreover, we study the bias correction performance by implementing Köppen–Geiger climate classification.

Our two major research questions are as follows:

Which bias correction methods of precipitation and temperature are able to reproduce climate classification based on the observed parameters in the 1961–1990 time period?

Which bias correction methods of precipitation and temperature are the most reliable for climate prediction over the whole pan-European domain?

This paper is organized as follows: In Section 2, a short description of the K-G classification, selected models and applied bias corrections are presented. In Section 3, the resulting climate classification with respect to the individual bias correction method is presented. Section 4 contains a discussion of our findings and Section 5 offers the conclusions we draw.

2. Materials and Methods

2.1. K-G Classification

Köppen and Geiger classified climate based on annual and monthly mean values of temperature and precipitation. Table 1 contains the methodology to calculate K-G climate zones in Europe based on [8].

Table 1. Key to calculate K-G zones in Europe and their third index. Pann is the accumulated annual precipitation. Pmin is the precipitation of the driest month. Psmin, Psmax, Pwmin and Pwmax are defined as the lowest and highest monthly precipitation values for the summer and winter half-years. Pth is the dryness threshold. Tann is the annual mean temperature, and the monthly mean temperatures of the warmest and coldest months are marked by Tmax and Tmin, respectively. The precipitation and temperature are given in mm and °C, respectively.

Type	Description	Criterion
B	Arid climates	$P_{ann} < 10\, P_{th}$
BS	Steppe climates	$P_{ann} > 5\, P_{th}$
BW	Desert climates	$P_{ann} \leq 5\, P_{th}$
C	Warm temperate climates	$-3\,°C < T_{min} < +18\,°C$
Cs	Warm temperate climates with dry summers	$P_{smin} < P_{wmin}, P_{wmax} > 3\, P_{smin}$ and $P_{smin} < 40$ mm
Cw	Warm temperate climates with dry winters	$P_{wmin} < P_{smin}$ and $P_{smax} > 10\, P_{wmin}$
Cf	Warm temperate climates, fully humid	neither Cs nor Cw
D	Snow climates	$T_{min} \leq -3\,°C$
Ds	Snow climates with dry summers	$P_{smin} < P_{wmin}, P_{wmax} > 3\, P_{smin}$ and $P_{smin} < 40$ mm
Dw	Snow climates with dry winters	$P_{wmin} < P_{smin}$ and $P_{smax} > 10\, P_{wmin}$
Df	Snow climates, fully humid	neither Ds nor Dw
E	Polar climates	$T_{max} < +10\,°C$
ET	Tundra climates	$0\,°C \leq T_{max} < +10\,°C$
EF	Frost climates	$T_{max} < 0\,°C$
	third index for C and D climate zones	
Type	Description	Criterion
a	Hot summers	$T_{mean} > 22°C$
b	Warm summers	not (a) and at least 4 $T_{mon} \geq +10\,°C$
c	Cool summers and cold winters	not (b) and $T_{min} > -38°C$
d	Extremely continental	like (c) and $T_{min} \leq -38°C$
	third index for B climate zone	
Type	Description	Criterion
h	Hot steppe/desert	$T_{ann} \geq +18\,°C$
k	Cold steppe/desert	$T_{ann} < +18\,°C$

The dryness threshold is calculated by

$$P_{th} = \begin{cases} 2\{T_{ann}\} & \text{if at least } \tfrac{2}{3} \text{ of the annual precipitation occurs} \in winter \\ 2\{T_{ann}\} + 28 & \text{if at least } \tfrac{2}{3} \text{ of the annual precipitation occurs} \in summer \\ 2\{T_{ann}\} + 14 & \text{otherwise} \end{cases}.$$

2.2. Datasets and Bias Corrections

For the analysis of bias correction influence on K-G zone distribution in Europe, we used simulations of five regional climate models from the ENSEMBLES project, as summarized in Table 2. The large scale forcing for two RCMs was taken from driving ARPÉGE GCM, and three of them were driven by ECHAM5-r3 GCM. The E-OBS version 10 gridded dataset of daily station observations with a spatial resolution of 0.25° in longitude and latitude was used as a reference dataset for validation and bias correction in the period from 1961 to 1990. Before the direct comparison of models and observations, the RCMs were interpolated from their native grids to the E-OBS 0.25 ° regular grid by the nearest neighbour remapping method.

Table 2. The institute, global climate models (GCMs), regional climate models (RCMs) and resolution of chosen models from the ENSEMBLES EU project.

	INSTITUTE/ REFERENCE	GCM	RCM	RESOLUTION
1	Centre National de Researches Météorologiques (CNRM)/ [33]	ARPÈGE	ALADIN	25 km
2	Danish Meteorological Institute (DMI)/ [34]	ARPÈGE	HIRHAM	25 km
3	Koninklijk Nederlands Meteorologisch Instituut (KNMI)/ [35]	ECHAM5-r3	RACMO2	25 km
4	Swedish Meteorological and Hydrological Institute (SMHI)/ [36]	ECHAM5-r3	RCA	25 km
5	International Centre for Theoretical Physics (ICTP)/ [37]	ECHAM5-r3	RegCM	25 km

The simulated climate zones were analysed in the Alps (AL), the British Isles (BI), Eastern Europe (EA), France (FR), the Iberian Peninsula (IP), the Mediterranean (MD), Mid-Europe (ME) and Scandinavian (SC) regions (Figure 1) specified in the framework of the PRUDENCE project [38].

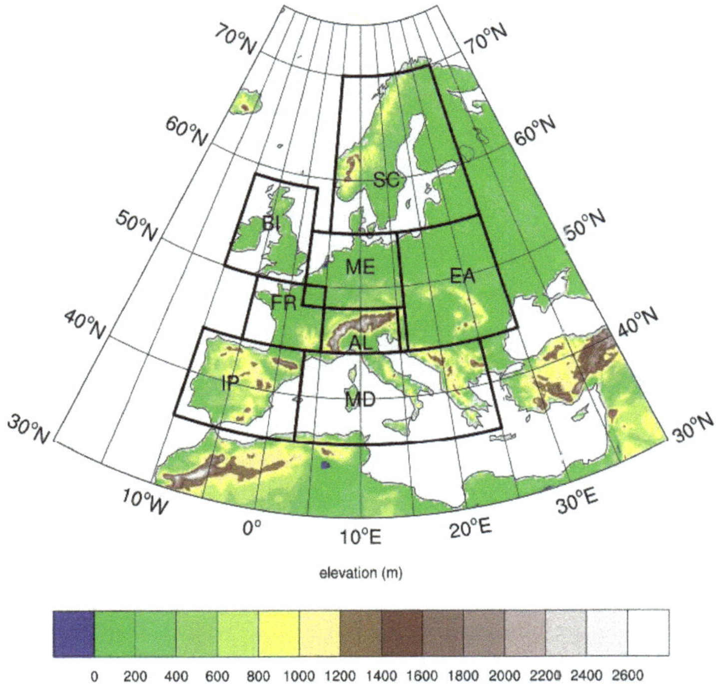

Figure 1. Subdomains based on the Prudence project: the Alps (AL), the British Isles (BI), Eastern Europe (EA), France (FR), the Iberian Peninsula (IP), the Mediterranean (MD), Mid-Europe (ME) and Scandinavia (SC).

Figure 2 demonstrates the simulated K-G zones without bias correction. Large differences can be seen between the simulated zones and between the simulated and observed zones. The distribution of K-G zones varied from the K-G zones based on the observed parameters in the case of ARPÈGE driven RCMs. HIRHAM RCM produced dryer climate zones in each region owing to the underestimated precipitation. It produced Csa and Csb zones instead of Cfb in France, Mid-Europe, Eastern Europe and in the Mediterranean and Dsb instead of Dfb in the Alps. Furthermore, the extension of BSk was extremely large in the Iberian Peninsula and in Eastern Europe. Both HIRHAM and ALADIN overestimated the Tundra climate (ET) zone in Scandinavia. In ALADIN simulation the Cfb zone was overestimated in the Iberian Peninsula and in the Italian Peninsula, while it was underestimated in Eastern Europe, in the Mediterranean and on the Western coast of France. The ECHAM5-r3 forced RCMs produced better K-G simulations but the RegCM simulated a wetter climate in the Iberian Peninsula and the Mediterranean, whilst RACMO2 and RCA produced drier climate zones in the Mediterranean and Eastern Europe. Each of them overestimated the ET zone in the Alps.

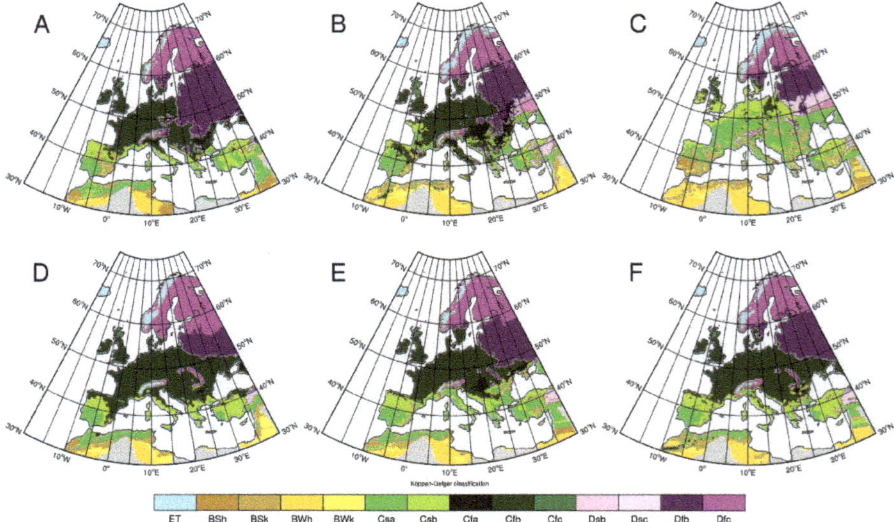

Figure 2. Simulated K-G climate classification according to E-OBS (**A**) and in ALADIN (**B**), HIRHAM (**C**), RegCM (**D**), RAMCMO2 (**E**) and RCA (**F**) without bias correction.

We applied the following bias correction methods: i) the empirical quantile mapping (eQM) of precipitation and temperature [15], ii) quantile mapping of precipitation and temperature based on a gamma + Generalized Pareto Distribution (gpQM) [39], iii) the power transformation of precipitation [40,41] the variance scaling of temperature [23], and iv) the local intensity scaling (LOCI) [42].

The daily mean precipitation and temperature values were used for the bias corrections.

2.2.1. Empirical Quantile Mapping

Empirical quantile mapping correction was used for correcting the nonparametric empirical cumulative distribution function in simulated daily data. This method calibrates the simulated Cumulative Distribution Function (CDF) by adding both the mean delta change and the individual delta changes in the corresponding quantiles to the observed quantiles. The implemented eQM obtained the correction function for 99 percentiles of observed and simulated distribution and linearly interpolated between two percentiles [15]. Outside the range of percentiles, e.g., for the 99th percentile, a constant correction was applied. In the case of precipitation, a 1 mm threshold value was considered so that the precipitation was redefined to zero if the value was less than 1 mm. We applied this bias correction with a 90-day moving window.

2.2.2. Quantile Mapping Based on Gamma + Generalized Pareto Distribution

Quantile mapping based on gamma + generalized Pareto distribution is a quantile mapping method similar to eQM but assumes that the observed and simulated precipitation density distribution are correctly approximated by gamma, and the temperature density distribution is correctly approximated by Gaussian distribution. Therefore, it uses theoretical distribution in the quantile mapping instead of empirical distribution. Due to the fact that gamma distribution is a light-tailed distribution, it is combined with a general Pareto distribution [39]. The observed and simulated quantiles were interpolated by inverse distance weighting. The 1 mm threshold cut off was also applied to precipitation in this approach. The gpQM bias correction with a 90-day moving window in the case of precipitation was used. Owing to the fact that the seasonal temperature density distribution cannot be approximated by Gaussian distribution in some places in Europe [43], the temperature gpQM correction produces a large number of infinitive values with a 90-day moving window. Therefore, gpQM was applied only in the case of precipitation.

2.2.3. Power Transformation of Precipitation

Power transformation of precipitation can be used for adjusting the variance statistics of precipitation. Simulated monthly precipitation is powered by a "b" value that guarantees the coefficient of variance (CV) of the simulated daily precipitation matches the CV of the observed daily precipitation. This power "b" value is estimated on a monthly basis using a 90-day window centred on the interval with a root-finding algorithm. Thereafter, the powered precipitation series is multiplied by the standard linear scaling parameter, which was calculated by dividing the monthly mean observed precipitation by the monthly mean powered simulated precipitation.

2.2.4. Variance Scaling of Temperature

Correspondingly, variance scaling of temperature corrects both the mean and variance values of temperature. In the first step, the temperature mean was corrected with the difference between the observed and simulated climatological monthly means. After that, the mean-corrected simulated temperature was shifted on a monthly basis to the zero mean. Thereafter, the standard deviation of the shifted temperature was scaled based on the ratio of the climatological monthly standard deviation of the observed and simulated data. Finally, the standard deviation corrected time series were shifted back using the corrected mean.

2.2.5. Local Intensity Scaling of Precipitation

The local intensity scaling correction corrects the mean as well as both the wet-day frequencies and wet-day intensities of precipitation. The frequency of wet-days in the case of observation considers those days when the precipitation value is higher than the 1 mm threshold.

The model's wet-day threshold was determined from the daily RCM precipitation series such that the threshold exceedance matched the wet-day frequency in the observed series. The scaling factor of this correction was calculated based on the ratio of the climatological monthly mean wet-day intensities between the observations and the RCMs with the adjusted wet-day thresholds. Subsequently, the simulated monthly precipitation values were adjusted with the model's wet-day threshold and multiplied by the scaling factor. Finally, the daily simulated precipitation was downscaled from the calibrated monthly scale such that the precipitation values were redefined to zero on those days when the observed precipitation was less than 1 mm.

The equations of bias correction methods are detailed in [44]. Bias corrections and K-G classification were implemented in Matlab by the MeteoLab [45] and Weaclim [46] toolboxes. The figures were created by NCAR Command Language [47].

3. Results

3.1. Empirical Quantile Mapping with a 90-Day Moving Window

The application of eQM bias correction with a 90-day moving window improved the climate classification. RCMs simulated appropriate climate zones in each region with the exception of HIRHAM (Figure 3). In the case of HIRHAM RCM, the climate zone simulation was improved in the Northern regions, e.g, in Scandinavia, the British Isles and Mid-Europe, but it still produced dryer climate zones in the Iberian Peninsula, the Mediterranean and Eastern Europe. In the other RCMs, the extension of climate zones differed from the observed ones mainly in the Iberia Peninsula, in the Alps, in the Mediterranean and in Eastern Europe. The difference in the frequency of the occurrence of the climate zones was only 1–2% between the RCMs with exception of HIRHAM in each region. For example, the occurrence of the Cfb zone in the Alps was 53%, 51%, 50% and 50% according to the ALADIN, RegCM, RACMO2 and RCA simulations, respectively.

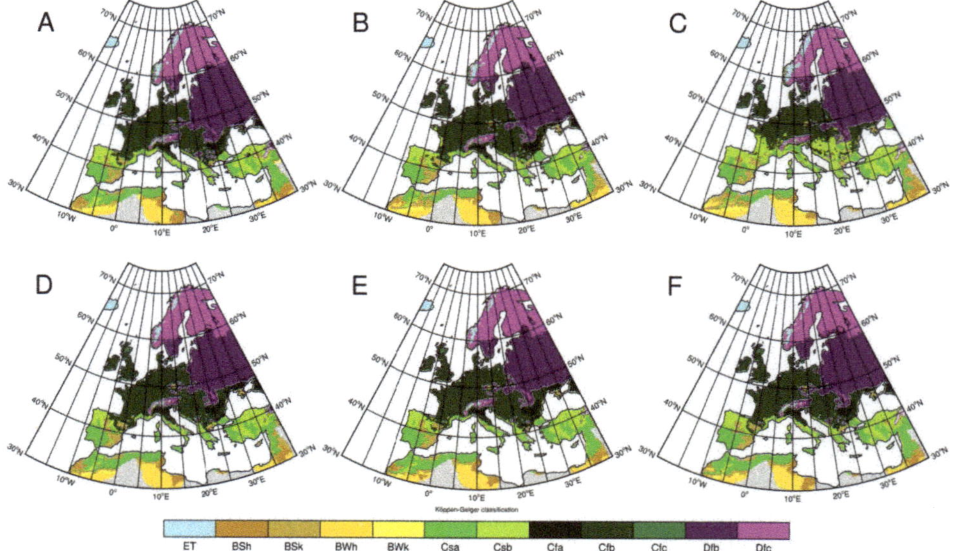

Figure 3. Simulated K-G climate classification according to E-OBS (**A**) and empirical quantile mapping (eQM) corrected precipitation and temperature with 90-day moving window in ALADIN (**B**), HIRHAM (**C**), RegCM (**D**), RAMCMO2 (**E**) and RCA (**F**).

The precipitation was mainly underestimated in each season except in the British Isles in winter (DJF) (Table 3a). The eQM with a 90-day moving window decreased the residual temperature bias in DJF but increased it in summer (JJA) in ALADIN, HIRHAM and RegCM in some regions (Table 3b).

Table 3. Residual bias of seasonal amount of simulated precipitation (a) and of seasonal mean of simulated temperature (b) in the case of eQM bias correction in eight different regions: the Alps (AL), the British Isles (BI), Eastern Europe (EA), France (FR), the Iberia Peninsula (IP), the Mediterranean (MD), Mid-Europe (ME) and Scandinavia (SC) in DJF and JJA The bias values are in % and in °C in the case of precipitation and temperature, respectively.

a)	ALADIN		HIRHAM		RegCM		RACMO2		RCA	
Region	**DJF**	**JJA**	**DJF**	**JJA**	**DJF**	**JJA**	**DJF**	**JJA**	**DJF**	**JJA**
AL	−2	−9	2	−23	−4	−6	−5	−6	−5	−7
BI	1	−2	2	−7	1	−3	0	−3	1	−2
EA	−8	−11	−4	−17	−7	−6	−8	−3	−9	−3
FR	−1	−5	0	−20	−6	0	−6	−5	−7	−4
IP	−6	−5	−8	−23	−8	−11	−8	−21	−9	−23
MD	−6	−8	0	−34	−4	−9	−4	−9	−5	−14
ME	−4	−9	−2	−12	−4	−3	−5	−3	−5	−2
SC	−6	−1	−3	−8	−3	−2	−2	−2	−4	1

−100 −50 −25 −10 10 25 50 100

Table 3. *Cont.*

b)	ALADIN		HIRHAM		RegCM		RACMO2		RCA	
Region	DJF	JJA	DJF	JJA	DJF	JJA	DJF	JJA	DJF	JJA
AL	0.2	0.1	0.3	−0.4	0.3	0.1	0.1	0.1	−0.1	0.0
BI	0.3	0.0	0.2	−0.3	0.2	0.0	0.1	0.1	0.1	0.1
EA	0.5	−0.2	0.4	−0.8	0.4	0.0	0.2	0.0	0.0	0.0
FR	0.3	0.0	0.2	−0.3	0.2	0.2	0.0	0.3	0.0	0.1
IP	0.1	0.5	−0.1	0.3	0.1	0.6	0.0	0.3	0.0	0.1
MD	0.1	0.6	0.1	−0.3	0.2	0.4	0.0	0.4	−0.1	0.2
ME	0.5	−0.2	0.5	−0.7	0.3	−0.1	0.2	0.0	0.0	0.0
SC	0.1	−0.6	0.3	−0.8	0.2	0.0	0.2	0.0	−0.1	0.0

5   2   1   0.5 −0.5   −1   −2   −5

3.2. Quantile Mapping Based on a Gamma + Generalized Pareto Distribution with a 90-Day Moving Window

Owing to the fact the seasonal temperature probability distribution does not fit a Gaussian distribution due to non-Gaussian tails occurrence, the gpQM bias correction with a 90-day moving window was implemented only on the precipitation data. The gpQM with a 90-day moving window bias-corrected precipitation was combined with eQM with 90-day moving window corrected temperature values for the calculation of K-G zones. The gpQM correction with a moving window also improved the climate classification, but it resulted in dryer climate zones in some regions compared to the eQM correction (Figure 4). The Csb and BSk ratio was larger in the Mediterranean and Eastern Europe, respectively, according to the gpQM in each RCM. Owing to gpQM method the extension of dry zones (Csa, Csb, Dsb), the Csb zone was predominant in France, Mid-Europe and the Mediterranean, and the BSk was overestimated in the Iberian Peninsula and Eastern Europe in HIRHAM model.

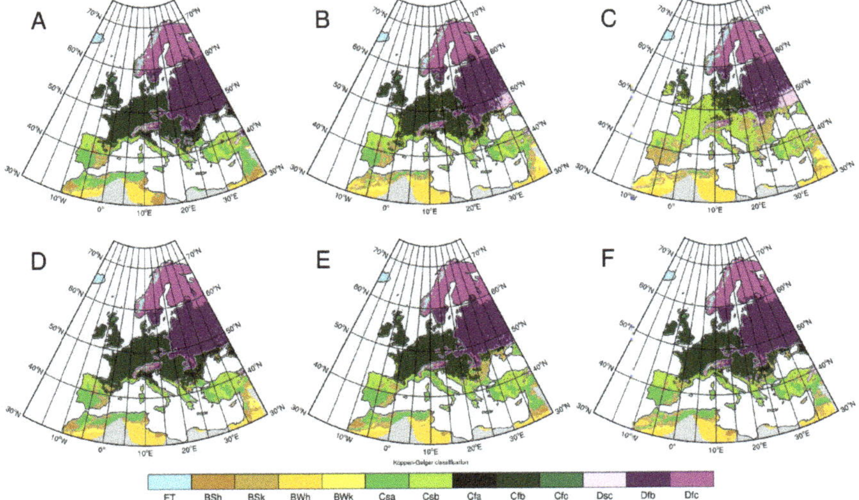

Figure 4. Simulated K-G climate classification according to E-OBS (**A**) and quantile mapping of precipitation and temperature based on a gamma + Generalized Pareto Distribution (gpQM) correction of precipitation and eQM correction of temperature with 90-day moving window in ALADIN (**B**), HIRHAM (**C**), RegCM (**D**), RAMCMO2 (**E**) and RCA (**F**).

The residual precipitation bias was variable. The precipitation was overestimated in some regions, mainly in DJF. Although eQM correction resulted in better K-G classification, the residual bias of gpQM correction was smaller in some regions (e.g., in the Mediterranean in the case of RegCM) (Table 4).

Table 4. Residual bias in the seasonal amount of simulated precipitation (a) in the case of gpQM bias correction in eight different regions: the Alps (AL), the British Isles (BI), Eastern Europe (EA), France (FR), the Iberian Peninsula (IP), the Mediterranean (MD), Mid-Europe (ME) and Scandinavia (SC) in DJF and JJA. The bias values are in %.

Region	ALADIN		HIRHAM		RegCM		RACMO2		RCA	
	DJF	JJA	DJF	JJA	DJF	JJA	DJF	JJA	DJF	JJA
AL	3	−12	0	−57	5	−11	0	−22	3	−10
BI	−2	−5	−3	−39	3	−2	3	−4	5	1
EA	3	−20	6	−46	3	−11	−2	−22	−3	−3
FR	5	−11	−6	−57	0	4	−1	−9	1	−3
IP	−11	−10	−33	−67	−3	−11	−3	−36	−5	−30
MD	−7	−15	−14	−68	0	−15	−4	−49	−3	−25
ME	8	−16	8	−42	5	−4	1	−9	1	4
SC	−7	−1	3	−19	−1	1	2	−2	−3	8

−100 −50 −2 −10 10 25 50 100

3.3. Power Transformation of Precipitation and Variance Scaling of Temperature

The power transformation of precipitation has been implemented in smaller domains in Europe, such as the basin of the river Meuse [40] and the mesoscale catchments of Sweden [30], where the precipitation is significant. In our work, the power value of precipitation was calculated with Brent's root-finding algorithm [48]. It is possible that the mean value of precipitation is near zero in the dryer regions. This zero mean value may have caused an invalid value in the coefficient of variation of precipitation that stopped the root-finding algorithm and produced incorrect K-G zones (this is not shown). To get around this issue, we applied two conditions before running the root-finding algorithm. The first condition was to ignore the RCM precipitation values if they were missing values. The second was to ignore the RCM precipitation values if their mean value was zero, as this causes an invalid value. Thanks to the above-mentioned conditions, the power transformation of precipitation combined with the variance scaling of temperature created the correct K-G classification in each RCM. Negligible differences were seen between the observed and simulated K-G zones (Figure 5). The difference in the frequency of occurrence of climate zones between observations and simulations was zero in each region with the exception of ALADIN. ALADIN simulated larger Cfb and smaller Csb extension in the Iberian Peninsula and in the Mediterranean regions where the difference from the observations was only 2%. Due to these facts, power transformation of precipitation and variance scaling of temperature appear to be the most suitable for climate classification in the whole pan-European domain.

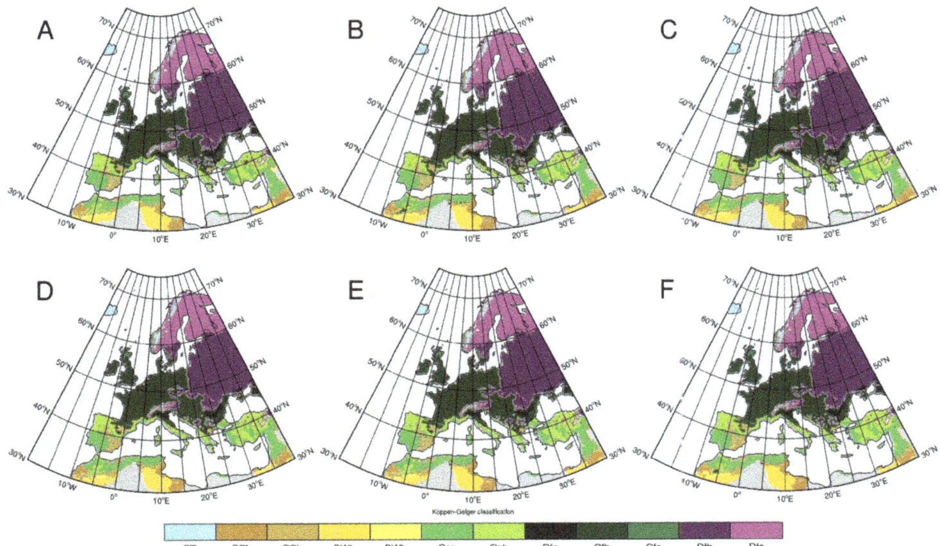

Figure 5. Simulated K-G climate classification according to E-OBS (**A**) and power transformation of precipitation and variance scaling of temperature correction in ALADIN (**B**), HIRHAM (**C**), RegCM (**D**), RAMCMO2 (**E**) and RCA (**F**).

The value of residual precipitation bias was similar in each RCM, with the exception of HIRHAM, where the residual bias values were zero (Table 5.). Furthermore, the bias was almost identical, except for HIRHAM, which means that power transformation is not dependent on the RCMs. The modelled temperature was almost commensurate with the observed data when variance scaling correction was implemented.

Table 5. Residual bias of seasonal amount of simulated precipitation in the case of power transformation of the precipitation bias correction method in eight different regions: the Alps (AL), the British Isles (BI), Eastern Europe (EA), France (FR), the Iberian Peninsula (IP), the Mediterranean (MD), Mid-Europe (ME) and Scandinavia (SC) in DJF and JJA. The bias values are in %.

	ALADIN		HIRHAM		RegCM		RACMO2		RCA	
Region	**DJF**	**JJA**	**DJF**	**JJA**	**DJF**	**JJA**	**DJF**	**JJA**	**DJF**	**JJA**
AL	0	−6	0	0	0	−6	0	−6	−1	−6
BI	−8	14	0	0	−8	14	−8	14	−8	14
EA	−9	−14	0	0	−9	−14	−9	−14	−9	−14
FR	−5	4	0	0	−5	3	−5	3	−5	4
IP	−12	8	0	0	−12	7	−12	7	−12	7
MD	−10	0	1	0	−10	4	−10	4	−10	4
ME	−9	−11	0	0	−9	−10	−9	−10	−9	−10
SC	−12	10	0	0	−12	10	−12	10	−12	9

3.4. Local Intensity Scaling of Precipitation and Variance Scaling of Temperature

Due to the fact that local intensity scaling correction can be applied only to precipitation, it was combined with variance scaling of temperature for the calculation of the K-G classification. Both corrections are distribution free and correct the diagnostics, as well as the mean. Owing to these facts, the difference between the observed and simulated zones was also negligible, only 1−2% (Figure 6). Apart from that, the RCMs resulted in very similar values.

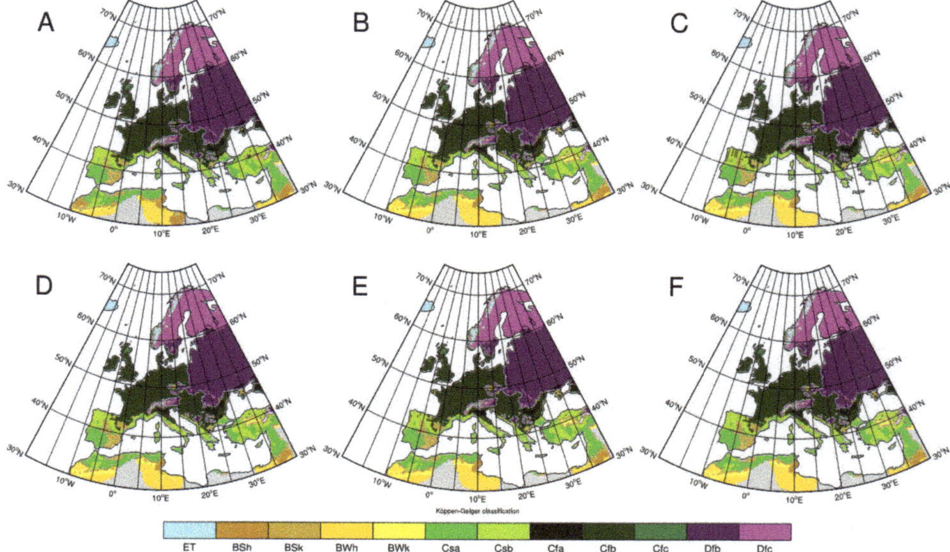

Figure 6. Simulated K-G climate classification according to E-OBS (**A**) and local intensity scaling of precipitation and variance scaling of temperature correction in ALADIN (**B**), HIRHAM (**C**), RegCM (**D**), RAMCMO2 (**E**) and RCA (**F**).

In the case of LOCI correction, the residual seasonal precipitation bias was the smallest compared to the other precipitation correction methods, except in the case of the HIRHAM RCM compared to power transformation of the precipitation method (Table 6). This caused a negative bias in both seasons in each RCM. Although the seasonal residual bias values were smaller than in the case of power transformation of precipitation, the minimum and maximum monthly precipitation values of the RCMs were closer to the observed minimum and maximum monthly precipitation values by power transformation in both seasons (not shown). These values determined the subtypes of the K-G zones.

Table 6. Residual bias of the seasonal amount of simulated precipitation in the case of local intensity scaling of the precipitation bias correction method in eight different regions: the Alps (AL), the British Isles (BI), Eastern Europe (EA), France (FR), the Iberia Peninsula (IP), the Mediterranean (MD), Mid-Europe (ME) and Scandinavia (SC) in DJF and JJA. The bias values are in %.

Regio	ALADIN		HIRHAM		RegCM		RACMO2		RCA	
	DJF	JJA	DJF	JJA	DJF	JJA	DJF	JJA	DJF	JJA
AL	−1	−1	−1	−1	−1	−1	−1	−1	−1	−1
BI	−2	−2	−2	−2	−2	−2	−2	−2	−2	−2
EA	−6	−1	−6	−2	−6	−1	−6	−1	−6	−2
FR	−2	−2	−2	−2	−2	−2	−2	−2	−2	−2
IP	−1	−2	−1	−2	−1	−2	−1	−2	−1	−2
MD	−2	−1	−2	−1	−2	−1	−2	1	−2	−1
ME	−5	−2	−6	−2	−5	−2	−6	−2	−6	−2
SC	−5	−2	−5	−2	−5	−2	−5	−2	−5	−2

−100 −50 −25 −10 10 25 50 100

3.5. Cross-Validation of Bias Corrections

Cross-validation was applied to test bias corrections. Due to the observed annual precipitation and temperature values being nearly stationary (not shown) in the 1961–2000 period, we applied a split-sample test (SST) as advocated by [49]. The parameters were split into calibration and test periods. The bias corrections were calibrated in the first twenty years of 1961–1980, and the corrections were implemented in the second twenty years of 1981–2000. The corrections were validated by the K-G zone simulation in the test period, and the results were compared with the K-G zones based on the observed data in the test period.

3.5.1. Validation of Empirical Quantile Correction

Figure 7 shows the simulated K-G distribution based on the validated eQM values of precipitation and temperature compared to K-G zones according to observed parameters in the test period. The ECHAM5-r3 RCMs produced similar results, whilst HIRHAM resulted in a drier climate in Eastern Europe and in the Mediterranean compared to the ALADIN model where the Csb zone was predominant. Each RCM significantly underestimated the BSk zone in the Iberian Peninsula compared to the K-G zones based on observations. In Scandinavia, the ET zone was overestimated with the exception of the ALADIN model, and the Dsc zone expanded in the RACMO2 and RCA models. Moreover, the DSb zone occurred in the RegCM model. In the ECHAM5-r3 driven RCMs, the Dfb zone shifted southwards in Southern Scandinavia. The ratio of the Dfb zone decreased in the Carpathians in Eastern Europe in each RCM. The Cfa zone diminished in the Eastern region of Eastern Europe in the HIRHAM, RACMO2 and RCA models. Moreover, the BSk zone occurred in Eastern Europe in ARPEGE-driven RCMs. The difference between the simulated and observed K-G zones was negligible in Mid-Europe and in the British Isles. The ECHAM5-r3 RCMs simulated a Csb zone in the Western region of France.

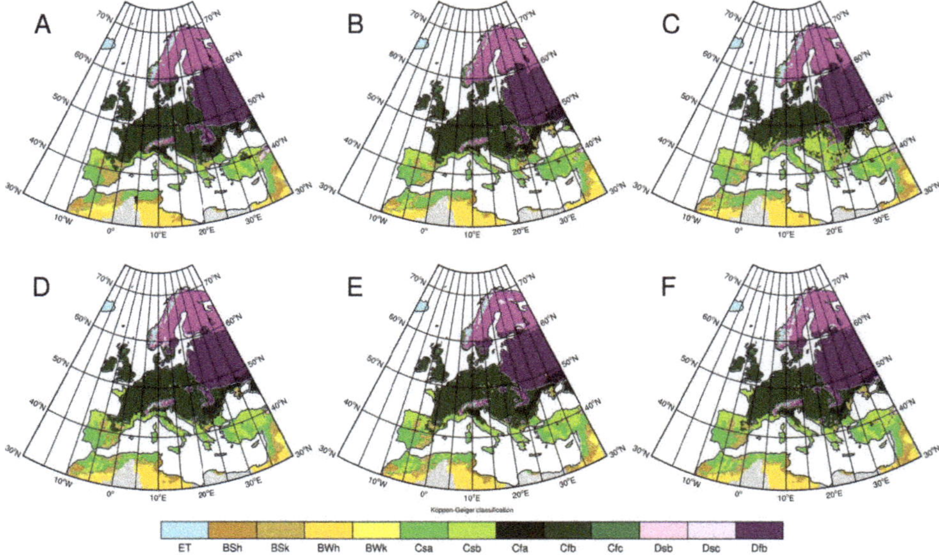

Figure 7. Simulated K-G climate classification according to E-OBS (**A**) and eQM corrected precipitation and temperature with 90-day moving window in ALADIN (**B**), HIRHAM (**C**), RegCM (**D**), RAMCMO2 (**E**) and RCA (**F**) in the test period in 1981−2000.

The residual bias of precipitation varied during the season and the eQM correction strongly depended on the regions and the RCMs (Table 7a). Larger residual bias was found in France, in the Iberian Peninsula and in the Mediterranean in each RCM with the exception of ALADIN. The residual bias of temperature was smaller than 1 °C, except in HIRHAM in Scandinavia in the DJF season (Table 7b).

Table 7. Residual bias of the seasonal amount of simulated precipitation (a) and of the seasonal mean of the simulated temperature (b) in the case of eQM bias correction in the test period in eight different regions: the Alps (AL), the British Isles (BI), Eastern Europe (EA), France (FR), the Iberia Peninsula (IP), the Mediterranean (MD), Mid-Europe (ME) and Scandinavia (SC) in DJF and JJA. The bias values are in % and in °C for precipitation and temperature, respectively.

a)	ALADIN		HIRHAM		RegCM		RACMO2		RCA	
Region	DJF	JJA	DJF	JJA	DJF	JJA	DJF	JJA	DJF	JJA
AL	−2	3	7	−26	3	−9	1	−2	3	−5
BI	−9	7	−10	0	−12	−7	−15	−9	−15	−11
EA	−3	5	2	−6	4	0	0	5	3	3
FR	−8	4	−6	−24	−17	−13	−17	−19	−18	−17
IP	9	3	5	−17	13	−1	14	−18	11	−21
MD	0	−5	7	−23	14	0	18	12	12	3
ME	−7	10	−5	−4	−7	−2	−14	−2	−8	−1
SC	−18	0	−15	−8	−14	−9	−11	−11	−13	−9

−100	−50	−25	−10	10	25	50	100

Table 7. *Cont.*

b)	ALADIN		HIRHAM		RegCM		RACMO2		RCA	
Region	DJF	JJA	DJF	JJA	DJF	JJA	DJF	JJA	DJF	JJA
AL	0.3	−0.4	0.3	−0.4	−0.2	−0.4	−0.5	−0.5	−0.6	−0.5
BI	0.5	−0.5	0.3	−0.7	−0.3	−0.3	−0.4	−0.2	−0.5	−0.4
EA	0.1	0.1	0.1	0.0	0.0	−0.5	−0.1	−0.6	−0.2	−0.6
FR	0.2	−0.9	0.2	−0.9	−0.6	−0.4	−0.7	−0.3	−0.9	−0.5
IP	−0.3	−0.2	−0.3	−0.4	−0.5	0.1	−0.5	0.1	−0.6	0.1
MD	0.2	0.2	0.2	−0.2	0.1	−0.5	−0.1	−0.7	−0.2	−0.6
ME	0.2	−0.3	0.3	−0.5	−0.2	−0.5	−0.3	−0.6	−0.4	−0.7
SC	0.9	0.4	1.1	−0.1	−0.1	−0.3	0.0	−0.4	−0.1	−0.2

| 5 | 2 | 1 | 0.5 | −0.5 | −1 | −2 | −5 |

3.5.2. Validation of Quantile Mapping Based on a Gamma + Generalized Pareto Distribution of Precipitation

Figure 8. demonstrates the simulated K-G zones according to the gpQM of precipitation and eQM of temperature combination during the test period. The gpQM of precipitation resulted in dryer climate zones compare to eQM in Mediterranean region, and a BSk zone was produced in the Southeastern area of Easter-Europe in each RCM except RegCM (Figure 8). The BSK zone dominantly decreased in the Iberian Peninsula in each RCM except HIRHAM. Moreover, significant extension of Csb was simulated in the HIRHAM model in France, Mid-Europe, EasternEurope, in the Mediterranean and in the Southwestern area of the British Isles. In this model, a larger area was covered by the BSk zone than in the other models in the Iberian Peninsula and in Eastern Europe.

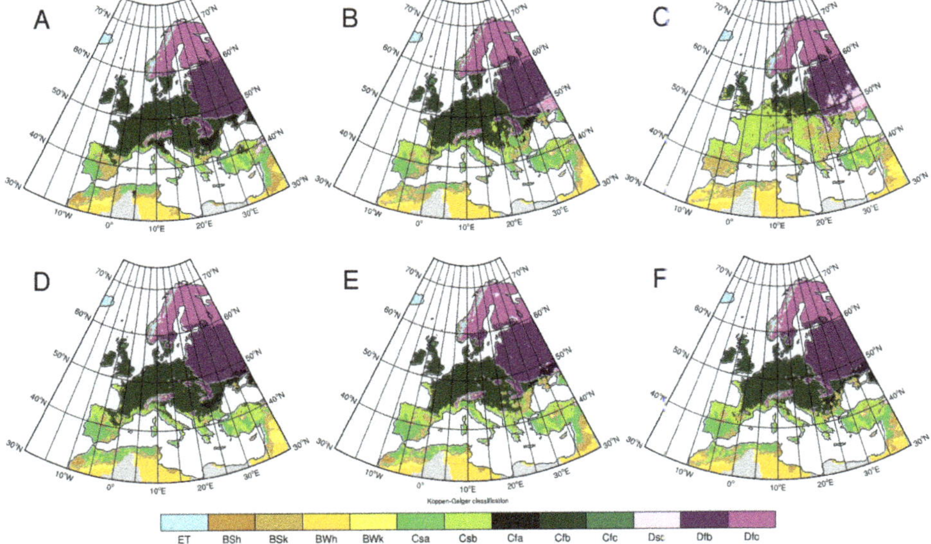

Figure 8. Simulated K-G climate classification according to E-OBS (**A**) and gpQM corrected precipitation and eQM corrected temperature with a 90-day moving window in ALADIN (**B**), HIRHAM (**C**), RegCM (**D**), RAMCMO2 (**E**) and RCA (**F**) in the test period in 1981−2000.

Although gpQM correction of precipitation resulted in a larger bias in K-G simulation in several regions compared to the eQM correction, the remained bias was smaller in France and in Scandinavia in ECHAM5-r3 forced RCMs (Table 8). Furthermore, gpQM produced a smaller residual bias in the RegCM model in the British Isles as well.

Table 8. Residual bias of seasonal amount of simulated precipitation in the case of gpQM bias correction in the test period in eight different regions: the Alps (AL), the British Isles (BI), Eastern Europe (EA), France (FR), the Iberian Peninsula (IP), the Mediterranean (MD), Mid-Europe (ME) and Scandinavia (SC) in DJF and JJA. The bias values are in %.

Region	ALADIN		HIRHAM		RegCM		RACMO2		RCA	
	DJF	JJA	DJF	JJA	DJF	JJA	DJF	JJA	DJF	JJA
AL	6	0	6	−59	11	−12	10	−20	12	−8
BI	−10	3	−13	−36	−9	−5	−11	−9	−10	−6
EA	12	−7	11	−40	14	−5	7	−15	9	4
FR	0	−7	−10	−62	−10	−10	−11	−23	−10	−17
IP	7	−2	−24	−64	21	0	20	−34	16	−33
MD	1	−14	−8	−62	17	−8	16	−42	15	−16
ME	7	2	3	−38	1	−1	−8	−7	−2	6
SC	−19	0	−11	−19	−10	−5	−7	−9	−11	0

−100 −50 −25 −10 10 25 50 100

3.5.3. Validation of Power Transformation of Precipitation and Variance Scaling of Temperature

The power transformation of precipitation and variance scaling of temperature bias corrections resulted in similar K-G zone distributions in each RCM (Figure 9). The extension of K-G zones was different between the RCMs and differed from the observed ones. The ECHAM5-r3 forced RCMs and HIRHAM simulated larger, while ALADIN resulted in a smaller ET fraction in Scandinavia. Moreover, ECHAM5-r3 forced RCMs simulated a large Dsc zone fraction in the Scandinavian mountains. The ratio of the Dfb zone decreased in the Carpathians in Eastern Europe in each RCM. The Cfa zone expanded in Eastern Europe according to ARPÉGE forced RCMs, whilst it decreased in ECHAM5-r3 driven models. In the Western part of France, the Csb zone was simulated with the exception of the ALADIN model. In the Mediterranean and in the Iberian Peninsula, the BSk zone was underestimated in each RCM.

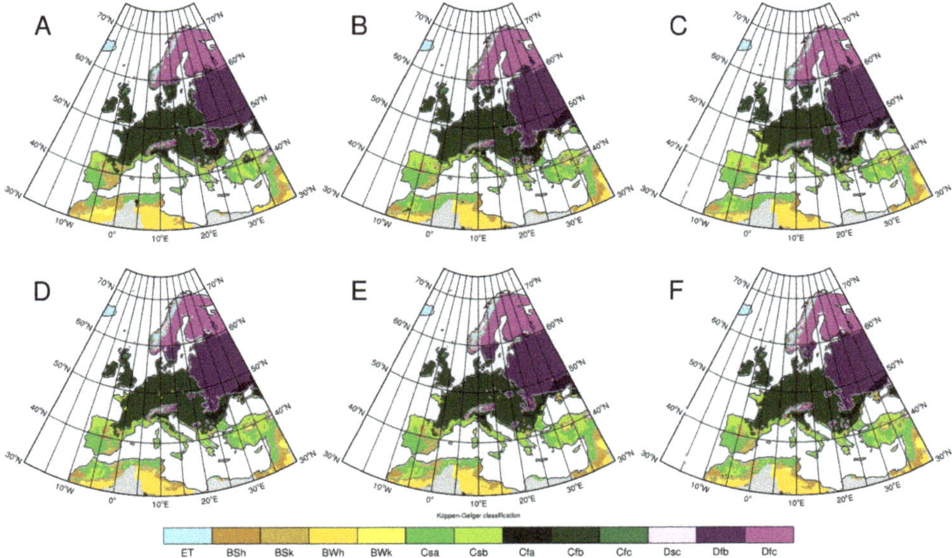

Figure 9. Simulated K-G climate classification according to E-OBS (**A**) and power transformation of precipitation and variance scaling of temperature correction in ALADIN (**B**), HIRHAM (**C**), RegCM (**D**), RAMCMO2 (**E**) and RCA (**F**) in the test period in 1981−2000.

Even though eQM produced a very small residual bias in ALADIN and RegCM in the Mediterranean and in RACMO2 in Eastern-Europe, the power transformation of precipitation better reproduced the K-G zone distribution. Furthermore, the power transformation of precipitation resulted in a smaller residual bias in Scandinavia, France and in the British Isles in both seasons, and in Mid-Europe in winter compared to eQM (Table 9a). The difference between the eQM and the variance scaling corrected temperature was negligible (Table 9b).

Table 9. Residual bias of seasonal amount of simulated precipitation (a) in the case of power transformation of precipitation and the seasonal mean of the simulated temperature, (b) in the case of variance scaling of the temperature bias correction methods in the test period in eight different regions: the Alps (AL), the British Isles (BI), Eastern Europe (EA), France (FR), the Iberia Peninsula (IP), the Mediterranean (MD), Mid-Europe (ME) and Scandinavia (SC) in DJF and JJA. The bias values are in % and in °C for precipitation and temperature, respectively.

a)	ALADIN		HIRHAM		RegCM		RACMO2		RCA	
Region	DJF	JJA	DJF	JJA	DJF	JJA	DJF	JJA	DJF	JJA
AL	1	16	4	−7	7	−2	8	7	6	5
BI	−8	9	−12	9	−12	−5	−15	−6	−15	−9
EA	5	20	4	11	11	8	11	8	14	9
FR	−7	10	−7	−9	−11	−12	−11	−14	−11	−11
IP	15	12	14	123	25	11	26	5	25	−4
MD	7	3	8	23	20	7	24	20	20	15
ME	−3	23	−4	8	−3	4	−8	3	−2	3
SC	−12	1	−12	1	−11	−6	−8	−9	−9	−8

−100   −50   −25   −10   10   25   50   100

Table 9. *Cont.*

b)	ALADIN		HIRHAM		RegCM		RACMO2		RCA	
Region	DJF	JJA	DJF	JJA	DJF	JJA	DJF	JJA	DJF	JJA
AL	0.1	−0.8	0.2	−0.8	−0.5	−0.4	−0.7	−0.6	−0.6	−0.6
BI	0.4	−0.6	0.2	−0.7	−0.6	−0.3	−0.6	−0.2	−0.6	−0.4
EA	−0.2	−0.1	−0.1	−0.1	−0.2	−0.5	−0.3	−0.5	−0.2	−0.6
FR	0.1	−1.1	0.1	−1.1	−0.7	−0.4	−0.9	−0.4	−0.9	−0.5
IP	−0.3	−0.5	−0.2	−0.6	−0.6	0.1	−0.6	0.1	−0.7	0.0
MD	0.1	−0.1	0.2	−0.4	0.0	−0.5	−0.1	−0.8	−0.2	−0.7
ME	−0.1	−0.6	0.1	−0.6	−0.4	−0.4	−0.5	−0.5	−0.4	−0.6
SC	0.8	0.4	0.9	0.0	−0.2	−0.2	−0.2	−0.3	0.0	−0.2

5   2   1   0.5   −0.5   −1   −2   −5

3.5.4. Validation of Local Intensity Scaling of Precipitation

The local intensity scaling of precipitation was also combined with the variance scaled of temperature to calculate the K-G zone. This combination produced a similar K-G distribution as the combination of power transformation of precipitation and variance scaling of temperature in some regions except in the HIRHAM RCM (Figure 10). Based on the LOCI bias correction, the Csb climate zone occurred in Eastern Europe, and the BSk zone decreased in the Eastern region of the Mediterranean. The extension of the Dsc zone in Scandinavia and the extension of the Csb zone in Western France decreased in the ECHAM5-r3 driven RCMs compared to the power transformation variance scaling bias correction combination.

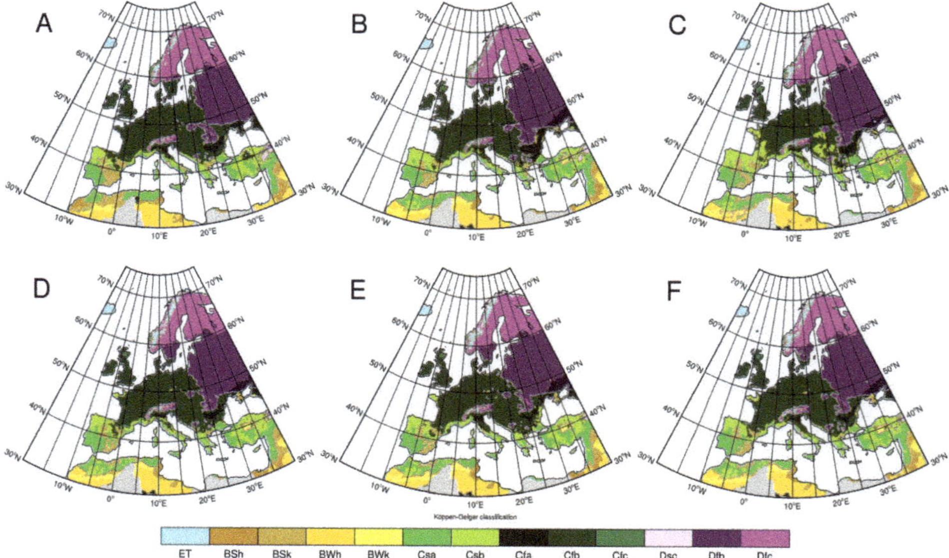

Figure 10. Simulated K-G climate classification according to E-OBS (**A**) and local intensity scaling of precipitation and variance scaling of temperature correction in ALADIN (**B**), HIRHAM (**C**), RegCM (**D**), RAMCMO2 (**E**) and RCA (**F**) in the test period in 1981−2000.

The residual bias of LOCI was extremely large in the HIRHAM model in the Iberian Peninsula in summer (Table 10). The residual bias increased in France compared to power transformation of the precipitation method. In contrast to this, the residual bias was smaller in Eastern Europe in the ECHAM5-r3 forced RCMs.

Table 10. Residual bias of the seasonal amount of simulated precipitation (a) in the case of local intensity scaling of the precipitation bias correction method in the test period in eight different regions: the Alps (AL), the British Isles (BI), Eastern Europe (EA), France (FR), the Iberia Peninsula (IP), the Mediterranean (MD), Mid-Europe (ME) and Scandinavia (SC) in DJF and JJA. The bias values are in %. The extremely large bias in the case of HIRHAM RCM in IP in the in JJA season is denoted by NA where the bias value is about 3×10^{12}.

	ALADIN		HIRHAM		RegCM		RACMO2		RCA	
Regio	DJF	JJA	DJF	JJA	DJF	JJA	DJF	JJA	DJF	JJA
AL	0	18	5	−3	7	−2	8	8	8	4
BI	−10	8	−14	9	−14	−7	−17	−9	−17	−12
EA	−1	19	−3	11	4	6	4	7	7	6
FR	−11	8	−11	−6	−15	−16	−16	−18	−16	−16
IP	13	14	11	NA	24	12	24	4	22	−4
MD	4	5	6	43	19	5	23	27	18	14
ME	−9	22	−9	6	−9	2	−14	0	−8	0
SC	−16	0	−16	−1	−15	−8	−13	−11	−14	−10

| −100 | −50 | −25 | −10 | 10 | 25 | 50 | 100 |

4. Discussion

The results confirmed our supposition that bias corrections have a significant effect on climate classification. The effect of the bias correction varied among the models and the regions of the model domains. Table 11 shows the differences between the observed and simulated Köppen−Geiger climate zones in each region. The results were received by the calculation of the number of grid points where the simulated and observed K-G zones were different in each region, and then this number was divided with the number of grid points of the regions. The eQM and gpQM resulted in the largest differences between the RCMs. These differences stemmed from the correction of precipitation. Simulated precipitation is very sensitive to the properties of a model, e.g., physical parameterization, surface properties, and resolution; hence, the distribution of precipitation varied among the RCMs. In the HIRHAM model, eQM and gpQM produced drier negative bias, i.e., dryer zones in almost the entire studied area. The dominance of these dry climate classes originated from the surface properties in the HIRHAM model, the 1 mm threshold value of precipitation and the correction method. Unlike the other RMCs, HIRHAM has only one soil moisture layer [50], which results in a smaller water-holding capacity, which probably causes a negative feedback effect on precipitation formation. Owing to the threshold value, most of the daily mean precipitation values were less than 1 mm, which were resized to zero. This threshold value also caused negative precipitation bias in the JJA season. Moreover, the eQM corrected the ranked category, but not the value of the variable. Hence, the precipitation (or temperature) values transformed into "very high" values correspond to what observations tell us about actual "very high "values [15]. Notwithstanding that the eQM is expected to be the best method according to some literature [15,51,52], but according to some studies, the distribution-based methods improve the RCMs [31,44,53]. The remaining large biases may originate from the weakness of linear extrapolation of the cumulative distribution of parameters.

ic

Table 11. Disagreement between observed and simulated K-G zones in eight different regions: the Alps (AL), the British Isles (BI), Eastern Europe (EA), France (FR), the Iberia Peninsula (IP), the Mediterranean (MD), Mid-Europe (ME) and Scandinavia (SC) and in the whole study area in DJF and JJA in the case of eQM-eQM, gpQM-eQM, power transformation of precipitation and variance scaling of temperature and LOCI and variance scaling of temperature bias correction combination. The values are in %.

DISAGREEMENT	AL	BI	EA	FR	IP	MD	ME	SC	Study Area
eQM-eQM									
ALADIN	8.9	1.5	15.5	8.8	12	17.8	2.3	4.5	9
HIRHAM	38.6	1.5	29.5	44.2	31.6	35.4	2.4	7.8	20
RegCM	9.4	1.3	10.3	1.5	16.2	17.2	0.7	6	8.4
RACMO2	8.9	2.2	12.9	1.5	12.9	19	0.7	6.4	9
RCA	10.3	2.5	8.6	1	14.2	11.7	0.9	10.5	9
Ensemble mean	15.2	1.8	15.4	11.4	17.4	20.2	1.4	7.0	
gpQM-eQM									
ALADIN	11.8	2.7	18.5	39.3	19.3	29.9	3.4	4.7	13.1
HIRHAM	73.9	46.9	47.4	99	51.2	55.8	46.9	8.9	38.2
RegCM	9.9	1.5	10.9	2.5	17.4	21.8	0.5	6.6	9.4
RACMO2	24.7	2.4	17.5	6.4	19.5	36.8	1.5	7.2	13.7
RCA	18.1	2.5	10	4.7	19.5	20.1	0.9	11.2	11.4
Ensemble mean	27.7	11.2	20.9	30.4	25.4	32.9	10.6	7.7	
power_variance									
ALADIN	1.2	0	0	0	2.9	3.8	0	0.2	0.8
HIRHAM	0	0	0	0	0	0	0	0.2	0.1
RegCM	0	0	0	0	0	0	0	0.2	0.1
RACMO2	0	0	0	0	0	0	0	0.2	0.1
RCA	0	0	0	0	1.1	0	0	0.2	0.2
Ensemble mean	0.2	0.0	0.0	0.0	0.8	0.8	0.0	0.2	
loci-variance									
ALADIN	0	0	0.3	1.2	2.7	1.1	0.8	0.3	0.7
HIRHAM	0	0	0.4	2.5	3.2	1.4	0.8	0.3	0.8
RegCM	0	0	0.3	1.2	2.7	1.1	0.8	0.3	0.7
RACMO2	0	0	0.3	1.2	2.9	3.4	0.8	0.3	0.9
RCA	0	0	0.3	1.2	2.6	1.2	0.8	0.3	0.7
Ensemble mean	0	0	0.3	1.4	2.8	1.6	0.8	0.3	

100	50	30	15	5

The model results corrected by the gpQM resulted in a similar climate classification to the eQM corrected simulations, regardless of the gpQM using gamma and generalized Pareto distributions. The remained bias can be explained by the fact that daily precipitation cannot be adequately expressed by gamma distribution for every region of Europe [54].

The power transformation of precipitation and the local intensity scaling of precipitation combined with the variance scaling of temperature performed correct K-G zone distribution with a negligible difference from the observed one. Furthermore, they resulted in very similar values in each of the RCMs. Their independence on the model and regions of the model domain can be explained by the fact that these are distribution-free correction approaches. Furthermore, they are also able to adjust the variance statistics of the precipitation time series, the simulated wet-day intensity, the wet-day frequency of precipitation and the variance and the mean values of temperature.

The bias correction methods were validated through a split-sample test by calculating the K-G zones in the 1981−2000 time period, except for the local intensity scaling of precipitation. According to the climate classification, the power transformation of precipitation and the variance scaling of temperature combination performed best in terms of K-G zones, despite the fact that the eQM bias correction methods had a smaller residual bias value in some RCMs, e.g., in ALADIN in the JJA season.

The bias correction methods were tested by the differential split-sample test in [44]. According to the statistical evaluation of the bias corrections in the test period, they found that the best method was distribution mapping based on gamma distribution, which was able to correct statistical moments other than means and standard deviations. Their findings presumably stemmed from their decision to choose smaller sized domains, in which only one European region was taken into account. We found the eQM and gpQM of precipitation had great limitations in the larger sized pan-European domain and produced incorrect climate classification in each RCM.

5. Conclusions

In this paper, the influence of bias corrections on K-G climate classification was investigated. Climate classification was calculated by eQM-corrected precipitation and temperature, by a combination of gpQM-corrected precipitation and eQM of temperature, by a combination of power transformation of precipitation and variance scaling of temperature, or by a combination of LOCI for precipitation and variance scaling for temperature. These bias correction methods were applied in five 25 km resolution ENSEMBLE RCMs in the historical time period of 1961−1990 and their results were compared with climate classification based on E-OBS-observed precipitation and temperature values to study their performance. The corrections were tested by a split-sample test, where the 1961−1980 period was training, and the corrections were validated in the 1981−2000 period. Subsequently, the climate classification was evaluated in eight individual subdomains: the Alps, the British Isles, Eastern Europe, France, the Iberian Peninsula, the Mediterranean, Mid-Europe and Scandinavia, defined according to the methodology devised for the PRUDENCE project.

When assessing the performance of the bias correction methods, we found similar results for eQM- and gpQM-corrected K-G classifications when daily data were used during the whole 30-year time period (not shown). Both of them were strongly dependent on the RCM, as the simulated climate zones varied between these RCMs. Moreover, the simulated climate zones significantly differed from the observed ones. These differences stemmed from the large bias in the seasonal precipitation amount. The 90-day moving window improved these correction methods. In comparison, a combination of LOCI and power transformation for precipitation with variance scaling of temperature, respectively, properly reproduced the climate zones by each of the RCMs in each region in the historical period. Furthermore, their test run contained the smallest differences from the observed K-G zones in most regions.

Our results suggest that the eQM and gpQM methods manifest a strong dependence on the spatial distribution of parameters, and this dependence causes a limitation in climate classification considering the large domain. Conversely, power transformation–and local intensity scaling of precipitation and variance scaling of temperature corrections–also generated a smaller bias between the simulated and observed parameters, except in HIRHAM in JJA, but their combination produced better results in climate classification for the whole European domain. This can be explained by the fact that they are distribution-free approaches.

This study is valid for Europe as a whole, since it was based on the E-OBS dataset with a resolution that may be coarser than that of some small regions studied in the quoted papers, where dense national datasets could be used. In the latter case, the statistical properties of the points reflect the smaller area and the results of the method evaluations could be different. It was beyond the scope of this study to devote itself to the several high-resolution gridded datasets that exist in Europe, but this will be the topic of future investigation using the next generation EURO-CORDEX regional climate model simulations.

Author Contributions: Writing–original draft preparation, B.S.-T.; review and editing, P.S., A.F. and J.M.

Funding: This research was funded by Ministry of Education, Youth and Sports of CR within the National Sustainability Program I (NPU I), grant number LO1415.

Acknowledgments: We are thankful for the E-OBS data set from the EU-FP6 project ENSEMBLES (http://www.ensembles-eu.org) and the data provided by the ECA&D project (http://www.eca.knmi.nl).We would like to thank Sixto Herrera García for his help in bias correction and MeteoLab application. We also thank the anonymous reviewer(s) for their comments.

Conflicts of Interest: The authors declare no conflict of interest. The funders had no role in the design of the study; in the collection, analyses, or interpretation of data; in the writing of the manuscript, or in the decision to publish the results.

References

1. Köppen, W. VersucheinerKlassifikation der Klimate, vorzugsweisenachihren Beziehungenzur Pflanzenwelt. *Geogr. Zeitschr.* **1900**, *6*, 593–611, 657–679.

2. Geiger, R. *Landolt-Börnstein—Zahlenwerte und FunktionenausPhysik, Chemie, Astronomie, Geophysik und Technik, alte Serie*; Ch. KlassifikationdefKlimatenach W. Köppen; Springer: Berlin/Heidelberg, Germay, 1954; Volume 3, pp. 603–607.

3. Trewartha, G.T. *An Introduction to Climate*; McGraw-Hill: New York, NY, USA, 1968; 408p.

4. Lohman, U.; Sausen, R.; Bengtsson, L.; Cubasch, U.; Perlwitz, J.; Roeckner, E. The Köppen climate classification as a diagnostic tool for general circulation models. *Clim. Res.* **1993**, *3*, 177–193. [CrossRef]

5. Kalvová, J.; Halenka, T.; Bezpalcová, K.; Nemešová, I. Köppen climate types in observed and simulated climates. *Stud. Geophys. Geod.* **2003**, *47*, 185–202. [CrossRef]

6. Chen, D.; Chen, H.W. Using the Köppen classification to quantify climate variation and change: An example for 1901–2010. *Environ. Dev.* **2013**, *6*, 69–79. [CrossRef]

7. Alessandri, A.; De Felice, M.; Zeng, N.; Mariotti, A.; Pan, Y.; Cherchi, A.; Lee, J.Y.; Wang, B.; Ha, K.J.; Ruti, P.; et al. Robust assessment of the expansion and retreat of Mediterranean climate in the 21st century. *Sci. Rep.* **2014**, *4*, 7211. [CrossRef] [PubMed]

8. Kottek, M.; Grieser, J.; Beck, C.; Rudolf, B.; Rubel, F. World Map of the Köppen-Geiger climate classification updated. *Meteorologische Zeitschrift* **2006**, *15*, 259–263. [CrossRef]

9. Rubel, F.; Kottek, M. Observed and projected climate shifts 1901–2100 depicted by world maps of the Köppen-Geiger climate classification. *Meteorologische Zeitschrift* **2010**, *19*, 135–141. [CrossRef]

10. Belda, M.; Holtanová, E.; Halenka, T.; Kalvová, J. Climate classification revisited: From Köppen to Trewartha. *Clim. Res.* **2014**, *59*, 1–13. [CrossRef]

11. Castro, M.; Gallardo, C.; Jylha, K.; Tuomenvirta, H. The use of climate-type classification for assessing climate change effect in Europe from an ensemble of nine regional climate models. *Clim. Chang.* **2007**, *81*, 329–341. [CrossRef]

12. Gallardo, C.; Gil, V.; Hagel, E.; Tejeda, C.; Castro, M. Assessment of climate change in Europe from an ensemble of regional climate models by the use Köppen-Trewartha classification. *Int. J. Climatol.* **2013**, *33*, 2157–2166. [CrossRef]

13. Trewartha, G.T.; Horn, L.H. *An introduction to Climate*, 5th ed.; McGraw-Hill: New York, NY, USA, 1980.

14. Ménard, R. Bias Estimation. In *Data Assimilation*; Lahoz, W., Khattatov, B., Menard, R., Eds.; Springer: Berlin/Heidelberg, Germany, 2010; pp. 113–135.

15. Déqué, M. Frequency of precipitation and temperature extremes over France in an anthropogenic scenario: Model results and statistical correction according to observed values. *Glob. Planet. Chang.* **2007**, *57*, 16–26. [CrossRef]

16. Sanchez-Gomez, E.; Somot, S.; Déqué, M. Ability of an ensemble of regional climate models to reproduce weather regimes over Europe-Atlantic during the period 1961–2000. *Clim. Dyn.* **2009**, *33*, 723–736. [CrossRef]

17. Farda, A.; Déqué, M.; Somot, S.; Horányi, A.; Spiridonov, V.; Tóth, H. Model ALADIN as regional climate model for central and eastern Europe. *Stud. Geophys. Geod.* **2010**, *54*, 313–332. [CrossRef]

18. Deser, C.; Phillips, A.; Bourdette, V.; Teng, H. Uncertainty in climate change projections: The role of internal variability. *Clim. Dyn.* **2012**, *38*, 527–546. [CrossRef]

19. Eden, J.M.; Widmann, M.; Grawe, D.; Rast, S. Skill, Correction, and Downscaling of GCM-Simulated Precipitation. *J. Clim.* **2012**, *25*, 3970–3984. [CrossRef]

20. Hawkins, E.; Sutton, R. The potential to narrow uncertainty in projections of regional precipitation change. *Clim. Dyn.* **2011**, *37*, 407–418. [CrossRef]

21. Allen, M.; Frame, D.; Kettleborough, J.; Stainforth, D. Model error in weather and climate forecasting. In *Predictability of Weather and Climate*; Palmer, T., Hagedorn, R., Eds.; Cambridge University Press: Cambridge, UK, 2006; pp. 275–294.

22. Hagemann, S.; Chen, C.; Haerter, J.O.; Heinke, J.; Gerten, D.; Piani, C. Impact of a Statistical Bias Correction on the Projected Hydrological Changes Obtained from Three GCMs and Two Hydrology Models. *J. Hydrometeorol.* **2011**, *12*, 556–578. [CrossRef]

23. Chen, J.; Brissette, F.P.; Leconte, R. Uncertainty of downscaling method in quantifying the impact of climate change on hydrology. *J. Hydrol.* **2011**, *401*, 190–202. [CrossRef]

24. Piani, C.; Haerter, J.O. Two dimensional bias correction of temperature and precipitation coupals in climate models. *Geophys. Res. Lett.* **2012**, *39*, L20401. [CrossRef]

25. Szabó-Takács, B.; Farda, A.; Zahradníček, P.; Štěpánek, P. Köppen-Geiger climate classification by different Regional Climate Models according to A1B SERES scenario in 21st century. In *Global Change: A Complex Challenge Conference Proceedings*; Urban, O., Šprtová, M., Klem, K., Eds.; Global Change Research Centre: Brno, Czech Republic, 2015; pp. 18–21.

26. Van der Linden, P.; Mitchell, J.F.B. (Eds.) *ENSEMBLES: Climate Change and Its Impacts: Summary of Research and Results from the ENSEMBLES Project*; Met Office Hadley Centre: Exeter, UK, 2009; 160p.

27. Nakićenovíc, N.; Swart, R. *Special Report on Emissions Scenarios: A Special Report of Working Group III of the Intergovernmental Panel on Climate Change*; Cambridge University Press: Cambridge, UK, 2000.

28. Solomon, S. Technical Summary. In *Climate Change the Physical Science Basis. Contribution of Working Group I to the Fourth Assessment Report of the Intergovernmental Panel on Climate Change*; Solomon, S., Qin, D., Manning, M., Chen, Z., Marquis, M., Averyt, K.B., Tignor, M., Miller, H.L., Eds.; Cambridge University Press: Cambridge, UK; New York, NY, USA, 2007.

29. Haylock, M.R.; Hofstra, N.; Klein Tank, A.M.G.; Klok, E.J.; Jones, P.D.; New, M. A European daily high-resolution gridded data set of surface temperature and precipitation for 1950–2006. *J. Geophys. Res.* **2008**, *113*, D20119. [CrossRef]

30. Teutschbein, C.; Seibert, J. Bias correction of regional climate model simulations for hydrological climate-change impact studies: Review and evaluation of different methods. *J. Hydrol.* **2012**, *456–457*, 12–29. [CrossRef]

31. Chen, J.; Brissette, F.P.; Chaumont, D.; Braun, M. Finding appropriate bias correction methods in downscaling precipitation for hydrologic impact studies over North America. *Water Resour. Res.* **2013**, *49*, 4187–4205. [CrossRef]

32. Fang, G.H.; Yang, J.; Chen, Y.N.; Zammit, C. Comparing bias correction methods in downscaling meteorological variables for a hydrologic impact study in an arid area in China. *Hydrol. Earth Syst. Sci.* **2015**, *19*, 2547–2559. [CrossRef]

33. Déqué, M.; Rowell, D.P.; Lüthi, D.; Giorgi, F.; Christensen, J.H.; Rockel, B.; Jacob, D.; Kjellström, E.; De Castro, M.; van den Hurk, B. An intercomparison of regional climate simulations for Europe: Assessing uncertainties in model projections. *Clim. Chang.* **2007**, *81* (Suppl. 1), 53–70. [CrossRef]

34. Christensen, J.H.; Christensen, O.B.; Lopez, P.; Van Meijgaard, E.; Botzet, M. *The HIRHAM4 Regional Atmospheric Model*; Scientific Report 96-4; The Danish Meteorological Institute: Copenhagen, Denmark, 1996.

35. Lenderink, G.; van der Hurk, B.; van Meijgaard, E.; van Ulden, A.P.; Cuijpers, J.H. *Simulation of Present-Day Climate in RACMO2: First Results and Model Developments*; Technical Report TR-252; Royal Netherlands Meteorological Institute: De Bilt, The Netherlands, 2003.

36. Kjellström, E.; Bärring, L.; Gollvik, S.; Hansson, U.; Jones, C.; Samuelsson, P.; Rummukainen, M.; Ullerstig, A.; Willén, U.; Wys, K. *A 140-Year Simulation of European Climate with the New Version of Rossby Central Regional Atmospheric Climate Model (RCA3)*; RCA Reports Meteorology and Climatology, 108; RCA: Norrköping, Sweden, 2005; 54p.

37. Giorgi, F.; Bi, X.; Pal, J. Mean, interannual variability and trends in regional climate change experiment over Europe. I. Present-day climate (1961–1990). *Clim. Dyn.* **2004**, *22*, 733–756. [CrossRef]

38. Christensen, J.H.; Carter, T.R.; Rummukainen, M.; Amanatidis, G. Evaluation the performance and utility of regional climate models: The PUDENCE project. *Clim. Chang.* **2007**, *81*, 1–6. [CrossRef]

39. Gutjahr, O.; Heinemann, G. Comparing precipitation bias correction methods for high-resolution regional climate simulations using COSMO-CLM: Effect on extreme values and climate change signal. *Theor. Appl. Climatol.* **2013**, *114*, 511–529. [CrossRef]

40. Leander, R.; Buishand, T.A. Resampling of regional climate model output for simulation of extreme river flows. *J. Hydrol.* **2007**, *332*, 487–496. [CrossRef]

41. Leander, R.; Buishand, T.A.; van den Hurk, B.J.J.M.; de Wit, M.J.M. Estimated changes in flood quantiles of the river Meuse from resampling of regional climate model output. *J. Hydrol.* **2008**, *351*, 331–343. [CrossRef]

42. Schmidli, J.; Frei, C.; Vidale, P.L. Downscaling from GCM precipitation: A benchmark for dynamical and statistical downscaling methods. *Int. J. Climatol.* **2006**, *26*, 679–689. [CrossRef]

43. Ruff, T.W.; Neelin, J.D. Long tails in regional surface temperature propability distributions with impactions for extremes under global warming. *Geophys. Res. Lett.* **2012**, *39*, L04704. [CrossRef]

44. Teutschbein, C.; Seibert, J. Is bias correction of regional climate model (RCM) simulations possible for non-stationary conditions? *Hydrol. Earth Syst. Sci.* **2013**, *17*, 5061–5077. [CrossRef]

45. MeteoLab. Meteorological Machine Learning Toolbox for Matlab. Available online: http://grupos.unican. es/ai/meteo/MeteoLab.html (accessed on 25 July 2014).

46. Weaclim. Available online: https://www.mathworks.com/matlabcentral/fileexchange/10881-weaclim (accessed on 27 April 2006).

47. The NCAR Command Language (Version 6.4.0) [Software]. UCAR/NCAR/CISL/TDD: Boulder, CO, USA, 2017. Available online: http://dx.doi.org/10.5065/D6WD3XH5 (accessed on 28 February 2017).

48. Brent, R.P. An algorithm with guaranteed convergence for finding a zero of a function. *Comput. J.* **1971**, *14*, 422–425. [CrossRef]

49. Klemeš, V. Operational testing of hydrological simulation models. *Hydrol. Sci. J.* **1986**, *31*, 13–24. [CrossRef]

50. Jacob, D.; Bärring, L.; Christensen, O.B.; Christensen, J.H.; de Castro, M.; Déqué, M.; Giorgi, F.; Hagemann, S.; Hirschi, M.; Jones, R.; et al. An inter-comparison of regional climate models for Europe: Model performance in present-day climate. *Clim. Chang.* **2007**, *81*, 31–52. [CrossRef]

51. Sennikovs, J.; Bethers, U. Statistical downscaling method of regional climate model results for hydrological modelling. In Proceedings of the 18th world IMACS/MODSIM Congress, Cairns, Australia, 13–17 July 2009.

52. Michelangeli, P.A.; Vrac, M.; Loukos, H. Probabilistic downscaling approaches: Application to wind cumulative distribution function. *Geophys. Res. Lett.* **2009**, *36*, L11708. [CrossRef]

53. Piani, C.; Haerter, J.O.; Coppola, E. Statistical bias correction for daily precipitation in regional climate models over Europe. *Theor. Appl. Climatol.* **2010**, *99*, 187–192. [CrossRef]

54. Vlček, O.; Radan, H. Is daily precipitation gamma-distributed?: Adverse effect of an incorrect use of the Kolmogorov-Smirnov test. *Atmos. Res.* **2009**, *93*, 759–766. [CrossRef]

Article

Analysis of Climate Change in the Caucasus Region: End of the 20th–Beginning of the 21st Century

Alla A. Tashilova [1,*], Boris A. Ashabokov [1,2], Lara A. Kesheva [1] and Nataliya V. Teunova [1]

[1] Federal State Budgetary Institution «High-Mountain Geophysical Institute», Lenin av., 2, 360030 Nalchik, Russia; ashabokov.boris@mail.ru (B.A.A.); kesheva.lara@yandex.ru (L.A.K.); n.teunova@gmail.com (N.V.T.)

[2] Institute of Computer Science and Problems of Regional Management, Kabardino-Balkarian Research Center, Russian Academy of Sciences, I. Armand str., 37a, 360000 Nalchik, Russia

* Correspondence: tashilovaa@mail.ru

Received: 15 December 2018; Accepted: 4 January 2019; Published: 10 January 2019

Abstract: The study of climate, in such a diverse climatic region as the Caucasus, is necessary in order to evaluate the influence of local factors on the formation of temperature and precipitation regimes in its various climatic zones. This study is based on the instrumental data (temperatures and precipitation) from 20 weather stations, located on the territory of the Caucasian region during 1961–2011. Mathematical statistics, trend analysis, and rescaled range Methods were used. It was found that the warming trend prevailed in all climatic zones, it intensified since the beginning of global warming (since 1976), while the changes in precipitation were not so unidirectional. The maximum warming was observed in the summer (on average by 0.3 °C/10 years) in all climatic zones. Persistence trends were investigated using the Hurst exponent H (range of variation 0–1), which showed a higher trend persistence of annual mean temperature changes ($H = 0.8$) compared to annual sum precipitations ($H = 0.64$). Spatial-correlation analysis performed for precipitations and temperatures showed a rapid decrease in the correlation between precipitations at various weather stations from $R = 1$ to $R = 0.5$, on a distance scale from 0 to 200 km. In contrast to precipitation, a high correlation ($R = 1.0$–0.7) was observed between regional weather stations temperatures at a distance scale from 0 to 1000 km, which indicates synchronous temperature changes in all climatic zones (unlike precipitation).

Keywords: temperature; precipitations; warming; Hurst exponent; persistence; spatial correlation; Caucasian region

---

1. Introduction

The problem of climate change is extremely urgent today. The global climate on our planet is changing rapidly. In this regard, an increasing number of studies are being devoted to this problem [1–17]. The Russia territory is more sensitive to the effects of climate than the Northern Hemisphere and the rest of the globe. Throughout Russia, the average growth rate of average annual air temperature has been 0.46 °C/10 years in 1976–2017. This is 2.5 times the growth rate of global temperature over the same period: 0.18 °C/10 years, and more than 1.5 times the average warming rate of surface air over the Earth's land: 0.28 °C/10 years (estimates according to the Hadley Center and the University of East Anglia: HADCRUT4, CRUTEM4) [18].

Observations of regional climate show that atmospheric phenomena are more significant and variable in regional rather than globally. Many factors affect the climatic features of the south of Russia, including zonal and altitudinal zonality. The geographic position in temperate latitudes contributes to the formation of a moderately continental type of climate, while the Caucasus Mountains serve as a climate cliff between the temperate and subtropical zones. The Caucasus region has significant impact on climatic features, which is supported by air masses, bringing the Mediterranean warm moist air.

An important factor is the difference in altitudes from the Caspian lowland (-28 m from sea level) to the peaks of the North Caucasus, with the highest point in Europe—Mount Elbrus with a height of 5642 m. According to the nature of the relief, the North Caucasus is usually divided into three zones: the plain (Black Sea zone, steppe, Caspian zone), with a height above sea level (a.s.l.) of less than 500 m a.s.l.; foothills (500–1000 m a.s.l.); mountain (>1000 m a.s.l.) and high-mountain (>2000 m a.s.l.). The issues of climate change in areas of the National Park "Prielbrusye" (the high-mountain zone) and the Sochi National Park (the Black Sea zone) are especially important, since they can be beyond landscapes causing disturbance of the ecosystem balance [19–25].

An important aspect of this region is an assessment of the regional response of the mountain climate against the backdrop of global warming, to study the glaciers deglaciation mechanisms. As research continues into climate of the Caucasus region, it becomes apparent that unfortunately, the historical information about climate fluctuations in the high mountains of the Caucasus is very scarce and not systematized. Due to the lack of long-term observations in mountainous areas, some authors [26] restored to the meteorological regime of the corresponding area according to NCEP/NCAR reanalysis, and corrected the information using data from individual instrumental observations. Others [27] restored to the series of temperatures and precipitation at the meteorogical stations Teberda (1280 m a.s.l., Teberda state biosphere reserve) and Terskol (2144 m a.s.l., Elbrus national park) of the Caucasus region, using dendroclimatological methods.

In the first case [26], for the restored meteorological regime in the Caucasus from 1948 to 2013, fragmentary observation materials of the first Elbrus expedition in July–August 1934–1935 [28], and of the second Elbrus expedition in 1957–1959, and in 1961–1962 (Institute of Geography, USSR Academy of Sciences) [29] were used. Analysis of the recovered data for the period 1948–2013 have shown that in the area of Mount Elbrus during the warm season, the positive trend does not go beyond the limits of natural variability, and the change in the average annual temperature is characterized as a stable value "-0.01 °C/10 years".

According to the results from the dendrological analysis [28], it was concluded that the mountain landscapes of Teberda and Terskol in 1960–2005 was characterized by relatively stable climatic conditions. In general, in this area there is a tendency to a slight increase in the air temperature in individual months, and to an increase in annual precipitation. However, under the conditions of an extremely rare network of meteorological observations, it is difficult to reliably determine the causes of this phenomenon, and attribute the differences due to the influence of local factors or the lack of representativeness of weather stations.

The series thus restored have one significant drawback for the study, they cover different time intervals, and the statistical characteristics of such meteorological series cannot be compared to each other. In addition, reanalysis of the data obtained using satellite meteorology can lead to heterogeneity of the series until the mid 70s.

The only way to somewhat reliably estimate regional climate change is by statistical analysis of long series of instrumental data covering the same time span, such as the approach in this paper.

The report from the Intergovernmental Panel on Climate Change (IPCC) of 31 March 2014 states that there are more significant climate changes on all continents, and spaces [30]. The observed effects of climate change have affected ecosystems of land and ocean, some sources of human livelihoods, water supply systems, agriculture, and human health. In this context, the study of climate and the identification of its possible consequences, have now become scientific problems that attract great attention from researchers around the world.

2. Materials and Methods

The focus of our study was climate of the Caucasian region (southern Russia), whose territory in the context of the article was limited to 41.28–47.14 degrees north latitude (°N) and 38.58–48.17 degrees east longitude (°E).

To study the climate in different regions of southern Russia, we used data from meteorological instrumental observations (1961–2011) by 20 weather stations, of the state observational network of Roshydromet and provided by the North Caucasian Administration for Hydrometeorology and Environmental Monitoring. (Table 1 and Figure 1). The data of the time series were homogeneous, throughout the period under study the location of the stations remained constant (outside populated areas), and the so-called urban warming did not affect them. The average, maximum, and minimum seasonal and annual temperatures in the south of Russia were investigated.

Table 1. Geographical location of weather stations inside the Caucasian region.

№ *n/n*	Weather Stations	Longitude (°N), Latitude (°E)	Height above the Sea Level, m (m a.s.l.)
	Plain stations (<500 m a.s.l.)		
1	Sochi (Black Sea zone)	43.35° N; 39.73° E	57
2	Krasnodar (steppe zone)	45.20° N; 38.58° E	26
3	Izobil'nyi (steppe zone)	45.22° N; 32.42° E	194
4	Mozdok (steppe zone)	43.44° N; 44.39° E	126
5	Prokhladnaya (steppe zone)	43.46° N; 44.05° E	198
6	Rostov-on-Don (steppe zone)	47.14° N; 39.44° E	64
7	Maykop (steppe zone)	44.37° N; 40.05° E	270
8	Derbent (Caspian zone)	42.04° N; 48.17° E	30
9	Kizlyar (Caspian zone)	43.51° N; 46.43° E	−17
10	Makhachkala (Caspian zone)	42.59° N; 47.31° E	173
11	Izberg (Caspian zone)	42.34° N; 47.45° E	21
	Foothill stations (500–1000 m a.s.l.)		
12	Stavropol	45.03° N; 41.58° E	540
13	Cherkessk	44.17° N; 42.04° E	526
14	Kislovodsk	43.54° N; 42.43° E	819
15	Nalchik	43.22° N; 43.24° E	500
16	Vladikavkaz	43.21° N; 44.40° E	680
17	Buinaksk	42.49° N; 47.07° E	560
	Mountain stations (1000–2000 m a.s.l.)		
18	Teberda	43.45° N; 41.73° E	1280
19	Akhty	41.28° N; 47.44° E	1054
	High-mountain station (>2000 m a.s.l.)		
20	Terskol	43.15° N; 42.30° E	2144

In previous studies [31,32], trends in the amount of precipitation and daily maximums of precipitation were analyzed in the Caucasus region, an analysis of the temperature regime was added in this study. In the series of temperatures, averaged values, anomalies (deviations of the observed value from the norm), and trends for the four seasons and the calendar year (January–December) were considered. The climatic norm was considered to be the mean multi-year value of the considered climate variable for the base period of 1961–1990 [33]. The anomalies were calculated for each year as the difference between the current value and the norm of the corresponding climate variable (average 1961–1990). In the series of mean temperature and sum precipitation, the data were averaged within the calendar seasons of each year (the winter season included December of the previous year) and for the year as a whole. Maximum and minimum temperatures were defined as the largest and lowest values for a certain period (month). The absolute maximum (minimum) was the largest (smallest) value was observed at least once in a month. We used absolute maxima and minima for each month of the season during the period 1961–2011.

Figure 1. Geographical location of weather stations inside the Caucasian region.

Time series were investigated by statistical methods, as well as by means of the STATISTICA, SPSS 15.0 programs [34–36], spatial fields of distributions were constructed using the geoinformation system Golden Software Surfer 8 [37]. Linear trends characterizing the trend of the considered value over the entire observation period from 1961 to 2011, and from 1976 (the beginning of global warming) were built in Excel. The estimation of the linear trend coefficients was considered the least squares method degrees per decade, °C/10 years.

To accept the hypothesis regarding the presence of a statistically significant linear trend, a 95% significance level (α) was adopted and determined through a determination coefficient R^2 characterizing the share of the trend in the explained variance (D, %). Using the coefficient of determination R^2, it was possible to check the significance of the trend. For this, the F-criterion was determined:

$$F = \frac{R^2}{1 - R^2} \frac{n - k - 1}{k},\tag{1}$$

where k is the number of trend equation coefficients. The constructed linear regression trend was significant with a level of significance α, if the inequality held:

$$F > F_{1-\alpha;k;n-k-1}\tag{2}$$

Quantile $F_{1-\alpha;k;n-k-1}$ calculated in Excel by expressing $\text{FINV}(\alpha; k; n - k - 1)$.

The lower threshold value of the coefficient of determination, which determines the statistical significance of the trend at a 95% confidence interval, was $D = 8\%$ (for $n = 51$, 1961–2011).

Climate of the Caucasus region, which includes various climatic zones (Table 1 and Figure 1), was primarily determined by its position in the temperate latitudinal climate zone.

The North Caucasus mountain system prevents the movement of cold air masses from north to south, and warm from south-west and west to north-east and east. A complex local circulation is created in the mountains with the separation of the two temperate zone regions: the Atlantic-continental (plain, foothill) and the mountain (high-mountain). At the same time, atmospheric processes in the region are complicated by local factors, namely the complex orography of the North Caucasus. Due to the complexity of climate formation in such a complex orographically heterogeneous terrain, the correlation of meteorological parameters of stations located in different climatic zones was of interest.

The spatial structures of the air temperature fields and precipitation fields were analyzed from weather station data in different climatic zones, and spatial correlation relationships between them were determined depending on the scale of the distance between them.

The study assessed the persistence of climate change. As its integral characteristic, the rescaled range method (R/S analysis) and fractal properties of time series (Hurst exponent H), were used [38–44]. By using the rescaled range method for the first time, the British hydrologist Harold Hurst studied the rise of the Nile River, as well as the sequence of measurements of atmospheric temperature, rainfall, river flow parameters, thickness of annual wood growth rings, and other natural processes [38]. The method is based on the analysis of the range R of the meteorological parameter (the largest and smallest value in the segment under study), and the standard deviation S and its dependence on the period of the studied time t. The Hurst exponent can distinguish a random series from a nonrandom one, even if the random series is not Gaussian (that is, not normally distributed).

To calibrate the time series, Hurst introduced a dimensionless ratio by dividing the range R by the standard deviation S of the observations. The range of R_n is the difference between the maximum and minimum levels of accumulated deviation X_n.

$$R_n = \max\left(X_k - \frac{k}{n}X_n\right) - \min\left(X_k - \frac{k}{n}X_n\right), \tag{3}$$

where $X_n = x_1 + x_2 + \ldots + x_n, n \geq 1$;

    X_n—accumulated deviation in n steps (t periods);
    R_n—deviation range in n steps, where

$$S_n = \sqrt{\frac{\sum\limits_{k=1}^{n} x_k - \bar{x}^2 n}{n}}\text{—empirical standard deviation;}$$

$\bar{x}_n = X_n/n$—empirical average;
R_n/S_n—normalized range of accumulated sums $R_k, k \leq n$.

Based on the formula for Brownian motion, the system displacement (normalized range) Hurst proposed to calculate using the following relation:

$$R_n/S_n = (at)^H, \tag{4}$$

from where

$$H = \frac{\ln(R_n/S_n)}{\ln(at)}, \tag{5}$$

where H is the Hurst exponent, varying from 0 to 1; R_n/S_n is the rescaled range; t is the studied period, and a is a constant. The value of the coefficient H characterizes the ratio of the strength of the trend (deterministic factor) to noise level (random factor). The indicator H is a tool for determining the systems persistence behavior and gives an answer to the question of what the next value of the investigated series would be, more or less than the current value. The processes for which $H = 0.5$ have an independent data distribution, and are characterized by the absence of a trend (classical Brownian motion). Time sequences for which H is greater than 0.5 are classified as persistent, preserving the existing trend. If the increments were positive for some time in the past—there was an increase—then on the average there will be an increase, this corresponds to a good predictability of the series. Thus, for a process with $H > 0.5$ there is a tendency to increase in the future, the effect of long-term memory is preserved. The case $0 < H < 0.5$ is characterized by antipersistence and is characterized by an alternating tendency.

On a large empirical material, it was found that the Hurst index value of the series of various natural processes is grouped in the interval $H = 0.72$–0.74. [38,41]. The question of why this is so remains open. Note that our average value of the Hurst index for time series of temperatures $H = 0.74$, also fell in this interval.

3. Analysis and Discussion

Using the method of spatial correlations, the spatial relationships between surface air temperature and precipitation at individual stations of the studied region were determined. This described the general spatial patterns of temperature and precipitation in the region, and generated a preliminary assessment of the features of regional climate formation.

Reference [45] describes that in all climatic zones of the region, changes in mean annual temperatures, unlike precipitation, were synchronous in time (Figure 2). In addition to the main climate-forming factors (radiation and circulation), climate of the Caucasus region is greatly influenced by the relief of the terrain, the orography of the terrain, and the distance of weather stations from each other. Figure 2b shows that changes in the precipitation regime in different climatic zones, unlike changes in the temperature regime (Figure 2a), were not synchronous.

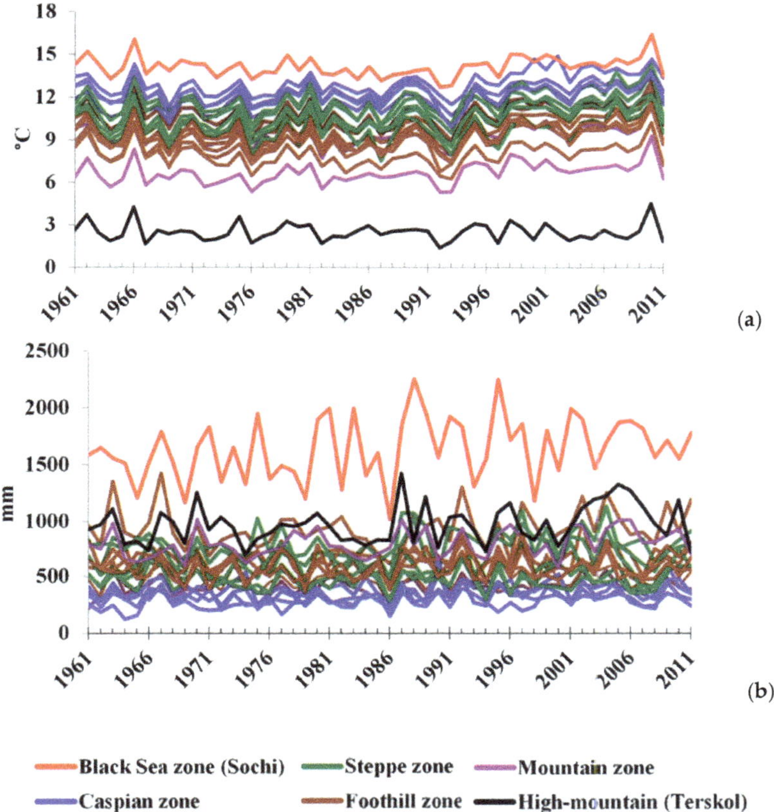

Figure 2. The course of average annual temperatures (**a**), and precipitation sums (**b**) according to data from 20 weather stations in southern Russia.

Apparently, this can be explained by influence of the same centers of low-frequency atmosphere variability, which by means of synoptic scale objects (cyclones, anticyclones, etc.), exert a distant influence on the climate in certain regions of Eurasia and are sources of anomalies in various meteorological fields.

The wide range of mean temperature dispersions and their rate of change in different climatic zones was also determined by the stations' location (from Rostov-on-Don in the north to Derbent in the south, and from Sochi in the west to Isberg in the east of the Caucasian region). Figure 2 show that the average annual temperature was maximum in Sochi (red line, t_{av} = 14.18 °C), and minimum in Terskol (black line, t_{av} = 2.5 °C). In all climatic zones, temperature changes occured synchronously, while temperature changes of stations within each climatic zone fit into their ranges.

The method of calculating spatial correlations was used to determine the spatial relationships of meteorological parameters (temperature, precipitation), measured at stations of different climatic zones of the southern Russia. Let x_i (j) be the average annual air temperatures or the precipitation sums at some station i, where $j = 1, \ldots N_i$, j is the year, N_i is the total number of measurements. Similar measurements at another station, l, respectively, x_l (k), where $l = 1, \ldots N_k$, i is the year, N_k is the total number of measurements. For joint analysis, it was necessary that $N_i = N_k$. The available series with average annual temperatures and precipitation sums of 20 stations satisfied this requirement.

The linear relationship between series x_i (j) and x_l (k) was quantitatively expressed by the Pearson correlation coefficient R_{ik}. Calculating all possible combinations of R_{ik}, we obtained matrices A_c, where $c = T$ (temperature) or $c = P$ (precipitation), symmetric with respect to the main diagonal, which consisted of units (correlation matrices). The latter can be associated with the same matrix Z of distances between each pair of matrices. The distance (L) between stations was calculated based on the geographical coordinates (latitude $\varnothing$, longitude λ, Table 1) of each of the station pairs according to the formula:

$$L = arccos(sin\varnothing 1 sin\varnothing 2 + cos\varnothing 1 cos\varnothing 2 cos\Delta\lambda) \tag{6}$$

Scatter diagrams of correlation matrices for air temperature and precipitation in the investigated region are presented in Figure 3. Each matrix elements was shown depending on the distance of the stations from each other, set in the distance matrices. Comparison of Figure 3a,b implied a significant difference in the correlation structure of the fields.

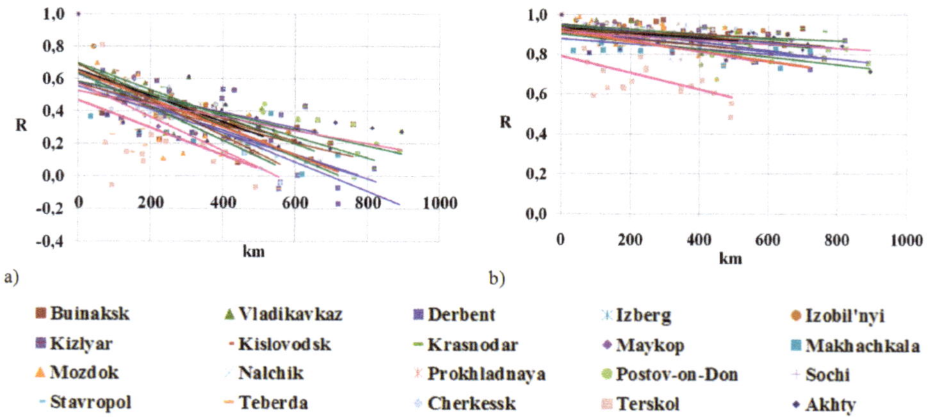

Figure 3. Spatial correlation of precipitation (**a**), and temperature (**b**) of the 20 weather stations, depending on the distance between them: the *x*-axis—distance between pairs of stations; and the *y*-axis—the correlations between the meteorological parameters (temperature, precipitation) of the 20 weather stations.

The spatial correlation of precipitation for all climatic zones tended to zero for scales of distances of the order of 600 km, and took negative values (from -0.04 to -0.2) for pairs of stations Derbent—Cherkessk, Derbent—Teberda, Derbent—Terskol, Derbent—Sochi, and Derbent—Maykop.

Such an inverse relationship between the precipitations of these stations can be explained, firstly, by the special geographical position of Derbent. Derbent is located on the western shore of the Caspian Sea (42.04° N; 48.17° E, 30 m a.s.l.), where the mountains of the Greater Caucasus are closest to the Caspian Sea, leaving only a narrow three-kilometer strip of plain, forming a semi-dry subtropical climate. Secondly, the great difference in altitude, and the remoteness of Derbent from other stations.

If we set the correlation threshold $R = 0.5$, it can be seen from Figure 3a that the correlation of precipitation decreased from $R = 0.7$ to $R = 0.5$ at a distance of 200 km, and it decreased further passing through $R = 0$ at a distance from 600 to 800 km.

The temperature correlations of different weather stations were quite high at distances from 0 to 1000 km, and lay mainly in a narrow range from $R = 1$ to $R = 0.7$ for plain, foothill, and mountain zones. The correlation between the temperatures of the Terskol (high-mountain zone) station with other stations was lower and varied from $R = 1$ to a threshold value $R = 0.50$ only for Terskol–Rostov-on-Don ($R = 0.554$) and Terskol–Derbent ($R = 0.485$).

Thus, the magnitude of temperature correlation depended mainly on the stations' location relative to height above sea level. In addition to this factor, the correlation of precipitation was determined by the geographical location of the weather stations (orography of the terrain) and the stations' remoteness from each other. The obtained values of the spatial correlation of temperature and precipitation explained the synchronous course of temperature variation in all climatic zones, and the regional precipitation regime in all climatic zones of southern Russia (Figure 2.)

Various estimates of the change in global surface air temperature have been given [1–5,46,47]. From the second half of the 20th century, and in the first decade of the 21st century, the rate of temperature growth on average has varied in the range $0.17 \pm 0.01\,°C$. According to our estimates, the warming trend across the Caucasus region corresponds to the overall trend of global temperature changes over the same period. Table 2 shows that the average annual temperature increased during the period 1961–2011 at $0.2\,°C/10$ years, while in 1961–1975 there was an insignificant negative trend, and since 1976 the temperature growth rate reached $0.43\,°C/10$ years with the trend contribution to the explained variance $D = 31.5\%$. The same trend was observed for all seasonal temperatures.

Table 2. Characteristics of seasonal and annual average air temperatures in the south of Russia.

Temperature, °C	Winter	Spring	Summer	Autumn	Annual
Average temperature, 1961–2011	−0.5	9.4	20.8	10.8	10.1
Standard deviation, 1961–2011	1.5	0.9	1.0	1.1	0.8
Anomalies, 1961–2011	0.3	0.1	0.4	0.1	0.2
The angular coefficient of the trend (1961–2011), °C/10 years (D^*, %)	0.22 (4.4%)	0.08 (1.6%)	**0.33 **** **(23%)**	0.15 (4.1%)	**0.2** **(13%)**
The angular coefficient of the trend (1961–1975), °C/10 years (D, %)	−1.34 (9.2%)	0.16 (0.5%)	0.07 (0.1%)	0.17 (0.7%)	−0.32 (3.3%)
The angular coefficient of the trend (1976–2011), °C/10 years (D, %)	0.38 (9.3%)	0.21 (6.1%)	0.65 (41%)	0.47 (18.4%)	0.43 (31.5%)

* D, the trend contribution to the explained variance. ** statistically significant trends are marked in bold ($n = 51$).

Negative trend of the average winter temperature in 1961–1975 ($-1.34\,°C/10$ years) since 1976 changed to a positive direction ($+0.38\,°C/10$ years). Interannual variability of temperature from all seasons was greatest in the winter season, $1.5\,°C$. The main reason for the large winter variability was the large temperature difference in winter between low and high latitudes. Since 1976, the growth rate of all other seasonal temperatures increased. If in the period 1961–2011 the trend of seasonal temperatures was statistically significant only in the summer season ($0.33\,°C/10$ years, $D = 23\%$), in the modern period, trends in autumn ($0.47\,°C/10$ years, $D = 18.4\%$) and winter ($0.38\,°C/10$ years, $D = 9.3\%$) were added to statistically significant trends. The maximum value of seasonal anomalies ($\Delta T = 0.4\,°C$) was also observed for summer temperatures. Since the mid 90s of the 20th century, there have been exceptionally positive anomalies in summer temperatures.

Next, we performed a comparative analysis of temperatures (average, maximum, minimum) within all climatic zones of the Caucasus region. The average values of mean annual temperatures, their standard deviation, characterizing the interannual variability, as well as the upper and lower boundaries of the intervals of mean annual air temperature in different climatic zones (at 95% confidence intervals) are presented in Table 3. The lower and upper limits of the confidence intervals of the mean annual temperature, according to data from the Black Sea, Caspian, steppe, piedmont, and mountain weather stations, intersect. The average annual temperature in Terskol ($2.5\,°C$, taking into account the interannual variability from $1.22\,°C$ to $3.78\,°C$) was significantly lower than the rest, which was explained by the high altitude zoning. This station was also characterized by the stability of the change in annual temperature ($0.01\,°C/10$ years). In our study, a high-altitude zone was distinguished (Terskol, 2144 m a.s.l.) with a negative trend of average annual temperature for the period 1961–2011 ($-0.01\,°C/10$ years), and for the years 1961–1975 ($-0.01\,°C/10$ years). Since 1976, the negative trend changed its direction to positive ($+0.05\,°C/10$ years).

Table 3. Characteristics of the temperature regime of surface air in the different climatic zones of southern Russia.

Temperature	Black Sea Zone (Sochi)	Steppe Zone	Caspian Zone	Foothill Zone	Mountain Zone	High-Mountain (Terskol)
Average temperature t_{av} (σ), °C						
Annual temperature (st. deviation)	14.18 (0.72)	10.82 (0.92)	12.38 (0.82)	9.24 (0.91)	8.05 (0.78)	2.5 (0.64)
Upper bound *	15.62	12.67	14.02	11.05	9.6	3.78
Lower bound	12.74	8.98	10.73	7.43	6.49	1.22
Rates of change of average temperature, °C/10 year $(D, \%)$						
Annual						
(a) 1961–2011	0.06 (2%)	**0.25 (14%)**	**0.21 (17%)**	**0.23 (17%)**	**0.17 (11%)**	−0.01 (0%)
(b) 1961–1975	−0.3 (4%)	−0.03 (2%)	−0.37 (5%)	−0.45 (7%)	−0.5 (6%)	−0.2 (2%)
(c) 1976–2011	0.31 (19.6%)	0.48 (28%)	0.42 (30%)	0.5 (33%)	0.38 (24.7%)	0.05 (0.7%)
Winter	−0.04 (0.2%)	0.3 (6%)	0.18 (4%)	0.27 (6%)	0.14 (2%)	−0.02 (0%)
Spring	−0.03 (0.2%)	0.09 (2%)	0.11 (4%)	0.1 (2%)	0.06 (1%)	−0.07 (1%)
Summer	0.28 (19%)	0.34 (20%)	0.26 (18%)	0.37 (22%)	0.35 (30%)	0.29 (25%)
Autumn	0.07 (1%)	0.16 (4%)	0.15 (4%)	0.19 (6%)	0.13 (3%)	−0.07 (1%)

* the upper (lower) boundary of the mean temperature ($t_{mean} \pm 2\,\sigma$) at 95% confidence interval; ** statistically significant trends are marked in bold ($n = 51$).

Changes in the average annual air temperature in different climatic zones of southern Russia were also represented by three periods: 1961–2011, 1961–1975, and 1976–2011 in Table 3. During the period 1961–2011 in all climatic zones of southern Russia, with the exception of the Black Sea zone (Sochi) and the high-mountain zone (Terskol), an average annual temperature increased from 0.17 °C/10 years in the mountain zone to 0.26 °C/10 years in the steppe zone. In the Black Sea zone, the rate of change in the mean annual temperature was 0.06 °C/10 years, and 0.01 °C/10 years in the high-mountain zone, which characterizes a stable temperature regime in these zones. According to the station Makhachkala (the Caspian region), it also received insignificant rates of change in the average annual temperature by 0.08 °C/10 years. The stability of average annual temperatures (0.08 °C/10 years) in Makhachkala was observed against the background of an increase in absolute maximums of temperatures, and an equally significant decrease in absolute minimums of temperatures.

It is probably due to regional features of the terrain: large water bodies (Black Sea, Caspian Sea) and snow massifs in the high-mountain zone that smooth out the amplitude of the change in the mean annual temperature.

During periods of seasonal average temperatures in all climatic conditions from 1961 to 2011, a stable pattern of temperature growth rates was observed in summer seasons: from 0.26 °C/10 years ($D = 18\%$) in the Caspian zone to 0.37 °C/10 years ($D = 22\%$) in the foothill zone.

The results of calculations of the Hurst exponent for determining the trend-persistence of precipitation and temperature series are presented in Table 4. Table 4 shows that the persistence indicators for temperature trends significantly exceeded the values for precipitation, and it characterizes the persistence and long-term changes in the temperature regime. The highest persistence trends have been observed at average annual, summer temperatures ($H = 0.80$), and autumn temperatures ($H = 0.73$). The annual, summer ($H = 0.75$), and autumn maximums ($H = 0.70$), in addition to spring minimum temperatures ($H = 0.72$) have had persistent trends. Since fractality is associated with determinism [41], it can be assumed that the summer warming observed in recent decades is a consequence of the coordinated influence of a number of climate-forming factors.

Table 4. Hurst exponent for climatic characteristics according to the data from 20 weather stations in the southern Russia.

Meteoparameters	Hurst Exponent (Standard Deviation), H (δ)				
	Winter	Spring	Summer	Autumn	Annual
1	2	3	4	5	6
Precipitation total	0.63 (0.06)	0.62 (0.07)	0.61 (0.07)	**0.68 (0.06)**	0.64 (0.09)
Daily maximums of precipitation	0.62 (0.07)	0.59 (0.08)	**0.63 (0.06)**	**0.63 (0.07)**	0.62 (0.06)
Average temperature	0.70 (0.05)	0.64 (0.08)	**0.80 (0.05)**	0.73 (0.05)	0.80 (0.06)
Maximum temperature	0.66 (0.07)	0.64 (0.06)	**0.75 (0.05)**	0.70 (0.08)	0.75 (0.04)
Minimum temperature	0.68 (0.08)	**0.72 (0.08)**	0.68 (0.09)	0.61 (0.08)	0.67 (0.08)

Previous studies [31] have found that, with the exception of the steppe and Black Sea zones, seasonal increase in precipitation amounts prevail. At the same time, an autumn increase in precipitation amounts was observed at all stations without exception. For annual sums of precipitation, in 27% of the steppe stations studied and in 10% of foothill stations, they decreased, for all other meteorological stations an increase was observed in annual sums of precipitation. In this case, Table 4 shows that in the autumn, for all stations the highest Hurst exponent value was observed both for the sum of precipitation and for the daily maximums $H = 0.63$, which characterized the persistence of the identified trends for a long period.

We used the data from Tables 3 and 4 to visualize changes in summer temperature regimes with indicators characterizing the statistical stability and persistence of the detected changes. Figure 4 shows the rates of change in the mean annual temperature (°C/10 years) with the coefficient of determination (D, %) for different climatic zones, against the background of the distribution fields of the Hurst exponent (H). It is seen from Figure 4 that the increase in average summer temperatures is not only distinguished by its high values in all climatic zones, but by the highest Hurst coefficients that characterize the persistence of the obtained trends. In the summer season, Hurst's exponent had a small spread of values from $H = 0.79$ in the steppe and Caspian zones, to $H = 0.82$ in the foothill, mountainous areas, from which a steady increase in average summer temperatures should be expected. The highest Hurst exponents characterized the process of growth of these temperatures as persistent, having a long-term memory, with a high probability of continuing in the future.

Thus for average, maximum, and minimum temperatures, the Hurst exponent lies in the range $0.5 < H < 1$, and all processes belong to the class of persistent ones that preserve the existing trends. Such a long-term memory takes place regardless of time scale. All annual changes are correlated with all future annual changes. The existing trend of rising surface air temperatures will continue in the future, at least for the next 50 years (for the period analyzed using the rescaled range method). Since fractality is related to determinism [34,36], it can be assumed that the increase in temperature observed in recent decades is a consequence of the coordinated effect of a variety of climate-forming factors, both natural and anthropogenic.

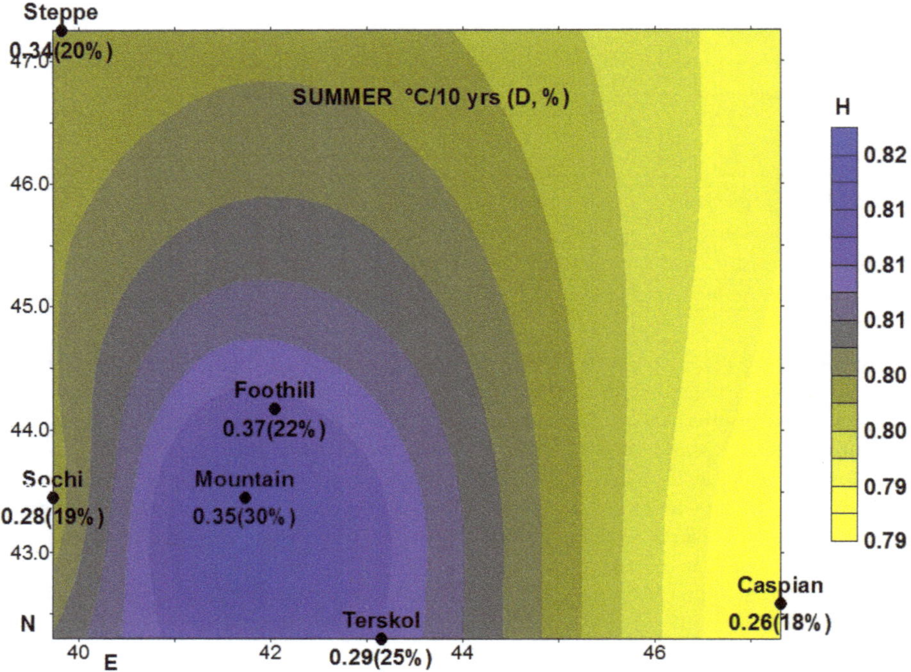

Figure 4. The rates of change in the average temperature (°C/10 years) with the coefficients of determination (*D*, %) in the summer season for different climatic zones, against the background of the trend-persistence field (the Hurst exponent *H* in color).

4. Conclusions

This study allowed us to approach the question different climatic zones sensitivity (plains, foothills, mountains) to climate change, and whether warming differs in mountains vs. plains. This is important because changes in temperature and precipitation in the mountains lead to significant changes in the hydrological cycle, in particular, to the observed process of deglaciation of the Caucasus glaciers. Furthermore, climate change in the areas of intensive farming in the south of Russia (plain, foothill) must be taken into account when solving problems in agriculture.

According to our study results, and taking into account previous studies of climate change in the Caucasus region, it became clear that the warming tendency prevails in all climatic zones with some features. Changes in the precipitation regime are not so unidirectional.

1. In the period 1961–2011, the average annual temperature in the entire territory of the southern Russia increased by 0.2 °C/10 years (*D* = 13%) with the most persistent trend being during the summer season (0.33 °C/10 years, *D* = 23%).

2. In all climatic zones of southern Russia, with the exception of the Black Sea zone (Sochi) and the high-mountain (Terskol), average annual and season temperatures have increased during 50 years of observation. In Sochi and Terskol, a statistically significant increase was observed only at average summer temperatures.

Since the beginning of global warming (since 1976), there has been the significant increase in the growth of average, maximum, and minimum temperatures in all climatic zones.

3. A change in the precipitation regime does not manifest itself as clearly as changes in temperature. During all seasons, increase and decrease in seasonal precipitation amounts occurred, but statistically insignificant. In all climatic zones, an increase in the amount of precipitation was observed in the autumn season, which was statistically significant in the steppe region.

4. Based on the study of the fractal properties of time series of precipitation and air temperature in the surface layer of the atmosphere in all climatic zones of southern Russia, it is shown that the Hurst exponent of the temperature (H = 0.74) trend significantly exceeds the Hurst exponent of the trend of precipitation (H = 0.63). It characterizes the persistence and long-term changes in the temperature regime. Of these, the trends of the average annual, summer (H = 0.80), and autumn (H = 0.73) temperatures are allocated. The changes in the maximum annual, summer (H = 0.75) and autumn (H = 0.70) temperatures, as well as minimum temperatures in spring (H = 0.72. High Hurst performance characterizes the process of increasing temperatures as constant, in contrast to increasing precipitation, having a long-term memory and with a high probability of continuation in the future.

5. Spatial correlation analysis performed for mean temperatures and precipitation sums of all climatic zones showed a high correlation (R = 1.0–0.7) between average temperatures of different climatic zones at distances of 0–1000 km between stations, and a decrease in correlation from 1 to 0.5 between precipitation sums at a scale of distances from 0 to 200 km.

Author Contributions: A.A.T. have formulated the statement of the problem and the main directions of research. Conclusions on the results of research are formulated. The final version of the article was edited before publication. B.A.A. have developed the algorithm necessary to solve the task. The analysis of the dynamics of changes in the temperature and precipitation regime was carried out. The first version of the article was formed before publication. L.A.K. time series of meteorological parameters were computed by methods of mathematical statistics. The obtained statistics of the meteorological parameters are formed into tables, their description is given. N.V.T. an array of data was prepared, a check was made for the homogeneity of time series. Graphs with linear trends, graphs with anomalies of meteoparameters are constructed.

Funding: This research received no external funding.

Acknowledgments: The authors are grateful to the reviewers for comments on the work, which led to a better presentation of the material.

Conflicts of Interest: The authors declare no conflicts of interest.

References

1. Hansen, J.; Sato, M.; Ruedy, R.; Lacis, A.; Oinas, V. Global warming in the twenty-first century: An alternative scenario. *Proc. Natl. Acad. Sci. USA* **2000**, *97*, 9875–9880. [CrossRef] [PubMed]
2. Jones, P.D.; Moberg, A. Hemispheric and large-scale surface air temperature variations: An extensive revision and an update to 2001. *J. Clim.* **2003**, *16*, 206–223. [CrossRef]
3. Hansen, J.; Ruedy, R.; Sato, M.; Lo, K. Global surface temperature change. *Rev. Geophys.* **2010**, *48*, RG4004. [CrossRef]
4. Broecker, W. When climate change predictions are right for the wrong reasons. *Clim. Change* **2017**, *142*, 1–6. [CrossRef]
5. Met Office. *Climate: Observation. Projections and Impacts (Russia) 2011*; Met Office: Exter, UK, 2011; 140p.
6. Bulygina, O.N.; Razuvaev, V.N.; Korshunova, N.N.; Groisman, P.Y. Climate variations and changes in extreme climate events in Russia. *Environ. Res. Lett.* **2007**, *2*, 045020. [CrossRef]
7. Gruza, G.V.; Rankova, E.Y. Detection of climate change: State, variability and extremes of climate. *Meteorol. Hydrol.* **2004**, *4*, 50–66.
8. Gruza, G.V.; Rankova, E.Y. Assessment of oncoming climate changes in the territory of the Russian Federation. *Meteorol. Hydrol.* **2009**, *11*, 15–29.
9. Gruza, G.V.; Rankova, E.Y. *Observed and Expected Climate Change in Russia*; FGBU "VNIIGMI-WDC": Moscow, Russia, 2012; 198p.
10. Ashabokov, B.A.; Bischokov, R.M.; Zherukov, B.K.; Kalov, K.M. *Analysis and Forecast of Climatic Changes of the Precipitation and Air Temperature Regime in Different Climate Zones of the North Caucasus*; KBNC RAN: Nalchik, Russia, 2008; 182p.
11. Mokhov, I.I. *Features of the Formation of Summer Heat 2010 On The European Territory of Russia in the Context of General Climate Changes And Its Anomalies Izvestiya of the Russian Academy of Sciences. Physics of the Atmosphere and Ocean*; Nauka Publishers: Moscow, Russia, 2011; Volume 47, pp. 709–716.

12. IPCC. Summary for Policy-makers of the Special Report. In *Managing the Risks of Extreme Events and Disasters to Advance Climate Change Adaptation*; Field, C.B., Barros, V., Stocker, T.F., Qin, D., Dokken, D.J., Ebi, K.L., Mastrandrea, M.D., Mach, K.J., Plattner, G.-K., Allen, S.K., et al., Eds.; Special Report of Working Groups I and II of the Intergovernmental Panel on Climate Change; The Intergovernmental Panel on Climate Change: Cambridge, MA, USA, 2012.

13. Rao, V.B.; Franchito, S.H.; Gerólamo, R.O.P.; Giarolla, E.; Ramakrishna, S.S.V.S.; Rao, B.R.S.; Naidu, C.V. Himalayan warming and climate change in India. *Am. J. Clim. Change* **2016**, *5*, 558–574. [CrossRef]

14. Van Wijngaarden, W.A.; Mouraviev, A. Seasonal and annual trends in australian minimum/maximum daily temperatures. *Open Atmos. Sci. J.* **2016**, *10*, 39–55. [CrossRef]

15. Stefanidis, S.; Stathis, D. Spatial and temporal rainfall variability over the Mountainous Central Pindus (Greece). *Climate* **2018**, *6*, 75. [CrossRef]

16. Bhuyan, M.D.I.; Islam, M.M.; Bhuiyan, M.E.K. A trend analysis of temperature and rain-fall to predict climate change for north-western region of Bangladesh. *Am. J. Clim. Change* **2018**, *7*, 115–134. [CrossRef]

17. Hasanean, H.; Almazroui, M. Rainfall: Features and variations over Saudi Arabia. A review. *Climate* **2015**, *3*, 578–626. [CrossRef]

18. Federal Service for Hydrometeorology and Environmental Monitoring (Roshydromet) (Ed.) *A Report on Climate Features on the Territory of the Russian Federation in 2017*; Roshydromet: Moscow, Russia, 2018; 69p, ISBN 978-5-906099-58-7. Available online: http://climatechange.igce.ru/index.php?option=com_docman&Itemid=73&gid=27&lang=ru (accessed on 30 April 2018).

19. Rybak, O.O.; Rybak, E.A. Climate change in the south of Russia: Tendencies and possibilities for prediction. *Sci. J. Kuban State Univ.* **2015**, *111*, 3–18.

20. Abshaev, M.T.; Malkarova, A.M.; Borisova, N.V. On Climate Change Trends in the North Caucasus. In Proceedings of the World Conference on Climate Change, Moscow, Russia, 29 September 2003; Abstracts of the World Conference on Climate Change; pp. 365–366.

21. Ashabokov, B.A.; Bischokov, R.M.; Fedchenko, L.M.; Kalov, K.M.; Bogachenko, E.M. *Analysis and Forecast of the Climate in the Kabardino-Balkarian Republic*; KBSAA: Nalchik, Russia, 2005; 150p.

22. Ilyin, Y.P.; Repetin, L.N. The Age-Old Changes in Air Temperature in the Black Sea Region and Their Seasonal Features. In *Ecological Safety of the Coastal and Shelf Zone: Collection of Scientific Papers*; Marine Hydrophysical Institute of Russian Academy of Sciences: Sevastopol, Russia, 2006; pp. 433–438.

23. Gruza, G.V.; Meshcherskaya, A.V. Changes in the Climate of Russia During the Period of Instrumental Observations. In *An Assessment Report on Climate Change and Their Consequences on the Territory of the Russian Federation*; Gruza, G.V., Zaitsev, A.S., Karol, I.L., Kattsov, V.M., Anisimov, O.A., Anokhin, Y.A., Boltneva, L.I., Vaganov, E.A., Zolotokrylin, A.N., et al., Eds.; Roshydromet: Moscow, Russia, 2008; 288p.

24. Atakuyev, Z.K.; Bekkiyev, M.Y.; Dokukin, M.D.; Kalov, K.M.; Kalov, R.K.; Tashilova, A.A.; Khatkutov, A.V. *Influence of Climate Changes on Dynamics of Glaciers, Environmental Processes and Recreational Complexes near the Mt. Elbrus*; Doklady Adygskoy (Cherkesskoy) Mezhdunarodnoy Akademii Nauk (AMAN): Cheresskoy, Russia, 2016; Volume 18, pp. 61–79.

25. Rybak, E.A.; Rybak, O.O.; Zasedatelev, Y.V. Complex geographical analysis of the greater Sochi region on the Black Sea coast. *GeoJ.* **1994**, *34*, 507–513.

26. Toropov, P.A.; Mikhalenko, V.N.; Kutuzov, S.S.; Morozova, P.A.; Shestakova, A.A. Temperature and radiation regime of glaciers on slopes of the Mount Elbrus in the ablation period over last 65 years. *Ice Snow* **2016**, *56*, 5–19. [CrossRef]

27. Solomina, O.; Dolgova, E.; Maksimova, O. Tree-ring based hydrometeorological reconstructions in Crimea, Caucasus and Tien-Shan. *M. SPb. Nestor History* **2012**, *232*.

28. Baranov, S.; Pokrovskaya, T. *The Work of the Meteorological Group of the Elbrus Expedition 1935. Proceedings of the Elbrus Expedition 1934 and 1935 M-L*; Academy of Sciences of the USSR: Moscow, Russia, 1936; 350p.

29. Voloshina, A.P. *Radiation Conditions During the ablation Period. The Icing of Elbrus*; Moscow State University: Moscow, Russia, 1968; 326p.

30. Pachauri, R.K.; Meyer, L.A. *Climate Change: Synthesis Report*; Contribution of Working Groups I, II and III to the Fifth Assessment Report of the Intergovernmental Panel on Climate Change; IPCC: Geneva, Switzerland, 2014; 151p.

31. Ashabokov, B.A.; Tashilova, A.A.; Kesheva, L.A.; Taubekova, Z.A. Trends in precipitation parameters in the climate zones of Southern Russia (1961–2011). *Rus. Meteorol. Hydrol.* **2017**, *42*, 150–158. [CrossRef]

32. Tashilova, A.A.; Kesheva, L.A.; Taubekova, Z.A.; Pshikhacheva, I.N. Analysis of the dynamics of the regime of total and maximum daily precipitation according to the data of the Buinaksk weather station (1961–2011). *Izvestiya Vuzov. North-Caucasian Region. Natural Sciences. No.2 (180), Rostov-on-Don, Russia* **2014**, 48–53.

33. 16th Session of the Commission for Climatology (CCl-16), Heidelberg, Germany, 3–8 July 2014. Available online: www.ccl-16.wmo.int (accessed on 1 June 2018).

34. Isayev, A.A. *Statistics in Meteorology and Climatology*; Izdatelstvo MSU: Moscow, Russia, 1988; 248p.

35. Byuyul, A.; Tsefel, P. *SPSS: The Art of Processing Information. Statistic Data Analysis*; SPb.: St. Petersburg, Russia; DiaSoft YUP: Moscow, Russia, 2002; 608p.

36. Borovikov, V. *Statistica. The Art of Data Analysis on the Computer*, 2nd ed.; SPb.: St. Petersburg, Russia, 2003; 688p.

37. Silkin, K.Y. *Geoinformation System Golden Sortware Surfer 8*; Publishing and Printing Center of Voronezh State University: Voronezh, Russia, 2008; 66p.

38. Hurst, H.E.; Black, R.P.; Simaika, Y.M. *Long-Term Storage: An Experimental Study*; Constable: London, UK, 1965; 240p.

39. Mandelbrot, B.B. The fractal geometry of trees and other natural phenomena. Buffon Bicentenary Symposium on Geometrical Probability. In *Lecture Notes in Biomathematics*; Miles, R., Serra, J., Eds.; Springer: New York, NY, USA, 1978; Volume 23, pp. 235–249.

40. Mandelbrot, B.B. *The Fractal Geometry of Nature*; W. H. Freeman and Company: New York, NY, USA, 1982; 656p.

41. Feder, J. *Fractals*; Trans. from English; Mir: Moscow, Russia, 1991; 254p.

42. Solntsev, L.A.; Iudin, D.I.; Snegireva, M.S.; Gelashvili, D.B. Fractal Analysis of the Secular Variation of the Average Air Temperature in Nizhny Novgorod. *Bull. Nizhny Novgorod Univ. Named Lobachevsky* **2007**, *4*, 88–91.

43. Butakov, V.; Grakovsky, A. Estimation of the stochasticity level of intermediate series of arbitrary origin using the Hurst Index. *Comput. Modell. New Technol.* **2005**, *9*, 27–32.

44. Lovejoy, S.; Schertzer, D. Multifractals and rain. In *New Uncertainty Concepts in Hydrology and Hydrological Modelling*; Kundzewicz, Z.W., Ed.; Cambridge University Press: Cambrigde, UK, 1995; Volume 199, pp. 109–113.

45. Ashabokov, B.A.; Tashilova, A.A.; Kesheva, L.A.; Teunova, N.V.; Taubekova, Z.A. Climatic changes in mean values and extremes of near-surface air temperature in the south of European Russia. *Fundam. Appl. Clim.* **2017**, *1*, 5–19. [CrossRef]

46. Federal Service for Hydrometeorology and Environmental Monitoring. *Roshydromet's Second Assessment Report on Climate Change and its Consequences on the Territory of the Russian Federation*; Rosgidromet: Moscow, Russia, 2013; p. 1009.

47. Gruza, G.V.; Rankova, E.Y.; Rocheva, E.V.; Samokhina, O.F. Features of the surface temperatures anomalies over the globe in the 2016. *Fundam. Appl. Clim.* **2017**, *1*, 124–130. [CrossRef]

Article

Assessing Heat Waves over Greece Using the Excess Heat Factor (EHF)

Konstantia Tolika

Department of Meteorology and Climatology, School of Geology, Aristotle University of Thessaloniki, Thessaloniki, 54124, Greece; diatol@geo.auth.gr; Tel.: +30-2310-998404

Received: 23 November 2018; Accepted: 29 December 2018; Published: 7 January 2019

Abstract: Heat waves are considered one of the most noteworthy extreme events all over the world due to their crucial impacts on both society and the environment. For the present article, a relatively new heat wave index, which was primarily introduced for the study of extreme warming conditions over Australia (Excess Heat Factor (EHF, hereafter)), was applied over Greece (eastern Mediterranean) for a 55-year period in order to examine its applicability to a region with different climatic characteristics (compared to Australia) and its ability to define previous exceptional heat waves. The computation of the EHF index for the period 1958–2012 demonstrated that, during the warm period of the year (June, July, August, and September (JJAS)), Greece experiences approximately 20 days per year with positive anomalous conditions (EHF > 0) with positive statistically significant trends for all stations under study. Moreover, an average of 128 spells with a duration of 3 to 10 consecutive days with positive EHF values were found during the examined 55-year period. As the duration of the spell was extended, their frequency lessened. Finally, it was found that the EHF index not only detected, identified, and described efficiently the characteristics of the heat waves, but it also provided additional useful information regarding the impact of these abnormal warming conditions on the human ability to adapt to them.

Keywords: temperature; heat wave; excess heat factor; acclimatization; Greece

1. Introduction

Heat waves have been a phenomenon of great worldwide interest due to their substantial societal and environmental impacts. These impacts intensify the necessity of measuring, studying, and even predicting these extreme hot conditions especially in the impacted communities and the affected regions [1] because remarkably warmer weather can have a direct negative effect on health, especially for the vulnerable elderly population [2–5]. There is also a global demonstration that extreme temperatures are highly correlated with human mortality [6–8], making heat waves one of the natural hazards with the greatest percentage of casualties [3,9]. Langlois et al. [10] mention that even though people tend to adapt and acclimatize themselves to temperature changes, if this change is sudden and abrupt, it can then cause certain heat-related diseases or even death.

Nevertheless, there is no single and standard definition of the physical nature of a heat wave and their overall description remains quite broad [11]. Heat waves are usually described as periods of exceptionally hot weather. However, the intensity of this temperature rise as well as the duration of the extreme warm consecutive days and the time of year that they occur are important aspects necessary to categorize a hot event as a heat wave. In general, a heat wave is an acute period of extreme warmth during the summer months, whereas the respective hot periods during winter are referred to as warm spells [12].

In order to define suitable metrics for waves, scientists have instituted either absolute or relative approaches [13]. Even though experts differ in the selection of thresholds and duration, the first approach is based on the meteorological/climatological values of certain parameters, such as daily mean temperature, maximum and minimum temperatures, temperature indices, duration, and relative humidity, whereas the relative approach also incorporates human acclimatization to weather and uses more human-related bioclimatic indices (e.g., [14–17]). Thus, the diverse definitions of a heat wave mainly depend on the scope that is being studied. If the climatological-statistical characteristics of these extreme hot events are of primary interest to the researcher, then straightforward metrics are being used. On the other hand, if the study is more human-centered, then the impact of the heat wave on people's health is the main drive and different approaches are used [11,18]. It should also be mentioned that due to the fact that most of the heat wave indices are developed for a specific use and a specific target group or sector, they are most of the time not flexible and cannot be applied to different regions or for different purposes [1].

Moreover, since temperature is increasing on a global scale, the interest concerning heat waves is also increasing as they are expected to become more frequent, more intense, and of longer duration [19,20]. Especially with respect to the Mediterranean region, which will probably experience a much larger number of heat waves in the future, particularly during the summer months (e.g., [21–25]), the need to define these extreme events efficiently becomes more and more urgent, due to their severe impacts on several aspects of human lives [26–28]. In Greece, which is the center of interest in the present study, heat waves have been analyzed by several researchers using different approaches, methodologies, and metrics either from a statistical or a more bioclimatological point of view (e.g., [24,29–33]).

However, in this study, an attempt was made to carry out an in-depth analysis of Greece's heat waves with a relatively new index, developed primarily for assessing heat waves in Australia [34] but which had recently been applied to the Czech Republic [35] and the Balkan Peninsula (Romania), where Greece is also located [3]. This index, defined as the excess heat factor (EHF) and described in detail in the next paragraph, is actually a set of indices, whose major advantage is that it combines both the statistical and the human-impact aspects of the heat wave. Moreover, with respect to temperature, not only maximum but also minimum temperatures were used for their definition. Adding Tmin on a heat wave index is not only climatologically tempting [1], but high minimum temperature values intensify the heat wave conditions, also increasing the degree of heat stress [34,36]. In addition, Karl and Knight [37] underlined that no relief from high minimum temperatures, for more than three consecutive days, could have crucial impacts on human health. Finally, another advantage of the excess heat factor index is that it takes into consideration not only the temperature conditions of the specific day but also of the previous two ones, which can intensify or reduce the heat wave's magnitude [1].

In the next section of this study, the methodology for the EHF index computation is analyzed as well as the data that are being used. Moreover, the statistical characteristics and results of the index are presented for the stations under study as well as the assessment of the EHF's ability to define and describe two representative heat waves (July 1987 and July 2007) that occurred in Greece during the past few years. Finally, the conclusions derived from the study as well as a literature discussion of them can be found in the last station of this research article.

2. Materials and Methods

The identification and determination of heat waves over the study area (Greece) was achieved using the computation of the excess heat factor (EHF) index, which provides a measure of the environmental temperature load [10] and the intensity of a potential heat wave [12]. The EHF is the product of the multiplication of two other excess heat indices (EHIs), namely, EHI_{accl} and EHI_{sig}.

$$EHF = EHI_{sig} \times \max\{1, EHI_{accl}\} \tag{1}$$

EHI_{accl} is defined as:

$$EHIaccl = \left[\frac{T_i + T_{i-1} + T_{i-2}}{3} \right] - \left[\frac{T_{i-3} + \ldots T_{i-32}}{30} \right], \tag{2}$$

where the first term of Equation (2) is the average daily Tmean for a three-day period and the second term is the average daily Tmean of the preceding thirty days. As proposed by Nairn et al. [34], this is an acclimatization index and its positive (negative) values are related to hot (cold) weather conditions. It determines a period of heat that is warmer than the recent past [10], and it should be highlighted that this index is not influenced by the potential general warming trend [34]. This index describes an important factor of the influence of heat to the population because, even though humans tend to acclimatize themselves to their environmental local climate according to the temperature variations throughout the year, they may be unprepared to an abrupt temperature rise above that of the recent past [12]. Thus, positive EHI_{accl} values indicate a lack of acclimatization to the warmer temperatures which may result in negative health impacts.

The second term of the EHF equation is the significance index (EHI_{sig}) defined by the following equation:

$$EHIsig = \left[\frac{T_i + T_{i-1} + T_{i-2}}{3} \right] - T_{95}, \tag{3}$$

which is calculated by the difference of the average daily Tmean for a three-day period minus the 95th percentile of the daily Tmean. The percentile is computed for a reference period of 30 years (1971–2000) using the daily values of the mean temperatures for all days throughout the year. A heat wave occurs when EHI_{sig} is positive, while the comparison with the 95th percentile measures the statistical significance of the heat event [30]. The authors also underline the fact that, since T_{95} is computed for a fixed climatological period, the EHI_{sig} (contrary to the EHI_{accl}) is expected to become more extreme under a general warming trend.

Therefore, the excess heat factor expresses the long-term temperature anomalies amplified by the short-term ones [10], and the days with positive EHF values indicate heat wave conditions, while the higher the values of the index, the more intense is the heat wave. However, according to Perking and Alexander [1], a heat wave episode will be defined when at least three consecutive days present EHF values above zero.

Finally, daily Tmean in all the above indices should be computed by averaging the Tmin and Tmax daily values since the diurnal temperature variation is highly associated with the ability of the biological systems to recover from high heat load. Hence, for this study, the daily Tmin and Tmax time series, derived from 14 meteorological stations, were used for a 55-year period starting from 1958 to 2012, for the computation of the Tmean values. Except for the data for the Thessaloniki station which were available from the meteorological station of the Aristotle University of Thessaloniki (AUTh), the remaining station data were provided by the Hellenic National Meteorological Service (HNMS). These data were proven to be homogenous according to the Alexandersson test [38] and had no gaps. The geographical distribution of the station locations is presented in Figure 1.

It is worth mentioning at this point that although abnormally warm conditions may occur even during the winter months, it was decided in this case to compute the EHF index for the hottest period of the year, that is, June, July, August, and September (the JJAS period) since summer heat waves tend to be more intense with severe impacts for humans during these months. Finally, adopting the definition used by Perkins and Alexander [1] who mentioned that a heat wave occurs when the abnormally hot conditions (EHF > 0) persist for more than three consecutive days, the duration of spells longer than 3 days was calculated for the 14 stations under study during the 55-year time period.

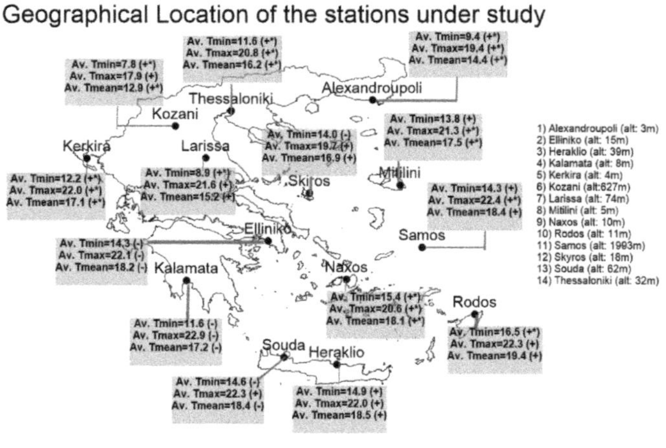

Figure 1. Geographical distribution of the stations under study. For each station, the altitude where it is located can be found on the right of the map. For each station, the average Tmin, Tmax, and Tmean values for the period 1958–2012 are provided. The sign of the trend of these time series is found in the brackets (). The asterisk indicates the statistical significance of the trends at the 95% level of significance.

3. Results

3.1. Statistical Analysis and Aspects of the EHF Index

Primarily required for the definition of the EHI_{sig} index, the 95th percentile of the Tmean was computed for a standard period of 30 years from 1971 to 2000. It was found that the T_{95} values varied from 24.8 °C to 28.7 °C. The lowest percentile values were observed over Kozani, a station in the west continental part of Greece, followed by Alexandroupoli (25.4 °C) in the north. Kozani is a typical continental station, of quite high altitude (400 m), and that is the main reason for the low Tmean 95th percentile values found. On the other hand, the second minimum of Alexandroupoli can be explained by the fact that this station is found in the northeastern part of Greece, and even though it is a coastal station, it presents an intense continental influence especially during winter, which explains the low 95th percentiles. The highest values were found in Elliniko in central Greece (28.5 °C) and the maximum was found in Samos, an island station over the southeastern Aegean Sea (28.7 °C) (Figure 2).

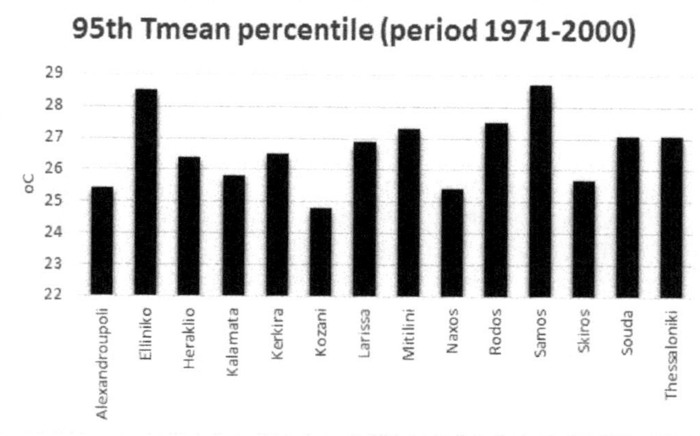

Figure 2. The 95th Tmean percentiles for the stations under study for the period 1971–2000.

As mentioned in the previous section, the analysis focused on the hot period of the year (JJAS) for the years from 1958 to 2012. The average number of days in the hot period of the year in Greece when the EHF index presented positive values varied from 17.9 (Larissa) to 23.9 (Kozani). This indicates that for approximately 20 days per year the stations under study experienced heat wave conditions (Table 1). In addition, although Kozani was the station with the lowest T_{95} percentile value, generally characterized by a relatively colder climate in comparison to the rest of the stations, it showed the highest average positive EHF days.

Table 1. Average, maximum and trends of the days with positive EHF values for June, July, August, and September (JJAS) for the period 1958–2012. All the trends were found statistically significant at the 95% level according to Kendall's tau test.

	Average Positive EHF Days	Max. Positive EHF Days	Trend Positive EHF Days
Alexandroupoli	23.0	72 (2012)	+0.76
Elliniko	21.7	76 (2011)	+0.41
Heraklio	18.5	51 (2010)	+0.41
Kalamata	22.8	80 (2012)	+0.65
Kerkira	21.5	71 (2012)	+0.56
Kozani	23.9	77 (2008)	+0.79
Larissa	17.9	56 (2012)	+0.44
Mitilini	21.0	63 (2007)	+0.63
Naxos	22.8	89 (2010)	+1.14
Rodos	19.4	50 (2012)	+0.34
Samos	21.0	67 (2012)	+0.89
Skiros	19.5	52 (2007)	+0.41
Souda	17.9	49 (2012)	+0.45
Thessaloniki	21.9	63 (2012)	+0.58

Moreover, the results regarding the year with the maximum number of days with EHF > 0 indicated that, for all of the stations, they occurred at the end of the examined time period, most of them being in 2012. It seems that during that year, Greece was characterized, in summer, by very intense warm conditions that lasted up to 80 days (Kalamata). However, the absolute maximum was observed in Naxos (89 days) two years earlier (2010). These abnormally hot days tended to become more frequent throughout the examined period, since positive trends were found in all the stations ranging from +3.4 days/decade in Rodos to +11.4 days/decade in Naxos (Table 1). The smaller trend in Rodos could be attributed to the geographical position of the station, in the northwest part of the island, which is highly influenced by the Etesian winds during the summer. The maximum in Naxos could also be related to the location of the station, which is more "protected" from the Etesian winds. Regardless of the trend values, it is worth mentioning that after the application of Kendall's tau test at a significance level of 95%, all of them were found statistically significant. This comes as a robust indication that the days of abnormally increased temperatures do significantly increase during the examined period and it is not just a random rise (Table 2). This finding encouraged the application of another statistical analysis, based on the application of the Mann–Kendall t test method [39], in order to identify breakpoints on the EHF time series. The results from this test are presented in Figure 3. It can be clearly seen from the normal curve in all stations that the time series of the positive EHF days present a statistically significant positive trend that exceeds the statistical significance level (95%) during the last years of the examined period. In addition, according to the criteria of this test [39], in all stations under study, an abrupt change (breakpoint) of the specific parameter is observed (a clear "X" shape between the normal and the retrograde curve). The actual year of the breakpoint is not the same in every station but it can be placed from the mid-90s until the first years of the 21st century. More specifically, the earliest breakpoint is in Samos (1993) and the latest one is in the Thessaloniki station (2002) (Figure 3).

Apart from the examination of the number of days with positive EHF values, the study of the spells of positive EHF is also included (Table 2). These heat wave spells were classified in four classes (1st class: 3 to 10 days, 2nd class: 11 to 20 days, 3rd class: 21 to 30 days, and 4th class: >30 days). In addition, the maximum spell duration was computed in order to provide a magnitude of the most extreme heat waves in terms of duration.

As expected, the most frequent spells are the ones belonging to the first class, with an average number of 122.7 spells during the years of study. The most heat waves with 3 to 10 days duration were found for the Kalamata station (146 spells) followed by the Kozani station (140 spells). These "shorter" heat waves were less frequent for two stations in the north of Greece, namely, Alexandroupoli and Kerkira with 106 and 105 spells, respectively. As the duration of the heat waves becomes longer, their frequency decreases. For the second heat wave class, the highest number of spells was found in Alexandroupoli and Elliniko (26 spells) and the lowest one was found in Larissa and Souda (15 spells), whereas for the third class the frequency of the heat waves did not exceed 7 (heat waves with a duration from 21 to 30 days) which was recorded at the Mitilini station over the eastern Aegean Sea. Regarding heat waves with a duration longer than 30 days, none were observed in the Heraklio, Larissa, Skiros, and Souda stations during the 55-year time period. On the other hand, seven (7) such heat waves were found for the Naxos station in the central Aegean Sea and five (5) were found in Kerkira in the Ionian Sea. Finally, calculating the maximum heat wave in each station during the examined period, it should be noted that during the year 2012, Kerkira and Kozani experienced 58 and 57 consecutive days, respectively, of abnormal hot conditions with positive EHF values. The maximum for this parameter (the duration of the maximum spell) is found in general over continental stations, over the western parts of the country, where the Etesian winds lack influence, during the summer months. Conversely, the minimum is observed for island stations where the sea probably plays an important role in cooling (temperature drop), especially during the night. It should also be highlighted that for most of the stations, this extremely long heat wave was detected during the last three years of the study period (2010, 2011, and 2012), especially the last year. This is in agreement with the general finding that 2012 is considered, on a planetary scale, as one of the warmest years all over the world according to the World Meteorological Organization (WMO).

Table 2. Number of heat wave spells (consecutive days with EHF values > 0.0) during the study period 1958–2012 for the 14 stations under study.

	3–10 Days	11–20 Days	21–30 Days	>30 Days	Maximum Spell Duration
Alexandroupoli	106	26	5	4	45 days (2012)
Elliniko	127	26	2	1	37 days (2011)
Heraklio	119	21	1	0	30 days (2010)
Kalamata	146	20	2	2	57 days (2012)
Kerkira	105	23	3	5	58 days (2012)
Kozani	140	23	5	1	32 days (2008)
Larissa	117	15	3	0	28 days (2012)
Mitilini	116	20	7	1	32 days (2011)
Naxos	106	19	2	7	57 days (2010)
Rodos	130	20	0	1	33 days (2012)
Samos	124	19	5	1	39 days (2012)
Skiros	134	18	2	0	25 days (2010)
Souda	132	15	0	0	18 days (1999)
Thessaloniki	116	22	5	1	38 days (2012)

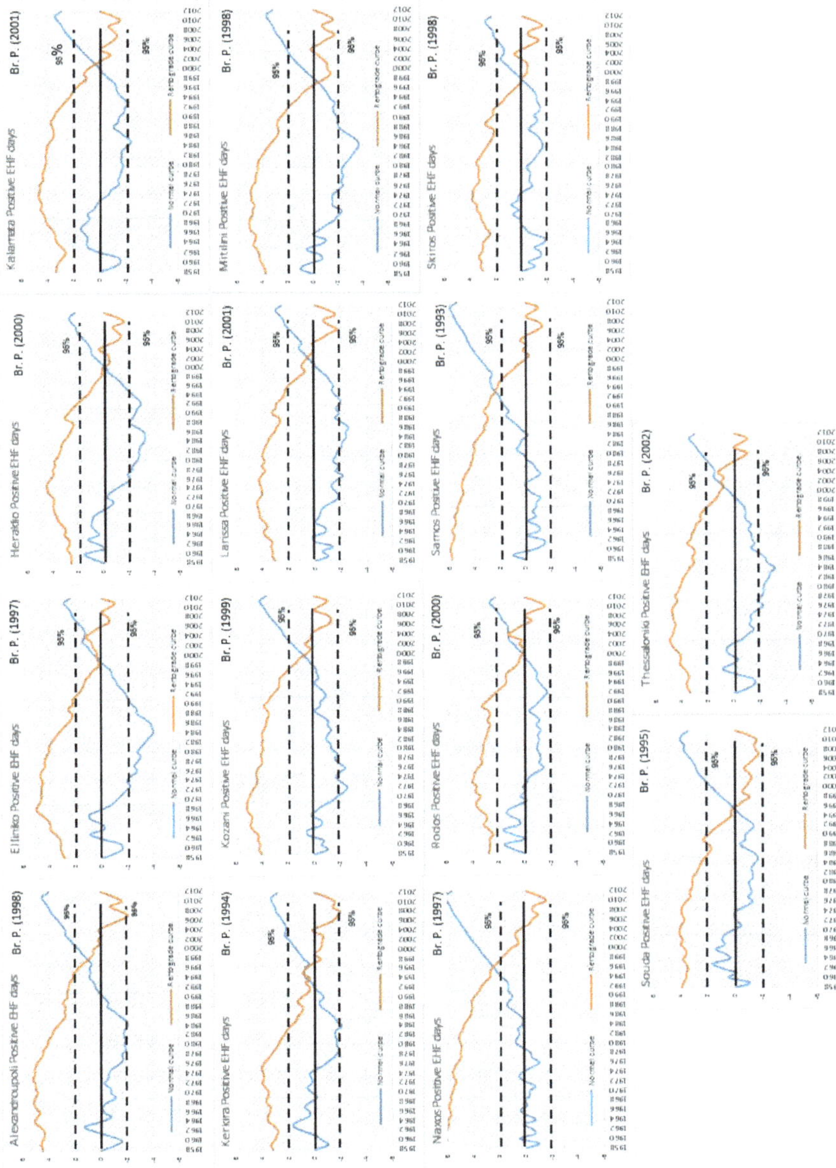

Figure 3. Mann–Kendall t test results on the statistically significant trends, at the 95% level, of the EHF positive days. In the upper right corner of each diagram, the year of the breakpoint (abrupt climatic change) is indicated in the brackets.

3.2. Examination of the EHF Index in "Capturing" Two Characteristic Heat Waves in Greece

3.2.1. The Heat Wave of the Year 1987

Listed as one of the major natural disasters by Berz [40], the heat wave of the summer of 1987 in Greece resulted in over a thousand deaths all over the country due to heat strokes, heat exhaustion, and heart-related conditions [16,30]. As a result, the heat wave of 1987 has been the subject of several studies, all agreeing that even though higher absolute Tmax values had been recorded before, the duration of these very intense and severe hot conditions was the main characteristic of this specific heat wave [29]. Using two physiological discomfort indices, Giles et al. [41] confirmed that this heat wave was continuous. For nine consecutive days, from 19 to 27 July [29], Tmax values exceeded 40 °C for most areas of Greece [41]; minimum temperatures were also relatively high and the days of the heat waves were characterized by a rather small diurnal range [29]. Matzarakis and Mayer [30] also mention that the thermal indices used in their study showed a very high thermal stress on people. Especially in the case of Athens, each afternoon was considered as "extremely hot" [31] and the heat wave was found to be more intense in northern Greece, especially in Thessaloniki [31,41].

Because of the above characteristics of the heat wave of 1987, this study now attempted to investigate if the EHF index was efficient enough to detect the heat wave, to identify it, and to provide a thorough analysis of this extremely hot summer in Greece. For this purpose, the acclimatization and the significance indices (EHI_{accl} and EHI_{sig}) as well as the EHF index were computed during the month of July 1987 for the 14 stations under study in comparison with the daily Tmean of that month and the Tmean95 of the reference period 1971–2000 (Figure 4). For all the stations, a gradual increase of the Tmean daily values was observed, rising above the Tmean95 value from 17 to 18 July. The peak of these daily values was found during days 25 to 27 of the month, and they dropped again below the 95th percentile at the end of the month (from 29 to 30 July 1987). If the study of the heat wave was based only on the daily mean temperatures that exceeded the Tmean95, then the duration of the heat wave would be defined from days 18 to 30 of the month. However, using the EHF index, additional and more detailed information was provided.

More specifically, EHI_{accl} began to have positive values much sooner, on 12–13 July, meaning that the averaged three-day temperature was higher than the recent past, according to the index definition. This indicated that there was now a lack of human acclimatization to the upcoming warmer conditions, which could result in an adverse impact on their everyday life and health [12]. Thus, even though the "actual" heat wave had not started, this rising index (EHI_{accl}) indicated that humans were not able to physically adapt to this warming, which could be used as a useful alert for heat wave policy management measures. The second computed index (EHI_{sig}) turned positive for several days (in general on 20 July). This was the starting point of the heat wave, since by definition, the EHI_{sig} values should be positive in order to consider the temperature conditions abnormally hot, higher than the 95th percentile Tmean. Both the indices dropped below zero (yellow line in Figure 4) at the end of July, either on day 29 or 30 of the month. In most of the stations, these were the same days that the daily Tmean also fell below the 95th percentile value. Overall, the examination of the combined EHF index showed that the heat wave of July 1987 started on day 20 of the month (in most of the stations under study) ending on day 30 when the EHF values were again negative. An interesting finding of the application of this index was that the peak of the heat wave was not placed on the same day that the actual daily mean temperature values reached their maximum, but one or two days later. In addition, it is worth mentioning that during this specific heat wave, not only were the temperatures exceptionally high (indicated by the EHI_{sig}) but, due to the rapid temperature increase, people did not have the chance to acclimatize themselves to these new "much hotter" conditions (positive EHI_{accl}), resulting in harmful effects on their health. Moreover, the previous studies' finding that the heat wave was more intense in northern Greece was also detected from the index application that showed higher values for the stations in the northern parts of the country. It is obvious that the definition and the description of this heat wave using the EHF match the previous findings, making it a suitable index for the study of heat waves.

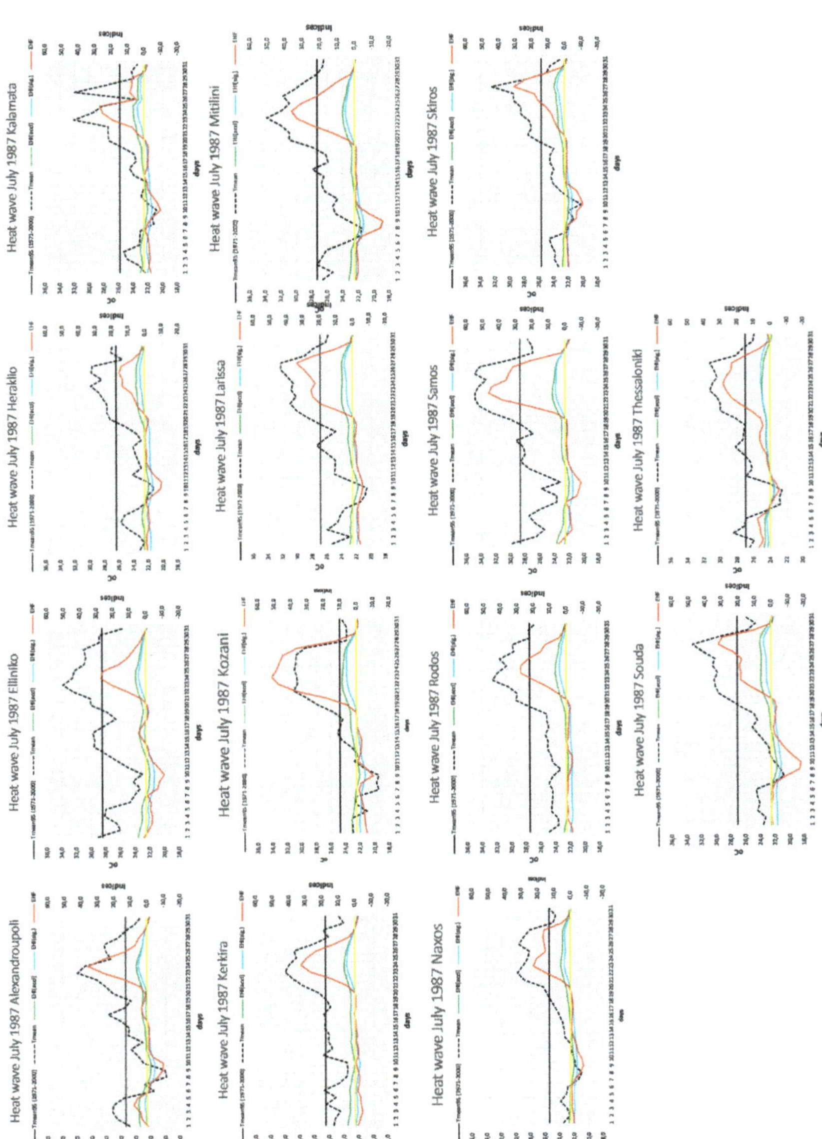

Figure 4. Day-to-day representation of the heat wave of July 1987 in Greece. The daily Tmean values as well as the Tmean95 percentile (reference period 1971–2000) are plotted in reference to the left y-axis (°C) and the three indices in reference to the right y-axis (°C). The yellow line represents zero and refers to the right y-axis.

3.2.2. The Heat Wave of the Year 2007

Occurring twenty years after the previous heat wave, the extreme hot conditions experienced during the summer of 2007 have been the subject of several climatological studies. Greece has experienced record-breaking temperatures in most of its regions [33] and the most extreme maximum temperatures appeared during the last days of each of the summer months [24], mainly between days 21 and 29. Theoharatos et al. [42] mentioned that especially in July, the daily Tmax values repeatedly exceeded 40 °C, while the regions where the population experienced discomfort (high discomfort index (DI) values) were Thessalia, Sterea Ellada, and western Pelloponisos. The impacts of this heat wave were substantial with an increase in forest fires, changes in the hydrological balance, and large losses in the agricultural and energy sectors [24,43]. Similarly to the previous paragraph, the three examined indices were computed on a daily basis during July 2007 in order to evaluate their ability to capture and describe the extreme temperatures in Greece during that year (Figure 5).

In most of the stations used in this study, the extreme hot conditions seemed to have started from days 15 to 17 of the month when the daily Tmean exceeded the 95th percentile. However, in most of the stations, the days approximately from 7 to 10 July also surpassed the Tmean95. According to the actual index values, the onset of the heat wave was determined one or two days later from the dates of the Tmean > 95th percentile, 16–18 July, and continued until the end of the month. This was the starting point where the EHI_{sig} was positive, a necessary condition for the determination of a heat wave. Regarding the acclimatization index, it started to present positive values on the same dates as the EHI_{sig}. Yet, in some stations (Souda, Skiros, Naxos, Mitilini, Alexandroupoli, Heraklio, Kalamata), it went negative sooner than the actual ending of the heat wave. This means, according to Nairn and Fawcett [12] that in some cases of longer heat waves, there may be some human adaptation to the extreme hot conditions, decreasing their impact on human health. For example, in the case of Skiros, EHI_{accl} dropped below zero on 28 July, whereas the end of the heat wave occurred two days later. This characteristic of the heat wave, due to its duration, was not observed in the previous heat wave case (July 1987).

Finally, one of the main issues observed in the computation of the EHF index was its magnitude which differed substantially from one station to the other. This means that even though all of Greece experienced heat wave conditions, its intensity varied considerably (Figure 5). The maximum index values were found for the stations of Kozani, Kerkira, Larissa, and Samos and the lowest ones were found in Rodos. On the other hand, the findings from all stations were in agreement that the peak (the highest EHF value) was found either on day 25 or 26 of the month and that the heat wave then started to wane.

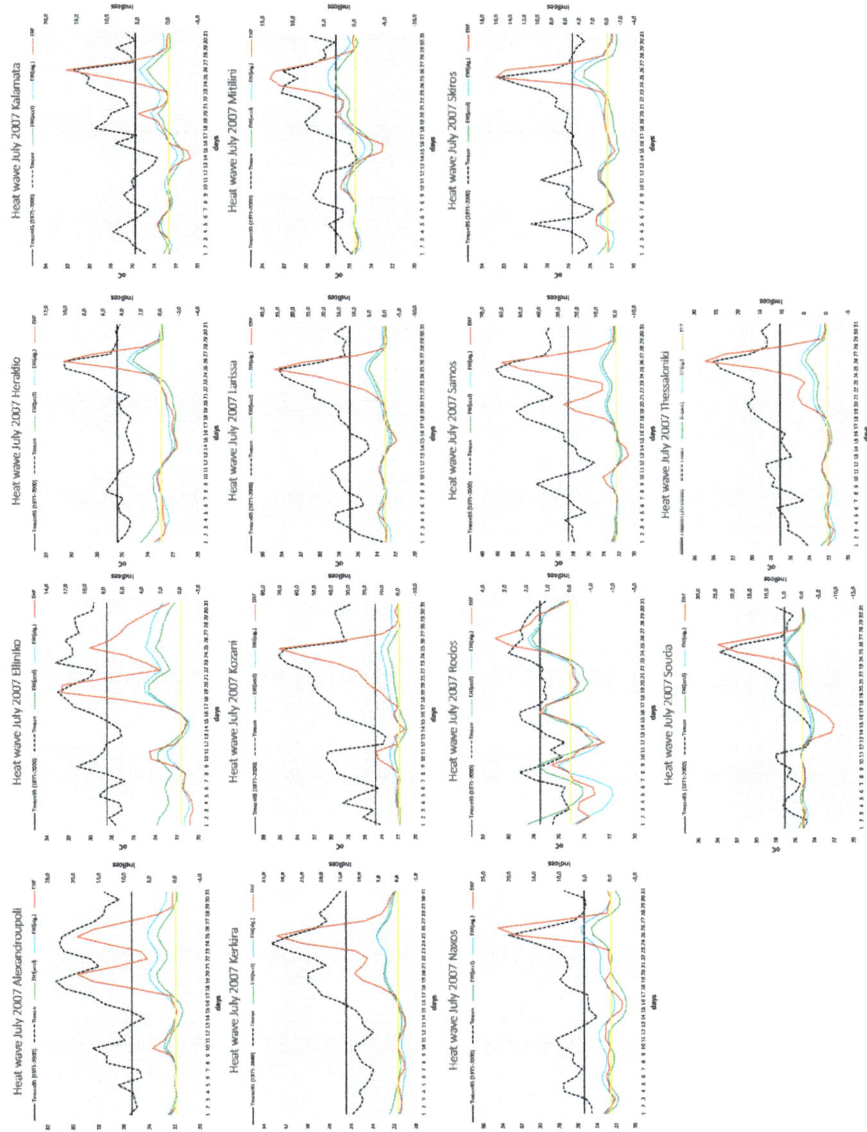

Figure 5. Same description as in Figure 4 but for the heat wave of July 2007 in Greece.

4. Discussion and Conclusions

The statistical characteristics of the combined heat wave index (excess heat factor (EHF)), as well as its ability to identify and describe efficiently previously recorded heat waves over Greece, was the main objective of the present study. Daily maximum and minimum temperature time series from 14 stations distributed over the geographical Greek region were employed for a 55-year period (1958–2018). This index introduced relatively recently [34] had, to the author's knowledge, never been used before in Greece. Although it was primarily proposed for the monitoring of heat waves in Australia [10,12,34], it was also used during 2018 for heat wave detection in another Balkan country, Romania [3].

Apart from the fact that it is the first time that the EHF index is being used over the center of interest, the main reasons for its selection are the following:

- It incorporates not only Tmax but also Tmin values which is an important parameter that should be taken under consideration especially if the heat wave study is more human-centered. Temperature rise is undoubtedly related to human health; however, the overnight high temperature impedes the night discharge and results in excess heat stress and an enhancement of the heat wave conditions [1,44–46].
- It measures daily temperature values for a three-day period (average) rather than the single temperature of one day, making it more sensitive to temperature changes as it considers also previous day conditions. [1].
- The first term of the EHF (EHI_{sig}) provides a measure of the statistical significance of the heating as it is compared with a fixed percentile value (95th percentile of a defined period which, in this case, was chosen to be 1971–2000).
- The second term (EHI_{accl}) compares the examined warm conditions with the recent past (previous 30 days) providing an indication of the people's acclimatization ability to this unusual heat. Conversely to the EHI_{sig}, EHI_{accl} does not change under a general climate-change warming [12].
- Overall, the application of the EHF can provide information both about the statistical characteristics of a heat wave but also about its effect on humans.

The computation of the EHF index for the period 1958–2012 demonstrated that, during the warm period of the year (JJAS), Greece experienced approximately 20 days per year with positive anomalous conditions (EHF > 0) with no discrepancies worth mentioning among the stations. In addition, these days tended to become more frequent (with positive statistically significant trends for all stations under study), agreeing with the general consideration of an increase in extreme hot events in the future over the Mediterranean [20]. The years that had the largest number of days with positive EHF values were the last ones of the time period used (2007 to 2012). Up to 89 days in Naxos, for 2010, were observed, where it seemed that most days of the summer season were characterized as abnormally hot.

However, since, according to Perkins and Alexander [1] a heat wave is defined when at least three or more consecutive days have positive anomalies, the spells of these days were also computed according to the index values. The selected stations presented an average of 128 spells with a duration of 3 to 10 days during the examined 55-year period. As the duration of the spell was extended, their frequency lessened. Four stations, mainly island ones, did not present any heat waves longer than 30 consecutive days, whereas others such as Kozani and Kerkira which are located at the northwestern part of Greece experienced an intense heat wave in 2012 lasting 58 and 57 days, respectively.

Apart from the significant insight into the heat wave statistical characteristics obtained from the EHF, an attempt was made in the study to assess its ability to record previous heat waves since the index was primarily introduced for another area with very different climatic characteristics from Greece. After its application to the daily temperature data for July 1987 (one of the most intense heat waves recorded in Greece), it was found that the index identified very efficiently both the duration and the intensity of the heat wave. In addition, the acclimatization index showed that the heating conditions were quite rapid, not allowing people to adapt to them a few days earlier than the defined start of the heat wave. The index was also able to capture the fact that the heat wave was more intense in the northern parts of the country where the EHF values were higher. Conversely, regarding the second heat wave case of July 2007, the computation of the EHF index made it clear that the heating conditions had a different intensity level over Greece since its values differed substantially among the stations. Moreover, the lack of human acclimatization started in this case on the same dates as the beginning of the main heat wave event (a few days of positive value index were also observed during the first days of July). However, the EHI_{accl} turned negative, for several stations, before the end of the heat wave, meaning that people started to adapt to these extreme conditions due to the longer duration of this specific heat wave.

Overall, the main conclusion of this study is that the EHF index applies not only to the detection and analysis of heat waves in Greece, but it also provides information about the conditions that may or may not have an impact on human health and well-being. Future work includes plans to examine other years which, according to the EHF, seemed to have been extremely hot with extended heat waves such as in the year 2012.

Funding: This research received no external funding.

Conflicts of Interest: The author declares no conflict of interest.

References

1. Perkins, S.E.; Alexander, L.V. On the Measurement of Heat Waves. *J. Clim.* **2013**, *26*, 4500–4517. [CrossRef]
2. Poumadère, M.; Mays, C.; Le Mer, S.; Blong, R. The 2003 heat wave in France: Dangerous climate change here and now. *Risk Anal.* **2005**, *25*, 1483–1494. [CrossRef] [PubMed]
3. Piticar, A.; Croitoru, A.E.; Ciupertea, F.A.; Harpa, G.V. Recent changes in heatwaves and cold waves detected based on excess heat factor and excess cold factor in Romania. *Int. J. Climatol.* **2018**, *38*, 1777–1793. [CrossRef]
4. Vandentorren, S.; Bretin, P.; Zeghnoun, A.; Mandereau-Bruno, L.; Croisier, A.; Cochet, C.; Ribéron, J.; Siberan, I.; Declercq, B.; Ledrans, M. August 2003 Heat Wave in France: Risk Factors for Death of Elderly People Living at Home. *Eur. J. Public Health* **2006**, *16*, 583–591. [CrossRef] [PubMed]
5. Díaz, J.; Jordán, A.; García, R.; López, C.; Alberdi, J.; Hernández, E.; Otero, A. Heat waves in Madrid 1986–1997: Effects on the health of the elderly. *Int. Arch. Occup. Environ. Health* **2002**, *75*, 163–170. [CrossRef] [PubMed]
6. Barnett, A.G.; Hajat, S.; Gasparrini, A.; Rocklov, J. Cold and heat waves in the United States. *Environ. Res.* **2012**, *112*, 218–224. [CrossRef] [PubMed]
7. D'Ippoliti, D.; Michelozzi, P.; Marino, C.; De'Donato, F.; Menne, B.; Katsouyanni, K.; Kirchmayer, U.; Analitis, A.; Medina-Ramón, M.; Paldy, A.; et al. The impact of heat waves on mortality in 9 European cities: Results from the EuroHEAT project. *Environ. Health* **2010**, *9*, 37. [CrossRef]
8. Baccini, M.; Biggeri, A.; Accetta, G.; Kosatsky, T.; Katsouyanni, K.; Analitis, A.; Anderson, H.R.; Bisanti, L.; D'ippoliti, D.; Danova, J.; et al. Heat Effects on Mortality in 15 European Cities. *Epidemiology* **2008**, *19*, 711–719. [CrossRef]
9. EM-DAT. The International Disaster Database: Center for Research on the Epidemiology of Disasters—CRED. Available online: https://www.emdat.be/ (accessed on 7 January 2019).
10. Langlois, N.; Mason, K.; Nairn, J.; Roger, W. Using the Excess Heat Factor (EHF) to predict the risk of heat related deaths. *J. Forensic Legal Med.* **2013**, *20*, 408–411. [CrossRef]
11. Smith, T.T.; Zaitchik, B.F.; Gohlke, J.M. Heat waves in the United States: Definitions, patterns and trends. *Clim. Chang.* **2013**, *118*, 811–825. [CrossRef]
12. Nairn, J.; Fawcett, R. The excess heat factor: A metric for heatwave intensity and its use in classifying heatwave severity. *J. Environ. Res. Public Health* **2015**, *12*, 227–253. [CrossRef] [PubMed]
13. Smouer-Tomic, K.E.; Kuhn, R.; Hudsonh, A. Heat wave hazards: An overview of Heat Wave impacts in Canada. *Nat. Hazards* **2003**, *28*, 463–485.
14. Matzarakis, A.; Mayer, H.; Iziomon, M. Applications of a universal thermal index: Physiological equivalent temperature. *Int. J. Biometeorol.* **1999**, *43*, 76–84. [CrossRef] [PubMed]
15. Matzarakis, A.; De Rocco, M.; Najjar, G. Thermal bioclimate in Strasbourg—The 2003 heat wave. *Theor. Appl. Climatol.* **2009**, *98*, 209–220. [CrossRef]

16. Matzarakis, A.; Nastos, P.T. Human-biometeorological assessment of heat waves in Athens. *Theor. Appl. Climatol.* **2011**, *105*, 99–106. [CrossRef]

17. Nastos, P.T.; Matzarakis, A. The effect of air temperature and human thermal indices on mortality in Athens, Greece. *Theor. Appl. Climatol.* **2012**, *108*, 591–599. [CrossRef]

18. WMO; WHO. Heatwaves and Health Guidance on Warming-System Development. Available online: https://public.wmo.int/en/projects (accessed on 29 December 2018).

19. Meehl, G.A.; Tebaldi, C. More Intense, More Frequent, and Longer Lasting Heat Waves in the 21st Century. *Science* **2004**, *305*, 994–997. [CrossRef]

20. IPCC. Managing the risks of extreme events and disasters to advance climate change adaptation. In *A Special Report of Working Groups I and II of the Intergovernmental Panel on Climate Change*; Field, C.B., Barros, V., Stocker, T.F., Qin, D., Dokken, D.J., Ebi, K.L., Mastrandrea, M.D., Mach, K.J., Plattner, G.-K., Allen, S.K., et al., Eds.; Cambridge University Press: Cambridge, UK; New York, NY, USA, 2012; 582p.

21. Xoplaki, E.; Gonzalez-Rouco, J.F.; Luterbacher, J.; Wanner, H. Mediterranean summer air temperature variability and its connection to the large-scale atmospheric circulation and SSTs. *Clim. Dyn.* **2003**, *20*, 723–739. [CrossRef]

22. Kostopoulou, E.; Jones, P.D. Assessment of climate extremes in the eastern Mediterranean. *Meteorol. Atmos. Phys.* **2005**, *89*, 69–85. [CrossRef]

23. Della-Marta, P.M.; Haylock, M.R.; Luterbacher, J.; Wanner, H. Doubled length of western European summer heat waves since 1880. *J. Geophys. Res.* **2007**, *112*, D15103. [CrossRef]

24. Tolika, K.; Maheras, P.; Tegoulias, I. Extreme temperatures in Greece during 2007. Could this be a "return to the future"? *Geophys. Res. Lett.* **2009**, *36*. [CrossRef]

25. Ouzeau, G.; Soubeyroux, J.-M.; Schneider, M.; Vautard, R.; Planton, S. Heat waves analysis over France in present and future climate: Application of a new method on the EURO-CORDEX ensemble. *Clim. Serv.* **2016**, *4*, 1–12. [CrossRef]

26. Salata, F.; Golasi, I.; Petitti, D.; de Lieto Vollaro, E.; Coppi, M.; de Lieto Vollaro, A. Relating microclimate, human thermal comfort and health during heat waves: An analysis of heat island mitigation strategies through a case study in an urban outdoor environment. *Sustain. Cities Soc.* **2017**, *30*, 79–96. [CrossRef]

27. Roldán, E.; Gómez, M.; Pino, M.R.; Pórtoles, J.; Linares, C.; Díaz, J. The effect of climate-change-related heat waves on mortality in Spain: Uncertainties in health on a local scale. *Stoch. Environ. Res. Risk Assess.* **2016**, *30*, 831–839. [CrossRef]

28. Miron, I.J.; Linares, C.; Montero, J.C.; Criado-Alvarez, J.J.; Díaz, J. Changes in cause-specific mortality during heat waves in central Spain, 1975–2008. *Int. J. Biometeorol.* **2015**, *59*, 1213–1222. [CrossRef]

29. Prezerakos, N. A contribution to the study of the extreme heat wave over the South Balkans in July 1987. *Meteorol. Atmos. Phys.* **1989**, *41*, 261–271. [CrossRef]

30. Matzarakis, A.; Mayer, H. The extreme heat wave in Athens in July 1987 from the point of view of human biometeorology. *Atmos. Environ.* **1991**, *25*, 203–211. [CrossRef]

31. Giles, B.D.; Balafoutis, C.J. The Greek heatwaves of 1987 and 1988. *Int. J. Climatol.* **1990**, *10*, 505–517. [CrossRef]

32. Balafoutis, C.; Makrogiannis, T. Analysis of a heat wave phenomenon over Greece and it's implications for tourism and recreation. In Proceedings of the First International Workshop o2n Climate, Tourism and Recreation, Halkidiki, Greece, 5–10 October 2001; pp. 113–121.

33. Founda, D.; Giannakopoulos, C. The exceptionally hot summer of 2007 in Athens, Greece—A typical summer in the future climate? *Glob. Plan. Chang.* **2009**, *67*, 227–236. [CrossRef]

34. Nairn, J.; Fawcett, R.; Ray, D. Defining and predicting excessive heat events: A national system. In Proceedings of the Modelling and Understanding High Impact Weather: Extended Abstracts of the Third CAWCR Modelling Workshop, Melbourne, Australia, 30 November–2 December 2009; Volume 17, pp. 83–86.

35. Urban, A.; Hanzlíková, H.; Kyselý, J.; Plavcová, E. Impacts of the 2015 heat waves on mortality in the Czech Republic—A comparison with previous heat waves. *Int. J. Environ. Res. Public Health* **2017**, *14*, 1562. [CrossRef]

36. Trigo, R.; Garia-Herrera, R.; Diaz, J.; Trigo, I.; Valente, M. How exceptional was the early August 2003 heatwave in France? *Geophys. Res. Lett.* **2005**, *32*, L10701. [CrossRef]

37. Karl, T.R.; Knight, R.W. The 1995 Chicago Heat Wave: How likely is a recurrence? *Bull. Am. Meteorol. Soc.* **1997**, *78*, 1107–1120. [CrossRef]

38. Alexandersson, H. A homogeneity test applied to precipitation data. *J. Climatol.* **1986**, *6*, 661–675. [CrossRef]

39. Sneyers, R. *Sur L'analyse Statistique des Series D'observations*; OMM Publ. No. 415. Note Technique 143; OMM: Geneve, Switzerland, 1975; 192p.

40. Berz, G. List of Major Natural Disasters, 1960–1987. *Nat. Hazards* **1998**, *1*, 97–99. [CrossRef]

41. Giles, B.D.; Balafoutis, C.; Maheras, P. Too hot for comfort: The heatwaves in Greece in 1987 and 1988. *Int. J. Biometeorol.* **1990**, *34*, 98–104. [CrossRef] [PubMed]

42. Theoharatos, G.; Pantavou, K.; Spanou, A.; Katavoutas, G.; Makrygiannis, G.; Mavrakis, A. Discomfort index levels during the heatwaves of June and July 2007 in Greece. In Proceedings of the 11th International Conference on Environmental Science and Technology, Chania, Greece, 3–5 September 2009; pp. 946–952.

43. Katavoutas, G.; Theoharatos, G.; Flocas, H.A.; Asimakopoulos, D.N. Measuring the effects of heat wave episodes on the human body's thermal balance. *Int. J. Biometeorol.* **2009**, *53*, 177–187. [CrossRef] [PubMed]

44. Pattendend, S.; Nikiforov, B.; Armstrong, B.G. Mortality and temperature in Sofia and London. *Epidemiol. Community Health* **2003**, *57*, 628–633. [CrossRef]

45. Pirard, P.; Vandentorren, S.; Pascal, M.; Laaidi, K.; Le Tertre, A.; Cassadou, S.; Ledrans, M. Summary of the mortality impact assessment of the 2003 heat wave in France. *Eurosurveillance* **2005**, *10*, 153–156. [CrossRef] [PubMed]

46. Nicholls, N.; Skinner, C.; Loughnan, M.; Tapper, N. A simple heat alert system for Melbourne, Australia. *Int. J. Biometeorol.* **2008**, *52*, 375–384. [CrossRef]

Article

Statistical Analysis of Recent and Future Rainfall and Temperature Variability in the Mono River Watershed (Benin, Togo)

Agnidé Emmanuel Lawin [1,2,*], Nina Rholan Hounguè [2], Chabi Angelbert Biaou [3] and Djigbo Félicien Badou [1,2]

[1] Laboratory of Applied Hydrology, National Water Institute, University of Abomey-Calavi, Cotonou 01 BP 526, Benin; fdbadou@gmail.com
[2] West African Science Service Centre on Climate Change and Adapted Land Use (WASCAL), Cotonou 01 BP 526, Benin; rholhu@yahoo.fr
[3] Laboratoire Eaux Hydro-Systèmes et Agriculture (LEHSA), Institut International d'Ingénierie de l'Eau et de l'Environnement, Ouagadougou 01 BP 594, Burkina Faso; angelbert.biaou@2ie-edu.org
* Correspondence: ewaari@yahoo.fr; Tel.: +229-975-818-09

Received: 17 November 2018; Accepted: 27 December 2018; Published: 6 January 2019

Abstract: This paper assessed the current and mid-century trends in rainfall and temperature over the Mono River watershed. It considered observation data for the period 1981–2010 and projection data from the regional climate model (RCM), REMO, for the period 2018–2050 under emission scenarios RCP4.5 and RCP8.5. Rainfall data were interpolated using ordinary kriging. Mann-Kendall, Pettitt and Standardized Normal Homogeneity (SNH) tests were used for trends and break-points detection. Rainfall interannual variability analysis was based on standardized precipitation index (SPI), whereas anomalies indices were considered for temperature. Results revealed that on an annual scale and all over the watershed, temperature and rainfall showed an increasing trend during the observation period. By 2050, both scenarios projected an increase in temperature compared to the baseline period 1981–2010, whereas annual rainfall will be characterized by high variabilities. Rainfall seasonal cycle is expected to change in the watershed: In the south, the second rainfall peak, which usually occurs in September, will be extended to October with a higher value. In the central and northern parts, rainfall regime is projected to be characterized by late onsets, a peak in September and lower precipitation until June and higher thereafter. The highest increase and decrease in monthly precipitation are expected in the northern part of the watershed. Therefore, identifying relevant adaptation strategies is recommended.

Keywords: Mono River watershed; trend analysis; climate

---

1. Introduction

Modifications in the climate as a result of both natural and anthropogenic processes have raised considerable concerns (such as more frequent and intense rainfall, droughts, dry spells, violent winds, etc.), as they induce adverse impacts on several development sectors [1].

In recent decades, weather and climate extremes such as droughts, heat waves, wild fires, floods and storms have increased in frequency and intensity in several regions of the world. In fact, Vincent et al. [2] noticed that the percentage of warm nights is increasing while that of cold nights is decreasing in South-America. In addition, the US Climate Change Science Program underscored the fact that heavy precipitations have become more frequent and intense in Northern America [3]. In central Asia, Savitskaya [4] reported that, during the last 50 years, there was high variability in the pattern of precipitation, whereas winter has become warmer in the entire region.

Furthermore, in tropical Africa a significant increase in temperature, about 0.15 °C per decade, was detected over the period 1979–2010 [5]. Consequently, high fatality rates are recorded in developing countries because of their high reliance on natural resources and their limited coping capacities [6]. Several authors highlighted that, since the 1970s, the number of natural disasters (flood, drought, windstorm, epidemic and famine) has been increasing in sub-Saharan Africa [7–9]. In 2012, central and western Africa were hit by severe floods which affected 1,538,242 people and caused 340 deaths as of September of that year. Moreover, flood events of 2010 have been recorded in West Africa as one of the most disastrous during the last decade. In 2010 only, Benin lost about USD 262 million [10], whereas Togo recorded about USD 43.934 million as damage and loss in the same year [11].

Thus, there is a need to carry out future climate analysis in order to foresee potential hazards and ultimately to develop appropriate strategies to combat them. According to the fifth assessment report AR5, "global surface temperature change for the end of the 21st century is likely to exceed 1.5 °C relative to 1850–1900 for all RCP scenarios except RCP2.6" [12]. However, it is clear that climate-change impacts will be time and location specific [13]. Therefore, undertaking climate projection at regional and local level will contribute to more accurate and relevant actions towards human security.

As in many other watersheds in the world, climate trend analyses have been carried out in the Mono watershed [14–18]. Ntajal et al. [18] noted that, over the period 1961–2013, at local scale, there is a significant decreasing trend in rainfall at the station of Sokode (upstream), while an insignificant increase in rainfall is observed in the downstream (Atakpame, Sotouboua, Aklakou and Tabligbo). The same assessment was conducted by Amoussou [15] for the period 1961–2000 using a cubic spatial interpolation for rainfall data in the watershed. The results showed an overall decreasing trend of rainfall. So far, however, there are few climate related studies which have addressed future climate projection in the Mono River watershed, despite the fact that people living at the downstream usually experience flood events during rainy seasons. The divergence noted among the results of previous studies makes it important to keep assessing climate trend in the watershed, mainly over a recent period. Therefore, this study aims at assessing current and future climate change in the Mono River watershed.

2. Materials and Methods

2.1. Study Area

The Mono River watershed occupies an area of 27,822 km^2 shared between two West-African countries, Togo and Benin. Specifically, it is located between the latitudes 06°16′ N and 09°20′ N, and the longitudes 0°42′ E and 2°25′ E (Figure 1). The major part of the basin lies in Togo's territory, totaling 21,750 km^2, whereas that of Benin stretches to 6072 km^2. The river serves as natural border between the two countries in the southern part. The climate is tropical (two rainy seasons and two dry seasons) downstream and subequatorial (one rainy season and one dry season) upstream.

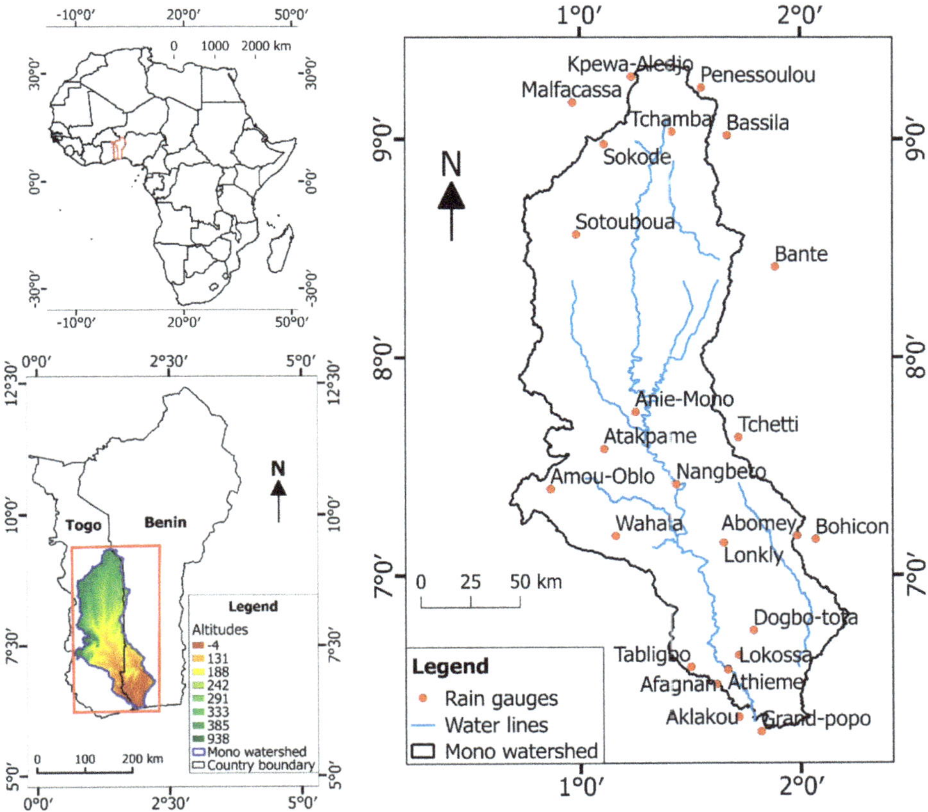

Figure 1. Location of Mono watershed and considered rain gauges.

2.2. Data Used

Both observation and projection data were used. Daily observed rainfall and temperature were collected from meteorological institutes of Benin and Togo (DMN, Direction de la Météorologie Nationale) for the period 1981–2010. That period is the current normal used for climatological analysis, and this study aims at taking it into account, as previous studies have already accounted for other normals [15,17,18]. As presented by Figure 1, rainfall data were collected at 24 rain gauges within and around Mono watershed (not farther than 25 km). As for temperature data, they were collected from three synoptic stations located in the watershed: Tabligbo, Atakpamé and Sokodé.

Projected data were provided by the regional climate model (RCM), REMO (Table 1). Rainfall and temperature data were extracted for the period 2018–2050 and under two representative concentration pathways: RCP4.5 (intermediate pathway) and RCP8.5 (most extreme pathway). Akinsanola et al. [19] have already noticed that REMO fairly simulates rainfall in West Africa and concluded that it can be used for future climate projections in the region. However, the use of a multi-model ensemble approach (more than one model) is recommended in order to better estimate the actual climate and improve the robustness of climate change projections [20–22].

Table 1. Characteristics of the regional climate model (RCM), REMO.

Model Name	Institute	Driven Model
REMO2009	Helmholtz-Zentrum Geesthacht, Climate Service Center, Max Planck Institute for Meteorology	Max Planck Institute—Earth System Model running on low resolution grid (MPI-ESMLR)

These data were accessed online (https://www.cordex.org) in the context of the Coordinated Regional Climate Downscaling Experiment (CORDEX) over Africa at 0.44° resolution.

2.3. Methods

2.3.1. Trend Analysis

Trends and breakpoints were assessed using the non-parametric test of Mann-Kendall first, followed by the Pettitt and SNH tests The Mann-Kendall test is used in order to establish whether there is a trend (increasing or decreasing) in the time series. It is done with a confidence level of 95%, and the hypotheses are:

H_0: there is no trend in rainfall time series;
H_1: there is a trend in rainfall time series.

For both Pettitt and SNH tests, significance level $\alpha = 0.05$, and the hypotheses are:

H_0: there is no change in annual rainfall data;
H_1: there is a date at which there is a change in the data.

Moreover, in order to assess the trend of annual rainfall at watershed scale, and because rainfall data are not measured in every single grid of the watershed, spatial interpolation was required. In the scope of this study, ordinary kriging (OK) was chosen over other methods—such as arithmetic mean, Thiessen polygon, inverse distance weighting—because (i) it takes into account not only the distance between observation stations and estimation point but also the distance between stations taken two by two; (ii) it is a stochastic method which provides the best linear unbiased predictions; (iii) the interpolation error can be estimated [23]. Nonetheless, one of the limitations of kriging is that it is not suitable when there are few observation points. Kriging was basically developed for geostatistics purposes [24] but is widely used in climatology. It is worth noting that the 'backbone' of kriging is the variogram which explains the variance of the studied variable with respect to distance between observation points. Equation (1) presents the formula of variogram

$$\gamma(h) = \frac{1}{2N(h)} \sum_{i=1}^{N(h)} (z(p_i) - z(p_i + h))^2, \tag{1}$$

where $\gamma(h)$ is the variogram, $N(h)$ the number of coupled points separated by the distance h, $z(p_i)$ the observed rainfall at location p_i, and $z(p_i + h)$ the observed rainfall at location $p_i + h$.

Furthermore, considering the fact that rainfall regime in Mono watershed is not homogenous, rainfall trend analysis is carried out with respect to three latitude-based regions, as done in previous studies [15,18]. The regions are defined as follows: latitude < 7, 7 ≤ latitude ≤ 8 and latitude > 8. Hereinafter, these regions are respectively referred to as southern part, central part and northern part of the Mono watershed.

Analysis of temperature trends over the watershed was conducted on the arithmetic mean from the stations of Tabligbo, Sokodé and Atakpamé.

2.3.2. Interannual Rainfall Variability Analysis

The standardized precipitation index (SPI) [25] is a tool recommended by the World Meteorological Organization (WMO) and widely used for quantifying the precipitation deficit over different timescales (3 to 48 months). For the selected timescale, rainfall records are fitted with a probability distribution which is then transformed into a normal distribution so that the mean SPI for the location and desired period is zero. Hence, this method improves the common anomaly method (Equation (2)), which does not take into account the fact that rainfall is typically not normally distributed for a cumulative period of 12 months or less.

$$I(i) = \frac{x_i - \overline{x_m}}{\sigma},$$

(2)

where $I(i)$, x_i, $\overline{x_m}$ and σ are respectively the standardized index of year i, the value for the year i, the average and the standard deviation of the time series.

In the present study, the SPI 12 (for 12 month's timescale) is used to assess rainfall deficit or excess on a yearly basis. Moreover, SPI 12 is the one recommended for watershed analysis [26]. Table 2 presents the guidelines for analyzing SPI values [25,26].

Table 2. Standardized precipitation index (SPI) values and their meanings.

SPI Value	Meaning
2.0 and plus	Extremely wet
1.5 to 1.99	Very wet
1.0 to 1.49	Moderately wet
−0.99 to 0.99	Near normal
−1 to −1.49	Moderately dry
−1.5 to −1.99	Severely dry
−2 and less	Extremely dry

2.3.3. Future Climate Analysis

Raw outputs from RCMs must be corrected prior to local impact studies because of the bias they encompass. There are several bias correction methods but in this study, the methods of delta, linear scaling and empirical quantile mapping (EQM) are used because they have produced satisfactory results in previous studies carried out in similar climatic regions [27–29]. The results of Ntcha M'po et al. [27], Essou and Brissette [30] and Speth et al. [31], who bias-corrected REMO data in the Ouémé watershed (Benin), guided the choice of correction methods in this study. Rainfall was corrected with delta method in the south, multiplicative scaling in the central part and EQM in the north. As for temperature data, they were corrected using only EQM.

2.3.4. Percentage of Relative Change in Rainfall Seasonal Cycle

It is computed using Equation (3).

$$P_{c,m} = \frac{R_{proj,m} - R_{obs,m}}{R_{obs,m}} \times 100$$

(3)

where, $P_{c,m}$, $R_{proj,m}$ and $R_{obs,m}$ are respectively the percentage of change for mth month, average projected rainfall for month m, and the average rainfall of month m during observation period.

3. Results and Discussion

3.1. Trends and Interannual Variability in Rainfall

3.1.1. Present Rainfall Pattern

The pattern of monthly rainfall varies from one part of the watershed to the other. The southern and northern parts are respectively characterized by bimodal and unimodal rainfall regime, whereas, a transitory (neither bimodal nor unimodal) regime was found in the central region (Figure 2).

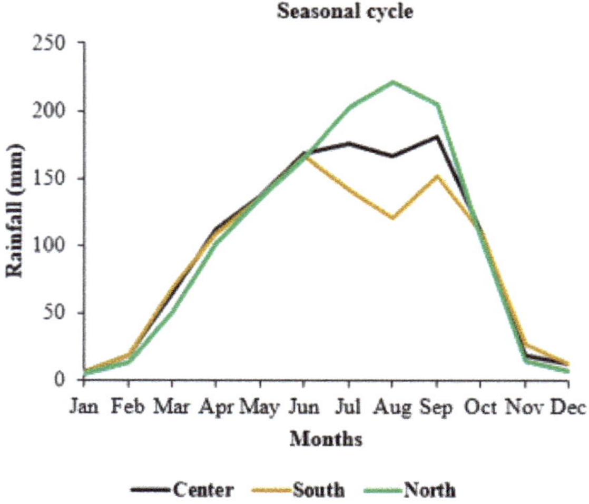

Figure 2. Rainfall seasonal cycles in Mono watershed.

In the south, two peaks are respectively recorded in June and September, while the unique peak in the northern part occurs in August. On the other side, the rainy season in the central part lasts from March to September. These results are in line with previous research findings [15,17,18]. Thus, the bimodal or unimodal characteristics of the precipitation regime seem not to change over the recent periods compared to the historical period.

The results of the Mann-Kendall test (Table 3) underscored the fact that rainfall in the three regions of Mono watershed had an increasing trend during the period 1981 to 2010.

Table 3. Results of break-point detection and Mann-Kendall test on observed rainfall.

Region	Break-Point Detection			Mann-Kendall Test	
	Tests	**Break-Point**	*p*-**Value**	*p*-**Value**	**Sen's Slope**
South	Pettitt	2001	0.0695	0.0022	9.99
	SNH	2001	0.0497		
Centre	Pettitt	1987	0.3113	0.0385	7.56
	SNH	1983	0.0217		
North	Pettitt	1987	0.0952	0.0295	6.49
	SNH	1983	0.0025		

As presented in Table 3 and Figure 3, the SNH test detected break points in the time series while the Pettitt test did not.

The change noted in the center and the north may be related to the well-known 1970s and 1980s droughts which affected many West-African countries and was documented by several authors such as Le Barbé and Lebel [32], Le Barbé et al. [33] and Le lay and Galle [34].

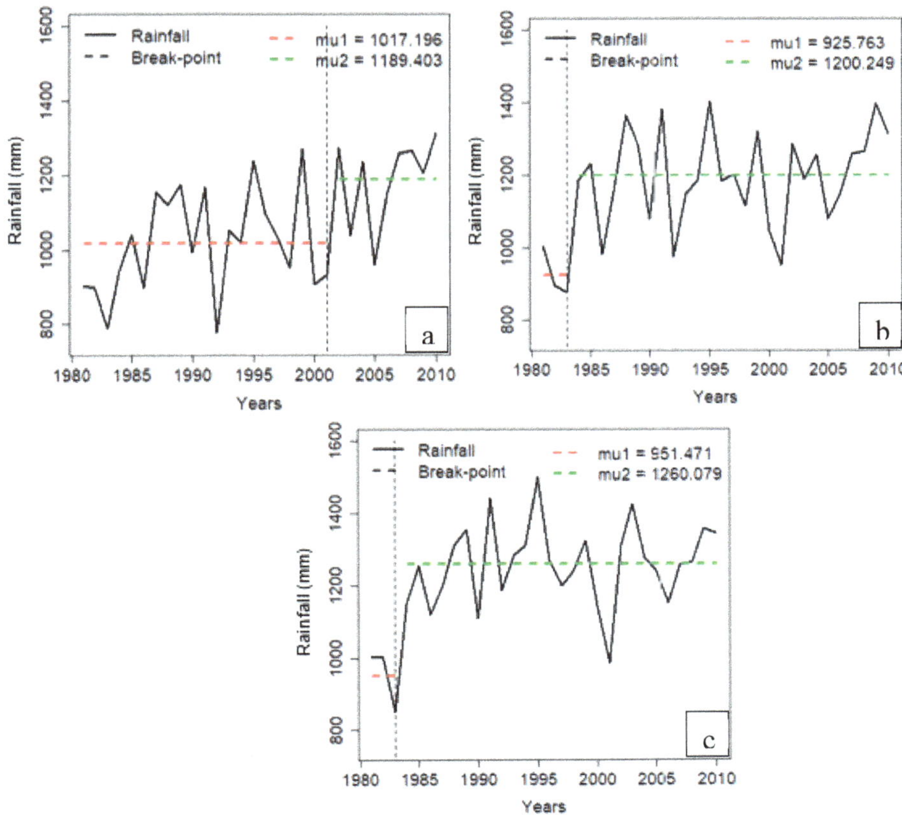

Figure 3. Standardized Normal Homogeneity (SNH) test on observed annual rainfall time series in the south (**a**), center (**b**) and north (**c**).

Figure 4 presents the results of the SPI computation.

According to the baseline period, the longest dry period is 1981–1986 in the south, 1981–1983 in the central part and 1981–1984 in the north. In addition, the driest year is 1992 in the south, and 1983 for both the center and north. As for years of highest excess, it is 2010 in the south and 1995 in central and northern parts. It is worth noting that, in the three regions, the longest dry period falls in the 1980s drought events. After this specific period no extreme drought occurred (except in 1992 in the south), and there were more years above normal than below. Overall, the period after 1990 is characterized by more wet years, and it explains the trends highlighted by statistical tests performed above. Similar results have been reported in other watersheds in West Africa by several authors, such as Adeyeri et al. [35] in Komadugu-Yobe basin, Nicholson et al. [36] over West Africa, Ozer et al. [37] over the Sahelian region, and by Lawin [38] and Attogouinon et al. [39] in the upper Ouémé river valley. However, this shift to wetter condition is region dependent, because other studies reported a decreasing trend in rainfall patterns over West-Africa [40,41].

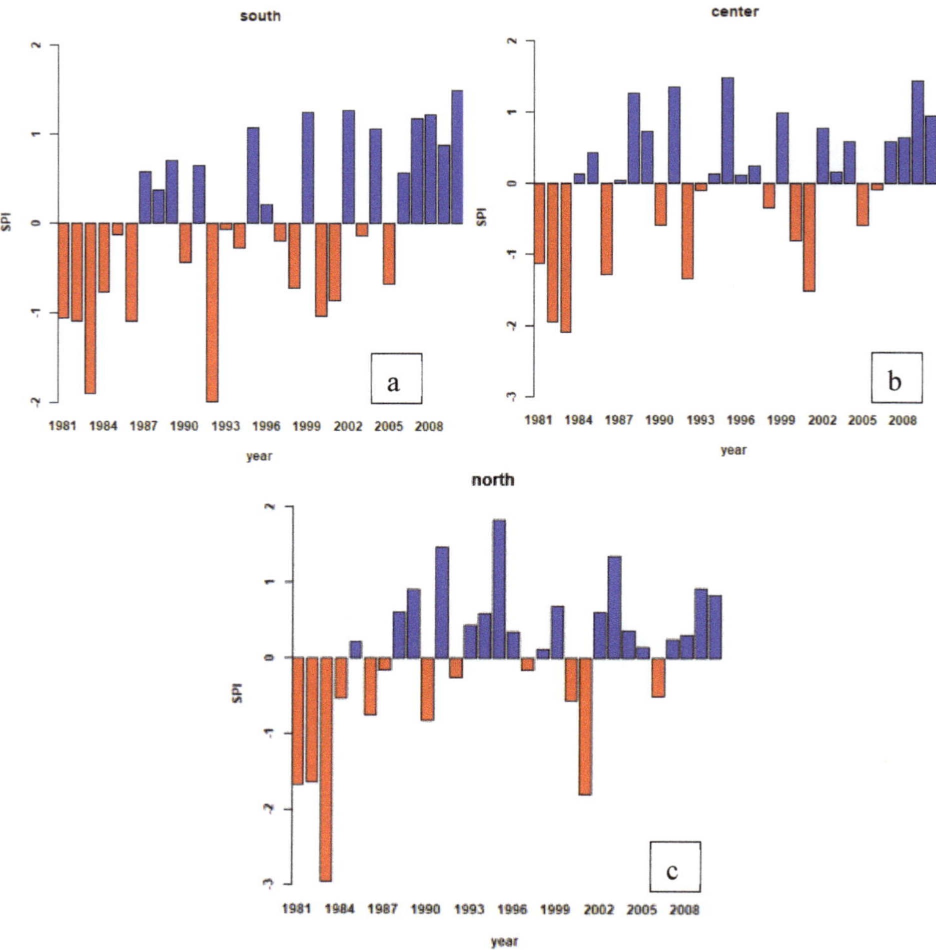

Figure 4. Standardized precipitation index of observed annual rainfall in the south (**a**), center (**b**) and north (**c**).

3.1.2. Future Rainfall Pattern

The results of break-point detection and the Mann-Kendall test performed on annual rainfall under RCP4.5 and RCP8.5 are summarized in Table 4.

The results of homogeneity tests and the Mann-Kendall test suggest that, all over the watershed, there is neither break-point nor a linear trend in annual rainfall time series, under emission scenarios RCP4.5 and RCP8.5. Statistically, rainfall time series are homogenous and present no trend. Nonetheless, some variabilities are observed. Specifically, in the northern part and for RCP8.5, rainfall might decrease by 2035 and increase thereafter. The increase in annual precipitation over recent decades in Mono river watershed seems not to be maintained in the future. Future pattern of rainfall may be marked by high variabilities. Such an absence of significant trend in rainfall is also reported by N'Tcha M'Po et al. [40] in Ouémé river basin by 2050, using a REMO model. Similarly, Lawin et al. [42] reported no trend in rainfall pattern in the Imbo north plain region in Burundi under central Africa climatology using an ensemble of eight regional climate models.

Table 4. Results of break-point detection and Mann-Kendall test on annual rainfall under RCP4.5 and RCP8.5.

Scenario	Region	Break-Point Detection			Mann-Kendall Test	
		Result	Break-Point	p-Value	p-Value	Sen's Slope
RCP4.5	South	Pettitt test	2021	0.621	0.914	−0.097
		SNH test	2049	0.942		
	Center	Pettitt test	2033	0.346	0.285	2.293
		SNH test	2033	0.525		
	North	Pettitt test	2031	0.129	0.091	4.427
		SNH test	2031	0.155		
RCP8.5	South	Pettitt test	2029	0.0731	0.588	−1.510
		SNH test	2024	0.842		
	Center	Pettitt test	2029	0.673	0.653	−1.341
		SNH test	2029	0.712		
	North	Pettitt test	2041	0.324	0.394	3.296
		SNH test	2041	0.104		

The SPI computed for each region emphasizes this variability (Figure 5).

Under RCP4.5, the number of projected deficit years in the watershed increases slightly from south to north. The years 2020 and 2024 are projected to be extremely wet in the south, whereas 2031 is expected to be extremely dry in the north. However, under RCP8.5, the projected number of deficit years decreases from south to north. The year 2033 is projected to be extremely dry in the north, and again, years 2020 and 2024 for this scenario are expected to be extremely wet in the south. The agreement of both scenarios for years 2020 and 2024 shows that those years will potentially be characterized by extreme precipitations. Furthermore, from south to north, RCP8.5 projects more extremely wet years than RCP4.5.

Furthermore, the pattern of the seasonal cycle of rainfall is expected to undergo some modifications. In the southern part (Figure 6a), rainfall seasonal cycle is projected to keep its bimodal pattern under RCP4.5 and RCP8.5. In addition, both scenarios project almost the same pattern. As in the normal period, the first peak is recorded in June but with a slightly lower amount. The second peak, which normally occurs in September, is expected to extend to October with a higher value.

In the central and northern parts (Figure 6b,c), and under RCP4.5 and RCP8.5, the rainfall regime is projected to be characterized by late onsets and lower precipitation until June, compared to observations, and higher thereafter. Both scenarios converge on the fact that rainfall peak will probably occur in September. The northern part is expected to keep its unimodal pattern under both scenarios, whereas a shift from a transitional regime to a unimodal one is expected in the central region.

Projected future rainfall regimes under the two scenarios are quite similar, apart from in August, where RCP4.5 predicted a slightly larger amount than RCP8.5, and an inverse proportion was predicted in June

Moreover, on a monthly scale, Figure 7 depicts how the rainfall seasonal cycle is expected to change under RCP4.5 and RCP8.5 compared to the baseline period.

Under RCP4.5, the relative change in monthly rainfall varies from −5.5% to 8.4% in the southern part, −29.9% to 22.2% in the central part and −39% to 91.4% in the northern part. For RCP8.5, the expected change ranges from −3.5% to 5.8% in the southern part, −55% to 20% in the central part and −64% to 85.9% in the northern part. Therefore, the biggest changes (both increase and decrease) in monthly rainfall are expected in the northern part of the watershed, regardless of the scenario considered. According to RCP4.5 and with respect to the observation period, the highest rainfall decrease during the period 2018–2050 is expected to occur in February, whereas January and November will record the highest increases.

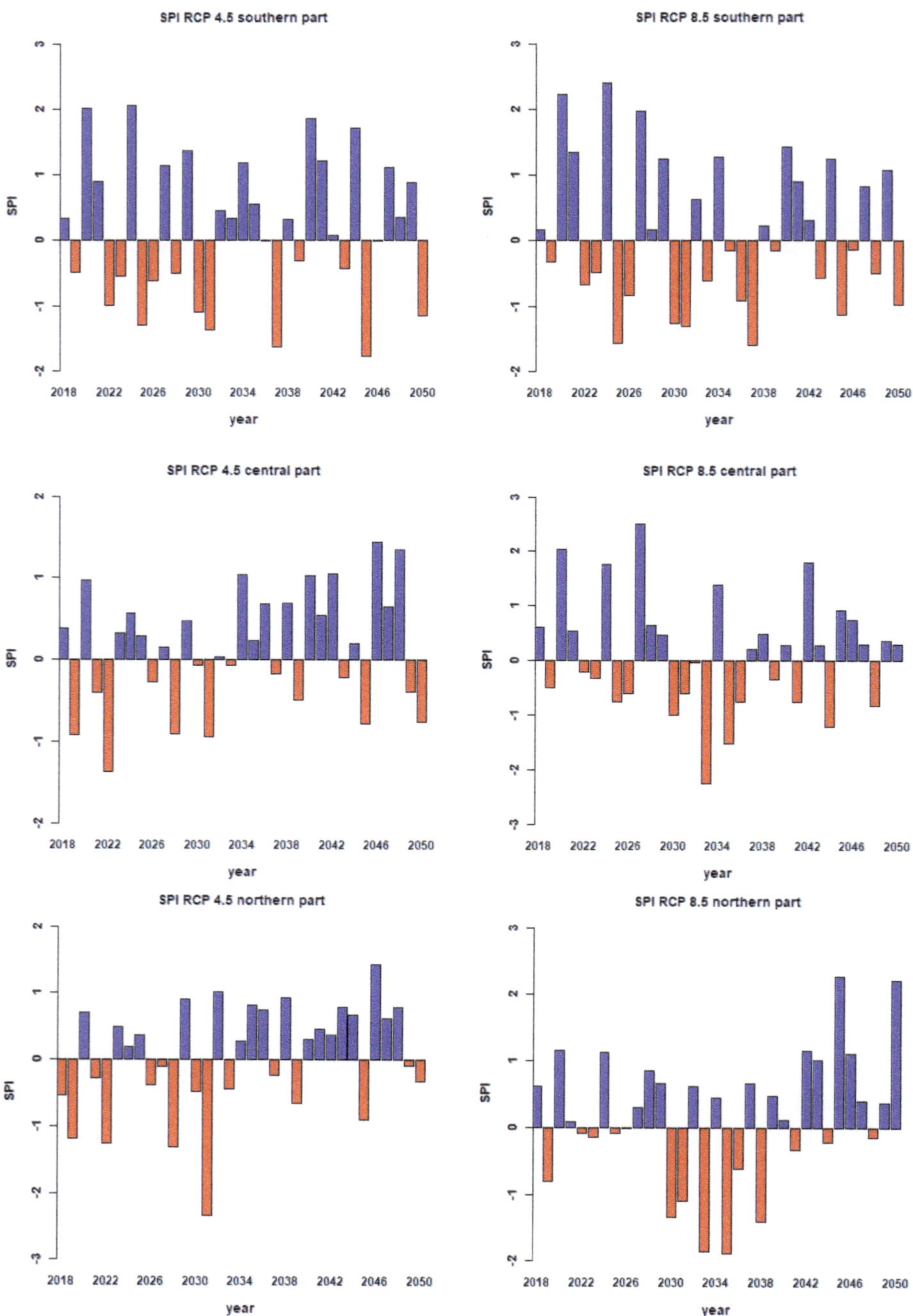

Figure 5. Standardized precipitation index of annual rainfall under RCP4.5 (**left panel**) and RCP8.5 (**right panel**).

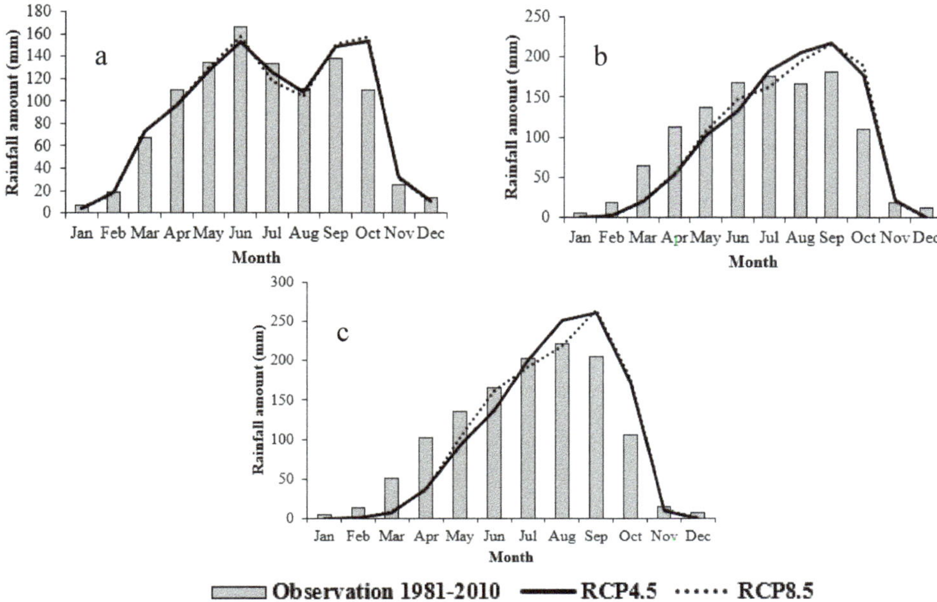

Figure 6. Seasonal cycles of rainfall in the south (**a**), center (**b**) and north (**c**) under RCP4.5 and RCP8.5.

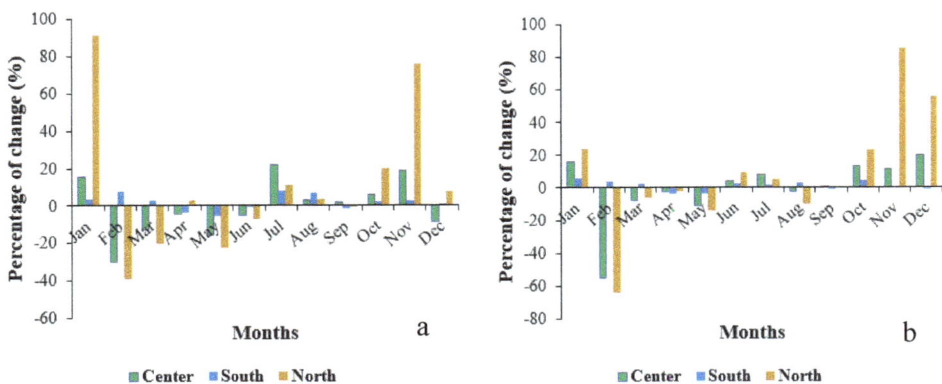

Figure 7. Expected change in rainfall seasonal cycles under RCP4.5 (**a**) and RCP8.5 (**b**).

Under RCP8.5, the highest decrease is projected to affect rainfall in February, whereas the highest increase is expected in November. Thus, the two scenarios project more than 70% increase of rainfall in the month of November by 2050, compared to the observation period. In addition, the highest increase is projected by RCP4.5 and the highest decrease by RCP8.5. Globally substantial changes are expected prior to and at the end of rainy seasons.

3.2. Temperature Trend

3.2.1. Present Temperature Change

The two homogeneity tests performed on annual temperature revealed the presence of break-point in the time series (Figure 8).

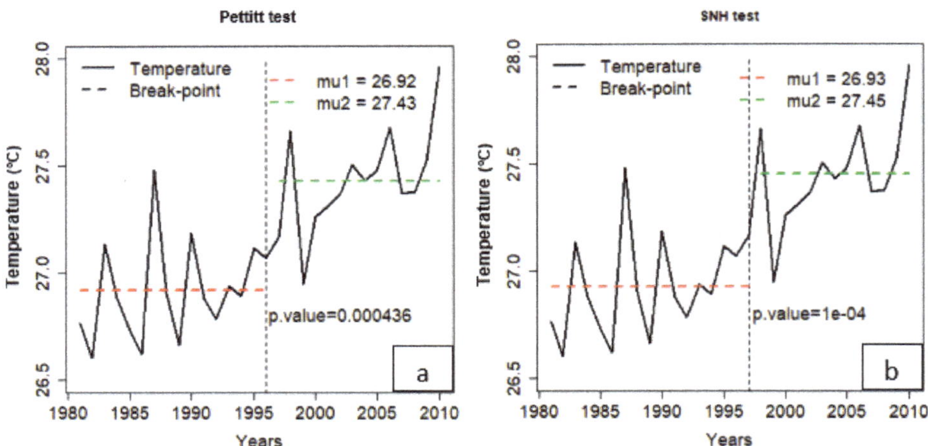

Figure 8. Pettit test (**a**) and SNH test (**b**) performed on annual temperature over the period 1981–2010.

The Pettitt test indicates that since 1996 mean annual temperature has increased by 0.51 °C in Mono watershed compared to the period 1981–1995, whereas the results of the SNH test implies an increase of 0.52 °C from the period 1981–1996 to 1997–2010. In addition, the Mann-Kendall test suggested a significant increasing trend (p-value = 3.457×10^{-6} and $\tau = 0.6$).

Anomalies computation revealed that, from 1981 to 1997, temperature was globally below and near normal, but since 1998 it has stayed above the normal (Figure 9).

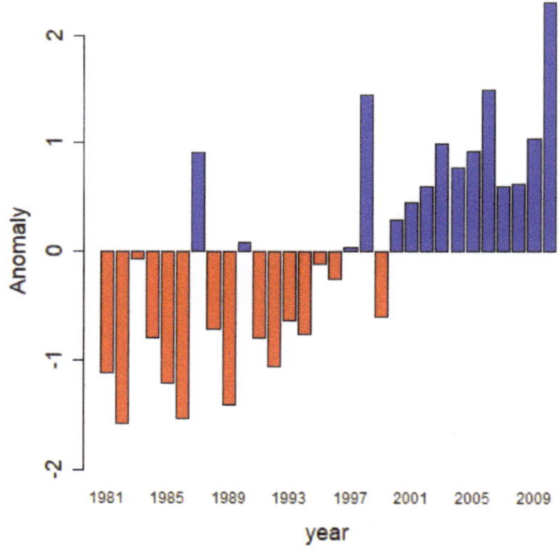

Figure 9. Anomaly of mean annual temperature in Mono watershed.

It therefore corroborates the outputs from homogeneity tests. In addition, these results are in line with previous studies which noted similar increasing trend of temperature in West Africa (Badjana [43] in the Kara river basin of Togo, Kabo-Bah et al. [44] in the Ghana part of Volta river basin, Oguntunde et al. [45] at Ibadan and Collins [5] over the West-African region).

3.2.2. Future Temperature Change

Under RCP4.5, the homogeneity tests detected break-points at different dates, but they got an agreement under RCP8.5 (Figure 10). Under the intermediate scenario, the Pettitt test indicated that, from the period 2018–2031 to 2032–2050, temperature may increase by 0.36 °C. On the other side, SNH test projected an increase of 0.39 °C from 2018–2027 to 2028–2050. For the high pathway scenario, both tests agreed on an increase of 0.87 °C from the period 2018–2038 to 2039–2050.

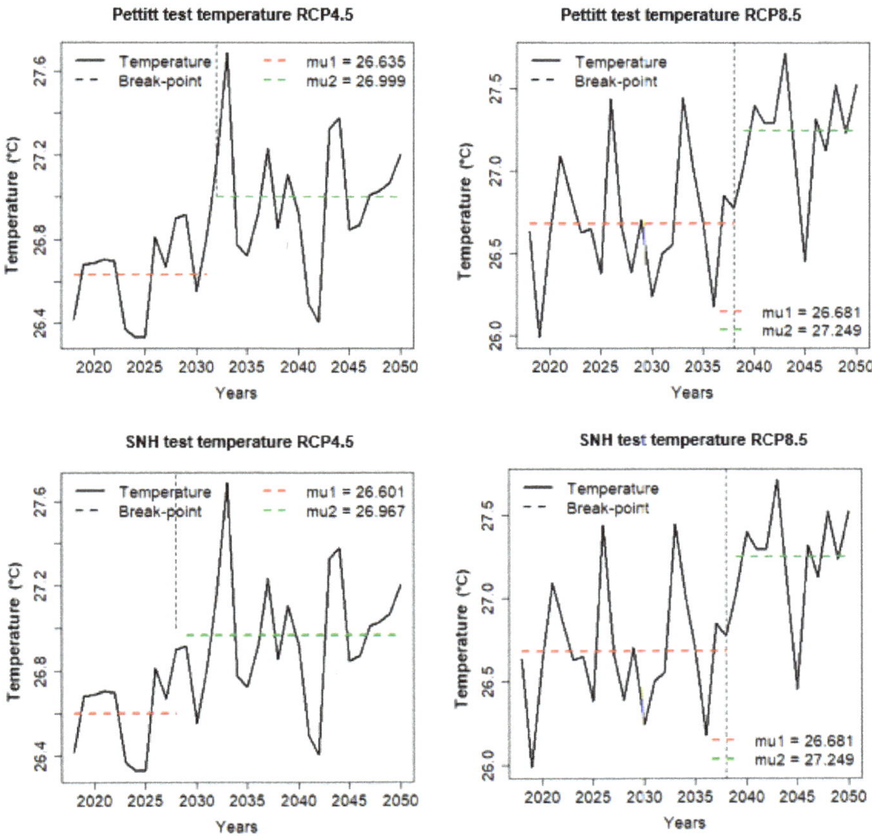

Figure 10. Pettitt test (**upper panel**) and SNH test (**lower panel**) performed on temperature under RCP4.5 and RCP8.5.

Regardless of the scenario used, an overall significant increasing trend in temperature is expected by 2050 (Table 5).

Figure 11 depicts anomalies of temperature under RCP4.5 and RCP8.5.

For both emission scenarios, the number of years above normal is higher compared to the number of years below normal. Therefore, future climate in Mono river watershed is projected to be warmer by 2050. Such an increasing trend by 2050 has been reported by Nelson et al. [46] in Togo. Similarly, Oyerinde [47] reported a consistently increasing trend in temperature over the Niger River Basin, 5% to 10% under RCP4.5 and 5% to 20% under RCP8.5, using an ensemble model from eight regional climate models. In the Massili basin of Burkina Faso, Bontogho [29] reported an increase in temperature by 1.8 °C (RCP4.5) and 3.0 °C (RCP8.5) from 1971 to 2050 using the regional model HIRHAM5. Badou [48] reported a temperature increase of up to 0.48 °C under RCP4.5 and up to 0.45 °C under RCP8.5 using the REMO model in the Benin part of Niger River basin. Overall, all models and scenarios considered by several authors converge to a moderate to high increase of temperature all over the world.

Table 5. Results of Mann-Kendall test on annual temperature under RCP4.5 and RCP8.5.

Scenario	Break-Point Detection			Mann-Kendall Test	
	Result	Break-Point	*p*-Value	*p*-Value	Sen's Slope
RCP4.5	Pettit test	2031	0.0039	0.0003	0.017
	SNH test	2027	0.0103		
RCP8.5	Pettitt test	2038	0.0082	0.0009	0.028
	SNH test	2038	0.0029		

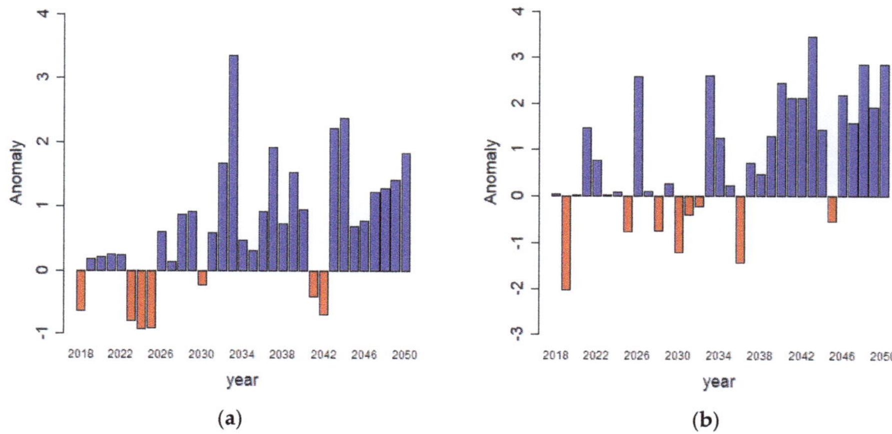

Figure 11. Temperature anomaly under RCP4.5 (**a**) and RCP8.5 (**b**).

4. Conclusions

This paper examined the trend in annual rainfall and annual temperature of Mono river watershed over the observation period 1981–2010 and by 2050 using the regional model REMO under RCP4.5 and RCP8.5. It also assessed the monthly pattern of rainfall over the same periods. During the last three decades, rainfall and temperature have been increasing all over the Mono River watershed. By 2050 and under emission scenarios RCP4.5 and RCP8.5, annual rainfall is projected to be characterized by high variability, whereas a significant increasing trend is projected for annual temperature (warmer future climate). For each of the three defined regions (south, center and north), and under both emission scenarios, the seasonal cycle of rainfall is expected to change: In the southern part, the first peak is projected to reduce slightly, whereas the second peak is expected to increase and shift to October; however, in the central and northern parts, it is expected that there will be late onset of rainfall and higher peaks. In addition, the seasonal cycle of rainfall in the central part is expected to shift from a transitional regime to a unimodal one. Moreover, on a monthly scale, the northern part of the watershed is expected to record the highest increase and decrease in rainfall regardless of the emission scenario considered. These considerable changes in the monthly rainfall of the northern part are expected to occur globally in the dry season, thus indicating potential extreme events. Considering the projected trends and patterns for rainfall and temperature over Mono watershed by 2050, it is recommended that experts identify and implement relevant adaptation strategies.

Author Contributions: A.E.L., N.R.H., D.F.B. and C.A.B. designed the study, developed the methodology and wrote the original manuscript. N.R.H. performed the field work, data collection and computer analyzis, meanwhile C.A.B. contributed to results analysis and interpretation. Overall the authors contributed equally to this paper.

Funding: This research was funded by the West African Science Service Center for Climate Change and Adapted Land use (WASCAL).

Acknowledgments: Authors thank the German Federal Ministry of Education who funded the Master Degree of Nina Rholan Houngue through the West African Science Service Center for Climate Change and Adapted Land use (WASCAL). We thank the ESGF grid (http://esg-dn1.nsc.liu.se/esgf-web-fe/) which provided the CORDEX-Africa future climate projections.

Conflicts of Interest: The authors declare no conflict of interest.

References

1. Intergovernmental Panel on Climate Change (IPCC). *Climate Change 2007: Synthesis Report. An Assessment of the Intergovernmental Panel on Climate Change*; IPCC: Valencia, Spain, 2007.
2. Vincent, L.A.; Peterson, T.C.; Barros, V.R.; Marino, M.B.; Rusticucci, M.; Carrasco, G.; Ramirez, E.; Alves, L.M.; Ambrizzi, T.; Berlato, M.A.; et al. Observed Trends in Indices of Daily Temperature Extremes in South America 1960–2000. *J. Clim.* **2005**, *18*, 5011–5024. [CrossRef]
3. Climate Change Science Program (CCSP). *Weather and Climate Extremes in a Changing Climate. Regions of Focus: North America, Hawaii, Caribbean, and U.S. Pacific Islands*; Karl Thomas, R., Meehl Gerald, A., Miller Christopher, D., Hassol Susan, J., Waple Anne, M., Murray, W.L., Eds.; Department of Commerce, NOAA's National Climatic Data Center: Washington, DC, USA, 2008.
4. Savitskaya, D. Statistical picture of climate changes in Central Asia: Temperature, precipitation, and river flow. In Proceedings of the International Environmental Modelling and Software Society (iEMSs) 2010 International Congress on Environmental Modelling and Software Modelling for Environment's Sake, Ottawa, ON, Canada, 5–8 July 2010; Swayne, D.A., Yang, W., Voinov, A.A., Rizzoli, A., Filatova, T., Eds.; iEMS: Manno, Switzerland, 2010.
5. Collins, J.M. Temperature Variability over Africa. *J. Clim.* **2011**, *24*, 3649–3666. [CrossRef]
6. Ahmed, S.A.; Diffenbaugh, N.S.; Hertel, T.W. Climate volatility deepens poverty. *Environ. Res. Lett.* **2009**, *4*. [CrossRef]
7. Bhavnani, R.; Vordzorgbe, S.; Owor, M.; Bousquet, F. *Disaster Risk Reduction in the Sub-Saharan Africa Region*; Commission of the African Union: Washington, DC, USA, 2008.
8. Badou, D.F.; Kapangaziwiri, E.; Diekkrüger, B.; Hounkpè, J.; Afouda, A. Evaluation of recent hydro-climatic changes in four tributaries of the Niger River Basin (West Africa). *Hydrol. Sci. J.* **2017**, *62*, 715–728. [CrossRef]
9. Hounkpè, J.; Diekkrüger, B.; Badou, D.F.; Afouda, A.A. Change in Heavy Rainfall Characteristics over the Ouémé River Basin, Benin Republic, West Africa. *Climate* **2016**, *4*, 15. [CrossRef]
10. United Nations Development Programme (UNDP). *Inondations au Bénin: Rapport D'évaluation des Besoins Post Catastrophe*; World Bank: Cotonou, Benin, 2011.
11. United Nations Development Programme (UNDP). *Evaluation des Dommages, Pertes et Besoins de Reconstruction Post Catastrophes des Inon-Dations de 2010 au Togo*; World Bank: Lomé, Togo, 2010.
12. Intergovernmental Panel on Climate Change (IPCC). *Climate Change 2013: The Physical Science Basis. Working Group I Contribution to the IPCC Fifth Assessment Report*; Stocker, T.F., Qin, D., Plattner, G.-K., Tignor, M., Allen, S.K., Boschung, J., Nauels, A., Xia, Y., Bex, V., Midgley, P.M., Eds.; IPCC: Cambridge, UK, 2013.
13. Intergovernmental Panel on Climate Change (IPCC). *Climate Change 2014: Synthesis Report. Approved Summary for Policymakers*; Pachauri, R.K., Meyer, L., IPCC, Eds.; IPCC: Cambridge, UK; Geneva, Switzerland, 2014.
14. Klassou, S.D. Evolution Climato-Hydrologique Récente et Conséquences sur L'environnement: L'exemple du Bassin Versant du Fleuve Mono (Togo-Bénin). Ph.D. Dissertation, Bordeaux III, Université Michel de Montaigne, Pessac, France, 1996.
15. Amoussou, E. Variabilité Pluviométrique et Dynamique Bassin Versant du Complexe la Gunaire Mono-Ahémé-Couffo (Afrique de l'Ouest). Ph.D. Dissertation, Université de Bourgogne, Dijon, France, 2010.
16. Gbeyetin, F.J. Inondations Dans la Basse Vallée du Mono: Typologie et Manifestations. Master's Thesis, Université d'Abomey-Calavi, Cotonou, Benin, 2014.
17. Kissi, A.E. *Flood Vulnerability Assessment in Downstream Area of Mono Basin*; Université de Lomé: Lomé, Togo, 2014.

18. Ntajal, J.; Lamptey, B.L.; Sogbedji, M.J.; Wilson-bahun, K.K. Rainfall trends and flood frequency analyses in the lower Mono River basin in Togo, West Africa. *Int. J. Des. Ris. Red* **2017**, *23*, 93–103. [CrossRef]

19. Akinsanola, A.A.; Ogunjobi, K.O.; Gbode, I.E.; Ajayi, V.O. Assessing the Capabilities of Three Regional Climate Models over CORDEX Africa in Simulating West African Summer Monsoon Precipitation. *Adv. Meteorol.* **2015**, *2015*, 1–13. [CrossRef]

20. Lambert, S.J.; Boer, G.J. CMIP1 evaluation and intercomparison of coupled climate models. *Clim. Dyn.* **2001**, *17*, 83–106. [CrossRef]

21. Diallo, I.; Sylla, M.B.; Giorgi, F.; Gaye, A.T.; Camara, M. Multimodel GCM-RCM Ensemble-Based Projections of Temperature and Precipitation over West Africa for the Early 21st Century. *Int. J. Geophys.* **2012**, *2012*. [CrossRef]

22. Giorgi, F.; Coppola, E. Does the model regional bias affect the projected regional climate change? An analysis of global model projections A letter. *Clim. Chang.* **2010**, *100*, 787–795. [CrossRef]

23. Oliver, M.A.; Webster, R. Kriging: A method of interpolation for geographical information systems. *Int. J. Geogr. Inf. Syst.* **1990**, *4*, 312–332. [CrossRef]

24. Matheron, G. Principles of geostatistics. *Econ. Geol.* **1963**, *58*, 1246–1266. [CrossRef]

25. Mckee, T.B.; Doesken, N.J.; Kleist, J. The relationship of drought frequency and duration to time scales. In Proceedings of the Eighth Conference on Applied Climatology, Anaheim, CA, USA, 17–22 January 1993; pp. 179–184.

26. WMO. *Standardized Precipitation Index User Guide*; Svoboda, M., Hayes, M., Wood, D., Eds.; WMO: Geneva, Switzerland, 2012.

27. N'Tcha M'Po, Y.; Lawin, A.E.; Oyerinde, G.T.; Yao, B.K.; Afouda, A.A. Comparison of Daily Precipitation Bias Correction Methods Based on Four Regional Climate Model Outputs in Ouémé Basin, Benin. *Hydrology* **2016**, *4*, 58–71. [CrossRef]

28. Obada, E.; Alamou, E.A.; Chabi, A.; Zandagba, J.; Afouda, A. Trends and Changes in Recent and Future Penman-Monteith Potential Evapotranspiration in Benin (West Africa). *Hydrology* **2017**, *4*, 38. [CrossRef]

29. Bontogho, T.-N.P.E. Modeling a Sahelian Water Resource Allocation under Climate Change and Human Pressure: Case of Loumbila dam in Burkina Faso. Ph.D. Dissertation, Université d'Abomey-Calavi, Cotonou, Benin, 2015.

30. Essou, G.R.C.; Brissette, F. Climate Change Impacts on the Ouémé River, Benin, West Africa. *Earth Sci. Clim. Chang.* **2013**, *4*, 2–11. [CrossRef]

31. Speth, P.; Christoph, M.; Diekkrüger, B. *Impacts of Global Change on the Hydrological Cycle in West and Northwest Africa*; Springer: Berlin/Heidelberg, Germany, 2010; ISBN 9783642129568.

32. Le Barbé, L.; Lebel, T. Rainfall climatology of the HAPEX-Sahel region during the years 1950–1990. *J. Hydrol.* **1997**. [CrossRef]

33. Le Barbé, L.; Lebel, T.; Tapsoba, D. Rainfall Variability in West Africa during the Years 1950–1990. *J. Clim.* **2002**, *15*, 187–202. [CrossRef]

34. Le lay, M.; Galle, S. Variabilité interannuelle et intra-saisonnière des pluies aux échelles hydroloques. La mousson ouest- africaine en climat saoudien. *J. Sci. Hydrol.* **2005**, *50*, 209–224.

35. Adeyeri, O.E.; Lamptey, B.L.; Lawin, A.E.; Sanda, I.S. Spatio-Temporal Precipitation Trend and Homogeneity Analysis in Komadugu-Yobe Basin, Lake Chad Region. *J. Climatol. Weather Forecast.* **2017**, *5*, 1–12. [CrossRef]

36. Nicholson, S.E.; Some, B.; Kone, B. An Analysis of Recent Rainfall Conditions in West Africa, Including the Rainy Seasons of the 1997 El Nino and the 1998 La Nina Years. *J. Clim.* **2000**, *13*, 2628–2640. [CrossRef]

37. Ozer, P.; Erpicum, M.; Demaree, G.; Vandiepenbeeck, M. Discussion of "Analysis of a Sahelian annual rainfall index from 1896 to 2000; the drought continues" The Sahelian drought may have ended during the 1990s. *Hydrol. Sci. J.* **2003**, *48*, 489–496. [CrossRef]

38. Lawin, A.E. Analyse Climatologique et Statistique du Regime Pluviométrique de la Haute Vallee de L'ouémé à Partir des Données Pluviographiques AMMA-CATCH Bénin. Ph.D. Dissertation, Institut National Polytechnique de Grenoble, Université d'Abomey-Calavi, Cotonou, Benin, 2007.

39. Attogouinon, A.; Lawin, E.A.; N'Tcha M'Po, Y.; Houngue, R. Extreme Precipitation Indices Trend Assessment over. *Hydrology* **2017**, *4*, 36. [CrossRef]

40. N'Tcha M'Po, Y.; Lawin, E.; Yao, B.; Oyerinde, G.; Attogouinon, A.; Afouda, A. Decreasing Past and Mid-Century Rainfall Indices over the Ouémé River Basin, Benin (West Africa). *Climate* **2017**, *5*, 74. [CrossRef]

41. Kouakou, A.B.P.; Lawin, E.A.; Kamagaté, B.; Dao, A.; Savané, I.; Srohourou, B. Rainfall Variability across the Agneby Watershed at the Agboville Outlet in Côte d'Ivoire, West Africa. *Hydrology* **2016**, *3*, 43. [CrossRef]

42. Lawin, A.E.; Manirakiza, C.; Batablinlè, L. Trends and changes detection in rainfall, temperature and wind speed in Burundi. *J. Water Clim. Chang.* **2018**. [CrossRef]

43. Badjana, H.M. River Basins Assessment and Hydrologic Processes Modeling for Integrated Land and Water Resources Management (ILWRM) in West Africa. Ph.D. Dissertation, Université d'Abomey-Calavi, Cotonou, Benin, 2015.

44. Kabo-Bah, A.; Diji, C.; Nokoe, K.; Mulugetta, Y.; Obeng-Ofori, D.; Akpoti, K. Multiyear Rainfall and Temperature Trends in the Volta River Basin and their Potential Impact on Hydropower Generation in Ghana. *Climate* **2016**, *4*, 49. [CrossRef]

45. Oguntunde, P.G.; Abiodun, B.J.; Olukunle, O.J.; Olufayo, A.A. Trends and variability in pan evaporation and other climatic variables at Ibadan, Nigeria, 1973–2008. *Meteorol. Appl.* **2012**, *19*, 464–472. [CrossRef]

46. Nelson, G.C.; Rosegrant, M.W.; Palazzo, A.; Gray, I.; Ingersoll, C.; Robertson, R.; Tokgoz, S. *Food Security, Farming, and Climate Change to 2050: Scenarios, Results, Policy Options*; International Food Policy Research Institute: Washington, DC, USA, 2010.

47. Oyerindé, G.T. Climate Change in the Niger River Basin on Hydrological Properties and functions of Kainji Lake, West Africa. Ph.D. Dissertation, Université d'Abomey-Calavi, Cotonou, Benin, 2016.

48. Badou, D.F. Multi-Model Evaluation of Blue and Green Water Availability under Climate Change in Four-Non Sahelian Basins of the Niger River Basin. Ph.D. Dissertation, University of Abomey-Calavi, Cotonou, Benin, 2016.

Article

Multi-Model Forecasts of Very-Large Fire Occurences during the End of the 21st Century

**Harry R. Podschwit [1],*, Narasimhan K. Larkin [2], E. Ashley Steel [3] and Alison Cullen [4]
and Ernesto Alvarado [5]**

[1] College of the Environment Special Programs, Quantitative Ecology & Resource Management (QERM),
   University of Washington, Seattle, WA 98195, USA
[2] Pacific Wildland Fire Sciences Laboratory, U.S. Forest Service, 400 N. 34th St #201, Seattle, WA 98103, USA;
   larkin@fs.fed.us
[3] Department of Statistics and School of Aquatic and Fishery Sciences, University of Washington, Seattle,
   WA 98195, USA; easteel@uw.edu
[4] Daniel J. Evans School of Public Policy and Governance, University of Washington, Seattle, WA 98195, USA;
   alison@uw.edu
[5] Pacific Wildland Fire Sciences Laboratory, School of Environmental and Forest Sciences,
   University of Washington, Seattle, WA 98195, USA; alvarado@uw.edu
* Correspondence: harryp@uw.edu; Tel.: +1-206-849-4225

Received: 9 November 2018; Accepted: 13 December 2018; Published: 19 December 2018

Abstract: Climate change is anticipated to influence future wildfire activity in complicated, and potentially unexpected ways. Specifically, the probability distribution of wildfire size may change so that incidents that were historically rare become more frequent. Given that fires in the upper tails of the size distribution are associated with serious economic, public health, and environmental impacts, it is important for decision-makers to plan for these anticipated changes. However, at least two kinds of structural uncertainties hinder reliable estimation of these quantities—those associated with the future climate and those associated with the impacts. In this paper, we incorporate these structural uncertainties into projections of very-large fire (VLF)—those in the upper 95th percentile of the regional size distribution—frequencies in the Continental United States during the last half of the 21st century by using Bayesian model averaging. Under both moderate and high carbon emission scenarios, large increases in VLF frequency are predicted, with larger increases typically observed under the highest carbon emission scenarios. We also report other changes to future wildfire characteristics such as large fire frequency, seasonality, and the conditional likelihood of very-large fire events.

Keywords: mega-fires; Bayesian-model averaging; model uncertainty; climate-fire models

---

1. Introduction

Although representing only a small fraction of the total number of fires, very-large fires (VLFs) are events often associated with dramatic economic, human health, and environmental risks that are unlike most other wildfires. The most salient and immediate economic impacts are suppression costs and property losses (Barrett 2018, [1]), which are often relatively large in VLFs compared to other smaller events (González-Cabán 1983 [2], Stephens et al., 2014 [3]). In addition to these direct costs, there is a suite of indirect economic impacts—such as damages from post-fire hazards, rehabilitation costs, lost tax and business revenue from community evacuations (Dale 2009 [4])—that are increasingly probable and costly in larger wildfires (Neary et al., 2003 [5], Peppin et al., 2011 [6], Beverly and Bothwell 2011 [7], Beverly et al., 2011 [8]). VLFs have the potential to burn large areas of vegetation and emit tremendous quantities of smoke within a short duration of time, which can adversely impact air quality for months at a time (Stephens et al., 2014 [3]), even at long distances from any active burning (Forster et al., 2001 [9], Val Martin et al., 2013 [10]). The large areas of active burning and sudden increase in air pollutants pose numerous risks to public health (Reid et al., 2016 [11]) and safety (Achtemeier 2009 [12], Stephens et al., 2014 [3]); and hospital admissions and treatment costs are expected to increase during VLFs (Moeltner et al., 2013 [13]). Although there are some ecological benefits of fire, VLFs have also been associated with significant, deleterious, and sometimes irreversible environmental changes. These include the production of environmental conditions conducive to the establishment of invasive species (Crawford et al., 2001 [14]), loss of ecosystem services (Rocca et al., 2014 [15]), and long-term modifications to forest structure (Haffey et al., 2018 [16]). Given the disproportionate costs VLFs pose to economic, social and environmental values, there is a broad need across disciplines, to better understand patterns and trends of their occurrence to mitigate future hazards. There is even greater urgency for this given that multiple lines of evidence suggest that VLF frequency has increased (Williams 2013 [17], Dennision et al., 2014 [18], Barbero et al., 2014 [19]) and will continue to increase into the future (Stavros et al., 2014 [20], Barbero et al., 2015 [21]). Still, there are several challenges to obtaining reliable predictions of future wildfire activity, many of which are related to various kinds of scientific unknowns or uncertainties. These uncertainties can come from many sources (Chen et al., 2018 [22]) and can be associated with a particular quantity of interest—what will the future frequency of very-large fire events be?—or associated with model structures—how are environmental conditions related to very-large fire frequency? The latter is referred to as model or structural uncertainties, which can have significant effects on the conclusions one draws from an analysis (Morgan et al., 1992 [23], Syphard et al., 2018 [24]). In the context of forecasting future wildfire activities, these structural uncertainties can arise from the selection of vegetation models (Sitch et al., 2008 [25], Syphard et al., 2018 [24]), assumed anthropogenic effects (Westerling et al., 2011 [26]), as well as greenhouse gas emmissions and their effects on the environment.

Structural uncertainties associated with the characteristics of the future environment are commonly accounted for in long-term climate impact studies through the use of multiple General Circulation Models (GCMs). Each GCM predicts climatological responses based on unique assumptions regarding chemical and physical interactions between a suite of factors including land, water, atmosphere, and the cryosphere. These models can be used to forecast future climate by forcing the models to observe historical atmospheric conditions and running the models forward using representative concentration pathways (RCPs) as plausible carbon emission scenarios. There are four future RCP scenarios, which are labeled RCP 8.5, RCP 6, RCP 4.5, and RCP 2.6; each assumes various levels of fossil fuel use and economic activity. The suffix labels correspond to the approximate 2100 radiative forcing levels. For instance, the high-emission (RCP 8.5) scenario corresponds to an approximate radiative forcing increase of 8.5 W/m^2 by 2100 compared to pre-industrial conditions. Since the choice of GCM and RCP can be thought of as competing plausible models of the future environment, and the precise climatological response and quantities of greenhouse gases that will be emitted and sequestered prior to 2100 are unknown, it makes sense to interpret them as structural uncertainties (Taylor et al., 2012 [27]).

In addition to the structural uncertainty arising from relating greenhouse gas emission scenarios to climatological impacts, structural uncertainty further arises when relating climatological variables to an impact of interest. In the context of fire, these relationships could include weather patterns such as temperature, precipitation, atmospheric moisture, winds, and clouds. Identifying and describing the relationship between these variables and VLFs is an immensely complex and subjective process, as there can be many competing hypotheses. For instance, temperature controls landscape flammability, is associated with thunderstorm activity and by extension ignition frequency (Flannigan et al., 2009 [28]), mediates tree mortality through drought (Allen et al., 2010 [29]) and insect pests (Bentz et al., 2010 [30]), and influences the length of the snow-free season (Westerling 2016 [31]). Additionally, the timing and amount of precipitation can also influence wildfire behavior in parallel with temperature by controlling the availability of fine fuels (Meyn et al., 2007 [32]), fuel moisture

(Flannigan et al., 2016 [33]), and distribution of flammable species (Bradley et al., 2016 [34]). Multiple weather variables may adequately measure a common phenomenon associated with wildfire risk, such as drought (Zargar et al., 2011 [35]), resulting in highly correlated covariates that when utilized in wildfire risk prediction, produce models with near-identical goodness-of-fit. Hence, although statistical models can be a useful tool for representing and identifying the relative importance of relationships between the environment and fire, they are still only approximations to reality. Confounding of unmeasured variables like vegetation management (Holsinger et al., 2016 [36]) may influence the predicted importance of some weather variables, which can make model selection challenging. In some cases, the same suite of covariates can be used to make predictions with multiple mathematical representations (Mallick and Gelfand 1994 [37]), presenting an additional level of uncertainty that is easily overlooked. Given the frequency with which we face structural uncertainties when modeling highly complex phenomena like wildfire, it is extremely risky to select any single model as an approximation of this phenomena, and a more robust approach should explore results from multiple models (Morgan et al., 1992 [23], Littel et al., 2011 [38]).

Bayesian model averaging is flexible and a commonly used method of accounting for structural uncertainties like these, lessening many of the risks of traditional model selection techniques and improving performance across a variety of metrics (Raftery and Zheng 2003 [39]). Within this framework, model weights, which are assumed to be uncertain, are used to combine covariates or predictions from multiple sources into a single probability distribution. The uncertainties in the model weights are represented using the posterior, which is a probability distribution representing the belief in the model parameters conditional on the observed data. While posteriors provide a natural framework for interpreting uncertainties, in practice, closed form expressions of the posterior are non-trivial and direct calculation is often impossible. Hence, simulation methods like Markov Chain Monte Carlo are typically used to generate samples from the distribution, which are in turn used to approximate the quantities that are of interest to the analyst (Fragoso et al., 2018 [40]). Computational barriers to Markov Chain Monte Carlo techniques have diminished greatly since they were first introduced, and a range of recently developed software options such as JAGS (Plummer 2003 [41]), Stan (Carpenter et al., 2017 [42]), and Integrated Nested Laplace Approximations (Rue et al., 2009 [43]) have facilitated the application of these methods in novel and previously infeasible contexts (Monnahan et al., 2017 [44]).

Hence, in this paper, we account for both kinds of structural uncertainty—uncertainty from the climate models and uncertainty from the choice of VLF models—using Bayesian model averaging to generate predictions of event frequency in the last half of the 21st century in the Continental United States. In Section 2, we present the methods used to produce robust predictions of future wildfire activity using GCMs and multiple fire occurrence models. In Section 3, the results of this analysis are available and demonstrate that increases in VLF activity should be expected in many regions in the Continental United States at the end of the century. In Section 4, we close the paper with a discussion of the implications of this analysis to decision-makers and researchers.

2. Methods

2.1. Fire Occurrence Data

Data from the Monitoring Trends in Burn Severity (MTBS) project [45], which describes individual fire size and severity based on changes in satellite imagery, are used to measure monthly fire occurrence. The original dataset included all detected fire events within the continental United States for the years 1984–2015 and is further filtered to remove all events 404 hectares or smaller, or that were non-wildfire. The filtered data are grouped into 18 regions with broadly similar climate and vegetation characteristics using a geospatial dataset of ecosystem divisions (Figure 1, Bailey 2016 [46]). For each region, two binary time series were constructed: one representing large fire (LF) occurrence, and another representing very-large fire (VLF) occurrence. The LF occurrence time series, $X_{LF,t}$, reports "1" if at least 1 event of at least 404 hectares is recorded during that month; otherwise, it reports

"0". The VLF occurrence time series, $X_{VLF,t}$, records "1" if at least 1 VLF—one that exceeds the 95th percentile of the region's filtered MTBS burn area records (Table 1)—is recorded during that month and region; otherwise, it reports "0". The Marine Division had one fire event and the Subtropical Regime Mountain had no VLF events during 1984–2005, and both were dropped from further consideration, yielding a total of 16 independent ecoregional analyses. All of the time series are split into a tuning dataset (1984–2005) and a training dataset (2006–2015).

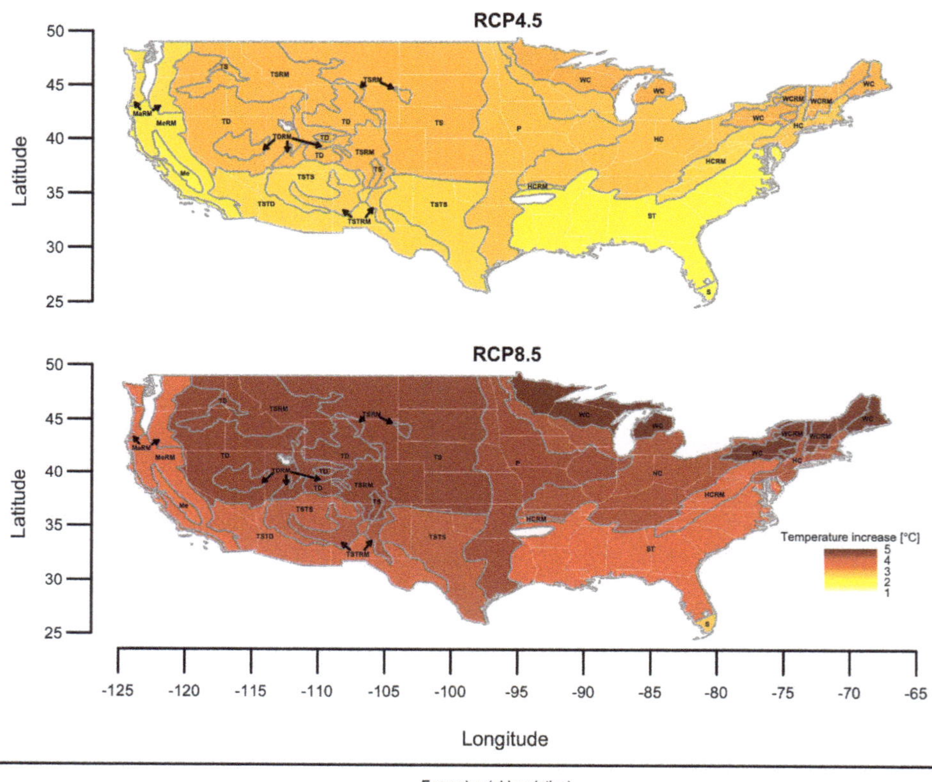

Ecoregion (abbreviation)

Warm Continental Division (WC)	Mediterranean Division (Me)
Marine Regime Mountains Redwood Forest Province (MaRM)	Hot Continental Regime Mountains (HCRM)
Temperate Steppe Regime Mountains (TSRM)	Temperate Desert Regime Mountains (TDRM)
Temperate Steppe Division (TS)	Subtropical Division (ST)
Prairie Division (P)	Tropical/Subtropical Steppe Division (TSTS)
Hot Continental Division (HC)	Tropical/Subtropical Desert Division (TSTD)
Temperate Desert Division (TD)	Tropical/Subtropical Regime Mountains (TSTRM)
Warm Continental Regime Mountains (WCRM)	Savanna Division (S)
Mediterranean Regime Mountains (MeRM)	

Figure 1. Map of multi-model mean temperature changes between 1955–2005 and 2050–2099 over the relevant Bailey's divisions under the representative concentration pathway (RCP) 4.5 (top) and RCP 8.5 emission scenarios (bottom). Regions with insufficient fire occurrence data for analysis are colored white.

Table 1. Very-large fire cutoffs, the number of fires, large fire months, and very-large fire months for two time periods: 1984–2005 and 2006–2015.

Domain	Division	95th Size Percentile (ha)	Total # of Fires		# of Large Fire Months		# of Very-Large Fire Months	
			'84-'05	'06-'15	'84-'05	'06-'15	'84-'05	'06-'15
Dry	Temperate Desert (TD)	16,007	1623	833	119	55	22	19
	Temperate Desert Regime Mountains (TDRM)	11,765	121	78	52	31	5	4
	Temperate Steppe (TS)	11,664	464	347	119	70	13	13
	Temperate Steppe Regime Mountains (TSRM)	19,853	798	541	101	56	14	13
	Tropical/Subtropical Desert (TSTD)	11,616	365	254	94	57	10	9
	Tropical/Subtropical Regime Mountains (TSTRM)	13,169	168	143	71	45	7	7
	Tropical/Subtropical Steppe (TSTS)	10,243	388	546	126	79	10	16
Temperate	Hot Continental (HC)	6180	267	75	67	41	11	2
	Hot Continental Regime Mountains (HCRM)	4586	169	45	30	27	4	1
	Marine Regime Mountains Redwood Forest Province (MaRM)	21,776	136	130	48	33	6	5
	Mediterranean (Me)	10,980	149	64	82	35	6	3
	Mediterranean Regime Mountains (MeRM)	17,772	799	374	143	60	16	17
	Prairie (P)	6707	155	275	47	56	5	9
	Subtropical (ST)	5908	431	312	149	82	16	14
	Subtropical Regime Mountains (SRM)	3927	4	16	3	12	0	1
	Warm Continental (WC)	6466	73	20	40	12	1	4
Humid	Savanna (S)	20,623	89	45	44	24	5	2

2.2. Meteorological Covariates

Regional averages of gridded weather variables from the University of Idaho gridMET dataset [47] are used to calculate 12 weather predictors that will provide coarse scale environmental descriptions during each month between 1984–2015. Of these 12 weather predictors, four are measures of temperature; six are measures of moisture levels, and two measure wind characteristics. The four temperature metrics are based on monthly space-time averages of daily average temperature, which are calculated by dividing the sum of the daily maximum and minimum values by two (Weiss et al., 2005 [48]). The quantity hereafter referred to as seasonality measures intra-annual temperature variability by normalizing monthly temperature averages by the mean and standard deviation of all 360 measurements in the most recent 30 years of data (e.g., 1986–2015). The inter-annual temperature variability is captured with a quantity referred to as the departure from normal, which instead normalizes by the mean and standard deviation of 30 measurements in the most recent 30 years of data that correspond to same month as the raw measurement. The remaining temperature metrics are the rolling 12-month minimum and maximum temperature, which will record extreme temperature events that have potential for delayed impacts on wildfire activity. The six moisture level metrics are average specific humidity and precipitation totals over five time periods (1, 3, 6, 12 and 24-month time windows). In addition to a simple space-time average of wind speed at 10 m, the maximum daily space-time average each month was also included as a covariate of the fire occurrence probabilities.

2.3. Probability Estimation Trees

Two quantities are estimated for each month and region, the probability that at least one LF (>404 hectares) occurs and the probability that a VLF occurs conditional on the occurrence of at least one LF. These probabilities are estimated using multi-model averages of a flexible and powerful type of binary classifier known as a probability estimation tree (PET). PETs use decision-tree structures to recursively divide the data with binary splits, eventually grouping all the data into mutually exclusive categories or leaves. With respect to the response, the splits create increasingly homogeneous clusters of observations, which also occupy an increasingly specific portion of the covariate space. Within the context of this analysis, we have 12 meteorological predictors available to form these categories, so that months—in which certain fire events did or did not occur—can be grouped into categories describing broadly similar environmental conditions. Prediction is performed by using the relevant covariates to identify the appropriate category, and taking the empirical frequency of the binary responses in that category as a probability estimate (Provost and Domingos 2000 [49]). While it is well known that predictions based on individual decision tree algorithms can be highly variable with significant levels of structural instability (Wang et al., 2016 [50]), these pathologies are often lessened through the use of model averaging (Provost and Domingos 2000 [49]). To that end, a suite of 100 PETs are generated for each region and for both probabilities of interest; and we will hereafter refer to each collection of 100 PETs as a 'forest'. Each individual PET within a forest is generated stochastically by applying the C4.5 learning algorithm without pruning (Quinlan 1993 [51]; Provost and Domingos 2000 [49]) to a random sample of the training dataset via the Roughly Balanced Bootstrapping algorithm (Hido et al., 2009 [52]). The LF forests and the conditional VLF forests are constructed somewhat differently in that the LF forests sample from all months in the training dataset, while the VLF forests are based only on samples of months in which at least one LF has occurred. In other words, the LF forests will discriminate between LF and no-fire months, and the VLF forests will discriminate between LF and VLF months. Identification of important predictors within each forest are assessed using two summary statistics: (1) the frequency that a predictor is present in the PETs; and (2) the frequency that a predictor is used in the first split of the PETs. The former identifies the frequency with which a given meteorological predictor is used at all within the forest, and the latter identifies the frequency with which a meteorological predictor is the best determinant of the response on a randomly generated dataset.

2.4. Multi-Model Very-Large Fire Predictions

The structural uncertainties arising from training PETs to observed meteorological data are compounded by structural uncertainties arising from the application of these models to long-term climate forecasts. To improve the quality of the probability estimates and VLF occurrence forecasts, multi-model averages of fire event probabilities are used to integrate both sources of structural uncertainty: the choice of the PET and selection of the climate model. The final probability estimates are assumed to be an average of predictions from all combination pairs of the 100 PETs within each forest and 13 modeled weather datasets; a total of 1300 individual predictions for each region, month, and probability of interest. The modeled weather data used to make PET predictions come from the second version of regional Multivariate Adaptive Constructed Analogs (MACA) [47] dataset that was trained on gridMET, and downscaled with 13 GCMs: bcc-csm1-1-m, BNU-ESM, CanESM2, CCSM4, CNRM-CM5, CSIRO-Mk3-6-0, GFDL-ESM2M, HadGEM2-ES365, inmcm4, IPSL-CM5A-MR, MIROC5, MRI-CGCM3, NorESM1-M.

The multi-model averages are calculated using both unweighted and weighted approaches. The unweighted approach assigns equal weight to each individual prediction and assumes that each PET and climate model is equally credible. The weighted approach assigns unequal weights to predictions from each pairing of PETs and climate models, to try to bias-correct the multi-model averages and optimize predictive performance. The final weight applied to each individual prediction is the product of two independent components: a climate model weight and a PET weight. For a specified region and month, we can write the estimated probabilities as,

$$\bar{p}_{LF} = \sum_{i=1}^{100} \sum_{j=1}^{13} u_{LF,i} v_j p_{LF,i,j},$$

$$\bar{p}_{VLF} = \sum_{i=1}^{100} \sum_{j=1}^{13} u_{VLF,i} v_j p_{VLF,i,j}.$$

Here, $u_{*,i}$ represents the weight applied to the predictions from PET i, v_j represents the weight applied to predictions utilizing climate model j, and $p_{*,i,j}$ is the prediction obtained from PET i utilizing climate model j. We estimate the weight components using a fully Bayesian approach that incorporates fire occurrence and modeled climate forcings from 1984–2005, as well as probabilistic representations of possible parameter estimates. Via Bayes rule, we know that the posterior of the model weight components, $\theta = \vec{u}_{LF}, \vec{u}_{VLF}, \vec{v}$, is proportional to the product of the likelihood and prior probability distributions. The likelihood component represents the probability of observing the fire occurrence time series, $X = \vec{X}_{LF}, \vec{X}_{VLF}$, assuming that they were generated from a Bernoulli process parameterized with our weighted multi-model averages:

$$X_{LF,t} \sim \text{Bernoulli}(\bar{p}_{LF,t}),$$

$$X_{VLF,t} \sim \text{Bernoulli}(\bar{p}_{VLF,t} X_{LF,t}).$$

The prior component, $p(\theta)$, which is a probability density function representing our a priori belief regarding the parameter values, is defined using independent Dirichlet priors with uninformative concentration parameters:

$$\vec{u}_{LF}, \vec{u}_{VLF}, \vec{v} \sim \text{Dirichlet}(\vec{1}).$$

The posterior was approximated using Just Another Gibbs Sampler (JAGS) software (Plummer 2003 [41]) and the runjags package in R (R Development Core Team 2008 [53]). An initial set of 30,000 samples were generated from three parallel Markov Chain Monte Carlo chains using a burn-in interval of 10,000 steps, adaptive phase of 10,000 steps, and thinning interval of 100. Calculations were performed on a MacBook Pro (Quanta Computer, Inc, Shanghai, China) with a 2.7 GHz Intel Core i7 processor (Hillsboro, Oregon, USA). Convergence was monitored visually, and also via the calculation of the potential scale reduction factor, using the range of the central 90th percentile of the marginal posteriors as a test statistic (Brooks and Gelman 1998 [54]). We assume that the second half of the chain has approximately converged if the maximum potential scale reduction factor fell below 1.01. If the chain had not converged, then it was continued in batches of 1000 iterations until the maximum potential scale reduction factor was less than 1.01 (Gelman and Shirley 2011 [55]). To provide guidance to future analysts looking to perform similar analyses, an informal computational comparison of JAGS (Version 4.3.0) and Stan software (Version 2.17.3) was completed, which is described in the supplementary materials. The final VLF probabilities can then be calculated by averaging both probabilities of interest—either with point estimate averages of the model weights or using the full posterior in the weighted approach—and applying Bayes rule (Figure 2). The resulting distribution of VLF probability time series is then used to estimate the changes in VLF frequency in the future by finding the difference between the expected number of VLFs in the historical climate (1956–2005), and the expected number of VLFs in the future (2050–2099) under moderate and severe warming scenarios, RCP 4.5 and RCP 8.5, respectively.

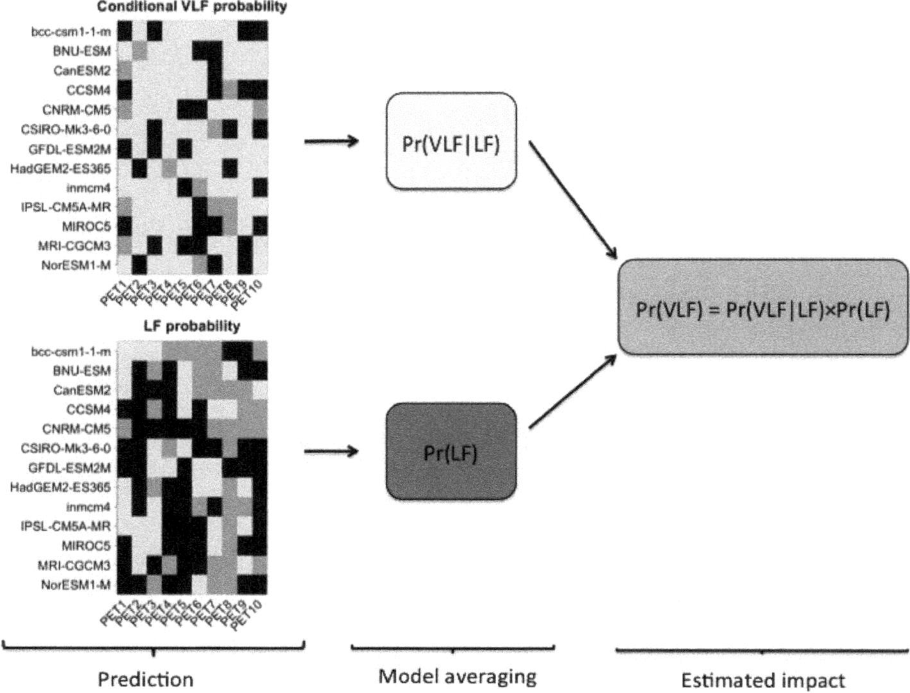

Figure 2. Workflow of very-large fire probability calculations. In the prediction phase, probability estimates are calculated for every combination pair of climate model and probability estimation tree. The model-averaging phase combines these predictions into probability estimates using either a weighted or unweighted averages. The final phase uses Bayes rule to calculate the very-large fire probability as the product of both components from the model-averaging phase.

2.5. Ensemble Assessment

The ability of the multi-model averages to estimate observed VLF frequencies was quantified over the temporal range of the extant MTBS fire occurrence record at three non-overlapping time periods. Two of the three time periods correspond to the tuning (1984–2005), training (2006–2015) datasets that were used to bias correct and fit the initial suite of PETs respectively. Additionally, a testing dataset independent of the information used to build the multi-model averages was constructed using 2016 MTBS occurrence data. For each time period, a sample of 100,000 probability time series were drawn from the relevant multi-model average posterior, which were then used to simulate the distribution of VLF counts predicted during that time period. Note that the 2016 fire data used to independently validate the multi-model averages represent an updated version of the MTBS data that was unavailable during the PET training and tuning stages, and that slight differences in the total number of large (>404 hectares) incidents between 1984–2015 were observed in the two versions. Specifically, the original MTBS dataset reported 10,295 large incidents between 1984–2015, while the updated version reported 10,298 large incidents during that same period.

3. Results

3.1. Important Predictors of Very-Large Fires

The diversity of predictors used in the PETs was high, and the important meteorological variables varied by region, the summary statistic, and the type of fire probability. Temperature metrics, in particular seasonality, are a commonly utilized weather predictor in LF forests, and in 10 of the 16 LF forests, seasonality is present in 90 or more of the PETs. On the other hand, while temperature metrics are frequently utilized when constructing PETs, they are not always the optimal splitting criterion. For instance, in the Savanna, Prairie, and Hot Continental Regime Mountains, precipitation metrics overwhelmingly replace temperature based metrics as the optimal discriminant of large and no fire months, and in other regions such as the Subtropical and Hot Continental Division, this designation is highly uncertain.

The importance of temperature metrics also varied by the type of fire probability considered, with temperature metrics more commonly identified as the optimal split criterion in LF forests compared to VLF forests. This sensitivity of PET structure to the type of fire probability could also arise in other ways. For example, in the Hot Continental Regime Mountains, the LF forest overwhelmingly relies on precipitation metrics for prediction, while the corresponding VLF forest utilizes no predictors and reports a constant conditional VLF probability. Similarly, in the Tropical/Subtropical Regime Mountains and Prairie divisions, conditional VLF forests tended to identify wind metrics as the optimal split criterion much more frequently than in the LF forests. Additionally, PET complexity tended to be lower in VLF forests than in the LF forests. The average number of variables used per PET, size, and the number of leaves were inflated in the latter, and weather invariant null models were only ever observed in the VLF forests. The variability in the optimal splitting criterion was also higher in the VLF forests, suggesting a relative lack of certainty regarding the optimal discriminant in conditional VLF probabilities compared to LF probabilities. Weighting did not appear to drastically influence the relative contribution of the weather predictors within each forest (Figure 3).

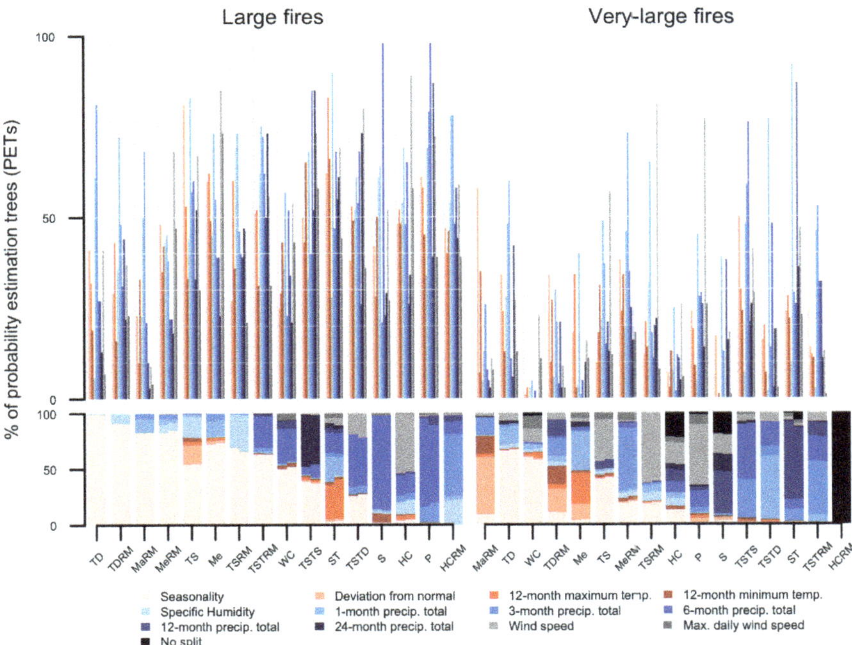

Figure 3. Summary statistics of each forest of probability estimation trees. The top panels show the percentage of probability estimation trees (PETs) in each forest that uses a particular weather predictor for at least one split. The bottom panels show the percentage of PETs for which a weather predictor is selected as the optimal splitting criterion. The relative contribution of each first-split variable under the unweighted (left) and weighted (right) averaging methods are displayed side-by-side.

3.2. Climate Change and Very-Large Fire Occurrence

In most divisions, the expected number of VLFs is predicted to increase in 2050–2099 compared to 1956–2005. The Marine Regime Mountains Redwood Forest division is predicted to have the largest absolute increase with about 13 additional fires per decade under the RCP 4.5 scenario and about 18 additional fires per decade under the RCP 8.5 scenario. For most of the regions under consideration, the average predicted increase ranges from near-zero to several additional VLF per decade relative to historical predictions. In some regions, like Mediterranean and Savanna divisions, the multi-model average predicts slight decreases in VLF activity. The largest absolute decrease occurred in Mediterranean California, which is predicted to have about one less VLF per decade relative to historical predictions under both RCP scenarios. In general, increases in VLF frequency are more severe under the RCP 8.5 scenario than under the RCP 4.5 scenario, although the sensitivity to RCP scenario varies by division (Figure 4). The largest absolute difference in average VLFs per decade between the RCP 8.5 and RCP 4.5 scenarios was in the Marine Regime Redwood Forest division, which had about 5 additional VLFs in the RCP 8.5 scenario. Although the Hot Continental Regime Mountains predicts a larger VLF count per decade under the RCP 4.5 scenario than the RCP 8.5, the difference is negligibly small. The median difference between the RCP 8.5 and RCP 4.5 scenario across the 16 ecoregions was 0.9 additional VLFs per decade under the RCP 8.5 scenario.

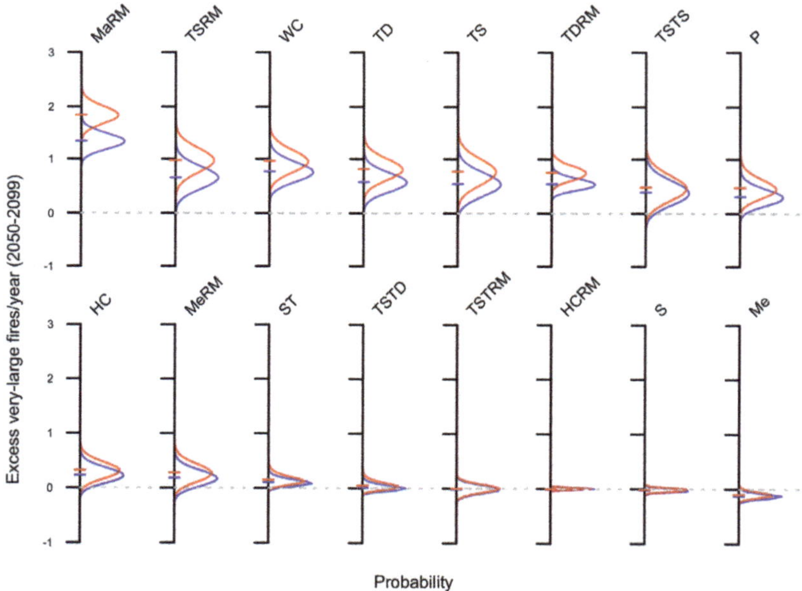

Figure 4. Kernel density estimates of the posterior and mean of the number of additional very-large fires per year relative to the 1956–2005 reference period per year by ecoregion under the representative concentration pathway (RCP) 4.5 (blue) and RCP 8.5 (red) scenarios arranged by magnitude of change. The excess very-large fire frequency is calculated by randomly sampling ($n = 10^6$) from the posterior of historical (1956–2005) and future (2050–2099) multi-model averages and calculating the difference.

Future changes in VLF frequency may or may not be uniformly distributed throughout the year. The largest absolute monthly changes in VLF frequency are observed in the Marine Regime Mountain Redwood Forest division during the summer months, while the shoulder months are not predicted to drastically differ from present day VLF frequency. In contrast to the predictions in the Marine Regime

Mountain Redwood Forest division, semi-uniform changes in VLF frequency are also predicted in some regions. For instance, the Subtropical division is predicted to have about 6-8 additional VLF events during the last half of the 21st century relative to the 1956–2005 reference period, but shows no strong preference as to what month these events will occur. For nearly all regions and months, VLF frequency is predicted to increase or show no change compared to historical reference conditions, with the Prairie division being an example of the former and the Hot Continental Regime Mountains the latter. The Mediterranean division is an exception to this pattern, as reductions in future very-large fire frequency are predicted from October to May (Figure 5).

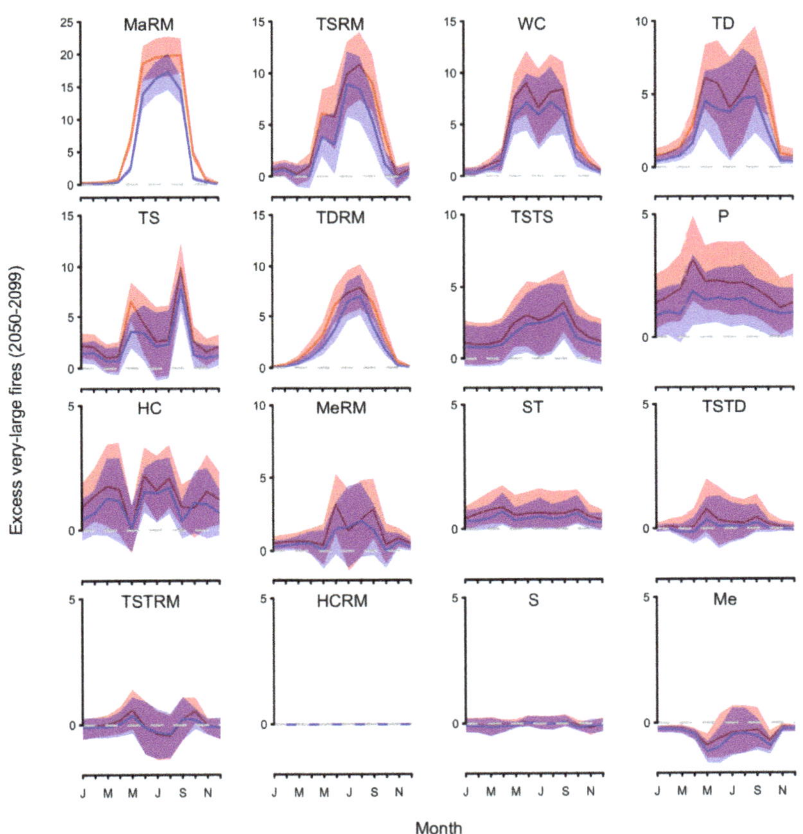

Figure 5. Predicted intra-annual changes in very-large fire frequency across sixteen biogeographical regions within the Continental United States under the RCP 4.5 (blue) and RCP 8.5 (red). The central 90th percentile and mean of the excess very-large fires are based on 1,000,000 random samples of the posterior multi-model average very-large fire probabilities from the historical (1956–2005) and future (2050–2099) scenarios.

The changes in overall VLF frequency are predicted to be a result of changes in both model components: the LF and conditional VLF occurrence probabilities. For divisions like Marine Regime Mountains Redwood Forest, both probabilities increase, implying that the LF months will become increasingly frequent and a larger proportion of the months classified as LF will become VLF months. Other divisions showed increases in only one of the model components. In the Temperate Steppe Regime Mountain division, only conditional VLF probabilities are predicted to increase, and in the Tropical/Subtropical Steppe division, only LF probabilities are anticipated to increase.

Significant decreases in the model components are only predicted in the Mediterranean and Tropical Subtropical Regime Mountains divisions, which respectively have decreases in the conditional VLF and LF probability components in 2050–2099 compared to 1956–2005 climate model forcings. The Mediterranean LF probability components are predicted to increase, while the conditional VLF probability component is expected to remain the same in the Tropical Subtropical Regime Mountains division. In general, the changes in model components are greater in the RCP 8.5 scenario compared to the RCP 4.5 scenario, although the differences between the two future scenarios were nearly imperceptible in some regions (Figure 6).

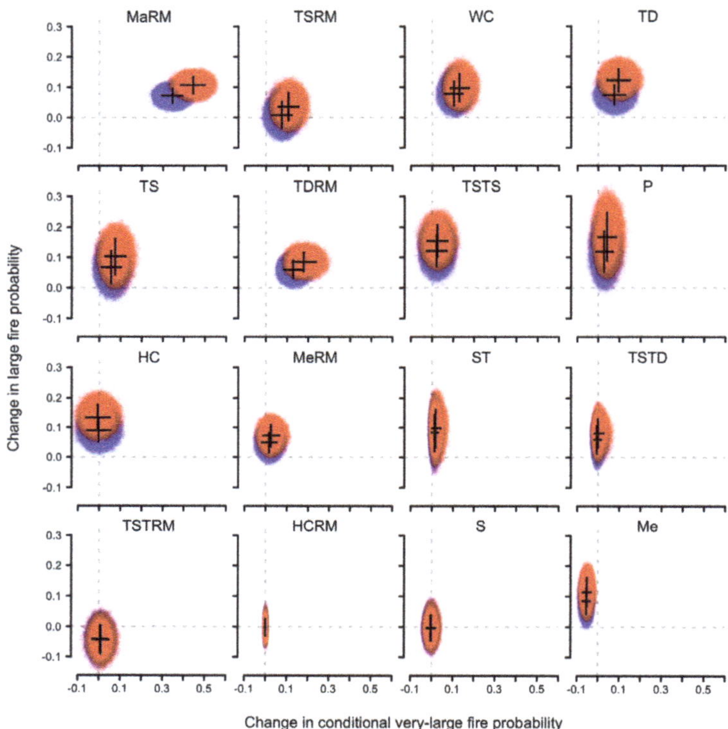

Figure 6. Simulated change in monthly multi-model large-fire and conditional very-large fire probability estimates across biogeographical divisions of the continental United States. The point cloud is a sample of 100,000 differences in average posterior probability components under the historical (1956–2005) and future scenarios (2050–2099); with the RCP 8.5 scenario colored red and RCP 4.5 colored blue. The solid black lines represent the central 90th percentile and the dashed lines are horizontal and vertical lines passing through the origin.

3.3. Ensemble Assessment

The proportion of simulated VLF counts equal to or below the observed values varied by region and time period, and the frequency with which this quantity fell within the central 95th percentile of the simulated VLF counts informs us of the overall quality of the multi-model average forecasts. Using this performance metric, the highest ensemble quality occurs in regions where the central 95th percentile of simulated VLF counts covers the observed VLF counts in all three time periods, which was observed in the Marine Regime Mountains Redwood Forest, Prairie, Hot Continental, Temperate Desert Regime Mountains, and Savanna divisions. In as many regions, this quantity fell in the central 95th percentile for the testing and tuning time periods only, or in the testing and training time periods only. This was observed in the Temperate Steppe Regime Mountains, Temperate Steppe, Temperate Desert, Mediterranean Regime Mountains, and Tropical/Subtropical Steppe divisions. Predictive performance was occasionally poor in the tuning and training time periods, but good during the testing time period, as was observed in the Warm Continental, Subtropical, and Tropical/Subtropical Desert divisions. In the Mediterranean division, the ensembles performed well on the tuning and training time periods, but showed poor performance when predicting data they were not already optimized on. The lowest model quality was seen in the Tropical/Subtropical Regime Mountains, where observed VLF counts were covered by the central 95th percentile in the tuning time period only, and in the Hot Continental Regime Mountains, where the central 95th percentile of the simulated VLF counts never covered the observed quantity.

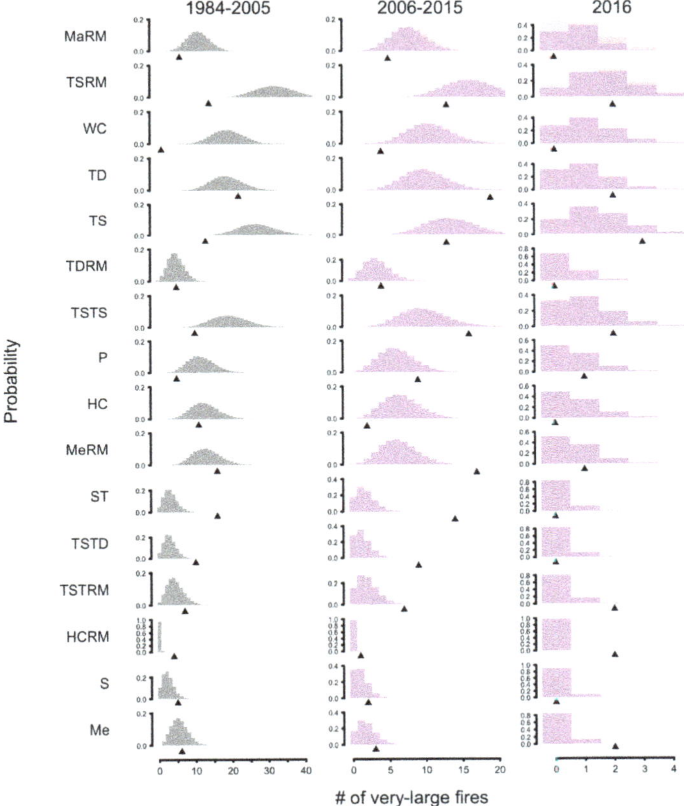

Figure 7. Simulated and actual very-large fire month counts for each region and three time periods: 1984–2005, 2006–2015, and 2016. A sample of 100,000 simulated very-large fire (VLF) counts are produced under historical (grey), RCP 4.5 (blue), and RCP 8.5 (red) scenarios by randomly selecting a VLF probability time series from the posterior and randomly generating a VLF occurrence time series. The observed VLF counts are represented with arrows.

Consistent underestimation, where the observed VLF count was equal to or greater than the median simulated VLF count in all three time periods, was reported in nine of the sixteen regions considered. The magnitude of these underestimates ranged from very minor, as in the Temperate Desert, to quite severe, as in the Hot Continental Regime Mountains. Consistent overestimates were much less frequently observed, with only the Marine Regime Mountains Redwood Forest and Warm

Continental divisions reporting observed VLF counts equal to or less than the median simulated VLF counts in every time period. Five regions had VLF counts that were located to the left or right of the median depending on the time period considered. The Temperate Steppe, Prairie, and Tropical/Subtropical Steppe divisions simulations tended to underestimate the true VLF counts, while the opposite was observed in Temperate Steppe Regime Mountains and Hot Continental divisions.

The simulated distributions did not appear to be strongly sensitive to the choice of RCP scenario during the temporal extent of the training (2006–2015) and testing (2016) time periods, as there are only slight differences between them during those times (Figure 7).

4. Discussion

4.1. Important Predictors of Very-Large Fires

Wildfire events are associated with a number of factors (Flannigan et al., 2009 [28]) that may vary in space (Stavros et al., 2014 [56], Barbero et al., 2015 [57], Arpaci et al., 2013 [58], Flannigan et al., 2006 [59]), and may reveal themselves only under certain conditions (Slocum et al., 2010 [60], Krueger et al., 2015 [61]); it should not then be unexpected that model variability can often be high. Attempts to identify any single factor as most closely associated with VLFs are frustrated by the complex behavior of wildfires, competition among models, data limitations, and diversity of performance criterion. Despite the ubiquity of structural and other uncertainties, the relative importance of various coarse scale meteorological factors to specific wildfire activities could be gauged by observing the frequency with which they were utilized to make predictions. In some cases, a meteorological variable could, with high confidence, be readily identified as important to predicting VLFs in a particular region. In the Temperate Desert division, seasonality was frequently utilized in PETs for both wildfire probabilities, and was also often identified as the optimal splitting criterion. More typically, however, some level of structural uncertainty was present and identifying a best predictor was not always as obvious. In the Subtropical division, LF forest, temperature and precipitation based variables were identified as the optimal splitting criterion with nearly equal frequency. In the Mediterranean Regime Mountains division, seasonality was frequently the optimal splitting criterion in the LF forest, but it was much less common in the VLF forest. Moreover, in the Mediterranean division, wind-based metrics were frequently utilized in LF forests in the Mediterranean forest, but not as a first-split in the PETs. Model variability could be particularly high in the LF forests in the Eastern Continental United States. Precipitation based variables were overwhelmingly preferred in extreme southern Florida and in the Appalachians, but wind based variables were preferred in the Hot Continental division; Temperature was slightly preferred in the Warm Continental division, and as already mentioned, the Subtropical region showed no strong preference with regard splitting criterion. Although some regions showed preferences for certain weather variables, model variability was fairly high in the VLF forests.

Although these structural uncertainties are sometimes obstacles to identifying important meteorological relationships with VLFs, they are also critical to understanding the true level of confidence we have in observed correlations and safeguard against overconfident conclusions. While clearly notable levels of model variability could be encountered across multiple factors, robust patterns and trends could still be inferred. For instance, we note that, in most of the West, with the exception of the Great Plains and the Tropical/Subtropical portions of the Southwest, temperature based metrics were often the best predictor of LFs and were commonly used in LF forests. In the remaining Western areas, temperature metrics were less useful and instead precipitation metrics were selected as the optimal splitting criterion. This apparent preference for precipitation based metrics over temperature based ones in these regions may be related to the characteristics of fuel-limited versus climate-limited fire regimes (Meyn et al., 2007 [32]), or due to a relative inability of seasonal temperature fluctuations to match wildfire activity compared to precipitation. The relative popularity of wind-based variables in very-large forests compared to very-large forests is also interesting, as wind has been reported to have variable influence on wildfire activity depending on fire size and geographic location (Slocum et al., 2010 [60]).

4.2. Climate Change and Very-Large Fire Occurrence

For both RCP scenarios and nearly all divisions, complex changes to wildfire activity are predicted that will result in an overall increase in the frequency of VLFs, which is largely consistent with many other projections (Flannigan et al., 2009 [28], Barbero et al., 2015 [21]). While overall increases in the frequency of these events are predicted using robust methods, the exact nature of these changes remain unclear. It is not certain, for instance, if the range of fire sizes will remain largely static in the future and only frequency of exceptionally large events will increase; or if the size distribution will shift, so that burn areas exceed historic records. These distinctions are important because the relative costs of these two competing possibilities are likely to vary across decision makers. The Mediterranean division was somewhat of an exception to the overall reported increases in VLF activity. Westerling et al. (2011 [26]) project either no change or modest increases in LF activity in much of lowland California, and large increases in mid- and high-elevation locations, which at first seems inconsistent with the predicted decrease in VLF activity, although there are a few explanations. Firstly, by considering a larger number of climate models and predictive models, the range of results in this analysis will be inherently more variable, and marginal results seen in other studies could emerge as significant when these structural uncertainties are incorporated. Secondly, as shown in this study, the environmental drivers of large and conditional VLF probabilities can vary, and differences regarding the definition of LF can result in variability amongst methodologies (Slocum et al., 2010 [60]). Thirdly, differences between the covariates considered and model structure are likely to alter the predictions across analyses. For instance, anthropogenic and vegetation effects on wildfire activity were omitted in this study, but are known to be an important influence of wildfire activity in California (Syphard et al., 2007 [62]) and elsewhere (Syphard et al., 2017 [63]).

The months in which VLF activity was historically highest may not necessarily apply in the second half of the 21st century, and noticeable changes in intra-annual patterns, usually increases, of VLF activity were predicted in most scenarios and regions. Some regions, like Temperature Desert Regime Mountains and Marine Regime Mountains Redwood Forests, are predicted to have increases in VLF frequency only during a limited portion of the year, while others, like the Subtropical division, are predicted to have a relatively uniform increase in VLF frequency throughout the year. Given that simultaneous increases in VLF probabilities are anticipated in multiple independent regions, it is likely that VLF activity will change in ways that will increase resource strain. Indeed, the results of this study suggest that, depending on the emission scenario, between 12–13 regions will have future VLF frequencies that exceed the historical record, and that intra-annual increases in VLF occurrence are often predicted during the same time of year in spatially distinct regions.

In addition to changes to intra-annual patterns of overall VLF frequency, it is important to acknowledge that the overall increases in VLF frequency are the product of two processes: changes in LF and conditional VLF probabilities. Any increase in VLF frequency is then the result of one of three scenarios: an increase in both probabilities, and increase in LF frequency only, or an increase in the frequency that LFs become VLFs. These specific changes in model components may be of particular relevance to firefighting, public health professionals, and other decision-makers who will—due to differences in the impacts of the events—react to no-fire, LF, and VLF months differently and require guidance regarding the characteristics of the novel future wildfire regimes. Reducing the uncertainty as to which emission scenario the future will resemble should also be a priority for decision-makers and researchers, as the predicted changes tend to be more exaggerated under the RCP 8.5 scenario compared to the RCP 4.5, which should influence adaptation and mitigation efforts of future wildfire impacts.

4.3. Caveats and Future Work

While the simultaneous acknowledgement of structural uncertainties in the climate models and PETs represents an interesting approach, there are still a number of uncertainties that were not addressed in this climate impact analysis. The limited availability of reliable and consistently recorded (e.g., satellite-based) measurements of wildfire activity (Taylor et al., 2013 [64]) and the inherent rarity of VLF events remain significant obstacles to validating predictions and estimating underlying model structures. The validation results should be considered as the current state of knowledge regarding the ensemble's predictive ability, and may change when more data becomes available in the future. If inter-annual variability in wildfire activity is high, then the validation results used in this study may be based on particularly predictable or unpredictable fire years, and therefore not be representative of the actual performance. Longer duration datasets would be preferred, and thirty year climatologies are often considered ideal (Arguez and Vose 2011 [65]), but the entire range of available burn area data only extends 33 years and it is unlikely that longer time scale meteorological associations with VLF activity will be accurately captured with the relative brevity of data (Westerling and Swetnam 2003 [66], Marlon et al., 2012 [67]). Moreover, if recent increases in VLF activity are indicative of a sudden a shift into overall wildfire patterns unlike what has been observed in the past, then forecasting future activities based on historical relationships could be inadequate. For instance, the two events occurring in the Hot Continental Regime Mountains in 2016 were quite unusual in historical terms, as only four VLF months were reported from 1984–2005, and only one VLF was reported from 2006–2015.

Data limitations may also be qualitative, and many of the remaining important structural uncertainties are due to unconsidered covariates, like vegetation changes, suppression effort, and population growth, that were not modeled due to data inavailability, practical considerations, and challenges related to predicting these quantities in the future. While the PETs used in this study produced a diverse suite of predictive models and are known to be highly unstable (Wang et al., 2016 [50]), there are many other lingering sources of structural uncertainty that could still be incorporated. For instance, generalized linear models could be used instead, which take a number of mathematical structures depending on the choice of link and response functions (Clyde 2003 [68]). Similarly, various data transformations could be used to generate competing models of the wildfire activities. Alternative models could be constructed that condense the two model components into VLF occurrence probabilities only, so that the event space of each month is purely binary. Instead of biogeographical classification of regions, the Continental United States could be partitioned using administrative or other boundaries to generate VLF predictions relevant to specific stakeholders. Hence, clearly a broad variety of other structural uncertainties still exist that could potentially influence predictions of future VLF frequency in the second half of the century.

It is important to understand that the VLF probabilities do not inform us as to what will actually happen, but rather communicates the degree of uncertainty about future outcomes conditional on carbon emission scenarios. For this reason, some tolerance to deviations between observed and expected VLF frequencies should be considered, as should the fact that the predictions were based on modeled climate data as opposed to direct observations. Still, in many regions, the ensemble performance was relatively adequate and the simulated distribution of fire counts covered the observations. Moreover, when deviations occurred, they tended to underestimate the future VLF counts. Hence, the overall claim that VLF counts will increase in the future under climate change is supported by the results of this study, as well as through the work of others (Stavros et al., 2014 [56], Barbero et al., 2015 [21]). Stochastic uncertainty will be critical when explicitly linking changes to VLF occurrence to human activities and for assessing the future levels of VLF simultaneity (Tedim et al., 2018 [69]) and is a factor that would be well addressed using the methods described here, but is beyond the scope of this paper. The inherent stochasticity of the PET construction process suggests that repeated applications of this methodology in the future may yield slight variations to the results presented here.

Interestingly, a standard factor analysis revealed that more than 86 percent of the variability in predicted probabilities could be attributed to variance amongst the PETs rather than variance amongst the climate models, and while the PETs are an inherently unstable choice of predictive model, this suggests that structural uncertainties should receive the attention of climate impact researchers in much the same way that the choice of climate model does. Further exploration of these structural uncertainties in climate impact analyses cannot be recommended enough in future analyses, as they inform not only of future impacts, but the reliability of these predictions, which can influence decision-maker behaviors in a variety of ways (Weber and Johnson 2009 [70]).

5. Conclusions

While the key conclusion from this research was that fires that were historically considered very-large and rare are likely to become increasingly frequent in most regions of the Continental United States at the end of the 21st century, there are also a number of other complexities in future wildfire activity that may be of further relevance to researchers and decision-makers. For instance, although temperature based metrics were often important for prediction, this analysis also found that the identification of important predictors could be highly uncertain across a number of factors, which should be ignored at one's peril. Moreover, even using the relatively simple probabilistic models we developed, rich details regarding future wildfire activities were constructed that reasonably matched observed fire frequencies and were dynamic in terms of intra-annual trends, fire frequency, simultaneous fire occurrence, and the readiness with which LFs become VLFs. Although overall increases are predicted, we also observed exceptions and regional variability. In the Northwestern United States, VLF frequencies were predicted to increase, with nearly two additional events per year, and increases close to one additional VLF per year were fairly commonly throughout much of the Continental United States as well. In rare instances, the potential for decreases in VLF activity was also reported, most surprisingly in Mediterranean California.

The cumulative impact of these changes are anticipated to affect decision-makers in various ways and the techniques described here have a number of benefits for addressing their needs. For instance, the presented Bayesian model averaging techniques avoids many of the risks of traditional model selection techniques that are especially dangerous when predicting complex phenomena such as wildfire. Moreover, this method simultaneously provides a natural method of calculating important event probabilities that are critical to informed decision-making. While uncertainty in climate models is well understood amongst climate impact researchers, these results highlight the hidden sources of structural uncertainty, and encourage the use of Bayesian model averaging to reconcile them into robust forecasts of future wildfire and other impacts resulting from climate change.

Author Contributions: Conceptualization, N.K.L.; Data curation, N.K.L.; Formal analysis, H.R.P., N.K.L. and E.A.S.; Funding acquisition, N.K.L.; Methodology, H.R.P., N.K.L., E.A.S., A.C. and E.A.; Project administration, N.K.L.; Resources, N.K.L.; Software, H.R.P.; Supervision, N.K.L. and E.A.S.: Visualization, H.R.P., N.K.L. and E.A.S.; Writing—original draft, H.R.P.; Writing—review & editing, H.R.P., N.K.L., E.A.S., A.C. and E.A.. H.R.P. contributed to the paper's methodology, software, formal analysis, writing—original draft preparation, writing—review and editing, and visualization. N.K.L. contributed heavily to the paper's conceptualization, methodology, formal analysis, resources, data curation, writing—review and editing, visualization, supervision, project administration, and funding acquisition. E.A.S. contributed to the methodology, formal analysis, writing—review and editing, visualization, and supervision. A.C. and E.A. contributed to the methodology, as well as the writing—review and editing.

Funding: This research was funded by Joint Fire Science Program project number 11-1-7-4 and by the Joint Venture Agreement 13-JV-11261987-094 between the USFS PNW Research Station and the University of Washington.

Acknowledgments: The authors would like to acknowledge John Abatzoglou and Renaud Barbero for their work that was done in parallel with this analysis. We would also like to thank the members of the AirFire team for their continued support and suggestions throughout this project's lifetime. We are also grateful to the contributions of three anonymous reviewers whose suggestions greatly enhanced the quality of this manuscript.

Conflicts of Interest: The authors declare no conflict of interest.

References

1. Barrett, K. The Full Community Costs of Wildfire. *Headwaters Economics* Available online: https://headwaterseconomics.org/wp-content/uploads/full-wildfire-costs-report.pdf (accessed on 14 December 2018).
2. González-Cabán, A. *Economic Cost of Initial Attack and Large-Fire Suppression*; USDA Forest Service General Technical Report PSW-068; U.S. Department of Agriculture, Forest Service, Pacific Southwest Forest and Range Experiment Station: Berkeley, CA, USA, 1983; 7p.
3. Stephens, S.L.; Burrows, N.; Buyantuyev, A.; Gray, R.W.; Keane, R.E.; Kubian, R.; Liu, S.; Seijo, F.; Shu, L.; Tolhurst, K.G.; et al. Temperate and boreal forest mega-fires: Characteristics and challenges. *Front. Ecol. Environ.* **2014**, *12*, 115–122. [CrossRef]
4. Dale, L. *The True Cost of Wildfire in The Western US*; Western Forestry Leadership Coalition: Denver, CO, USA, 2009.
5. Neary, D.G.; Gottfried, G.J.; Ffolliott, P.F. Post-wildfire watershed flood responses. In Proceedings of the 2nd International Fire Ecology Conference, American Meteorological Society, Orlando, FL, USA, 28 November–2 December 2003; Volume 65982.
6. Peppin, D.L.; Fulé, P.Z.; Sieg, C.H.; Beyers, J.L.; Hunter, M.E.; Robichaud, P.R. Recent trends in post-wildfire seeding in western US forests: Costs and seed mixes. *Int. J. Wildl. Fire* **2011**, *20*, 702–708. [CrossRef]
7. Beverly, J.L.; Bothwell, P. Wildfire evacuations in Canada 1980–2007. *Nat. Hazards* **2011**, *59*, 571–596. [CrossRef]
8. Beverly, J.L.; Flannigan, M.D.; Stocks, B.J.; Bothwell, P. The association between Northern Hemisphere climate patterns and interannual variability in Canadian wildfire activity. *Can. J. For. Res.* **2011**, *41*, 2193–2201. [CrossRef]
9. Forster, C.; Wandinger, U.; Wotawa, G.; James, P.; Mattis, I.; Althausen, D.; Simmonds, P.; O'Doherty, S.; Jennings, S.G.; Kleefeld, C.; et al. Transport of boreal forest fire emissions from Canada to Europe. *J. Geophys. Res. Atmos.* **2001**, *106*, 22887–22906. [CrossRef]
10. Val Martin, M.; Heald, C.L.; Ford, B.; Prenni, A.J.; Wiedinmyer, C. A decadal satellite analysis of the origins and impacts of smoke in Colorado. *Atmos. Chem. Phys.* **2013**, *13*, 7429–7439. [CrossRef]
11. Reid, C.E.; Brauer, M.; Johnston, F.H.; Jerrett; M; Balmes, J.R.; Elliott, C.T. Critical review of health impacts of wildfire smoke exposure. *Environ. Health Perspect.* **2016**, *124*, 1334. [CrossRef] [PubMed]
12. Achtemeier, G.L. On the formation and persistence of superfog in woodland smoke. *Meteorol. Appl.* **2009**, *16*, 215–225. [CrossRef]
13. Moeltner, K.; Kim, M.K.; Zhu, E.; Yang, W. Wildfire smoke and health impacts: A closer look at fire attributes and their marginal effects. *J. Environ. Econ. Manag.* **2013**, *66*, 476–496. [CrossRef]
14. Crawford, J.A.; Wahren, C.H.; Kyle, S.; Moir, W.H. Responses of exotic plant species to fires in Pinus ponderosa forests in northern Arizona. *J. Veg. Sci.* **2001**, *12*, 261–268. [CrossRef]
15. Rocca, M.E.; Miniat, C.F.; Mitchell, R.J. Introduction to the regional assessments: Climate change, wildfire, and forest ecosystem services in the USA. *For. Ecol. Manag.* **2014**, *327*, 8. [CrossRef]
16. Haffey, C.; Sisk, T.D.; Allen, C.D.; Thode, A.E.; Margolis, E.Q. Limits to Ponderosa Pine Regeneration following Large High-Severity Forest Fires in the United States Southwest. *Fire Ecol.* **2018**, *14*, 143–163.
17. Williams, J. Exploring the onset of high-impact mega-fires through a forest land management prism. *For. Ecol. Manag.* **2013**, *294*, 4–10. [CrossRef]
18. Dennison, P.E.; Brewer, S.C.; Arnold, J.D.; Moritz, M.A. Large wildfire trends in the western United States. *Geophys. Res. Lett.* **2014**, *41*, 2928–2933. [CrossRef]
19. Barbero, R.; Abatzoglou, J.T.; Steel, E.A.; Larkin, N.K. Modeling very large-fire occurrences over the continental United States from weather and climate forcing. *Environ. Res. Lett.* **2014**, *9*, 124009. [CrossRef]
20. Stavros, E.N.; Abatzoglou, J.T.; McKenzie, D.; Larkin, N.K. Regional projections of the likelihood of very large wildland fires under a changing climate in the contiguous Western United States. *Clim. Chang.* **2014**, *126*, 455–468. [CrossRef]
21. Barbero, R.; Abatzoglou, J.T.; Larkin, N.K.; Kolden, C.A.; Stocks, B. Climate change presents increased potential for very large fires in the contiguous United States. *Int. J. Wildl. Fire* **2015**, *24*, 892–899. [CrossRef]

22. Chen, J.; Brissette, F.P.; Poulin, A.; Leconte, R. Overall uncertainty study of the hydrological impacts of climate change for a Canadian watershed. *Water Resour. Res.* **2011**, *47*. [CrossRef]

23. Morgan, M.G.; Henrion, M.; Small, M. *Uncertainty: A Guide to Dealing With Uncertainty in Quantitative Risk and Policy Analysis*; Cambridge University Press: Cambridge, UK, 1992.

24. Syphard, A.D.; Sheehan, T.; Rustigian-Romsos, H.; Ferschweiler, K. Mapping future fire probability under climate change: Does vegetation matter? *PLoS ONE* **2018**, *13*, e0201680. [CrossRef] [PubMed]

25. Sitch, S.; Huntingford, C.; Gedney, N.; Levy, P.E.; Lomas, M.; Piao, S.L.; Betts, R.; Cias, P.; Cox, P.; Friedlingstein, P.; et al. Evaluation of the terrestrial carbon cycle, future plant geography and climate-carbon cycle feedbacks using five Dynamic Global Vegetation Models (DGVMs). *Glob. Chang. Biol.* **2008**, *14*, 2015–2039. [CrossRef]

26. Westerling, A.L.; Bryant, B.P.; Preisler, H.K.; Holmes, T.P.; Hidalgo, H.G.; Das, T.; Shrestha, S.R. Climate change and growth scenarios for California wildfire. *Clim. Chang.* **2011**, *109*, 445–463. [CrossRef]

27. Taylor, K.E.; Stouffer, R.J.; Meehl, G.A. An overview of CMIP5 and the experiment design. *Bull. Am. Meteorol. Soc.* **2012**, *93*, 485–498. [CrossRef]

28. Flannigan, M.D.; Krawchuk, M.A.; de Groot, W.J.; Wotton, B.M.; Gowman, L.M. Implications of changing climate for global wildland fire. *Int. J. Wildl. Fire* **2009**, *18*, 483–507. [CrossRef]

29. Allen, C.D.; Macalady, A.K.; Chenchouni, H.; Bachelet, D.; McDowell, N.; Vennetier, M.; Kitzberger, T.; Rigling, A.; Breshears, D.D.; Hogg, E.H.; et al. A global overview of drought and heat-induced tree mortality reveals emerging climate change risks for forests. *For. Ecol. Manag.* **2010**, *259*, 660–684. [CrossRef]

30. Bentz, B.J.; Régnière, J.; Fettig, C.J.; Hansen, E.M.; Hayes, J.L.; Hicke, J.A.; Kelsey, R.G.; Negrón, J.F.; Seybold, S.J. Climate change and bark beetles of the western United States and Canada: Direct and indirect effects. *BioScience* **2010**, *60*, 602–613. [CrossRef]

31. Westerling, A.L. Increasing western US forest wildfire activity: Sensitivity to changes in the timing of spring. *Philos. Trans. R. Soc. B* **2016**, *371*, 20150178. [CrossRef] [PubMed]

32. Meyn, A.; White, P.S.; Buhk, C.; Jentsch, A. Environmental drivers of large, infrequent wildfires: The emerging conceptual model. *Prog. Phys. Geogr.* **2007**, *31*, 287–312. [CrossRef]

33. De Groot, W.J.; Johnston, J.; Jurko, N.; Cantin, A.S. Fuel moisture sensitivity to temperature and precipitation: Climate change implications. *Clim. Chang.* **2016**, *134*, 59–71.

34. Bradley, B.A.; Curtis, C.A.; Chambers, J.C. Bromus response to climate and projected changes with climate change. In *Exotic Brome-Grasses in Arid and Semiarid Ecosystems of the Western US*; Springer International Publishing: New York, NY, USA, 2016; pp. 257–274.

35. Zargar, A.; Sadiq, R.; Naser, B.; Khan, F.I. A review of drought indices. *Environ. Rev.* **2011**, *19*, 333–349. [CrossRef]

36. Holsinger, L.; Parks, S.A.; Miller, C. Weather, fuels, and topography impede wildland fire spread in western US landscapes. *For. Ecol. Manag.* **2016**, *380*, 59–69. [CrossRef]

37. Mallick, B.K.; Gelfand, A.E. Generalized linear models with unknown link functions. *Biometrika* **1994**, *81*, 237–245. [CrossRef]

38. Littell, J.S.; McKenzie, D.; Kerns, B.K.; Cushman, S.; Shaw, C.G. Managing uncertainty in climate-driven ecological models to inform adaptation to climate change. *Ecosphere* **2011**, *2*, 1–19. [CrossRef]

39. Raftery, A.E.; Zheng, Y. Discussion: Performance of Bayesian model averaging. *J. Am. Stat. Assoc.* **2003**, *98*, 931–938. [CrossRef]

40. Fragoso, T.M.; Bertoli, W.; Louzada, F. Bayesian model averaging: A systematic review and conceptual classification. *Int. Stat. Rev.* **2018**, *86*, 1–28. [CrossRef]

41. Plummer, M. JAGS: A program for analysis of Bayesian graphical models using Gibbs sampling. In Proceedings of the 3rd International Workshop on Distributed Statistical Computing, Vienna, Austria, 20–22 March 2003.

42. Carpenter, B.; Gelman, A.; Hoffman, M.D.; Lee, D.; Goodrich, B.; Brubaker, M.; Guo, J.; Betancourt, M.; Li, P.; Riddell, A. Stan: A probabilistic programming language. *J. Stat. Softw.* **2017**, *76*. [CrossRef]

43. Rue, H.; Martino, S.; Chopin, N. Approximate Bayesian inference for latent Gaussian models by using integrated nested Laplace approximations. *J. R. Stat. Soc.* **2009**, *71*, 319–392. [CrossRef]

44. Monnahan, C.C.; Thorson, J.T.; Branch, T.A. Faster estimation of Bayesian models in ecology using Hamiltonian Monte Carlo. *Methods Ecol. Evol.* **2017**, *8*, 339–348. [CrossRef]

45. Monitoring Trends in Burn Severity. Available online: http://www.mtbs.gov (accessed on 21 November 2017).

46. Bailey, R.G. *Bailey's ecoregions and subregions of the United States, Puerto Rico, and the U.S. Virgin Islands*; Forest Service Research Data Archive: Fort Collins, CO, USA, 2016. Available online: https://doi.org/10.2737/RDS-2016-0003 (accessed on 14 December 2018).

47. Climatology Lab. Available online: http://www.climatologylab.org (accessed on 29 January 2018).

48. Weiss, A.; Hays, C.J. Calculating daily mean air temperatures by different methods: Implications from a non-linear algorithm. *Agric. For. Meteorol.* **2005**, *128*, 57–65. [CrossRef]

49. Provost, F.; Domingos, P. *Well-Trained PETs: Improving Probability Estimation Trees, CeDER Working Paper #IS-00-04, Stern School of Business*; New York University: New York, NY, USA, 2000.

50. Wang, H.; Yang, F.; Leu, Z. An experimental study of the intrinsic stability of random forest variable importance measures. *BMC Bioinf.* **2016**, *17*, 60. [CrossRef] [PubMed]

51. Quinlan, J.R. *C4.5: Programs for Machine Learning*; Elsevier: Amsterdam, The Netherlands, 2014.

52. Hido, S.; Kashima, H.; Takahashi, Y. Roughly balanced bagging for imbalanced data. Statistical Analysis and Data Mining: *ASA Data Sci. J.* **2009**, *2*, 412–426. [CrossRef]

53. Development Core Team R. A Language and Environment for Statistical Computing. 2008. Available online: http://www.R-project.org (accessed on 14 December 2018).

54. Brooks, S.P.; Gelman, A. General methods for monitoring convergence of iterative simulations. *J. Comput. Gr. Stat.* **1998**, *7*, 434–455.

55. Gelman, A.; Shirley, K. Inference from simulations and monitoring convergence. In *Handbook of Markov chain Monte Carlo*; CRC Press: Boca Raton, FL, USA, 2011; pp. 163–174.

56. Stavros, E.N.; Abatzoglou, J.; Larkin, N.K.; McKenzie, D.; Steel, E.A. Climate and very large wildland fires in the contiguous western USA. *Int. J. Wildl. Fire* **2014**, *23*, 899–914. [CrossRef]

57. Barbero, R.; Abatzoglou, J.T.; Kolden, C.A.; Hegewisch, K.C.; Larkin, N.K.; Podschwit, H. Multi-scalar influence of weather and climate on very large-fires in the Eastern United States. *Int. J. Climatol.* **2015**, *35*, 2180–2186. [CrossRef]

58. Arpaci, A.; Eastaugh, C.S.; Vacik, H. Selecting the best performing fire weather indices for Austrian ecoregions. *Theor. Appl. Climatol.* **2013**, *114*, 393–406. [CrossRef] [PubMed]

59. Flannigan Mike, D. Forest fires and climate change in the 21 st century. *Mitig. Adapt. Strateg. Global Chang.* **2006**, *11*, 847–859. [CrossRef]

60. Slocum, M.G.; Beckage, B.; Platt, W.J.; Orzell, S.L.; Taylor, W. Effect of climate on wildfire size: A cross-scale analysis. *Ecosystems* **2010**, *13*, 828–840. [CrossRef]

61. Krueger, E.S.; Ochsner, T.E.; Engle, D.M.; Carlson, J.D.; Twidwell, D.L. Soil Moisture Affects Growing-Season Wildfire Size in the Southern Great Plains. *Soil Sci. Soc. Am. J.* **2015**, *79*, 1567–1576. [CrossRef]

62. Syphard, A.D.; Radeloff, V.C.; Keeley, J.E.; Hawbaker, T.J.; Clayton, M.K.; Stewart, S.I.; Hammer, R.B. Human influence on California fire regimes. *Ecol. Appl.* **2007**, *17*, 1388–1402. [CrossRef] [PubMed]

63. Syphard, A.D.; Keeley, J.E.; Pfaff, A.H.; Ferschweiler, K. Human presence diminishes the importance of climate in driving fire activity across the United States. *Proc. Natl. Acad. Sci. USA* **2017**, 201713885. [CrossRef] [PubMed]

64. Taylor, S.W.; Woolford, D.G.; Dean, C.B.; Martell, D.L. Wildfire prediction to inform fire management: Statistical science challenges. *Stat. Sci.* **2013**, *28*, 586–615. [CrossRef]

65. Arguez, A.; Vose, R.S. The definition of the standard WMO climate normal: The key to deriving alternative climate normals. *Bull. Am. Meteorol. Soc.* **2011**, *92*, 699–704. [CrossRef]

66. Westerling, A.L.; Swetnam, T.W. Interannual to decadal drought and wildfire in the western United States. *EOS* **2003**, *84*, 545–555. [CrossRef]

67. Marlon, J.R.; Bartlein, P.J.; Gavin, D.G.; Long, C.J.; Anderson, R.S.; Briles, C.E.; Brown, K.J.; Colombaroli, D.; Hallett, D.J.; Power, M.J.; et al. Long-term perspective on wildfires in the western USA. *Proc. Natl. Acad. Sci. USA* **2012**, *109*, E535–E543. [CrossRef] [PubMed]

68. Clyde, M. Model averaging. *Subject. Object. Bayesian Stat.* **2003**, *25*, 320–326.

69. Tedim, F.; Leone, V.; Amraoui, M.; Bouillon, C.; Coughlan, M.R.; Delogu, G.M.; Fernandes, P.M.; Ferreira, C.; McCaffrey, S.; McGee, T.K.; et al. Defining extreme wildfire events: Difficulties, challenges, and impacts. *Fire* **2018**, *1*, 9. [CrossRef]
70. Weber, E.U.; Johnson, E.J. Decisions under uncertainty: Psychological, economic, and neuroeconomic explanations of risk preference. *Neuroeconomics* **2009**, *2009*, 127–144.

Article

Objective Definition of Climatologically Homogeneous Areas in the Southern Balkans Based on the ERA5 Data Set

Christos J. Lolis *, Georgios Kotsias and Aristides Bartzokas

Laboratory of Meteorology, Department of Physics, University of Ioannina, 45110 Ioannina, Greece; gkotsias@cc.uoi.gr (G.K.); abartzok@uoi.gr (A.B.)
* Correspondence: chlolis@uoi.gr; Tel.: +30-26510-08472

Received: 15 November 2018; Accepted: 5 December 2018; Published: 7 December 2018

Abstract: An objective definition of climatologically homogeneous areas in the southern Balkans is attempted with the use of daily $0.25° \times 0.25°$ ERA5 meteorological data of air temperature, dew point, zonal and meridional wind components, Convective Available Potential Energy, Convective Inhibition, and total cloud cover. The classification of the various grid points into climatologically homogeneous areas is carried out by applying Principal Component Analysis and K-means Cluster Analysis on the mean spatial anomaly patterns of the above parameters for the 10-year period of 2008 to 2017. According to the results, 12 climatologically homogenous areas are found. From these areas, eight are mainly over the sea and four are mainly over the land. The mean intra-annual variations of the spatial anomalies of the above parameters reveal the main climatic characteristics of these areas for the above period. These characteristics refer, for example, to how much warmer or cloudy the climate of a specific area is in a specific season relatively to the rest of the geographical domain. The continentality, the latitude, the altitude, the orientation, and the seasonal variability of the thermal and dynamic factors affecting the Mediterranean region are responsible for the climate characteristics of the 12 areas and the differences among them.

Keywords: Mediterranean climate; cluster analysis; objective classification; ERA5

1. Introduction

The Mediterranean Sea is located in a transitional climatic area between Europe, Africa, and Asia and its climate is widely known as "the Mediterranean climate". The Mediterranean climate is generally characterized by considerable seasonal variations in almost all climatic parameters, for example air temperature, cloudiness, precipitation, lightning activity, etc. [1–5]. Although these seasonal variations appear over the whole Mediterranean region, they present significant differences among the various subregions. These differences are connected to various atmospheric and geographical factors. The position and the variability of the large-scale atmospheric circulation systems (e.g., the subtropical anticyclone of the North Atlantic and the south Asian summer low) and the global atmospheric oscillations affecting the region are dominant atmospheric circulation factors connected to the significant spatial variability of climate within the Mediterranean region [6–8]. The North Atlantic Oscillation (NAO), which refers to the sea level pressure seesaw between the Icelandic low and the subtropical anticyclone of the North Atlantic, affects the atmospheric conditions over most of the Mediterranean region. Specifically, positive/negative values of the NAO index are generally connected to below/above normal cyclonic activity and precipitation over the Mediterranean [9,10]. The Arctic Oscillation (AO) is connected to the variation of the intensity of the polar low. Positive values of the AO index are generally associated with anticyclonic conditions and dryness over the Mediterranean, while negative AO index values are connected to the transfer of cold masses towards the south and the prevalence of cyclonic activity and above normal precipitation over the same region [11,12]. The North Sea–Caspian Pattern (NCP) is a pressure seesaw between the North Sea and the Caspian Sea. A positive phase of NCP is accompanied by northeasterly flow over a part of the Eastern Mediterranean (Balkans and Western Turkey), while a negative phase is accompanied by southwesterly flow over the same regions [13,14]. Furthermore, the spatial variability of climate characteristics is enhanced by the complicated geographical features of the region (coastline, relief, land–sea alteration), which affect significantly most of the climatic parameters including temperature, cloudiness, precipitation, and wind. For example, the windward or leeward character of a specific region, as well as its distance from the sea, plays an important role on its cloudiness and precipitation regimes. These regimes are strongly affected by the associated atmospheric humidity and static stability levels and the frequent prevalence of adiabatic sink (warming) or rise (cooling) [3,15]. Considering the above atmospheric and geographical features of the Mediterranean region and the associated high spatial variability of its climate, climatological studies, which can be carried out with the use of high-resolution meteorological data, would lead to interesting results referring to the detailed spatial characteristics of the Mediterranean climate. This is further supported by the fact that the future climatic changes, as they are forecasted by the climatic models, are also expected to be characterized by high spatial variability over the same region [16,17].

The climate of the Balkans, a Mediterranean subregion of great climatological interest, has been extensively studied by means of the spatial regimes of specific climatic parameters [18,19] and their connection to specific atmospheric circulation modes [20]. The impacts of the acting teleconnections, global warming, and local surface warming (land–atmosphere interactions) reveal high spatial complexity over the region. Most studies involve station meteorological data and/or Reanalysis data of spatial resolution up to $1° \times 1°$. Moreover, an attempt for a climate regionalization of the Balkan Peninsula has been made with the application of Cluster Analysis on station sea level pressure data over the region [21]. Nowadays, the recently introduced ERA5 high resolution meteorological data set [22] provides the scientific community with an extra useful tool for studying the spatial variability of climate over a region with relatively complex relief. The southern Balkans is such a region and the present study aims at examining this variability by defining climatologically homogeneous subregions within it, i.e., subregions with characteristic seasonal variations of the main climatic parameters. This is achieved with the application of a multivariate statistical scheme including Principal Component Analysis and Cluster Analysis on the high-resolution ERA5 meteorological data set. The temporal availability of the data set is at the moment restricted to the recent 10-year period of 2008 to 2017. Although this period is relatively short, not being sufficient for the full establishment of the statistical parameters connected to the climate of the region, the general characteristics of the spatial variability of climate, especially those that are significantly influenced by the complicated relief, can be highlighted by means of the high spatial resolution and the reliability of the recently introduced data set.

2. Materials and Methods

The data used in the present study are daily (00UTC and 12UTC) $0.25° \times 0.25°$ grid point values of air temperature (AT), dew point temperature (DP), zonal (ZW) and meridional (MW) wind components, Convective Available Potential Energy (CAPE), Convective Inhibition (CIN), and total cloud cover (TCC) for the southern Balkans area (19°–29° E, 34°–42° N) (Figure 1) for the 10-year period of 2008 to 2017, obtained from the ERA5 Reanalysis data set [22]. The selection of the above parameters has been made taking into account that their values over the examined area are directly connected to the climate of the region, by either determining its main characteristics (AT, DP, ZW, MW, and TCC) or being responsible for the in situ extreme precipitation events related to thunderstorms (CAPE and CIN). This is not the case for other parameters for example sea level pressure or geopotential height, which affect the climate characteristics of the region indirectly and remotely and have to be examined over a broader area. Also, the data corresponds to 00UTC and 12UTC hours in order to involve both midnight and midday atmospheric conditions, which are generally different especially during the warm period of the year, mainly because of the intense daytime land warming and the development of small-scale circulations (e.g., see breezes). ERA5 is a recently introduced ECWF data set, which provides hourly values of many atmospheric, land, and oceanic parameters at a horizontal resolution of 31 km on 137 levels from the surface up to 0.01 hPa (~80 km above the earth's surface). It combines large quantities of historical observations into global estimates with the use of advanced modeling and data assimilation procedures [22]. For each of the above parameters (AT, DP, ZW, MW, CAPE, CIN, and TCC) and time (00UTC and 12UTC), the 2008–2017 long-term mean spatial anomaly patterns are calculated for each of the 365 calendar days of the year. The spatial anomaly pattern of a specific parameter for a specific calendar day is calculated by subtracting the spatial average from the value of each grid point. Thus, a matrix containing all the long-term mean spatial anomaly patterns of the above parameters at 00UTC and 12UTC for the 365 calendar days of the year is constructed. Each column of the matrix corresponds to a specific parameter, a specific hour (00UTCor 12UTC) and a specific calendar date of the year, while each line corresponds to a specific grid point of the study area.

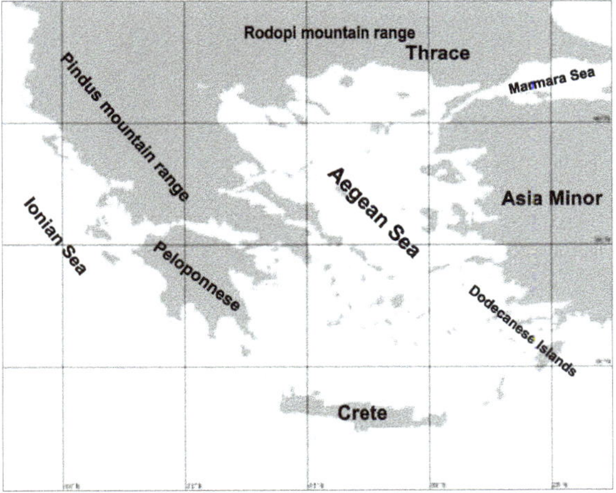

Figure 1. The geographical domain used.

Principal Component Analysis (PCA), with varimax rotation, is applied on the above matrix as a dimensionality reduction tool. PCA is a multivariate statistical method which projects a set of possibly correlated variables onto a set of uncorrelated variables, which are called principal components. Only the statistically significant components are used for the next step and their number is indicated by the SCREE plot and the physical hypostasis of the results [23,24]. Next, K-Means Cluster Analysis (CA) is applied on the time series of the standardized significant principal components in order to group grid points, and thus to define the areas with homogenous climate characteristics regarding the spatial anomalies of specific climatic parameters during specific sub-periods of the year. CA is a statistical method that classifies cases of a set of variables into objectively defined distinct and homogeneous clusters. The squared Euclidean distance is selected to be the measure of similarity, while the k-means technique succeeds in the continuous rearrangement of the cases in new clusters optimizing the final classification [25–27]. The optimum number of clusters is indicated by the distortion test [28]. For the grid points classified into each of the clusters, the mean intra-annual variations of all the climatic parameters are constructed. These intra-annual variations are smoothed by averaging the daily values over each of the 73 (365/5) 5-day periods of the year. In this way the main climate characteristics of the objectively defined areas regarding the magnitude of each climatic parameter, relatively to the spatial average, during the year, are revealed. The methodology scheme, which is followed in the present study and is described in the above paragraphs, is presented in Figure 2. Finally, a comparison between the ERA5 and ERA-Interim data sets is carried out for the common period of 2008 to 2017.

This comparison involves air temperature and total cloud cover, parameters which are connected to the most significant climate characteristics, and it is performed separately for the land and the sea areas. For this purpose, daily values of ERA-Interim $1° \times 1°$ grid point data of air temperature and total cloud cover are also used [29].

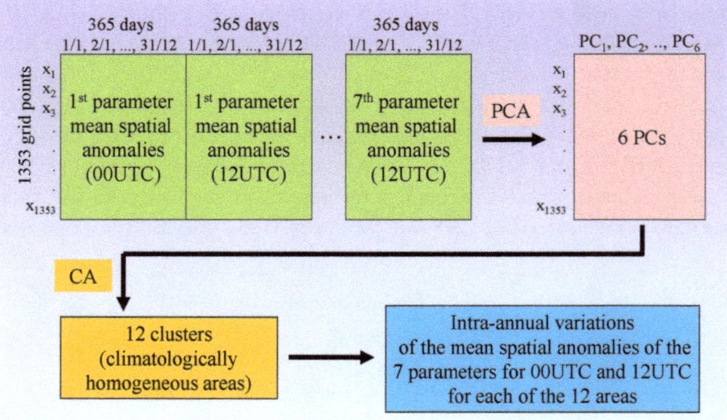

Figure 2. The methodology scheme used in the present study.

3. Results

The application of PCA leads to six PCs accounting for 74% of the total variance and the application of CA leads to 12 clusters (Figure 2). The 12 clusters correspond to specific geographical areas, which are presented in Figure 3. Each area is characterized by specific seasonal variations of the climatic parameters' spatial anomalies, which are presented in Figures 4–9. The main climatic characteristics of each cluster (subregion) are presented in the following paragraphs.

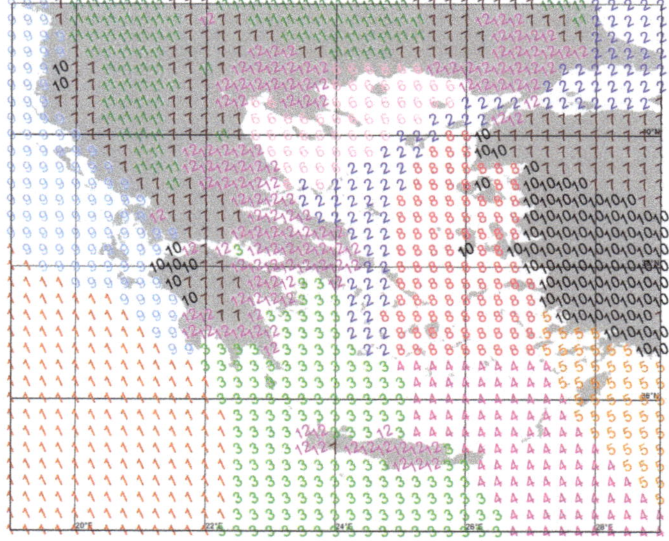

Figure 3. The spatial distribution of the 12 clusters.

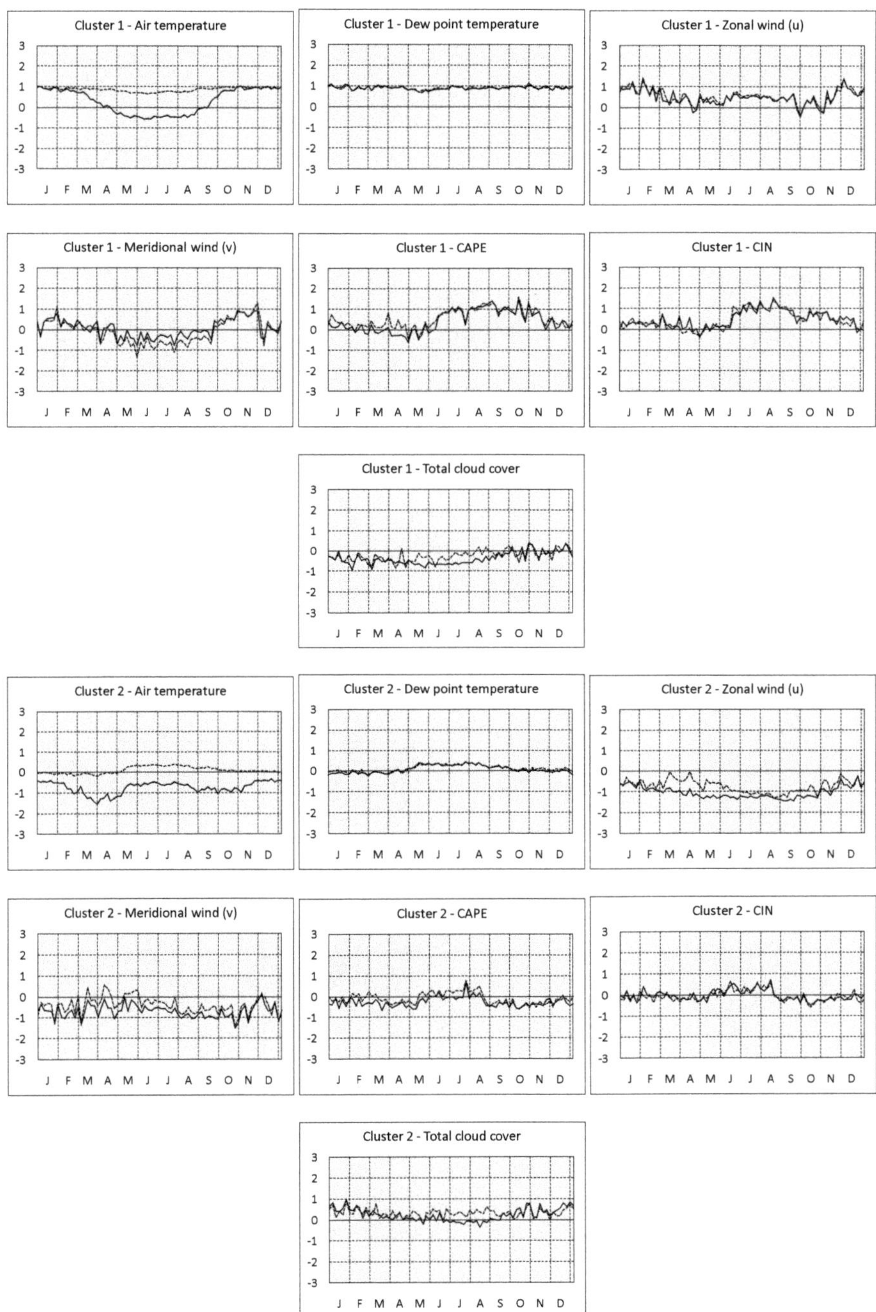

Figure 4. Mean intra-annual variations of the spatial anomalies of the meteorological parameters for the areas of clusters 1 and 2. Continuous and dashed lines correspond to 12UTC and 00UTC respectively.

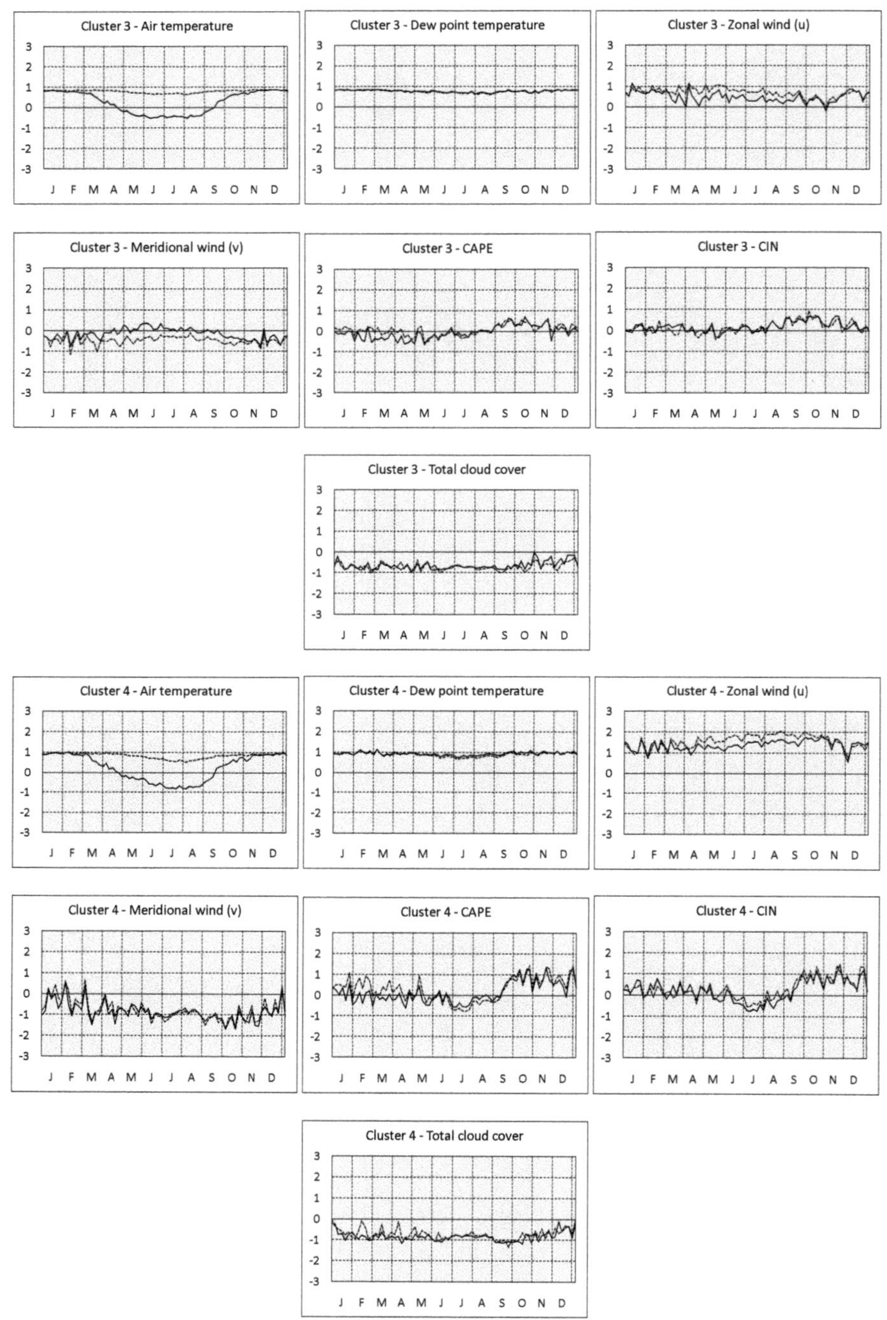

Figure 5. As in Figure 4, but for clusters 3 and 4.

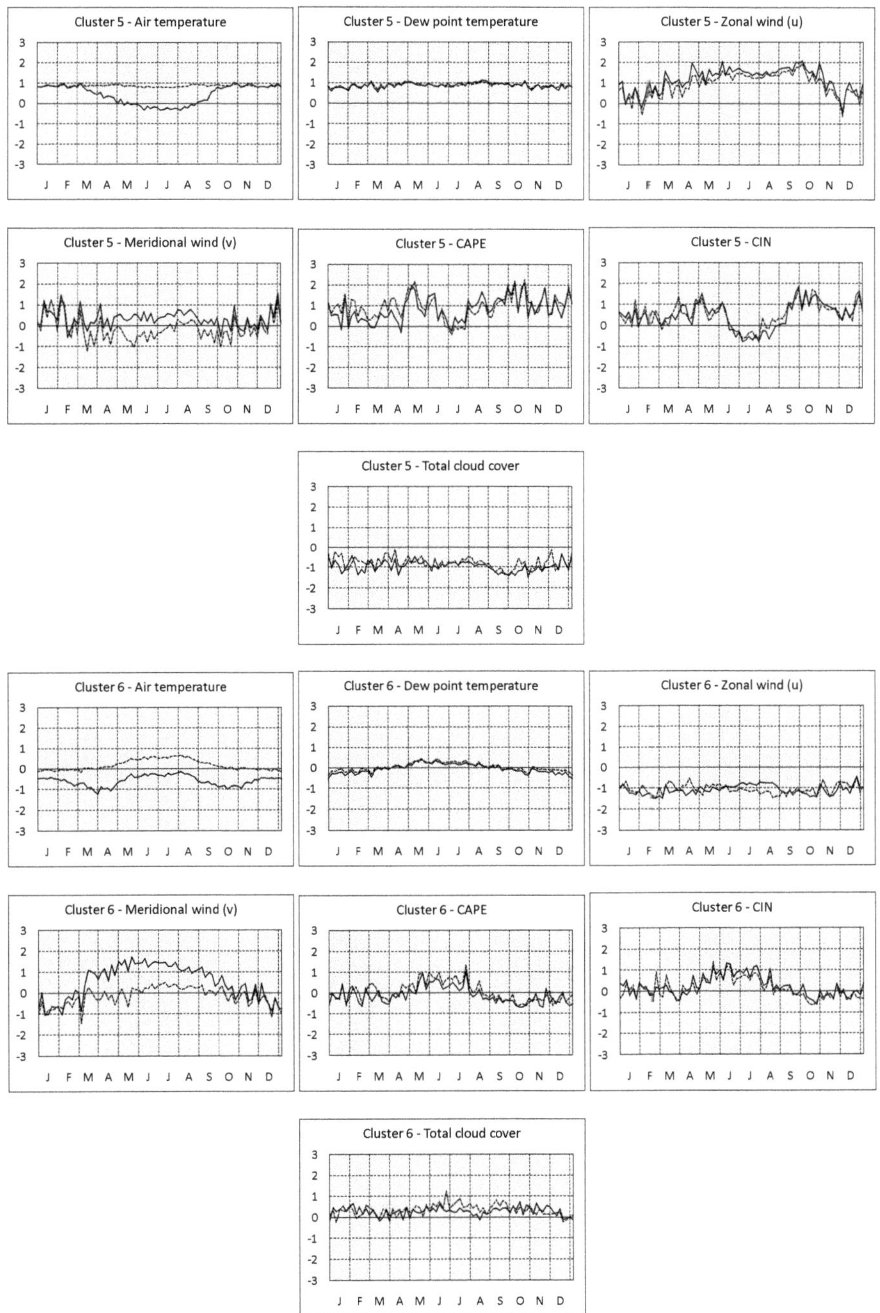

Figure 6. As in Figure 4, but for clusters 5 and 6.

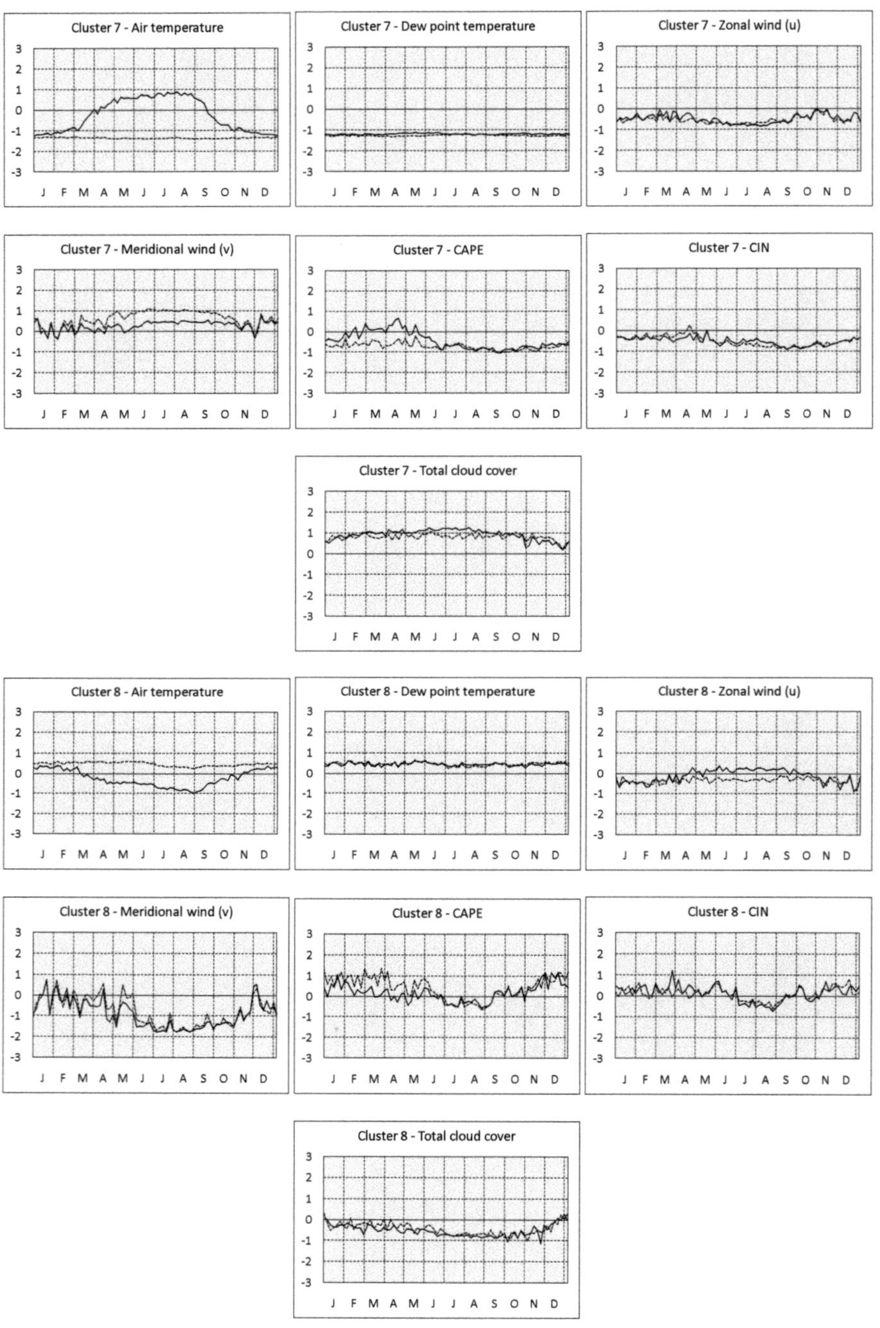

Figure 7. As in Figure 4, but for clusters 7 and 8.

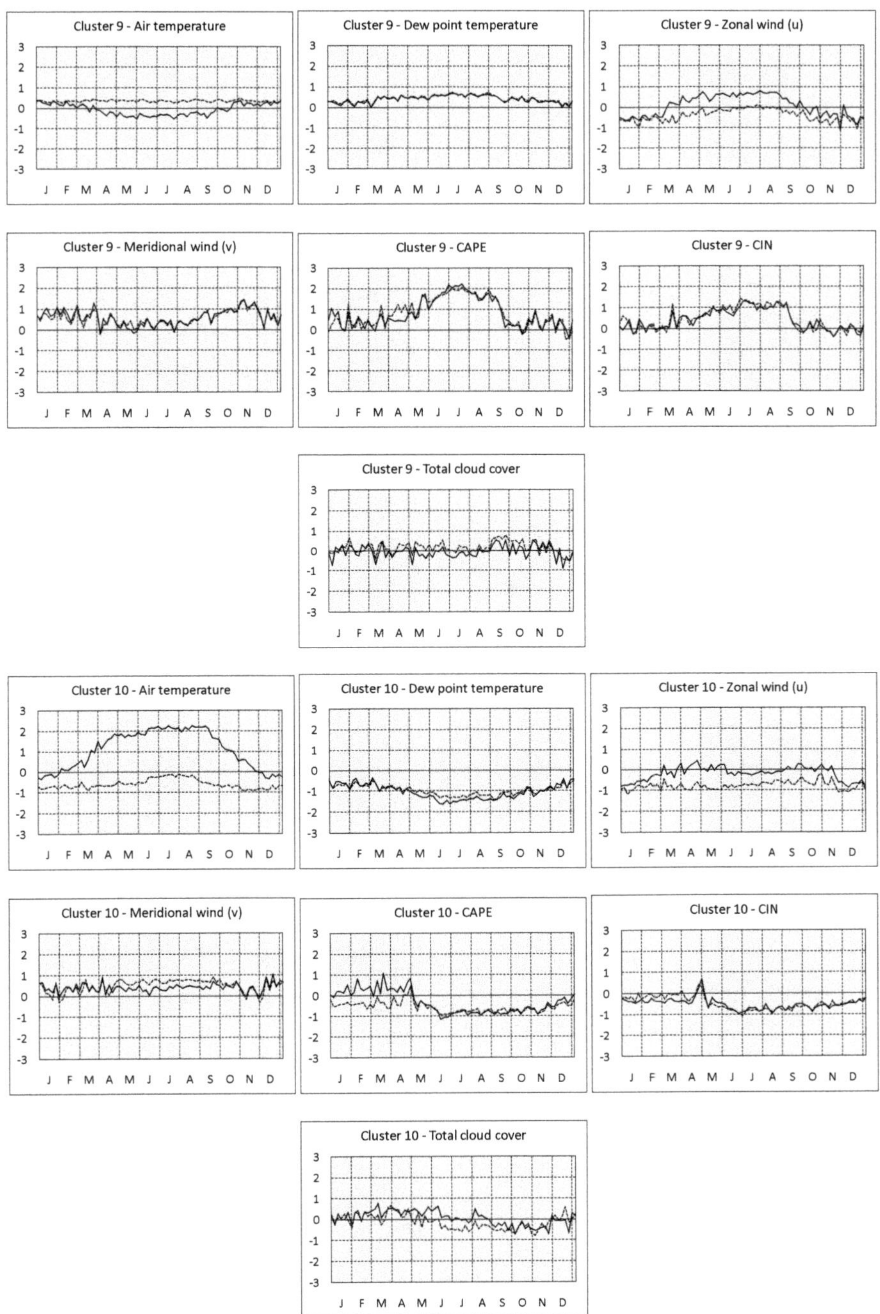

Figure 8. As in Figure 4, but for clusters 9 and 10.

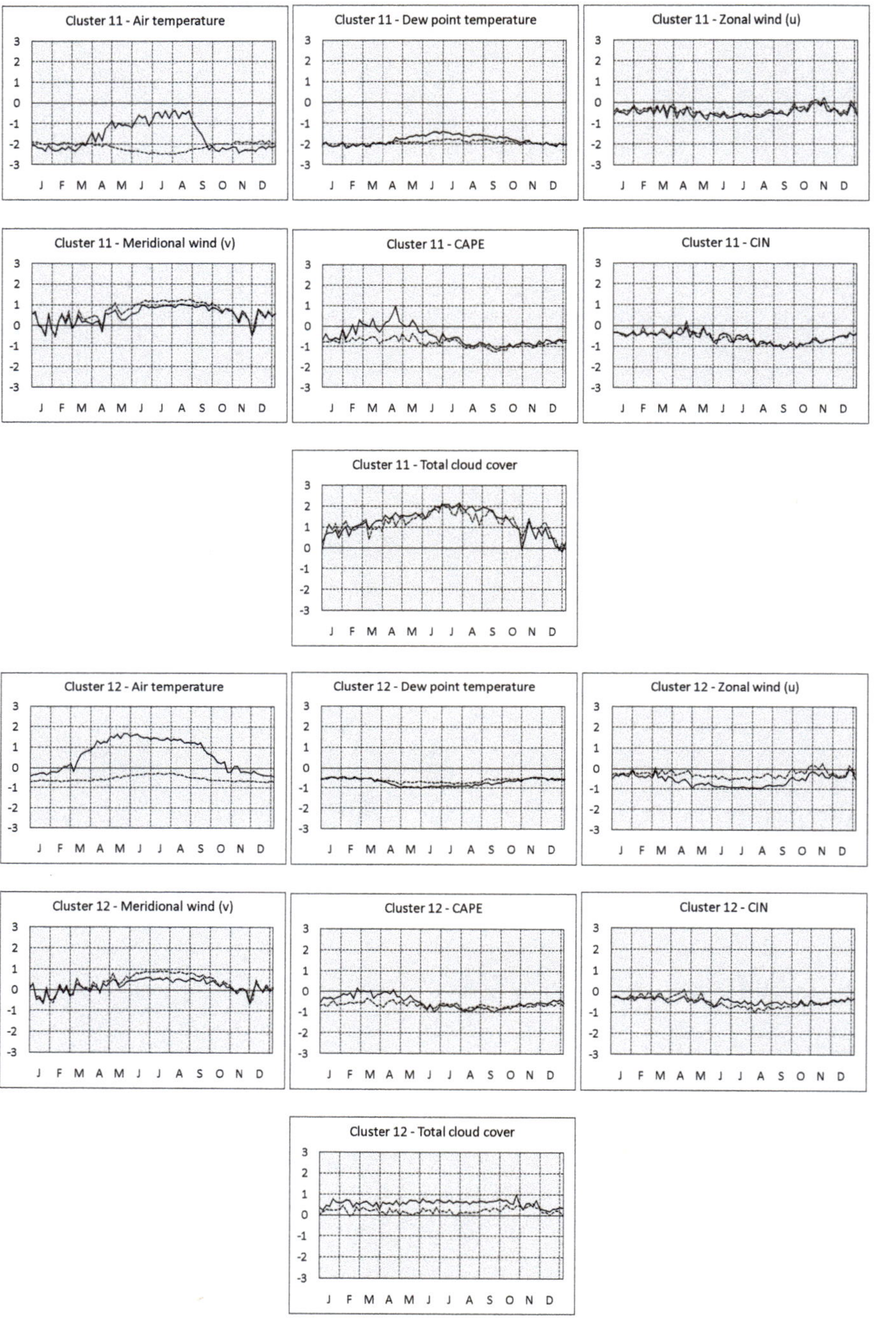

Figure 9. As in Figure 4, but for clusters 11 and 12.

Cluster 1 comprises the grid points of the southern Ionian Sea (Figure 3). The intra-annual variations of the anomalies of the climatic parameters in this area are presented in Figure 4. It is seen that 12UTC air temperature over the southern Ionian is higher than the spatial average from the middle of September to the middle of April by one standard deviation, while it is lower than the spatial average by half standard deviation during the rest of the year. This is not valid for 00UTC air temperature which is equally higher than the spatial average during the whole year. These seasonal variations of temperature anomalies are due to the fact that at night sea-surface is generally warmer than land during the whole year, while during daytime this is valid only for the cold period as for the warm period the high insolation leads to intense land warming and highest temperature anomalies over the land. For the rest of the parameters, there is not any notable difference between 00UTC and 12UTC intra-annual variations, except for cloud cover, which presents a small difference in summer. Dew point is about one standard deviation above the spatial average during the whole year. The zonal wind component is higher than the spatial average during almost the whole year with highest values in winter, while the meridional wind component is highest during late autumn. Both wind seasonal variations are in agreement with the high frequency of southwesterly winds associated with the passages of Mediterranean depressions during late autumn and winter. CAPE and CIN spatial anomalies are highest in summer and early autumn when they are approximately one standard deviation higher than the spatial average. This agrees with the findings of Lolis [30] for the spatial distribution of CAPE in the Mediterranean region. Finally, total cloud cover is generally lower than the spatial average during the whole year except autumn, while for late spring and summer 12UTC anomalies are considerably lower than 00UTC anomalies. The low cloud cover values relatively to the spatial average during late spring and summer are due to the convective cloud development over the land during the same period [1].

Cluster 2 contains a relatively narrow geographical zone mainly covered by the sea, which is extended from the western central Aegean to the Marmara Sea and the Black Sea, while it includes also a part of Eastern Thrace (Figure 3). According to the intra-annual variations of the anomalies (Figure 4), it appears that the nighttime temperature is near the spatial average during the whole year, while the daytime temperature is below the spatial average. The prevalence of low daytime temperature anomalies is probably a result of the frequent and/or strong northeast winds in both the cold and the warm period appearing in the anomalies of the wind components. The northeast winds over this area are mainly a result of stationary synoptic conditions, which in the warm period are associated with the prevalence of the etesian winds [31,32]. The CAPE and CIN anomalies do not present considerable variations during the year, while cloud cover is generally above the spatial average during the whole year. An exception exists for summer during daytime when the spatial maximum of cloud cover is located over the land because of the intense land warming there.

Clusters 3, 4, and 5 comprise the southern Aegean and the sea area south of Crete and the Dodecanese islands (Figure 3). According to the intra-annual variations of the spatial anomalies (Figures 5 and 6), the above area is characterized by a remarkable difference between the daytime and nighttime temperature anomalies during the warm period, which was the case for the previously analyzed neighboring sea area of cluster 1 (southern Ionian Sea). Other characteristics of the areas of clusters 3, 4, and 5 are the positive anomalies of the zonal wind component during the whole year, the autumn CAPE and CIN maxima and the negative anomalies of cloud cover which are mainly associated with the low latitude and the large distance from the land in summer and the absence of orographic effect in winter. The main differences among the areas of the above three clusters mainly refer to the sign and the magnitude of the meridional wind, CAPE and CIN anomalies. Specifically, one of the main differences is that the area of cluster 5 is characterized by high CAPE and CIN variations relatively to the other two areas, while two maxima appear instead of one, the first in spring and the second in autumn.

Cluster 6 corresponds to the northern Aegean (Figure 3). In this area, the 12UTC air temperature anomalies are remarkably lower than the 00UTC ones, not only during the warm period as is the case for the rest of the sea areas, but during the whole year (Figure 6). This can be attributed to the frequent advection of cold masses from the neighboring continental areas of the Balkans to the Aegean via the frequent northeasterly winds in winter. This justification is also supported by the negative anomalies of the zonal and meridional wind components in winter. In the warm period, it is seen that the daytime meridional wind component anomalies are remarkably higher than the nighttime ones, which can be attributed to the effect of the southerly sea breeze from the Aegean to the coasts of Macedonia and Thrace and the weak etesian winds (relative to the rest of the Aegean Sea). Regarding the rest of the parameters, cloud cover is slightly above the spatial average, while CAPE and CIN are higher than the spatial average during late spring and early summer.

Cluster 7 comprises inland areas which are mainly areas of intermediate altitudes between the plains and the mountainous ones (Figure 3). Such areas are shown over both the Balkan Peninsula and northwestern Asia Minor. According to the intra-annual variation of the anomalies (Figure 7), their climate is characterized by low temperatures during winter and high daytime and low nighttime temperatures during summer. The low nighttime temperature anomalies prevailing during the whole year are connected to the inland character of the areas favoring nighttime radiative cooling and the presence of high altitude mountainous areas in their vicinity favoring the development of mountain breezes. Another characteristic of cluster 7 is the high cloud cover values associated with the windward character of the regions favoring the formation of orographic clouds in the cold period and the inland character of the region favoring convective cloud development in the warm period [1].

Cluster 8 comprises the area of central and eastern Aegean Sea (Figure 3), which presents most of the climate characteristics found for the rest of sea areas. The remarkable difference between daytime and nighttime temperature anomalies during summer and the high dew point anomalies during the whole year are among these characteristics, while furthermore it has to be noted that the high negative meridional wind anomalies are in agreement with the fact that the north etesian winds are very frequent, persistent, and strong over this area [32].

Cluster 9 corresponds to the north Ionian Sea and the coasts of northwestern Greece and Albania (Figure 3). The main characteristics of the seasonal variations of the anomalies (Figure 8) are (i) the difference between daytime and nighttime air temperatures in summer, which has been found for most sea areas; (ii) the difference between daytime and nighttime zonal wind component anomalies in the warm period, which is due to the sea breeze circulation between the Ionian Sea and the Balkan Peninsula; and (iii) the broad and strong summer maximum of CAPE (approximately two standard deviations above the spatial average), which in agreement with the climatology of CAPE in the Mediterranean region [30].

Cluster 10 comprises the areas of northwestern Peloponnese and southwestern Asia Minor (Figure 3). These areas are characterized by very high daytime air temperature in the warm period of the year relative to the spatial average, low values of dew point during the whole year, low values of CAPE and CIN during summer and autumn, and approximately one standard deviation difference between daytime and nighttime zonal wind anomalies (Figure 8). The high 12UTC summer temperature anomalies are due to the intense land warming and the katabatic character of the etesian winds over the above areas.

Cluster 11 corresponds to the high altitude mountainous areas of the southern Balkans, which include the Pindus and Rodopi mountain ranges (Figure 3). The very low air temperature and dew point values and the very high values of cloudiness relatively to the spatial averages are the main climate characteristics of these areas. Also, the spring maximum of daytime CAPE is responsible for the frequent appearance of air mass thunderstorms during the same season, as this maximum is not accompanied by a corresponding maximum of CIN (Figure 9).

Cluster 12 comprises mainly the plains of eastern Greek mainland, Thrace, and Crete (Figure 3). The climate over these areas is relatively warm and dry during the whole year as it can be seen in the seasonal variations of temperature and dew point. Furthermore, upper air static stability appears to be stronger relatively to the other areas, while cloud cover is slightly higher than the spatial average, especially at noon (Figure 9).

Finally, in order to validate the results that are based on the ERA5 database, an attempt is made to compare them with that of the ERA-Interim database. The ERA-Interim data used refer to $1° \times 1°$

grid point values and the parameters of air temperature and total cloud cover (00UTC and 12UTC) are selected as an example. The 12 clusters are separated into two categories: land (clusters 7, 10, 11, and 12) and sea (clusters 1, 2, 3, 4, 5, 6, 8, and 9). For the integer coordinates of the two datasets (the common grid points) and for the above two parameters, the scatterplots of the spatial mean daily values are constructed and are presented for the land and sea clusters in Figures 10 and 11. According to the results, there is a very high linear correlation between the two datasets for both land and sea areas: R^2 is higher than 0.98 for air temperature, while it is equal to 0.92 for cloud cover. These values imply a high degree of covariability between the two data sets even for cloud cover, which is a very sensitive parameter to the variations of dynamic and geomorphologic factors. The agreement between the two data sets is also reflected to the mean intra-annual variations of the spatial anomalies of air temperature and total cloud cover for the 12 clusters. These variations have been calculated also for ERA-Interim data set for all clusters and the results are found to be similar to the ones of ERA5. The small differences found can be mainly attributed to the different resolution of the data sets. In Figure 12, an example of this comparison is presented for clusters 1 and 11, which correspond to sea and mountainous areas, respectively. As it can be seen, the intra-annual variations of air temperature and cloud cover for both 00UTC and 12UTC are quite similar and this is also supported by the corresponding correlation coefficients. For cluster 1, these correlation coefficients are 0.95 and 0.99 for 00UTC and 12UTC air temperature and 0.81 and 0.88 for 00UTC and 12UTC total cloud cover, respectively. The corresponding coefficients for cluster 11 are 0.94 and 0.90 for 00UTC and 12UTC air temperature and 0.89 and 0.98 for 00UTC and 12UTC total cloud cover, respectively. The high correlation coefficients enhance the reliability of the results and confirm the suitability of both data sets for such climatic studies.

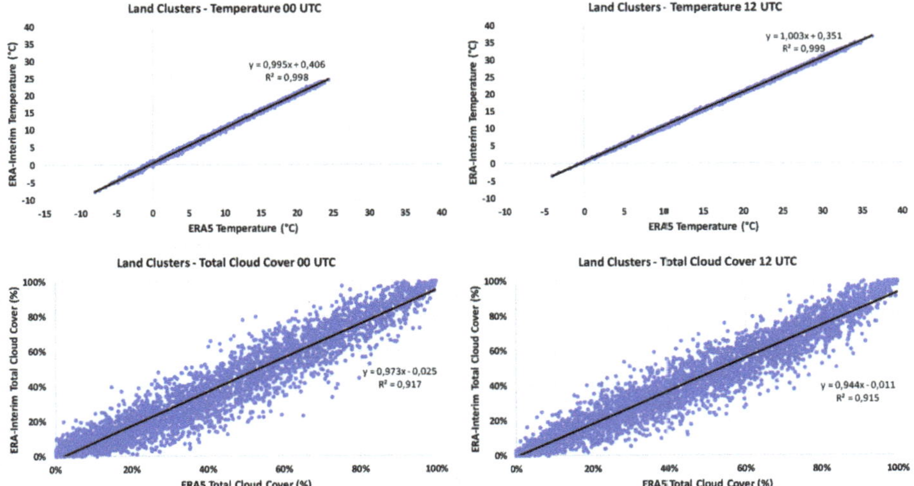

Figure 10. Scatterplots of the spatial mean values of ERA-Interim and ERA5 air temperature and total cloud cover for the areas of clusters 7 and 10–12 (land areas).

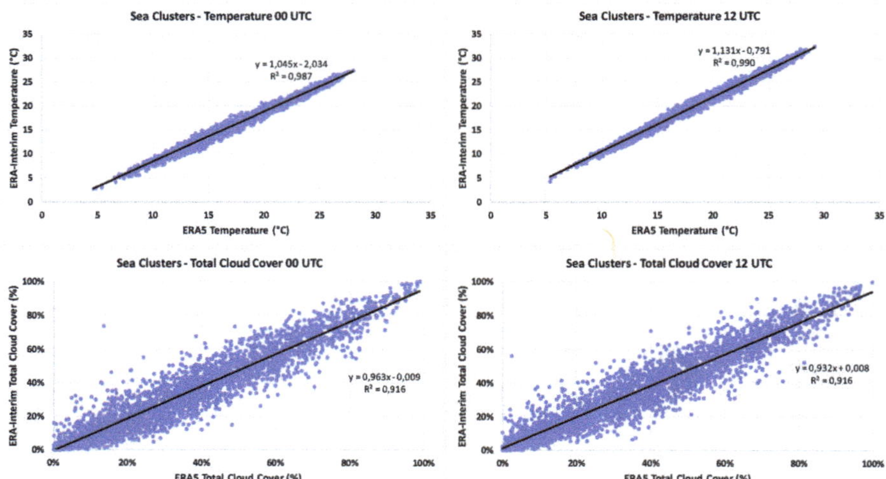

Figure 11. As in Figure 10, but for clusters 1–6 and 8–9 (sea areas).

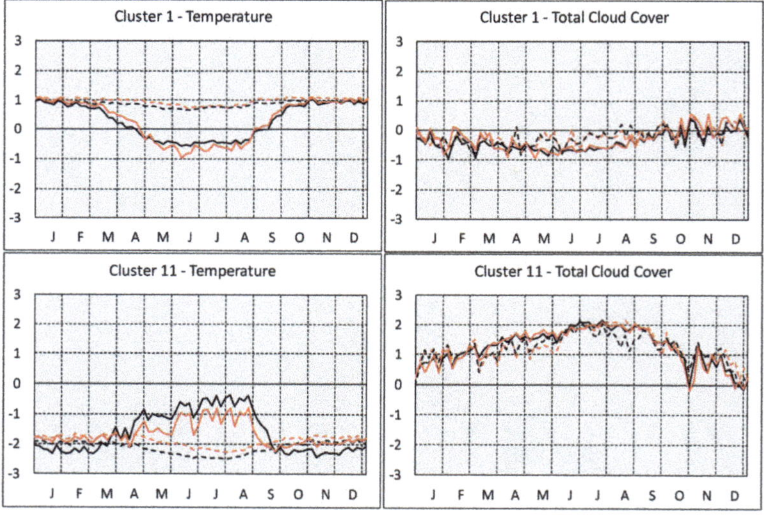

Figure 12. The mean intra-annual variations of the spatial anomalies of air temperature and total cloud cover for clusters 1 and 11. Black lines represent the ERA5 dataset while red lines represent the ERA-Interim dataset. Solid and dashed lines correspond to 12UTC and 00UTC, respectively.

4. Discussion

The division of the southern Balkans into 12 areas with characteristic homogeneous climatic characteristics confirms the significant spatial variability of climate, which is directly connected to the complicated coastline and relief of the greater region. For each area, the deviations of the climatic parameters from the spatial averages have to be examined and interpreted taking into account that some spatial averages can be affected by the portion between land and sea coverage of the geographical domain. The 12 areas can be generally distributed into two groups depending on whether they are sea or land areas. The first group consists of eight areas which are mainly over the sea, while the second group consists of the rest four areas which are over the land. The areas of each group present some common characteristics that they are connected to the effect of land or sea surface to the temperature, humidity, or stability characteristics of the overlying air, but they also present significant differences based on the specific geographical position, atmospheric circulation, orientation, latitude, and altitude characteristics of each area. The sea areas are generally characterized by (i) high dew point values due to the high evaporation rates which lead to saturation conditions and (ii) low summer daytime temperatures relatively to the land areas due to the high thermal capacity and conductivity of the water. Their differences refer mainly to the wind, cloud cover, and stability characteristics, which are significantly affected by the latitude and their location relatively to large-scale circulation systems (extend of the Azores high, Asian thermal low, etc.) [20,21] being responsible for the prevalence of specific wind regimes, for example the etesian winds [31]. On the other hand, the land areas are generally characterized by (i) high summer daytime temperatures relatively to the sea areas due to the low thermal capacity and conductivity of soil, (ii) low dew point values because of the lower evaporation rates relatively to the sea areas, and (iii) static instability maxima in spring associated with the high solar radiation and lapse rate values during this season. Their differences refer mainly to wind, static and dynamic instability, and cloud cover regimes, which are significantly affected by the orientation and the altitude of the areas via their effect on the orographic cloud development and the lifted condensation level [33]. The climatologically homogenous areas and their geographical borders present both similarities and differences compared with the results of the recent climate regionalization made by Nojarov [21], but it has to be taken into account that Nojarov has made the regionalization using sea level pressure, while in the present work a set of meteorological parameters including temperature, cloudiness, etc., over the study area is used.

Also, it has to be mentioned that the above characteristics of the 12 regions refer to the 10-year period of 2008 to 2017 and may present significant deviations from the corresponding climate characteristics of the past and the future decades. The derived 2008–2017 regional characteristics could simply be an expression of coinciding several dynamical factors (i.e., North Atlantic Oscillation, teleconnection with Indian monsoon, and climate change), which can sometimes hinder or amplify the derived spatial relationships/features. In order to detect such deviations, data availability for a very longer period and climate simulations for the future decades are needed [34–36].

The special type of climatic classification, which has been achieved in the present study with the use of the recently released ERA5 high resolution data set, can be considered as an initial attempt for climate classifications using high resolution grid point data. As mentioned above, such a classification can be significantly improved in the future when the database will have been extended to the past and a longer data period will be available (now only available from 2008). This will allow the incorporation of precipitation data in the classification, which is not used at the moment because of (i) the fact that they consist of forecast and not of analysis values and (ii) the relatively short 10-year period does not allow the establishment of the main statistical parameters, including the long-term averages. However, the study provides significant evidence about the spatial variability of the climate and it can be considered as a useful tool by the scientists dealing with the Mediterranean climate. Also, it has to be taken into account that the spatial variations of the associated parameters might change under the influence of the ongoing climatic change [37].

5. Conclusions

In the present work an objective definition of 12 climatologically homogeneous areas in the southern Balkans has been carried out for the 10-year period of 2008 to 2017 with the use of $0.25° \times 0.25°$ ERA5 meteorological data and the following main conclusions can be drawn.

1. The high resolution of the ERA5 data set reveal spatial variations in climate which are connected to the complicated relief of the region and cannot be adequately described with the use of a low resolution data set.
2. The geographical distribution of land and sea is one of the dominant factors affecting the definition and the geographical borders of the 12 areas.

3. The complicated geographical relief of the area being responsible for the windward or leeward character of the various subregions affects significantly the spatial distribution of humidity and cloud cover.

4. There are significant differences between nighttime and daytime cloud cover over the land areas. The low thermal capacity and conductivity of the soil allow the significant influence of daytime radiative heating of its surface on the temperature and static stability regimes of the lowest atmospheric layers, regulating convection and cloud development.

5. There are significant differences between nighttime and daytime wind regimes, especially near the coasts. These differences are associated with the development of diurnal small-scale circulations between land and sea (see breezes).

6. The main climatic characteristics of the 12 areas for the above period have been also confirmed with the use of the ERA-Interim data base. Strong similarity is found between the ERA5 and the ERA-Interim results. The small differences that exist are mainly associated to the different resolution of the data sets.

7. A time extension of the high resolution data set to the past would allow the full establishment of the statistical parameters associated with the climate of the region.

8. The effect of the future climate change on the characteristics of the 12 areas revealed in the present work can be examined using also the results of climate model simulations and it is an interesting subject for a future research work.

Author Contributions: Conceptualization, C.J.L.; Methodology, C.J.L. and G.K.; Data Curation, G.K.; Writing—Original draft preparation, C.J.L.; Writing—Review and editing, C.J.L., G.K., and A.B.; Visualization, G.K.; Supervision, A.B.

Funding: This research received no external funding.

Acknowledgments: Acknowledgement is made for the use of ECMWF's computing and archive facilities in this research, generated using Copernicus Climate Change Service information [2018].

Conflicts of Interest: The authors declare no conflicts of interest.

References

1. Ioannidis, E.; Lolis, C.J.; Papadimas, C.D.; Hatzianastassiou, N.; Bartzokas, A. On the intra-annual variation of cloudiness over the Mediterranean region. *Atmos. Res.* **2018**, *208*, 246–256. [CrossRef]

2. Galanaki, E.; Kotroni, V.; Lagouvardos, K.; Argiriou, A. A ten-year analysis of cloud-to-ground lightning activity over the Eastern Mediterranean region. *Atmos. Res.* **2015**, *166*, 213–222. [CrossRef]

3. Hatzianastassiou, N.; Papadimas, C.D.; Lolis, C.J.; Bartzokas, A.; Levizzani, V.; Pnevmatikos, J.D.; Katsoulis, B.D. Spatial and temporal variability of precipitation over the Mediterranean Basin based on 32-year satellite global precipitation climatology project data, part I: Evaluation and climatological patterns. *Int. J. Climatol.* **2016**, *36*, 4741–4754. [CrossRef]

4. Ziv, B.; Saaroni, H.; Alpert, P. The factors governing the summer regime of the eastern Mediterranean. *Int. J. Climatol.* **2004**, *24*, 1859–1871. [CrossRef]

5. Harpaz, T.; Ziv, B.; Saaroni, H.; Beja, E. Extreme summer temperatures in the East Mediterranean—Dynamical analysis. *Int. J. Climatol.* **2014**, *34*, 849–862. [CrossRef]

6. Davis, R.E.; Hayden, B.P.; Gay, D.A.; Phillips, W.L.; Jones, G.V. The North Atlantic subtropical anticyclone. *J. Clim.* **1997**, *10*, 728–744. [CrossRef]

7. Tyrlis, E.; Lelieveld, J.; Steil, B. The summer circulation over the eastern Mediterranean and the Middle East: Influence of the South Asian monsoon. *Clim. Dyn.* **2013**, *40*, 1103–1123. [CrossRef]

8. Cherchi, A.; Annamalai, H.; Masina, S.; Navarra, A. South Asian Summer Monsoon and the Eastern Mediterranean Climate: The Monsoon–Desert Mechanism in CMIP5 Simulations. *J. Clim.* **2014**, *27*, 6877–6903. [CrossRef]

9. Wanner, H.; Brönnimann, S.; Casty, C.; Gyalistras, D.; Luterbacher, J.; Schmutz, C.; Stephenson, D.B.; Xoplaki, E. North Atlantic Oscillation—Concepts and studies. *Surv. Geophys.* **2001**, *22*, 321–382. [CrossRef]

10. Nasr-Esfahany, M.A.; Ahmadi-Givi, F.; Mohebalhojeh, A.R. An energetic view of the relation between the Mediterranean storm track and the North Atlantic Oscillation. *Q. J. R. Meteorol. Soc.* **2011**, *137*, 749–756. [CrossRef]

11. Thompson, D.W.J.; Wallace, J.M. The Arctic Oscillation signature in the wintertime geopotential height and temperature fields. *Geophys. Res. Lett.* **1998**, *25*, 1297–1300. [CrossRef]

12. Grassi, B.; Redaelli, G.; Visconti, G. Arctic sea ice reduction and extreme climate events over the Mediterranean region. *J. Clim.* **2013**, *26*, 10101–10110. [CrossRef]

13. Kutiel, H.; Maheras, P.; Türkeş, M.; Paz, S. North Sea—Caspian Pattern (NCP)—An upper level atmospheric teleconnection affecting the eastern Mediterranean—Implications on the regional climate. *Theor. Appl. Climatol.* **2002**, *72*, 173–192. [CrossRef]

14. Kutiel, H.; Benaroch, Y. North Sea-Caspian pattern (NCP)—An upper level atmospheric teleconnection affecting the Eastern Mediterranean: Identification and definition. *Theor. Appl. Climatol.* **2002**, *71*, 17–28. [CrossRef]

15. Kotsias, G.; Lolis, C.J. A study on the total cloud cover variability over the Mediterranean region during the period 1979–2014 with the use of the ERA-Interim database. *Theor. Appl. Climatol.* **2018**, in press. [CrossRef]

16. Hertig, E.; Jacobeit, J. Considering observed and future non-stationarities in statistical downscaling of Mediterranean precipitation. *Theor. Appl. Climatol.* **2015**, *122*, 667–683. [CrossRef]

17. Paxian, A.; Hertig, E.; Seubert, S.; Vogt, J.; Jacobeit, J.; Paeth, H. Present-day and future Mediterranean precipitation extremes assessed by different statistical approaches. *Clim. Dyn.* **2014**, *44*, 845–860. [CrossRef]

18. Lolis, C.J. High-resolution precipitation over the southern Balkans. *Clim. Res.* **2012**, *55*, 167–179. [CrossRef]

19. Gatidis, C.; Lolis, C.J.; Lagouvardos, K.; Kotroni, V.; Bartzokas, A. On the seasonal variability and the spatial distribution of lightning activity over the broader Greek area and their connection to atmospheric circulation. *Atmos. Res.* **2018**, *208*, 180–190. [CrossRef]

20. Kostopoulou, E.; Jones, P.D. Comprehensive analysis of the climate variability in the eastern Mediterranean. Part I: Map-pattern classification. *Int. J. Climatol.* **2007**, *27*, 1189–1214. [CrossRef]

21. Nojarov, P.J. Genetic climatic regionalization of the Balkan Peninsula using cluster analysis. *J. Geogr. Sci.* **2017**, *27*, 43–61. [CrossRef]

22. ERA5 Data Documentation. Available online: https://www.ecmwf.int/en/forecasts/datasets/archive-datasets/reanalysis-datasets/era5 (accessed on 11 September 2018).

23. Jolliffe, I.T. *Principal Component Analysis*; Springer: New York, NY, USA, 1986.

24. Richman, M.B. Rotation of principal components. *J. Climatol.* **1986**, *6*, 293–335. [CrossRef]

25. Sharma, S. *Applied Multivariate Techniques*; Wiley: New York, NY, USA, 1995.

26. Davis, R.E.; Walker, D.R. An upper air synoptic climatology of the western United States. *J. Clim.* **1992**, *5*, 1449–1467. [CrossRef]

27. Kalkstein, L.S.; Nichols, M.C.; Barthel, C.D.; Greene, J.S. A new spatial synoptic classification: Application to air mass analysis. *Int. J. Climatol.* **1996**, *16*, 983–1004. [CrossRef]

28. Sugar, A.C.; James, M.G. Finding the number of clusters in a dataset: An information—Theoretic approach. *J. Am. Stat. Assoc.* **2003**, *98*, 750–763. [CrossRef]

29. Dee, D.P.; Uppala, S.M.; Simmons, A.J.; Berrisford, P.; Poli, P.; Kobayashi, S.; Andrae, U.; Balmaseda, M.A.; Balsamo, G.; Bauer, P.; et al. The ERA-Interim reanalysis: Configuration and performance of the data assimilation system. *Q. J. R. Meteorol. Soc.* **2011**, *137*, 553–597. [CrossRef]

30. Lolis, C.J. A climatology of convective available potential energy in the Mediterranean region. *Clim. Res.* **2017**, *74*, 15–30. [CrossRef]

31. Anagnostopoulou, C.; Zanis, P.; Katragkou, E.; Tegoulias, I.; Tolika, K. Recent past and future patterns of the Etesian winds based on regional scale climate model simulations. *Clim. Dyn.* **2014**, *42*, 1819–1836. [CrossRef]

32. Tyrlis, E.; Lelieveld, J. Climatology and dynamics of the summer Etesian winds over the eastern Mediterranean. *J. Atmos. Sci.* **2013**, *70*, 3374–3396. [CrossRef]

33. Saaroni, H.; Halfon, N.; Ziv, B.; Alpert, P.; Kutiel, H. Links between the rainfall regime in Israel and location and intensity of Cyprus lows. *Int. J. Climatol.* **2010**, *30*, 1014–1025. [CrossRef]

34. Spinoni, J.; Vogt, J.V.; Barbosa, P.; Dosio, A.; McCormick, N.; Bigano, A.; Füssel, H.-M. Changes of heating and cooling degree-days in Europe from 1981 to 2100. *Int. J. Climatol.* **2018**, *38*, e191–e208. [CrossRef]

35. Ruosteenoja, K.; Räisänen, P. Seasonal Changes in Solar Radiation and Relative Humidity in Europe in response to Global Warming. *J. Clim.* **2013**, *26*, 2467–2481. [CrossRef]

36. Rizou, D.; Flocas, H.A.; Hatzaki, M.; Bartzokas, A. A statistical investigation of the impact of the Indian monsoon on the eastern Mediterranean circulation. *Atmosphere* **2018**, *9*, 90. [CrossRef]
37. El-Samra, R.; Bou-Zeid, E.; Bangalath, H.K.; Stenchikov, G.; El-Fadel, M. Seasonal and Regional Patterns of Future Temperature Extremes: High-Resolution Dynamic Downscaling Over a Complex Terrain. *J. Geophys. Res. Atmos.* **2018**, *123*, 6669–6689. [CrossRef]

Article

Time Series Analysis of MODIS-Derived NDVI for the Hluhluwe-Imfolozi Park, South Africa: Impact of Recent Intense Drought

Nkanyiso Mbatha * and Sifiso Xulu

Department of Geography and Environmental Studies, University of Zululand,
KwaDlangezwa 3886, South Africa; xulusi@unizulu.ac.za
* Correspondence: mbathanb@unizulu.ac.za; Tel.: +27-035-902-6400

Received: 8 October 2018; Accepted: 14 November 2018; Published: 30 November 2018

Abstract: The variability of temperature and precipitation influenced by El Niño-Southern Oscillation (ENSO) is potentially one of key factors contributing to vegetation product in southern Africa. Thus, understanding large-scale ocean–atmospheric phenomena like the ENSO and Indian Ocean Dipole/Dipole Mode Index (DMI) is important. In this study, 16 years (2002–2017) of Moderate Resolution Imaging Spectroradiometer (MODIS) Terra/Aqua 16-day normalized difference vegetation index (NDVI), extracted and processed using JavaScript code editor in the Google Earth Engine (GEE) platform was used to analyze the vegetation response pattern of the oldest proclaimed nature reserve in Africa, the Hluhluwe-iMfolozi Park (HiP) to climatic variability. The MODIS enhanced vegetation index (EVI), burned area index (BAI), and normalized difference infrared index (NDII) were also analyzed. The study used the Modern Retrospective Analysis for the Research Application (MERRA) model monthly mean soil temperature and precipitations. The Global Land Data Assimilation System (GLDAS) evapotranspiration (ET) data were used to investigate the HiP vegetation water stress. The region in the southern part of the HiP which has land cover dominated by savanna experienced the most impact of the strong El Niño. Both the HiP NDVI inter-annual Mann–Kendal trend test and sequential Mann–Kendall (SQ-MK) test indicated a significant downward trend during the El Niño years of 2003 and 2014–2015. The SQ-MK significant trend turning point which was thought to be associated with the 2014–2015 El Niño periods begun in November 2012. The wavelet coherence and coherence phase indicated a positive teleconnection/correlation between soil temperatures, precipitation, soil moisture (NDII), and ET. This was explained by a dominant in-phase relationship between the NDVI and climatic parameters especially at a period band of 8–16 months.

Keywords: drought; NDVI; ENSO; wavelet; time series analysis; Hluhluwe-iMfolozi Park; Google Earth Engine

---

1. Introduction

Vegetation within protected areas such as game reserves provides wildlife and society with indispensable ecosystem goods and services [1] including food, medicinal resources, aesthetic value, and recreational opportunities [2]. However, inappropriate management and other disturbances affect the potential productivity and spatial extent of this resource [3]. Thus, any factor that poses a threat to vegetation and its associated benefits which could affect their productivity in the protected areas needs to be identified and monitored. One such threat is an increase in temperature above normal as well as a prolonged decline in precipitation and soil moisture, leading to extreme climatic events such as droughts, which severely affect vegetation productivity [4]. Drought-related impacts are becoming more multifaceted, as explained by their rapidly growing consequences in sectors such as recreation and tourism, agriculture, and energy [5].

The influence of drought on vegetation varies in the spatial and temporal scales, and these are projected to increase with climate change [6,7]. This behavior affects wildlife, particularly in semi-arid and arid environments where herbivory is strongly restricted by vegetation extent and water availability [8]. In the north-east part of KwaZulu-Natal, South Africa, for example, droughts are becoming a recurrent and prominent feature [9,10], affecting vegetation, water and wildlife resources notably in the Hluhluwe-iMfolozi Park (HiP), the oldest proclaimed game reserve in Africa, as reported in this paper. Furthermore, these impacts have potential consequences that could incapacitate this game reserve's support of its specialist grazers such as rhinos [11].

Understanding the association between vegetation productivity and climatic variables such as precipitation and temperature has, therefore, become a high priority. To address this, spatiotemporal tools that can integrate climate data with other information of interest are required. Remotely sensed data provide the opportunity to monitor vegetation dynamics in a systematic manner [12]. They play a growing role in drought detection and management as they afford up-to-date information over various time and geographic scales and complement alternative techniques such as field surveys [4] and interviews [13]. Remote sensing's systematic observation allows us to track vegetation conditions from the 1970s to the present [14] and provides the means to integrate the record with causal factors. This study investigates vegetative drought which is the vegetation stress as a function of moisture deficit [15].

Several drought studies based on satellite-derived measurements have exploited key indicators such as the (a) normalized difference vegetation index (NDVI), a ratio of the difference between the near-infrared and red bands of the spectrum over the sum of the near-infrared and red bands [16,17], which is a robust indicator of vegetation productivity [18]; (b) the normalized difference infrared index (NDII) which contain additional information on water availability in the soil for use by vegetation [19] as measured by the ratios of the near-infrared and short-wave infrared [20]; and (c) the evapotranspiration (ET) which includes both the loss of root zone soil water through transpiration (influenced by stomatal conductance), as well as evaporation from bare soil [21]. These studies have enhanced our understanding of how vegetation reacts to drought events over time [22–25].

Hitherto, numerous studies have explored vegetation changes using NDVI in response to climatic variability. Most have shown that vegetation is largely swayed by the El Niño/Southern Oscillation (ENSO) phenomenon and have been established to respond well to climatic variables [10,24,26,27]. These studies in different climatic regions have revealed climate-induced effects in key economic sectors such as agriculture [28] and forestry [24,29]. Most recently, Huang et al. [30] used MODIS-derived NDVI to demonstrate how vegetation responds to climate variation in the Ziya-Daqing basins of China. Their results showed that the trends of growing season NDVI were significant in the forest, grassland, and highlands of Taihang but insignificant in most plain drylands [27]. They also showed how grassland, as the primary vegetation on the Qinghai-Tibet Plateau, has been increasingly influenced by water availability due to droughts over the last decade.

Several factors make the HiP an ideal site for assessing the effects of drought on wildlife. First, Bond et al. [31] established that droughts largely influence the extent of grazing vegetation in the reserve. More recently, Xulu et al. [10] showed how the recent intense drought moderated the vegetation health of commercial plantations located ~70 km from the park. Second, the HiP is an important conservation area and ecotourism destination in South Africa [32], so the resultant socio-economic impacts of ecosystem changes are of great concern. In this study, therefore, we aim to evaluate the influence of climatic variability on vegetation in the game reserve over the period of 2002 to 2017. This is the first attempt to demonstrate the spatial dimension of the drought effects in the HiP using satellite data. We show how to construct a MODIS-derived NDVI time series in the GEE platform, and perform statistical tests to determine the causal influence of climatic variables in the reserve.

2. Materials and Methods

2.1. Study Area

This study was conducted at the Hluhluwe-iMfolozi Park, which covers 900 km^2 and extends between 28°00′ S and 28°26′ S, and 31°43′ E and 32°09′ E in the northern KwaZulu-Natal, South Africa (Figure 1). The reserve was established in 1895 and is managed by Ezemvelo KwaZulu-Natal Wildlife (EKZN Wildlife). The landscape undulates with an altitude ranging from approximately 50 to 500 m.a.s.l. and comprises a mixture of soil types resulting from topographic and climatic heterogeneity [33]. The terrain of the study area on the right side of Figure 1 was plotted using the Global Multi-resolution Terrain Elevation Data 2010 data set. The version of this data is the Breakline Emphasis, 7.5 arc-s, and is archived as USGS/GMTED2010 in the Google Earth Engine (GEE) JavaScript platform. Land cover classification was also performed in the GEE environment so as to show the types of vegetation cover in the HiP.

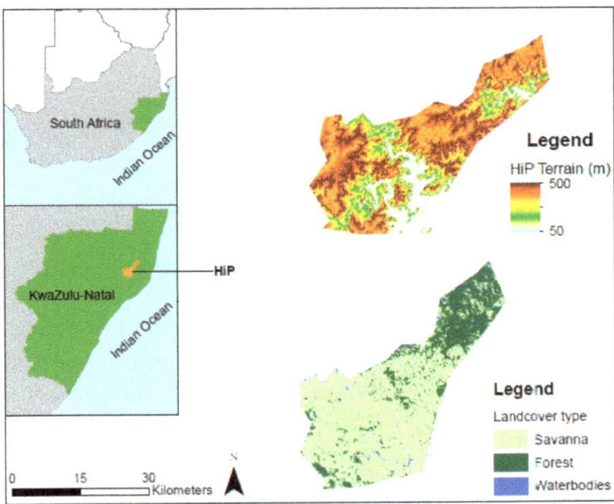

Figure 1. The study area showing the Hluhluwe-iMfolozi Park in the north-eastern part of South Africa.

There are two main rivers that pass through this nature park, namely the Black and White Umfolozi. The entire area of the park is fenced and borders on populated rural communities. Vegetation varies from semi-deciduous forests in the north of Hluhluwe to open savanna woodlands in the southern iMfolozi. Much of the area is dominated by woodland savarna interspersed with shrub thicket [34]. The northern part of the park has hilly terrain and is dominated by forest. The climate is subtropical with summer rainfall. It receives a mean annual rainfall ranging from 700 to 985 mm, much of it occurring between October and March [35]. The park supports approximately 1200 plant species, including 300 tree and 150 grass species.

2.2. Data

In order to investigate the variability of vegetation in the HiP in response to climatic conditions as well as the recent intense drought of 2014–2016, we opted to use the monthly averaged MODIS Terra/Aqua 16-day datasets measured for the period from 2002 to 2017 (16 years). With its considerable time resolution (about for images per month) compared to other satellites, MODIS images were the most appropriate for this study because of the size of the geographic area. The MODIS data used here are archived in the GEE as image collection. This data product is generated from a MODIS/MCD43A4 version 6 surface reflectance composite. More details about the MCD43A4 MODIS/Terra and Aqua nadir BRDF-adjusted reflectance daily level 3 global 500 m SIN grid V006 data can be found in a study by Schaaf et al. [36]. The data were extracted and processed using the JavaScript code editor in the GEE platform (https://earthengine.google.com/, Mountain View, CA, USA) (see Appendix A), which provides possibilities of parallel computing and large data processing for even very large study areas. For the purpose of this investigation, our main parameter is the NDVI, but we also considered other vegetation indices such as the Enhanced Vegetation Index (EVI), the Burned Area Index (BAI), and Normalized Difference Infrared Index (NDII). The BAI was also included in order to determine the possible vegetation burning activity, which may have been triggered by drier conditions associated with an intense drought period. NDII has been recently proven to be a robust indicator for monitoring the moisture content in the root-zone from the observed moisture state of vegetation [19,21]. These spectral indices were calculated using the formulas:

$$\mathrm{NDVI} = \frac{NIR - R}{NIR + R} \tag{1}$$

$$\mathrm{EVI} = 2.5\,\frac{NIR - R}{NIR - 6\,R - 7.5\,B + 1} \tag{2}$$

$$\mathrm{BAI} = \frac{1}{(0.1 + R)^2 + (0.06 + NIR)}$$
$$\mathrm{NDII} = \frac{NIR - SWIR1}{NIR + SWIR1} \tag{3}$$

where R, NIR, and $SWIR1$ are spectral bands in the blue (450–500 nm), red (600–700 nm), near-infrared (700–1300 nm), and shortwave infrared (1550–1750 nm) regions.

In this study, we derived the precipitation values averaged for the study area for the period from 2002 to 2017 using the Climate Engine Application (CEA, http://climateengine.org/, Moscow, ID, USA), while soil temperature monthly mean data was derived from the National Aeronautics and Space Administration (NASA, Washington, DC, USA): http://giovanni.gsfc.nasa.gov. Both the soil temperature and precipitation data are an output of the Modern Retrospective Analysis for the Research Application (MERRA-2) model [37]. The MERRA model is an American global reanalysis tool operating from 1979 onwards that is based on the NASA Goddard Earth Observation serving Data Assimilation System version 5 (GEOS-5). The MERRA-2 model data are given at a spatial resolution of $0.67° \times 0.50°$ at 1-hourly to 6-hourly intervals.

There is always an expected variability of surface water content due to changes in both weather and climatic conditions. Therefore, in a study such as this one, it is essential to always investigate the water lost to the atmosphere through both evaporation and transpiration. This can be an important process as it could explain details about vegetation water stress. Given that the study area is a remote area which does not have evaporation and/transpiration measurements records, we opted to use the Global Land Data Assimilation System (GLDAS) evapotranspiration (ET) data. The GLDAS system was designed to generate optimal fields of land surface and fluxes, and it is also capable of generating quality controlled, spatially and temporally consistent, terrestrial hydrological data including ET and other related parameters [38].

The ENSO phenomenon influences rainfall and temperature conditions largely over southern Africa [39,40]. Previous studies have demonstrated how vegetation responds significantly to ENSO [40] and the DMI [41] index as a measure of climatic conditions [42–44] in some parts of southern Africa. Thus, in order to investigate changes in vegetation in the HiP due to variability in climatic conditions, it is important to consider these climate indices. In this study, we used the Niño3.4 monthly mean time series retrieved from the National Oceanic and Atmospheric Administration (NOAA) website (https://www.esrl.noaa.gov/psd/gcos_wgsp/Timeseries/, Washington, DC, USA). The Niño3.4 index is calculated by taking the area-averaged sea-surface temperature (SST) within the Niño3.4 region, which is at 5° N–5° S longitude and 120° W–170° W latitude in the Pacific Ocean. On the other hand, the DMI is calculated by taking the difference between the SST anomalies in the western (50° E–70° E; 10° S–10° N) and eastern (90° E–110° E), (10° S–0° N) sectors of the equatorial Indian Ocean [41].

The DMI data were downloaded from the website: http://www.jamstec.go.jp/frcgc/research/d1/iod/iod/dipole_mode_index.html. The relevant time series of Niño3.4 and DMI are shown in Figure 2.

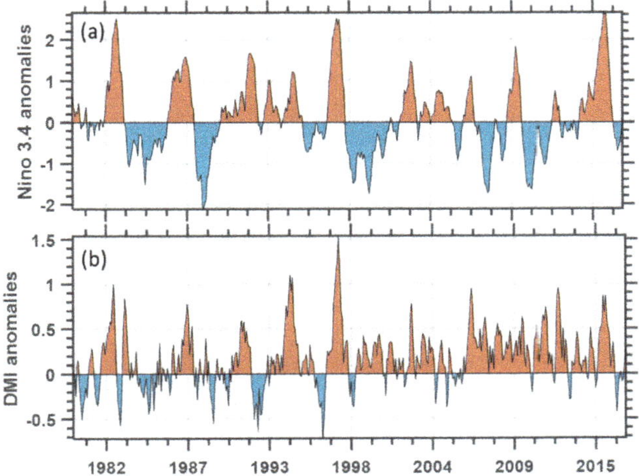

Figure 2. The standardized monthly Niño3.4 (**a**) and dipole mode index (DMI) (**b**) time series for the period from 1980 to 2017.

2.3. Multiple-Linear Regression

One of the principal objectives of this study is to quantify the effects of temperature, precipitation, ET, soil moisture at root-zone (NDII), ENSO and DMI on the NDVI as a surrogate for vegetation in the study area. Multiple-linear regression analysis (MLR), which is commonly used to explain the relationship between one continuous dependent variable and two or more independent variables, was employed. The MLR model output of a number n observations can be represented as

$$y_i = \beta_0 + \beta_1 x_{i2} + \cdots + \beta_p x_{ip} + \varepsilon_i \text{ where } i = 1, 2, 3, \ldots, n \tag{4}$$

where y_i is the dependent variable (NDVI in this case), x_{ip} represents the independent variables (soil temperature, precipitation, Niño3.4, and DMI in this case), β_0 is the intercept, and $\beta_1, \beta_2, \ldots \beta_p$ are the coefficients of the x terms. The term ε_i represents the error term, which the model always tries to minimize.

2.4. Mann–Kendall Test

It is always useful to assess the monotonic trends in a time series of any geophysical data. In this study, the Mann–Kendall test [45–47] was used. This is a non-parametric rank-based test method, which is commonly used to identify monotonic trends in a time series of climate data, environmental data, or hydrological data. Non-parametric methods are known to be resilient to outliers [48], hence it is desirable to choose such methods. Based on a study by Kendall [47] and recently by Pohlert [49] and others, the Mann–Kendall test statistic is calculated from the following formula:

$$S = \sum_{k=1}^{n-1} \sum_{j=k+1}^{n} sign\left(X_j - X_k\right) \tag{5}$$

where

$$sign(x) = \begin{cases} +1, & \text{if } x > 1 \\ 0, & \text{if } x = 0 \\ -1, & \text{if } x < 1 \end{cases} \tag{6}$$

The average value of S is $E[S] = 0$, and the variance σ^2 is given by the following equation:

$$\sigma^2 = \left\{ n(n-1)(2n+5) - \sum_{j=1}^{p} t_j(t_j - 1)(2t_j + 5) \right\}/18 \tag{7}$$

where t_j is the number of data points in the jth tied group, and p is the number of the tied group in the time series. It is important to mention that the summation operator in the above equation is applied only in the case of tied groups in the time series in order to reduce the influence of individual values in tied groups in the ranked statistics. On the assumption of random and independent time series, the statistic S is approximately normally distributed provided that the following z-transformation equation is used:

$$z = \begin{cases} \frac{S-1}{\sigma} & \text{if } S > 1 \\ 0 & \text{if } S = 0 \\ \frac{S+1}{\sigma} & \text{if } S < 1 \end{cases} \tag{8}$$

The value of the S statistic is associated with the Kendall

$$\tau = \frac{S}{D} \tag{9}$$

where

$$D = \left[\frac{1}{2}n(n-1) - \frac{1}{2}\sum_{j=1}^{p} t_j(t_j - 1) \right]^{1/2} \left[\frac{1}{2}n(n-1) \right]^{1/2} \tag{10}$$

In regards to the z-transformation equation defined above, this study considered a 5% confidence level, where the null hypothesis of no trend was rejected if $|z| > 1.96$. Another important output of the Mann–Kendall statistic is the Kendall τ term, which is a measure of correlation which indicates the strength of the relationship between any two independent variables. In this study, the Mann–Kendall test system summarized above was applied to the NDVI data by writing a piece of code in R-project and following the instructions by Pohlert [49].

The Mann–Kendall trend method can be extended into a sequential version of the Mann–Kendall test statistic which is called the Sequential Mann–Kendall (SQ-MK). This method was proposed by Reference [50], and it is used to detect approximate potential trends turning points in long-term time series. This test method generates two time series, a forward/progressive one ($u(t)$) and a backward/retrograde one ($u'(t)$). In order to utilize the effectiveness of this trend detection method, it is required that both the progressive and the retrograde time series are plotted in the same figure. If they happen to cross each other and diverge beyond the specific threshold (± 1.96 in this study), then there is a statistically significant trend. The region where they cross each other indicates the time period where the trend turning point begins [51]. Basically, this method is computed by using ranked values of y_i of a given time series (x_1, x_2, x_3, ..., x_n) in the analyses. The magnitudes of y_i, ($i = 1, 2, 3, ..., n$) are compared with y_i, ($j = 1, 2, 3, ..., j-1$). At each comparison, the number of cases where $y_i > y_j$ are counted and then donated to n_i. The statistic t_i is thereafter defined by the following equation:

$$t_i = \sum_{j=1}^{i} n_i$$

The mean and variance of the statistic t_i are given by

$$E(t_i) = \frac{i(i-1)}{4}$$

and

$$\text{Var}(t_i) = \frac{i(i-1)(2i-5)}{72}$$

Finally, the sequential values of statistic $u(t_i)$ which are standardized are calculated using the following equation:

$$u(t_i) = \frac{t_i - E(t_i)}{\sqrt{\text{Var}(t_i)}}$$

The above equation gives a forward sequential statistic which is normally called the progressive statistic. In order to calculate the backward/retrograde statistic values ($u'(t_i)$), the same time series ($x_1, x_2, x_3, \ldots, x_n$) is used, but statistic values are computed by starting from the end of the time series. The combination of the forward and backward sequential statistic allows for the detection of the approximate beginning of a developing trend. Additionally, in this study, a 95% confidence level was considered, which means critical limit values are ± 1.96. This method has been successfully utilized in studies of trends detection in temperature [52,53] and precipitation [51,53,54].

2.5. Wavelet Transforms and Wavelet Coherence

In this study, we opted to employ the wavelet transform analyses method [55] because of its ability to obtain a time–frequency representation of any continuous signal. Basically, the continuous wavelet transform (CWT) of a given geophysical (in this case) time series is given by transforming the time series into a time and frequency space. While there are several types of wavelets, the choice of the wavelet function is determined by the data series, of which, for geophysical data, the Morlet wavelet function has been shown to perform well [55–57]. Thus, the CWT [W_n (s)] for a given time series (x_n, $n = 1, 2, 3, \ldots, N$) with respect to wavelet Ψ_0 (η) is defined as:

$$W_n^X(s) = \sum_{n'-1}^{n-1} X_{n'} \Psi^* \left[\frac{(n'-n)}{s} \delta t \right] \tag{11}$$

where s is the wavelet scale, n' is the translated time index, n is the localized time index, and Ψ^* is the complex conjugate of the normalized wavelet. δt is the uniform time step (which is months in this case). The wavelet power is calculated from $|W_n(n)|^2$. In this study, the CWT statistical significance at a 95% confidence level was estimated against a red noise model [55,57]. Using a continuous wavelet transform analysis, it is also possible to quantify the relationship between two independent time series of the same time step δt. In this study, the aim was to quantify the relationship between NDVI averaged for the study area and selected climatic parameters. Following Grinsted et al. [57], for the two time series of X and Y, with different CWT $W_n^X(s)$ and $W_n^Y(s)$ values, the cross-wavelet transform $W_n^{xy}(s)$ is given by

$$W_n^{XY}(s) = W_n^X(s) W_n^{Y*}(s) \tag{12}$$

where "*" represents the complex conjugate of the Y time series. The output of the above equation can also assist in calculating the wavelet coherence. Basically, wavelet coherence is a measure of the intensity of the covariance of the two time series in a time–frequency domain. This is an important parameter because the cross-wavelet only gives a common power. Another important process is to calculate the phase difference between the two time series. Here, the procedure is to estimate the mean and confidence interval of the phase difference. Following a study by Grinsted et al. [57], we used the circular mean of the phase-over regions with relatively high statistical significance that are inside the cone of influence (COI) to quantify the phase relationship between any two independent time series. As defined in a study by Zar [58] and also later by Grinsted et al. [57], the mean circulation of a set of angles (a_i, $i = 1, 2, 3, \ldots, n$) can be defined by the following equation:

$$a_m = \arg(X, Y) \text{ with } X = \sum_{i=1}^{n} \cos(a_i) \text{ and } Y = \sum_{i=1}^{n} \sin(a_i) \tag{13}$$

Following the studies by Torrence et al. [55,57], the wavelet coherence between two independent time series can be calculated using the following equation:

$$R_n^2(s) = \frac{\left|S\left(s^{-1}W_n^{XY}(s)\right)\right|^2}{S\left(s^{-1}|W_n^X(s)|^2\right) \times S\left(s^{-1}|W_n^Y(s)|^2\right)} \tag{14}$$

where the parameter S is the smoothing operator defined by $S(W_n\ (s)) = S_{\text{scale}}\ (S_{\text{time}}\ (W_n(s)))$. The parameter S_{time} represents the smoothing in time. For further details about the theory of wavelet analyses, the reader is referred to [55,57,59].

3. Results and Discussion

To investigate whether the El Niño event of 2014–2016 can be classified as a strong El Niño event, a time series for the period from the beginning of the satellite era (1980) to 2017 was plotted (see Figure 2a). We also considered the DMI index (Figure 2b) as a measure of climatic conditions of the eastern part of southern Africa [43]. A general classification of ENSO events should contain 5 consecutive overlapping 3-month periods with SST anomalies below -0.5 for the La Niña events and above $+0.5$ for the El Niño events.

In Figure 2a,b, both the El Niño events and positive DMI are shaded in red, whereas La Niña and negative DMI are indicated in blue. To identify the strength of the ENSO events, the threshold is further broken down to weak (0.5–0.9 SST anomaly), moderate (1.0–1.4 anomaly), and strong ($\geq$1.5 anomaly) events. Figure 2a shows that the 2014–2016 El Niño was one of the strongest since the beginning of the record. Other notably strong El Niño occurrences were in 1982/1983, 1997/1998, and 2009/2010. On the other hand, there were many episodes of positive DMI, with one such event in the 2014–2016 period, which seems to be in phase with the recent strong El Niño of 2014–2016.

A composite of the NDVI index averaged for each year from 2002 to 2017 is shown in Figure 3. In this figure, regions where there are greener colors indicate higher NDVI values, whereas the brownish colors indicate low NDVI values. These results show that there seems to be a direct influence of the ENSO in the vegetation of the HiP, especially during strong El Niño years (2014–2016). It is evident that during El Niño years, there was a decline in NDVI values especially in the southern and western parts of the study area. This is presumably because the vegetation of the northern part of the HiP is dominated by a forest which is consist of indigenous trees which are believed to be drought resistant (see Figure 1). Additionally, the contributing factor could be that the eastern part of the HiP is benefiting from orographic lifting as it is situated in a high terrain (see Figure 1). The evidence of the influence of El Niño is more prominent during the strong El Niño years such as 2003 and the recent intense 2014–2016 drought period, as well as the 2008 non-ENSO drought period.

Figure 4a shows the deseasonalized monthly averaged MODIS NDVI time series for HiP from 2002 to 2017 (red line) plotted together with the 12 months running mean smooth trend (black dotted line). The monthly mean NDVI values plotted in Figure 4a were calculated by taking an averaged of MODIS images available in that month. In this study area, the MODIS satellite records four images per month. In general, there is a steady trend of NDVI measured at the HiP beside some anomalies observed in specific parts of the time series. This seems to be the case for southern Africa because other studies also indicated a steady trend for this region [10]. Remarkably, during the 2014–2016 period, a period that coincided with the recent intense El Niño, there was a sudden decrease in the NDVI values which reduced to the lowest minimum value of about 0.3 in November 2015. During this period, EVI values also decreased to minimum values of about 0.11 (results not shown here).

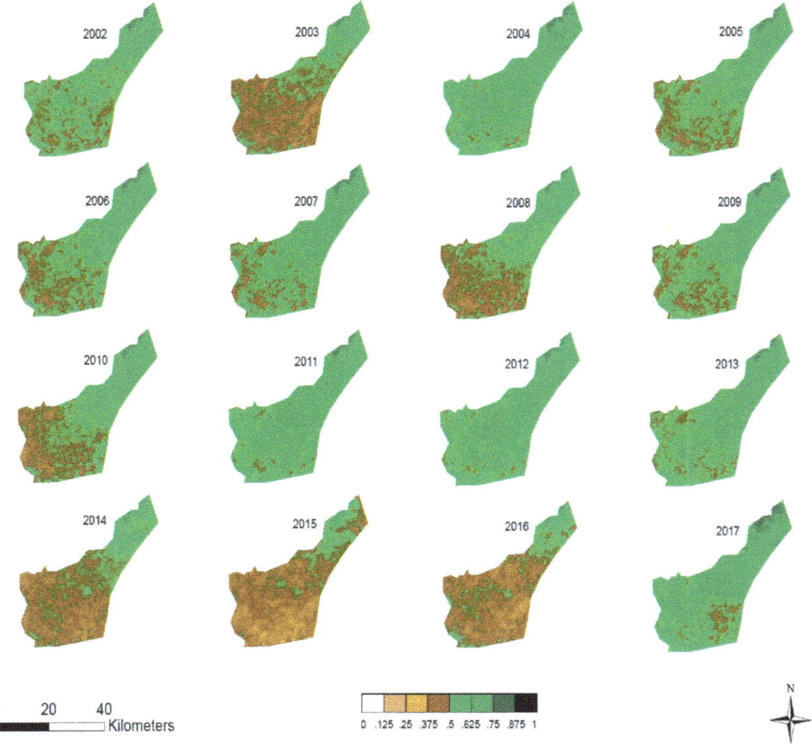

Figure 3. The spatiotemporal variability of normalized difference vegetation index (NDVI) at the Hluhluwe-iMfolozi Park for the period from 2002 to 2017. The scale represents the range of NDVI values from 0 to 1.

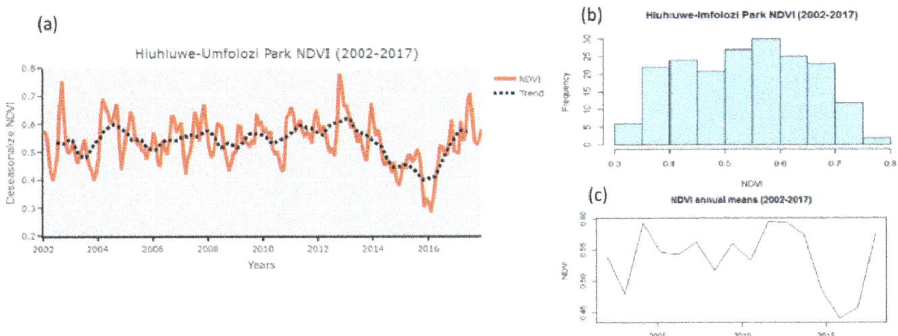

Figure 4. (**a**) The deseasonalized monthly mean NDVI time series for HiP. The continuous red line indicates the trend estimate and the dashed red lines show the 95% confidence interval for the trend based on resampling methods. (**b**,**c**) show the histogram and yearly mean time series, respectively.

A study by Mberego and Gwenzi [60] investigated the temporal patterns of precipitation and vegetation variability over Zimbabwe during extreme dry and wet rainfall seasons using data covering the period 1981–2005. Their NDVI time series indicated a steady trend over this period. However, it seemed to be strongly affected by severe dry conditions, an observation which is consistent with the results presented here. In this study, the deseasonalized monthly mean NDVI time series in Figure 4a (red line) indicates the possible response that corresponds to both dry and wet years, especially during the most recent strong El Niño events of 2003 and 2014–2016. In relation to the strength of the influence of El Niño in the south-western part of southern Africa, a study by Manatsa et al. [61] analyzed agricultural drought in Zimbabwe using the standardized precipitation index (SPI). They reported the 1991–1992 period as the period which experienced the most extreme drought conditions. A little later, observations by Mberego and Gwenzi [60] reported the year 2002–2003 as the drought period with the most prolonged time of relatively low NDVI values in their time series. While our study does indicate a significant influence of the 2003 El Niño event in the HiP NDVI values, the observations presented here indicate that 2014–2016 was the longest period with low NDVI values. Thus, 2014–2016 could be regarded as the most recent intense El Niño period, with a maximum effect on vegetation in the HiP. The NDVI values dropped from a value ~0.65 in November 2013 to 0.3 in November 2015. This is also verified by a much smoother representation of the NDVI in Figure 4c in which a reduction in NDVI values is observed. This reduction coincides with the most recent strong El Niño. Additionally, a reduction which coincides with the El Niño of 2003 (see Figure 4c). Another significant feature is a strong peak, which reaches ~0.8 just after the Irina tropical storm, which occurred in early March 2012 (see Figure 4).

Figure 5 shows monthly mean time series values plotted together with their corresponding 12-month running-mean smooth trend for NDVI (Figure 5a), EVI (Figure 5b), BAI (Figure 5c), soil temperature (Figure 5d), precipitation (Figure 5e), evapotranspiration (Figure 5f), and NDII (Figure 5g). These monthly mean values are plotted together with their respective smooth trends, which were calculated using the Breaks For Additive Season and Trend (BFAST) method, which is described in details by Verbesselt et al. [62,63]. Basically, the BFAST method integrates the decomposition of time series seasonal, trend, and remainder components of any satellite image time series, and can be applied to any other type of time series in the geosciences that deals with seasonal or non-seasonal time series. The period of the most recent intense drought (2014–2016) is indicated by the grey shaded box in each figure. In general, all the parameters show a seasonal cycle in terms of monthly means.

There is an expected resemblance between the NDVI and EVI observations in both the monthly mean time series and the smooth trend, with a clear indication of the effect of the 2014–2016 drought period. These observations are consistent with a study by Xulu et al. [10], who investigated the response of commercial forestry to the recent strong and broad El Niño event in a region which is 70 km south-east of the HiP. In their study, Xulu et al. [10] reported a significant decline of NDVI values which corresponded to the 2014–2016 El Niño years [10]. Although the influence of the 2014–2016 El Niño in the HiP seems to be the strongest, it follows the same pattern as that reported by Anyamba et al. [40] in their study of the influence of both El Niño and La Niña in the vegetation status over eastern and southern Africa. Considering the level of browning of vegetation demonstrated in Figure 3 for the years 2014–2016, it is necessary to consider the possible fire activity given the relatively dry conditions in the HiP. Figure 5c indicates that during the period of the intense drought of 2014–2016 there was an increase in fire incidences in the HiP. This is revealed by a rise in the BAI values of the smooth trend to its maximum level of approximately 50 in November 2015. During the 2014–2016 period, the HiP experienced an unprecedented decline in the total precipitation per month (see Figure 5d). During the same period, the soil temperature increased to its highest maximum (see Figure 5e). The GLDAS monthly mean ET time series shown in Figure 5f indicates a declining trend during the period 2014–2016, which indicated a possible vegetation stress. In order to investigate the moisture content at root-zone, the NDII index was used. The NDII (Figure 5g) indicates a similar pattern to that of the NDVI and EVI time series. It is observed in Figure 5a that the NDII had a steady trend (0.10) during the period 2002–2013 which was followed by a sudden decrease which reached a minimum value of −0.06 in November 2015.

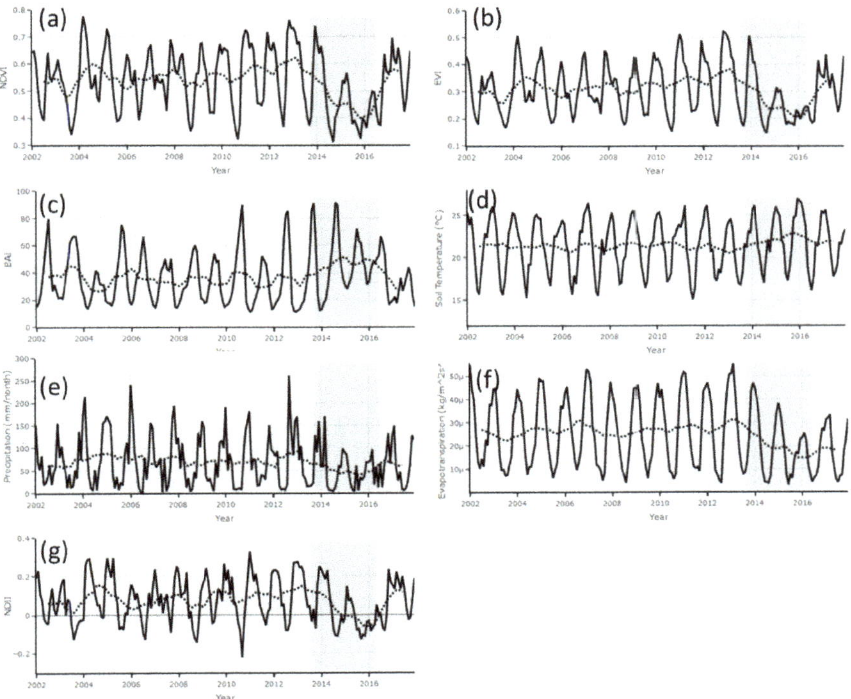

Figure 5. The monthly mean time series values of (**a**) NDVI, (**b**) Enhanced Vegetation Index (EVI), (**c**) Burned Area (BAI), and Modern Retrospective Analysis for the Research Application (MERRA-2) model soil temperature (**d**) and precipitation (**e**), Global Land Data Assimilation System (GILDAS) evapotranspiration (**f**) and Normalized Difference Infrared Index (NDII) (**g**). The dotted lines represent the 12-month smooth trends.

The 12-month running mean smooth trends extracted using the BFAST method for NDVI, EVI, and BAI plotted against anomalies of climatic forcers Niño3.4 and DMI are shown in Figure 6. This plot was constructed to investigate any possible 2-dimensional teleconnection between vegetation condition and the Niño3.4 and DMI climatic forcers, respectively. The panels on the left represent the vegetation indices versus Niño3.4, and the panels on the right show the vegetation indices versus DMI. Both the NDVI (Figure 6a) and EVI (Figure 6b) values show a fairly steady pattern for most parts of the time series, which vary between NDVI values of 0.50 and 0.60, and between EVI values of 0.28 and 0.34. However, both the NDVI and EVI values seem to be enhanced by the extreme amount of rainfall that was brought by the tropical cyclone Irina during early 2012 in the eastern part of southern Africa. In that year, NDVI values increased to a maximum value of approximately 0.62, whereas the more sensitive EVI index reached its maximum of approximately 0.38. The strong peaks that were observed during 2004 for both the NDVI and EVI time series correspond to the greening of vegetation in the HiP which was produced by heavy rainfall that was brought by tropical cyclone Elite in January 2004 [64]. NDVI values were observed to decrease sharply from late 2013 until they reached their minimum of approximately 0.40 in 2015 before beginning to recover to normal average conditions in 2017. This pattern is also depicted in the EVI time series and is directly linked to the stronger and more extensive 2014–2016 El Niño event. Similar results were also presented in a study by Xulu et al. [10], who investigated the influence of recent droughts on forest plantations in Zululand. The notable browning observed in Figure 3 for years 2014, 2015 and 2016, which was also revealed by the NDVI and EVI time series (Figure 5), seems to represent favorable conditions for biomass burning in the

HiP. This is revealed by the unprecedented sudden increase of BAI values to its highest maximum of approximately 50, which coincides with the enhancement of Niño3.4 during 2014–2016 (Figure 6e).

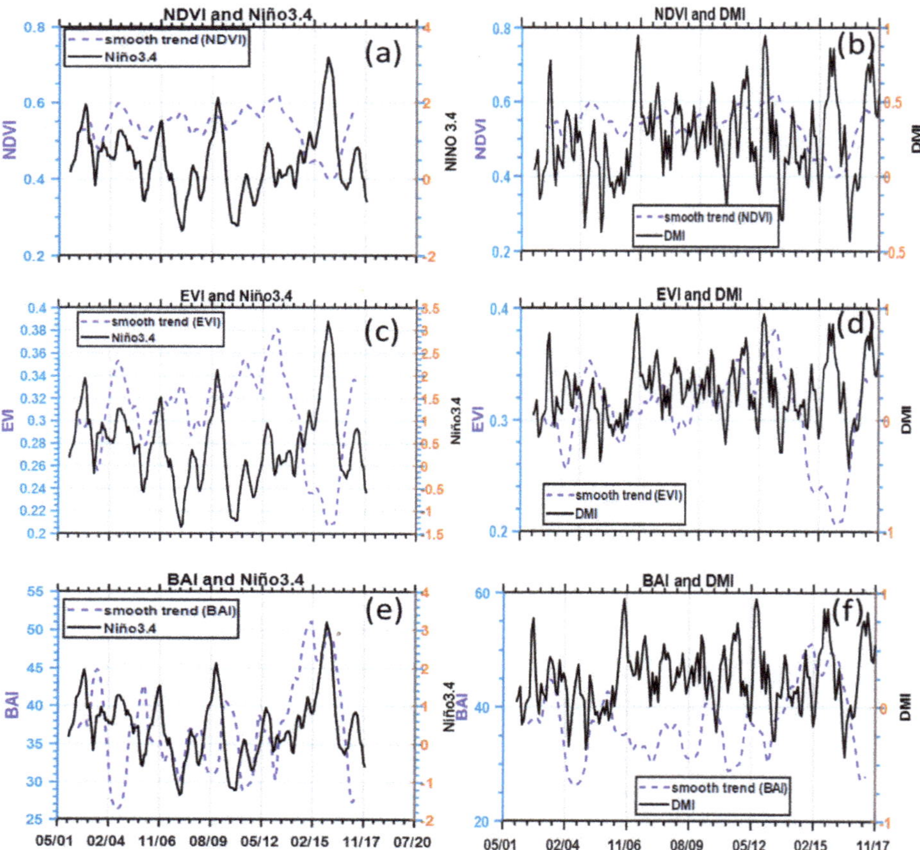

Figure 6. The (**a,b**) NDVI, (**c,d**) EVI and (**e,f**) BAI (blue dashed line) 12-month smooth trends versus Niño3.4 (left panels) and DMI (right panels) for the period from 2002 to 2017 for the HiP.

The DMI was highly variable compared to the Niño3.4 climatic forcer throughout the study period, with several distinctive positive DMI values that reached a maximum of just below 1.0. Remarkably, there is a strong peak that extends up to approximately 0.8 during the period band corresponding to 2014–2016 that coincided with the decline in NDVI and EVI and the increase of the ENSO and BAI time series. We note here that the widespread browning observed during the 2014–2016 drought period could have been accelerated by the fact that the climatic forcers, which are known to influence the south-eastern part of southern Africa, may have been in phase during this period. This, of course, needs further investigation; and is discussed below.

3.1. Correlations Statistics and Mann–Kendall Test

The Pearson correlation between NDVI and climatic variables used in this study for the whole study record was derived. Figure 7 shows the heat map which summarizes the linear relationships between all the parameters monitored in this study. In this figure, it is clear that there is a strong correlation between NDVI and Soil temperature ($r = 0.35$), precipitation ($r = 0.43$), ET ($r = 0.68$), and NDII ($r = 0.92$). On the other hand, there is a significance strong negative correlation between the NDVI and BAI, which is not surprising because greener vegetation reduces chances of biomass burning, while the possibility of the satellite detecting a charcoal signal from burnt vegetation during dry conditions is high. There is also a noteworthy negative ($r = -0.27$) correlation between NDVI and Niño3.4. The results shown in Figure 7 also reaffirm the strong relationship between soil temperature with a strong correlation coefficient of $r = 0.77$. Considering that Figure 2 indicates some episodes where a strong Niño3.4 peak is in phase with the DMI peaks, the noteworthy correlation of $r = 0.28$ between these two climatic indices seems to reaffirm this.

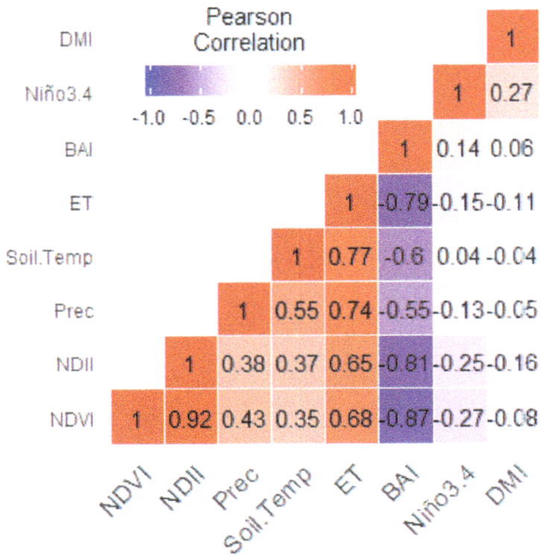

Figure 7. The heat map of Pearson correlation coefficients for NDVI, NDII, precipitation (Prec), soil temperature (Soil.Temo), ET, BAI, Niño3.4, and DMI.

Figure 8 shows the inter-annual variability of the Pearson linear correlation between the HiP NDVI values and parameters such as BAI, Soil Temp, Prec, Niño3.4, DMI, ET, and NDII for the period from 2002 to 2017. The correlation between NDVI and EVI was not analyzed because the two parameters closely resemble each other. In general, NDVI is positively correlated to soil temperature, precipitation, ET and NDII through the study period. The NDVI–NDII correlation was the strongest positive correlation with an average value of $r = 0.91$. This reaffirms the strong relationship between vegetation water stress and soil moisture at the root-zone. The NDVI–ET correlation was observed to be steady at an average correlation coefficient value of $r = 0.65$ during the period from 2002–2013. However, this linear relationship decreased to $r = 0.40$ and $r = 0.30$ in 2015 and 2016, respectively. A study by [65] also used MODIS NDVI and GILDAS evapotranspiration data in order to investigate the relationship between NDVI and evapotranspiration. In their study, they reported a steady positive inter-annual variability of correlation coefficients with an average value of $r = 0.58$. As expected, the NDVI–Niño3.4 correlation is dominated by negative values which are observed to decrease during the periods corresponding to El Niño years. This is consistent with previous studies such as those of Xulu et al. [10,40] who reported a significant influence of ENSO on the vegetation of southern Africa, especially the north-eastern part. Moreover, a salient observation is that the greatest minimum correlation recorded was in 2015, a year with a particularly strong El Niño. The negative correlation between DMI and NDVI also seems to be greater during the recent intense drought period, which could indicate that Niño3.4 and DMI were in phase during this time. The correlation between NDVI and the BAI is expected to be strongly negative as greening is not conducive to biomass burning. However, the results presented in Figure 8 indicate that there was a sudden increase in correlation between NDVI and BAI in 2015 before it returned to its average position in 2016 and 2017. Overall, the inter-annual variation of almost all the study parameters indicates a noticeable change during El Niño events in the years 2003 and more prominently during the 2014–2016 period.

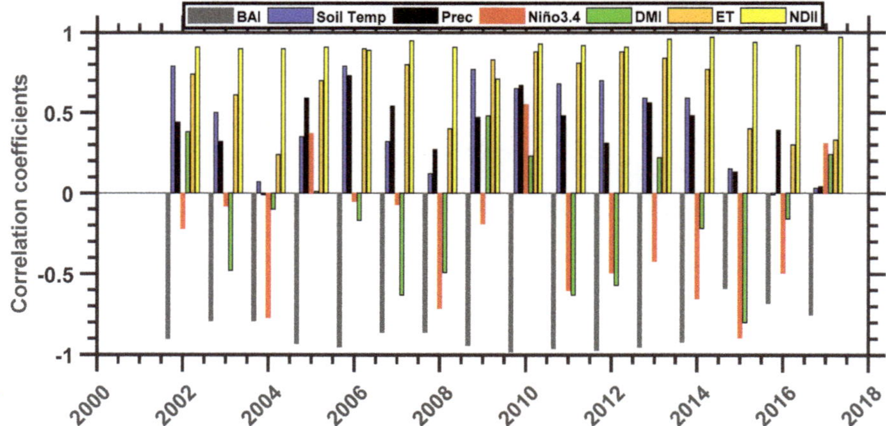

Figure 8. The inter-annual variability of linear correlations between NDVI and BAI, Soil Temp, Prec, Niño3.4, DMI, ET, and NDII for the period 2002 to 2017.

A comprehensive summary of the MLR analysis statistics encompassing NDVI, temperature, precipitation, Niño3.4, and DMI is shown in Table 1. It should be mentioned that the soil temperature, precipitation, ET, NDII, and Niño3.4 were used in this model because of their well-known possible influence on NDVI variability. The DMI climatic parameter was not used as an explanatory variable in the MLR model because of its weak correlation with the NDVI. The results in Table 1 reveal a statistically significant relationship between NDVI and soil temperature and between NDII and ET, with p-values of 0.000386, $<2.00 \times 10^{-16}$ and 0.000173, respectively. Both Precipitation and Niño3.4 indicate a statistically insignificant association with the NDVI because of p-values which are far greater than 0.05. A positive significant correlation between NDVI and Soil temperature, NDII, and ET, which is also represented as in Figures 7 and 8, indicates that soil moisture, soil temperature, and evapotranspiration play a significant role in vegetation health in the HiP. The significant but negative correlation between Niño3.4 and NDVI confirms the notion that ENSO variability plays a role in the climatic conditions of southern Africa [35,52].

Table 1. The output of the Multiple Linear Regression (MLR) model in which Normalized Difference Vegetation Index (NDVI) is a dependent variable and soil temperature, precipitation (Soil Temp), Niño3.4, Normalized Difference Infrared Index (NDII), Dipole Model Index (DMI), and Evapotranspiration (ET) are independent variables.

Variable	Estimate	Std. Error	t-Value	p-Value	Sig
Soil. Temp	-1.23×10^{-02}	3.39×10^{-03}	-3.615	0.000386	***
Prec	7.35×10^{-05}	1.72×10^{-04}	0.427	0.669736	
Niño3.4	-6.44×10^{-03}	7.62×10^{-03}	-0.845	0.399427	
NDII	$1.46 \times 10^{+00}$	7.60×10^{-02}	19.214	$<2.00 \times 10^{-16}$	***
ET	$4.03 \times 10^{+03}$	$1.05 \times 10^{+03}$	3.833	0.000173	***

Significance codes: 0 '***' 0.001 '**' 0.01 '*' 0.05 '.' 0.1 ' ' 1.

In this study, the Mann–Kendall trend test was used for the analysis of the trend in the HiP NDVI time series. The main advantage of this technique is that it provides a non-parametric test that does not require data to be normally distributed, and it is not dependent on the magnitude of data. Furthermore, this non-parametric test method has a low sensitivity to abrupt breaks in heterogeneous time series [66]. The Mann–Kendall test model was applied to the NDVI data, and the results are shown in Figure 9. In summary, the z-score and *p*-value for the entire NDVI time series period (2002–2017) were found to be −1.22 and 0.224, respectively. Both the z-score and the *p*-value seem to indicate that there was a downward but not significant trend in the NDVI data. The indication of an insignificant downward trend (negative z-score) presumably due to the unprecedented sudden reduction of the NDVI values which coincided with the 2014–2016 drought. In order to investigate the influence of drought conditions in the study area using the Mann–Kendall method, it is necessary to calculate the inter-annual variation of Mann–Kendall z-scores. These Mann–Kendall z-scores were calculated from monthly means for each year starting from 2002–2017.

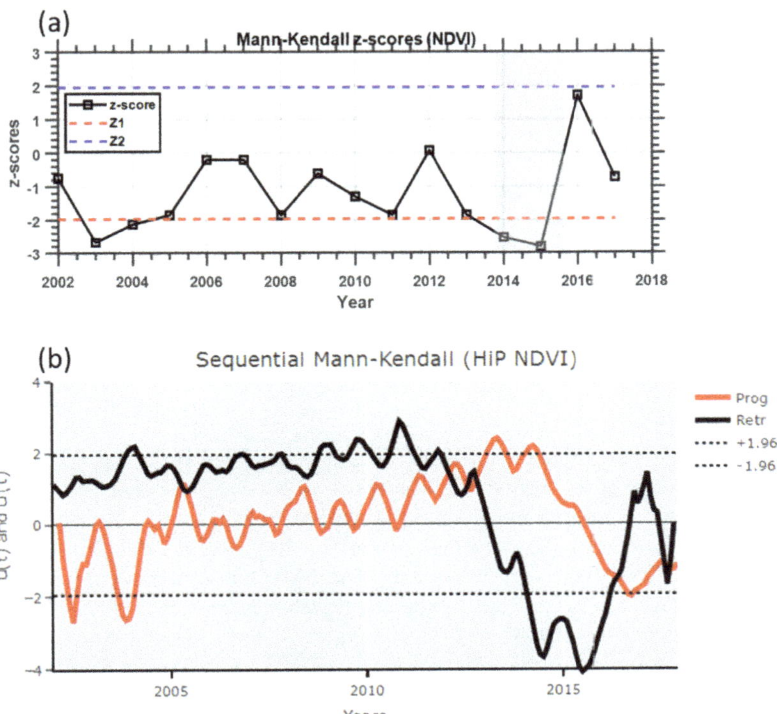

Figure 9. (**a**) The inter-annual variation of Mann–Kendall z-scores ($\alpha = 0.05$, Z1 = −1.96, Z2 = 1.96) for the HiP from 2002 to 2017. (**b**) Sequential statistics values of progressive (Prog) $u(t)$ (solid red line) and retrograde $u'(t)$ (black solid line) obtained by Sequential Mann-Kendall (SQ-MK) test for HiP monthly mean NDVI data for the period from 2002 to 2017.

Figure 9a shows the Mann–Kendall z-scores based on the 16 years of monthly average NDVI data for the game reserve. In general, it is expected that vegetation will respond to climate fluctuating conditions, and this is clearly depicted by significantly negative z-scores (less than Z1 = −1.96) during strong El Niño events (e.g., in 2003 and 2014/2015). The significant downward trend observed between 2014 to 2015 is the strongest such downward trend in the history of the MODIS NDVI data used in this study; it demonstrates a clear response of the vegetation of the reserve to the strong El Niño event of 2014–2016. Similar analysis and results comparable with those presented here were reported by Hou et al. [24] in their study on the inter-annual variability in growing-season NDVI and its correlation with climate variables in the south-western Karst region of China.

The sequential version of the Mann–Kendall test was applied to the NDVI monthly mean time series so as to determine the approximates time periods of the beginning of a significant trend.

Figure 9b shows the sequential statistic values of forward/progressive (Prog) $u(t)$ (solid red line) and retrograde (Retr) $u'(t)$ (black solid line) obtained by SQ-MK test for HiP monthly mean NDVI data for the period from 2002 to 2017. There is a noticeable statistically significant downward trend which seems to coincide with the 2003 and strongly the 2014–2016 strong El Niño event. These independent calculations are in agreement with the inter-annual variation of the Mann–Kendall *z*-scores results presented in Figure 9a. In the case that seems to be associated with the 2014–2016 strong El Niño events, there is an apparent downward trend (indicated by the retrograde) that begins in November 2012 and reaches the negative significant trend limit (-1.96) in April 2014. The retrograde statistic values stay in significant negative territories during the period from April 2014 to May 2016 before it starts to revert back to be within the 95% confidence level limits (±1.96). This trend is regarded as significant because the progressive and retrograde curves intersect each other (turning point) within the limits of the 95% confidence level. This significant trend turning point took place during November 2012. Another significant downward trend was observed in late 2003, with the significant trend turning point observed in June 2005.

The intensity of the 2014–2016 drought impact in the HiP has been identified to be identical to that of the early 1980s [11]. Some of the additional factors that reportedly intensified the impact of the 2014–2016 drought include the reduction in the grazing lawns, siltation of rivers, and the increasing number of carnivores [11]. The impact of the 2014–2016 drought did not only affect this natural protected area (HiP), but also the comical plantations which are situated at about 70 km southwest of the HiP [10], [67]. A study by Crous et al. [67] reported a large-scale dieback of *Eucalyptus grandis* × *E. urophylla* (SClone) in the Zululand coastal plain, KwaZulu-Natal, South Africa, during the recent intense drought. This was later supported by Xulu at al. [10], where they reported that the commercial forest of kwaMbonambi, northern Zululand suffered drought stress during 2015.

3.2. Wavelet Analyses

In order to analyze the localized variation of the spectral power within the time series, wavelet analyses, the most common tool for this purpose, was conducted. As mentioned earlier, the wavelet method assists by decomposing a time series into a time–frequency space, which makes it possible to determine the dominant modes of variability and how they vary in time. Figure 10 shows the normalized wavelet power spectra for the monthly mean NDVI, precipitation, soil temperature, DMI, Niño.3.4 NDII, and ET data. The results of the EVI wavelet analyses are not shown here because this time series is identical to that of the NDVI. In Figure 10, the blue color indicates low wavelet power, and the yellow color represents areas of high wavelet power. The horizontal axis is the time scale (in years) and the vertical axis is the period (in months). The thick black line represents the 95% confidence level. The areas of the wavelet power that are considered are those which are within the cone-of-influence (indicated by the solid "u" shaped line). The con-of-influence indicates areas where edge effects occur in the coherence data [55,57].

The NDVI of the HiP seems to follow the distinctive pattern of the seasonality of precipitation in the north-eastern part of South Africa. The region experiences rainfall during the summer period (December–February) and dry winter period (June–August). This is confirmed by a statistically significant peak observed at around the 12-month cycle (see Figure 10a), which seems to correspond with that of precipitation (Figure 10b). The wavelet power spectra of soil temperature (Figure 10c), NDII (Figure 10f) and ET (Figure 10g) also indicate a strong peak at around the 12-month cycle. This is plausible because wet seasons (summer in this case) lead to increased soil moisture and also create conditions of low evapotranspiration and thus accelerate the greening process in the HiP. It should also be noted that the NDVI wavelet power spectra have significant peaks showing the presence of the semi-annual oscillation (6 months), which is observed during the distinctive period from 2006–2007 to 2011–2012. The semi-annual oscillation observed during the 2006–2007 period is also apparent in the NDII wavelet power spectra. The results of the NDVI wavelet spectral presented here are remarkably similar to the findings of Azzali and Menenti [12], who used a Fourier transform-based technique and reported a substantial seasonal change in NDVI for southern Africa. The significant power of a period of 3–4 months that is observed during the distinctive period 2012–2013 and 2015–2016 in the precipitation power spectra is perhaps related to cyclone Irina in early 2012 and the most recent intense drought of 2014–2016. The wavelet power spectra of the DMI indicate a significant power peak of distinctive periods in the 3–20 months band primarily during the period between 2008 and 2013. On the other hand, the Niño3.4 power spectra exhibit significant power peaks in the 8–32 months band throughout the study period. It should be noted, however, that this frequency of occurrence of peaks observed in the Niño3.4 wavelet spectra is similar to that reported in the studies of Torrance and Compo [55] and also of Jevrejeva et al. [56], who used a much longer time series of the ENSO signal.

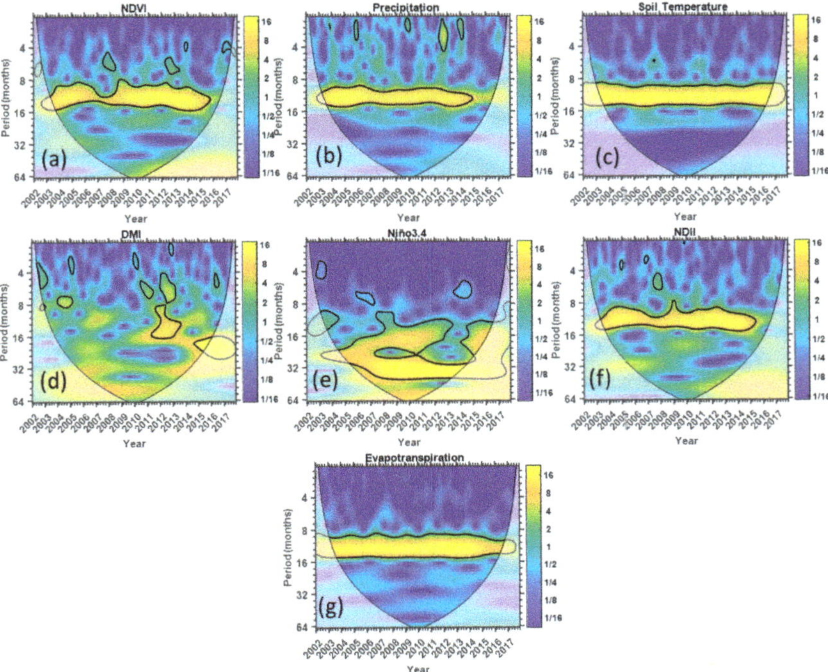

Figure 10. The normalized wavelet power spectra of monthly mean (**a**) NDVI, (**b**) precipitation, (**c**) Soil temperature, (**d**) DMI, (**e**) Niño3, (**f**) NDII, and (**g**) ET, plotted for the period from 2002 to 2017. The black lines which encircle the yellowish colors indicate the areas of significance at the 95% confidence level using the red noise model.

The wavelet coherence between NDVI–Niño3.4, NDVI–DMI, NDVI–precipitation, NDVI–soil temperature, NDVI–NDII, and NDVI–ET was investigated to determine whether NDVI significant wavelet spectra peaks observed at a given time correspond with those observed by the other parameters. Furthermore, the phase relationship between NDVI and the other parameters was calculated and superimposed graphically in Figure 11. The phase relationship is represented by arrows, where two cross-wavelet parameters are in phase if the arrows point to the right, anti-phase if the arrows point to the left, and NDVI leading or lagging if the arrows point upwards or downwards, respectively. The vectors were only plotted for areas where the squared coherence is greater or equal to 0.5. More details about these calculations can be found in References [56,57] and later by studies by Schulte et al. [59].

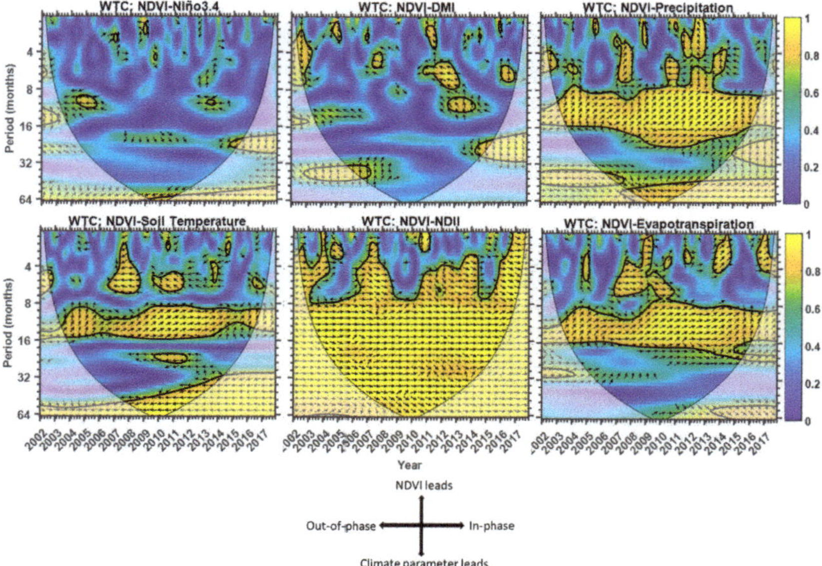

Figure 11. The squared cross-wavelet power spectra for NDVI–Niño3.4, NDVI–DMI, NDVI-precipitation, NDVI–soil temperature, NDVI–NDII, and NDVI–ET. The continuous black lines demarcate the areas of significance at the 95% confidence level using the red noise model. The arrows are vectors indicating the phase difference between the cross-wavelet parameters (see the legend in the bottom left corner).

The local wavelet coherence spectra together with their distinctive cross-spectra phase for NDVI–Niño3.4, NDVI–DMI, NDVI-precipitation, NDVI–soil temperature, NDVI–NDII, and NDVI–ET are shown in Figure 11. In general, all the wavelet coherence spectra indicate that Niño3.4, DMI, precipitation, soil temperature, NDII, and ET do have some degree of coherence with the HiP NDVI in a variety of both periods and timescales. However, it should be mentioned that because statistically, the significant correlation between any two variables being investigated could occur by chance, a significant commonality in a wavelet coherence spectra analysis does not necessarily imply interconnection. Moreover, there is a possibility of smaller areas of wavelet coherence occurring by chance, which would not indicate interconnection, whereas larger areas of significance are improbable due to chance. For this reason, further investigation is required in regard to a possible teleconnection between any two-time series.

A study by Torrance and Compo [55] investigated the periodicities present in a much longer time series (1871–1996) of Niño3.4 using Morlet wavelets and reported the domination of periods greater than 12 months, with some episodes of shorter periods also present in their spectra. In this study, the wavelet coherence between NDVI and Niño3.4 indicates smaller or no areas of high power significance, which is understandable because the 16-year monthly mean NDVI time series is dominated by periodicities of less than 16 months (Figure 10a) whereas the Niño3.4 wavelet spectra are dominated by periodicities greater than 12 months. Remarkably, there is a significant power at a period band of 22–27 months from 2014 to 2017 with cross-spectra phase pointing at the leading position for Niño3.4, which indicates that the recent strong El Niño event may have started first before the response of NDVI months after the El Niño.

The wavelet coherence between NDVI and DMI is observed to delineate some areas that have high significant power at periods of 2–16 months. It is also important to mention that there are significant peaks which are within the cone of influence at the period band 32–48 months during 2005–2007 and 2013–2014, respectively. The cross-wavelet phase during the years 2013–2014 indicates that the DMI was leading the NDVI. This significant peak seems to be similar to that observed in the Niño3.4–NDVI wavelet coherence spectra, which indicates that it is possible that the DMI and Niño3.4 time series were in phase during this period. If so, their joint effect could have maximized the browning observed during 2014–2016. The wavelet coherence between NDVI and precipitation, soil temperature, and ET indicates high significant power during most parts of the study record. In general, these spectra vectors are observed to have an in-phase relationship especially during the period band 8–18 months. This pattern is also observed in the distinctive periods which are less than 8 months especially for the period band 2006–2013. The NDVI and soil temperature wavelet coherence spectra delineate distinctive high power significance with an anti-phase relationship in a 2–8 months band during 2006–2014. Apart from the two distinctive period bands of 2004–2006 and 2015–2017 of high significant power during which the NDVI time series led the temperature time series during the period band 9–14 months, the annual cycle is dominated by the in-phase relationship. Both these scenarios indicate the possible teleconnection between the two time series. The dominant in-phase relationship in the NDVI–precipitation, NDVI–soil temperature, and NDVI–ET suggests that these parameters are positively correlated to the NDVI. This also indicates that the NDVI of the HiP follows the seasonal cycle of precipitation and temperature that is experienced in this region of southern Africa. As expected, the NDVI–NDII coherence spectra indicate a significant coherence at periods greater than 3 months, with a dominant in-phase relationship which indicates a strong correlation between NDVI and NDII. This is in agreement with the Pearson correlation coefficient results presented in Figures 7 and 8.

Overall, factors such as DMI, Niño3.4, precipitation, soil temperature, NDII, and ET are shown to influence NDVI at different distinctive periods and timescales. During the La Niña years, the relationship between NDVI and precipitation and temperature seemed to not indicate any alarming patterns. However, during strong El Niño years (especially broad and strong El Niño years such as the 2014–2016), intense droughts occur. This condition is associated with less humidity and cloud cover, which allows for more solar radiation reaching the ground and accelerated evapotranspiration, which impedes photosynthetic activity.

4. Conclusions

Time series analyses methods were employed in this study to investigate the basic structure variability and trend of the HiP NDVI and its response to the variability of climatic conditions. The results of this study indicate that drought stress reaction patterns of vegetation within HiP provide temporal responses to climate variability, suggesting a strong causal influence. Both the NDVI and EVI values, averaged over the study area, decreased suddenly during 2014–2016 to their greatest minima of approximately 0.28 and 0.11, respectively, in 2015. The linear relationship between climatic indices and NDVI indicated that precipitation soil temperature, soil moisture at root-zone (NDII), ET and to some extent ENSO play a significant role in the variability of vegetation health. The Pearson correlation r and MLR p-value for precipitation and ENSO were found to be 0.45 and 2.0×10^{-7}, and 0.27 and 8.4×10^{-4}, respectively. While some studies [17] reported temperature as the main meteorological parameter that influences vegetation, in this study, we conclude that the influence of precipitation on vegetation was more significant. Different areas of the HiP are affected differently by the strong El Niño signal because of the special variation of land cover. The southern part of the HiP was affected the most because it is dominated by savanna. On the other hand, the northern part of the HiP seems to not be affected presumably because land cover in this area is dominated by forests which are composed of trees which are drought resistant. Moreover, terrain appears to have additional influence on the state of vegetation in the reserve. For example, the lower NDVI values corresponded with the 2014–2016 drought period, particularly in the south-western (flat) part of the reserve, whereas the northern parts (hilly) seem to have benefited from orographic precipitation which promoted vegetation growth. Terrain is also assumed to restrict wildlife grazing in hilly parts of the reserve where stable NDVI are noticeable, placing more burden in flat areas that are accessible to most grazers.

The Mann–Kendall trend significance test and the sequential version of the Mann–Kendall test statistic revealed a significant decreasing pattern of NDVI during the extreme drought periods of 2003 and 2014–2016, with unprecedented lowest minimum values of NDVI detected in 2015. This study has also demonstrated how the wavelet coherence signal processing technique can serve in identifying periodicities in NDVI time series and can also help demonstrate the temporal response of vegetation status to environmental disturbances. The wavelet coherence power spectra indicate a strong influence of precipitation, soil temperature, soil moisture, and ET on the viability of NDVI. This was revealed by a dominant in-phase relationship between the climatic variables and NDVI, which suggests a positive correlation.

While the El Niño of 2014–2016 was both extended and strong, it is possible that its influence in the study area was also supported by a corresponding positive DMI peak which took place at the same time with the with the 2014–2016 El Niño period. It is, therefore, desirable to use the wavelet coherence technique and other methods to investigate the phase relationship between ENSO and DMI for determining the corresponding influence of rainfall in the north-eastern part of South Africa.

Finally, we conclude that the recent intense drought of 2014–2016 influenced the spatiotemporal pattern of the vegetation condition in the HiP. This holds more implications for the tourism potential of the HiP with attractive grazers such as white rhinos and buffalos that were reportedly affected by this event [11]. The results portend that the freely GEE-archived satellite data is a capable tool for monitoring droughts with a high temporal resolution across game reserves located in drought-prone areas of South Africa and other parts of the world.

Author Contributions: N.M. and S.X. designed the research. N.M. performed data analyses, visualization, and interpretation of results. N.M. and S.X. wrote the paper.

Funding: This research is funded by the National Research Foundation (NRF) of South Africa and the University of Zululand.

Acknowledgments: The authors would like to thank all personnel involved in the development of the Google Earth Engine system and climate engine. We also thank the providers of the important public data set in the Google Earth Engine, in particular, NASA, USGS, NOAA, EC/ESA, and MERRA-2 model developers.

Conflicts of Interest: The authors declare no conflict of interest.

Appendix A

Here we show the Google Earth Engine interactive development environment. The dark green area indicates the location of the Hluhluwe-iMfolozi Park in north-eastern South Africa. The NDVI time series, which is averaged for the study area, is shown in the console part of the GEE interactive development environment.

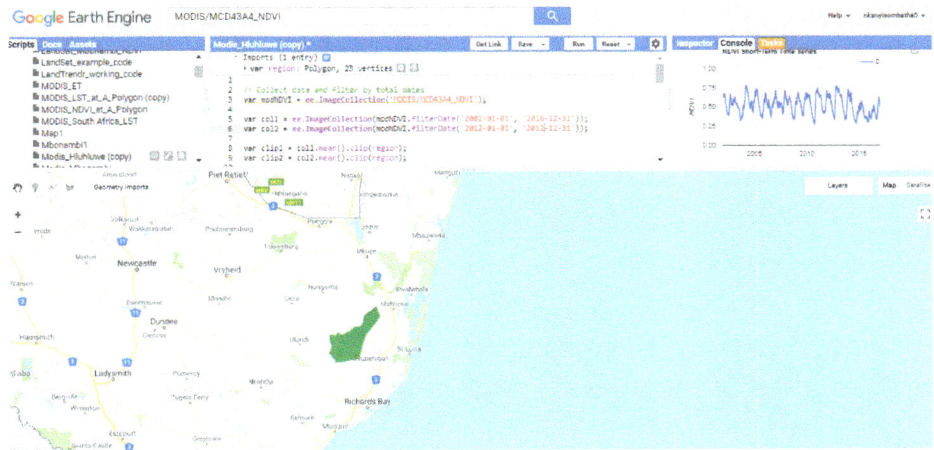

Figure A1. The Google Earth Engine interactive development environment.

References

1. Stolton, S.; Dudley, N.; Avcıoğlu Çokçalışkan, B.; Hunter, D.; Ivanić, K.Z.; Kanga, E.; Kettunen, M.; Kumagai, Y.; Maxted, N.; Senior, J.; et al. Values and benefits of protected areas. In *Protected Area Governance and Management*; Worboys, G.L., Lockwood, M., Kothari, A., Feary, S., Pulsford, I., Eds.; ANU Press: Canberra, Australia, 2015; pp. 145–168.
2. Kettunen, M.; ten Brink, P. *Social and Economic Benefits of Protected Areas: An Assessment Guide*; Routledge: Abingdon, UK, 2013.
3. White, R.P.; Murray, S.; Rohweder, M.; Prince, S.D.; Thompson, K.M. *Grassland Ecosystems*; World Resources Institute: Washington, DC, USA, 2000.
4. Sruthi, S.; Aslam, M.M. Agricultural drought analysis using the NDVI and land surface temperature data; a case study of Raichur district. *Aquat. Procedia* **2015**, *4*, 1258–1264. [CrossRef]
5. Wilhite, D.A.; Svoboda, M.D.; Hayes, M.J. Understanding the complex impacts of drought: A key to enhancing drought mitigation and preparedness. *Water Resour. Manag.* **2007**, *21*, 763–774. [CrossRef]
6. Myhre, G.; Shindell, D.; Bréon, F.M.; Collins, W.; Fuglestvedt, J.; Huang, J.; Koch, D.; Lamarque, J.F.; Lee, D.; Mendoza, B.; et al. Anthropogenic and Natural Radiative Forcing. In *Climate Change 2013: The Physical Science Basis. Contribution of Working Group I to the Fifth Assessment Report of the Intergovernmental Panel on Climate Change*; Tignor, K., Allen, M., Boschung, S.K., Nauels, J., Xia, Y., Bex, V., Midgley, P.M., Eds.; Cambridge University Press: Cambridge, UK; New York, NY, USA, 2013.
7. Clark, P.U.; Shakun, J.D.; Marcott, S.A.; Mix, A.C.; Eby, M.; Kulp, S.; Levermann, A.; Milne, G.A.; Pfister, P.L.; Santer, B.D.; et al. Consequences of twenty-first-century policy for multi-millennial climate and sea-level change. *Nat. Clim. Chang.* **2016**, *6*, 360. [CrossRef]
8. Duncan, J.M.; Biggs, E.M. Assessing the accuracy and applied use of satellite-derived precipitation estimates over Nepal. *Appl. Geogr.* **2012**, *34*, 626–638. [CrossRef]
9. Dube, L.; Jury, M. The nature of climate variability and impacts of drought over KwaZulu-Natal, South Africa. *S. Afr. Geogr. J.* **2000**, *82*, 44–53. [CrossRef]
10. Xulu, S.; Peerbhay, K.; Gebreslasie, M.; Ismail, R. Drought influence on forest plantations in Zululand, South Africa, using MODIS time series and climate data. *Forests* **2018**, *9*, 528. [CrossRef]
11. Whateley, A. The Impact of Drought in the Hluhluwe Imfolozi Park (HiP) South Africa. 2017. Available online: http://theconservationimperative.com/?p=235 (accessed on 10 June 2018).
12. Azzali, S.; Menenti, M. Mapping vegetation-soil-climate complexes in southern Africa using temporal Fourier analysis of NOAA-AVHRR NDVI data. *Int. J. Remote Sens.* **2000**, *21*, 973–996. [CrossRef]
13. Hassan, M.H.; Hutchinson, C. *Natural Resource and Environmental Information for Decision Making*; The World Bank: Washington, DC, USA, 1992.

14. Xie, Y.; Sha, Z.; Yu, M. Remote sensing imagery in vegetation mapping: A review. *J. Plant Ecol.* **2008**, *1*, 9–23. [CrossRef]

15. Rulinda, C.M.; Dilo, A.; Bijker, W.; Stein, A. Characterizing and quantifying vegetative drought in East Africa using fuzzy modelling and NDVI data. *J. Arid Environ.* **2012**, *78*, 169–178. [CrossRef]

16. Tucker, C.J. Red and photographic infrared linear combinations for monitoring vegetation. *Remote Sens. Environ.* **1979**, *8*, 127–150. [CrossRef]

17. Poveda, G.; Salazar, L.F. Annual and interannual (ENSO) variability of spatial scaling properties of a vegetation index (NDVI) in Amazonia. *Remote Sens. Environ.* **2004**, *93*, 391–401. [CrossRef]

18. Wang, J.; Rich, P.M.; Price, K.P.; Kettle, W.D. Relations between NDVI and tree productivity in the central Great Plains. *Int. J. Remote Sens.* **2004**, *25*, 3127–3138. [CrossRef]

19. Sriwongsitanon, N.; Gao, H.; Savenije, H.H.G.; Maekan, E.; Saengsawan, S.; Thianpopirug, S. Comparing the Normalized Difference Infrared Index (NDII) with root zone storage in a lumped conceptual model. *Hydrol. Earth Syst. Sci.* **2016**, *20*, 3361–3377. [CrossRef]

20. Sriwongsitanon, N.; Gao, H.; Savenije, H.H.G.; Maekan, E.; Saengsawan, S.; Thianpopirug, S. The Normalized Difference Infrared Index (NDII) as a proxy for soil moisture storage in hydrological modelling. *Hydrol. Earth Syst. Sci. Discuss.* **2015**, *12*, 8419–8457. [CrossRef]

21. Joiner, J.; Yoshida, Y.; Anderson, M.; Holmes, T.; Hain, C.; Reichle, R.; Koster, R.; Middleton, E.; Zeng, F.W. Global relationships among traditional reflectance vegetation indices (NDVI and NDII), evapotranspiration (ET), and soil moisture variability on weekly timescales. *Remote Sens. Environ.* **2018**, *219*, 339–352. [CrossRef]

22. Deshayes, M.; Guyon, D.; Jeanjean, H.; Stach, N.; Jolly, A.; Hagolle, O. The contribution of remote sensing to the assessment of drought effects in forest ecosystems. *Ann. For. Sci.* **2006**, *63*, 579–595. [CrossRef]

23. Fensholt, R.; Rasmussen, K.; Nielsen, T.T; Mbow, C. Evaluation of earth observation based long term vegetation trends—Intercomparing NDVI time series trend analysis consistency of Sahel from AVHRR GIMMS, Terra MODIS and SPOT VGT data. *Remote Sens. Environ.* **2009**, *113*, 1886–1898. [CrossRef]

24. Hou, W.; Gao, J.; Wu, S.; Dai, E. Interannual variations in growing-season NDVI and its correlation with climate variables in the southwestern karst region of China. *Remote Sens.* **2015**, *7*, 11105–11124. [CrossRef]

25. Xu, C.; Hantson, S.; Holmgren, M.; Nes, E.H.; Staal, A.; Scheffer, M. Remotely sensed canopy height reveals three pantropical ecosystem states. *Ecology* **2016**, *97*, 2518–2521. [CrossRef] [PubMed]

26. Forkel, M.; Carvalhais, N.; Verbesselt, J.; Mahecha, M.D.; Neigh, C.S.; Reichstein, M. Trend change detection in NDVI time series: Effects of inter-annual variability and methodology. *Remote Sens.* **2013**, *5*, 2113–2144. [CrossRef]

27. Liu, S.; Zhang, Y.; Cheng, F.; Hou, X.; Zhao, S. Response of grassland degradation to drought at different time-scales in Qinghai Province: Spatio-temporal characteristics, correlation, and implications. *Remote Sens.* **2017**, *9*, 1329.

28. Jiang, L.; Shang, S.; Yang, Y.; Guan, H. Mapping interannual variability of maize cover in a large irrigation district using a vegetation index–phenological index classifier. *Comput. Electron. Agric.* **2016**, *123*, 351–361. [CrossRef]

29. Cui, Y.P.; Liu, J.Y.; Hu, Y.F.; Kuang, W.H.; Xie, Z.L. An analysis of temporal evolution of NDVI in various vegetation-climate regions in Inner Mongolia, China. *Procedia Environ. Sci.* **2012**, *13*, 1989–1996. [CrossRef]

30. Huang, F.; Mo, X.; Lin, Z.; Hu, S. Dynamics and responses of vegetation to climatic variations in Ziya-Daqing basins, China. *Chin. Geogr. Sci.* **2016**, *26*, 478–494. [CrossRef]

31. Bond, W.J.; Archibald, S. Confronting complexity: Fire policy choices in South African savanna parks. *Int. J. Wildl. Fire* **2003**, *12*, 381–389. [CrossRef]

32. Nsukwini, S.; Bob, U. The socio-economic impacts of ecotourism in rural areas: A case study of Nompondo and the Hluhluwe-iMfolozi Park (HiP). *Afr. J. Hosp. Tour. Leis.* **2016**, *5*, 1–15.

33. Boundja, R.P.; Midgley, J.J. Patterns of elephant impact on woody plants in the Hluhluwe-imfolozi park, Kwazulu-Natal, South Africa. *Afr. J. Ecol.* **2010**, *48*, 206–214. [CrossRef]

34. Trinkel, M.; Ferguson, N.; Reid, A.; Reid, C.; Somers, M.; Turelli, L.; Graf, J.; Szykman, M.; Cooper, D.; Haverman, P.; et al. Translocating lions into an inbred lion population in the Hluhluwe-iMfolozi Park, South Africa. *Anim. Conserv.* **2008**, *11*, 138–143. [CrossRef]

35. Jolles, A.E.; Etienne, R.S.; Olff, H. Independent and competing disease risks: Implications for host populations in variable environments. *Am. Nat.* **2006**, *167*, 745–757. [CrossRef] [PubMed]

36. Schaaf, C.B.; Gao, F.; Strahler, A.H.; Lucht, W.; Li, X.; Tsang, T.; Strugnell, N.C.; Zhang, X.; Jin, Y.; Muller, J.P. First operational BRDF, albedo nadir reflectance products from MODIS. *Remote Sens. Environ.* **2002**, *83*, 135–148. [CrossRef]

37. Reinecker, M.M.; Suarez, M.J.; Gelaro, R.; Todling, R.; Bacmeister, J.; Liu, E.; Bosilovich, M.G.; Schubert, S.D.; Takacs, L.; Kim, G.K. MERRA: NASA's modern-era retrospective analysis for research and applications. *J. Clim.* **2011**, *24*, 3624–3648. [CrossRef]

38. Rodell, M.; Houser, P.R.; Jambor, U.E.A.; Gottschalck, J.; Mitchell, K.; Meng, C.J.; Arsenault, K.; Cosgrove, B.; Radakovich, J.; Bosilovich, M.; et al. The global land data assimilation system. *Bull. Am. Meteorol. Soc.* **2004**, *85*, 381–394. [CrossRef]

39. Kruger, A. The influence of the decadal-scale variability of summer rainfall on the impact of El Niño and La Niña events in South Africa. *Int. J. Climatol.* **1999**, *19*, 59–68. [CrossRef]

40. Anyamba, A.; Tucker, C.J.; Mahoney, R. From El Niño to La Niña: Vegetation response patterns over east and southern Africa during the 1997–2000 period. *J. Clim.* **2002**, *15*, 3096–3103. [CrossRef]

41. Saji, N.H.; Goswami, B.N.; Vinayachandran, P.N.; Yamagata, T. A dipole mode in the tropical Indian Ocean. *Nature* **1999**, *401*, 360. [CrossRef] [PubMed]

42. Reason, C.; Mulenga, H. Relationships between South African rainfall and SST anomalies in the southwest Indian Ocean. *Int. J. Climatol.* **1999**, *19*, 1651–1673. [CrossRef]

43. Reason, C. Subtropical Indian Ocean SST dipole events and southern African rainfall. *Geophys. Res. Lett.* **2001**, *28*, 2225–2227. [CrossRef]

44. Reason, C.; Rouault, M. ENSO-like decadal variability and South African rainfall. *Geophys. Res. Lett.* **2002**, *29*, 161–164. [CrossRef]

45. Mann, H.B. Nonparametric tests against trend. *Econometrica* **1945**, *1*, 245–259. [CrossRef]

46. Gilbert, R.O. *Statistical Methods for Environmental Pollution Monitoring*; John Wiley & Sons: Toronto, ON, Canada, 1987.

47. Kendall, M.G. *Rank Correlation Methods*; Griffin: London, UK, 1975.

48. Lanzante, J.R. Resistant, robust and non-parametric techniques for the analysis of climate data: Theory and examples, including applications to historical radiosonde station data. *Int. J. Climatol.* **1996**, *16*, 1197–1226. [CrossRef]

49. Pohlert, T. Non-Parametric Trend Tests and Change-Point Detection. 2018. Available online: https://cran.r-project.org/web/packages/trend/trend.pdf (accessed on 27 July 2018).

50. Sneyers, R. *On the Statistical Analysis of Series of Observations*; Technical Note No. 143, WMO No. 415; World Meteorological Organization: Geneva, Switzerland, 1991; p. 192.

51. Mosmann, V.; Castro, A.; Fraile, R.; Dessens, J.; Sanchez, J.L. Detection of statistically significant trends in the summer precipitation of mainland Spain. *Atmos. Res.* **2004**, *70*, 43–53. [CrossRef]

52. Chatterjee, S.; Bisai, D.; Khan, A. Detection of approximate potential trend turning points in temperature time series (1941–2010) for Asansol Weather Observation Station, West Bengal, India. *India Atmos. Clim. Sci.* **2014**, *4*, 64–69. [CrossRef]

53. Zarenistanak, M.; Dhorde, A.G.; Kripalani, R.H. Trend analysis and change point detection of annual and seasonal precipitation and temperature series over southwest Iran. *J. Earth Syst. Sci.* **2014**, *123*, 281–295. [CrossRef]

54. Soltani, M.; Rousta, I.; Taheri, S.S.M. Using Mann-Kendall and time series techniques for statistical analysis of long-term precipitation in Gorgan weather station. *World Appl. Sci. J.* **2013**, *28*, 902–908.

55. Torrence, C.; Compo, G.P. A practical guide to wavelet analysis. *Bull. Am. Meteorol. Soc.* **1998**, *79*, 61–78. [CrossRef]

56. Jevrejeva, S.; Moore, J.; Grinsted, A. Influence of the arctic oscillation and El Niño-Southern Oscillation (ENSO) on ice conditions in the Baltic Sea: The wavelet approach. *J. Geophys. Res. Atmos.* **2003**, *108*, 4677. [CrossRef]

57. Grinsted, A.; Moore, J.C.; Jevrejeva, S. Application of the cross wavelet transform and wavelet coherence to geophysical time series. *Nonlinear Proc. Geophys.* **2004**, *11*, 561–566. [CrossRef]

58. Zar, J.H. *Biostatistical Analysis*; Prentice Hall: Upper Saddle River, NJ, USA, 1999.

59. Schulte, J.A.; Najjar, R.G.; Li, M. The influence of climate modes on streamflow in the Mid-Atlantic region of the United States. *J. Hydrol. Reg. Stud.* **2016**, *5*, 80–99. [CrossRef]

60. Mberego, S.; Gwenzi, J. Temporal patterns of precipitation and vegetation variability over Zimbabwe during extreme dry and wet rainfall seasons. *J. Appl. Meteorol. Climatol.* **2014**, *53*, 2790–2804. [CrossRef]

61. Manatsa, D.; Mukwada, G.; Siziba, E.; Chinyanganya, T. Analysis of multidimensional aspects of agricultural droughts in Zimbabwe using the standardized precipitation index (SPI). *Theor. Appl. Climatol.* **2010**, *102*, 287–305. [CrossRef]

62. Verbesselt, J.; Hyndman, R.; Newnham, G.; Culvenor, D. Detecting trend and seasonal changes in satellite image time series. *Remote Sens. Environ.* **2010**, *114*, 106–115. [CrossRef]

63. Verbesselt, J.; Hyndman, R.; Zeileis, A.; Culvenor, D. Phenological change detection while accounting for abrupt and gradual trends in satellite image time series. *Remote Sens. Environ.* **2010**, *114*, 2970–2980. [CrossRef]

64. Fitchett, J.M.; Grab, S.W. A 66-year tropical cyclone record for south-east Africa: Temporal trends in a global context. *Int. J. Climatol.* **2014**, *34*, 3604–3615. [CrossRef]

65. Islam, M.M.; Mamun, M.M.I. Variations of NDVI and its association with rainfall and evapotranspiration over Bangladesh. *Rajshahi Univ. J. Sci. Eng.* **2015**, *43*, 21–28. [CrossRef]

66. Tabari, H.; Marofi, S.; Aeini, A.; Talaee, P.H.; Mohammadi, K. Trend analysis of reference evapotranspiration in the western half of Iran. *Agric. For. Meteorol.* **2011**, *151*, 128–136. [CrossRef]

67. Crous, C.J.; Greyling, I.; Wingfield, M.J. Dissimilar stem and leaf hydraulic traits suggest varying drought tolerance among co-occurring *Eucalyptus grandis* × *E. urophylla* clones. *South. For. J. For. Sci.* **2018**, *80*, 175–184. [CrossRef]

Article

Selecting and Downscaling a Set of Climate Models for Projecting Climatic Change for Impact Assessment in the Upper Indus Basin (UIB)

Asim Jahangir Khan [1,2,*] **and Manfred Koch** [1]

1   Department of Geohydraulics and Engineering Hydrology, University of Kassel, 34125 Kassel, Hessen, Germany; manfred_kochde@yahoo.de
2   Department of Environmental Sciences, COMSATS University Islamabad, Abbottabad Campus, University Road, Tobe Camp, Abbottabad, KP 22060, Pakistan
*   Correspondence: asimjkw@gmail.com or uk053114@student.uni-kassel.de; Tel.: +49-17631674283

Received: 26 September 2018; Accepted: 10 November 2018; Published: 14 November 2018

Abstract: This study focusses on identifying a set of representative climate model projections for the Upper Indus Basin (UIB). Although a large number of General Circulation Models (GCM) predictor sets are available nowadays in the CMIP5 archive, the issue of their reliability for specific regions must still be confronted. This situation makes it imperative to sort out the most appropriate single or small-ensemble set of GCMs for the assessment of climate change impacts in a region. Here a set of different approaches is adopted and applied for the step-wise shortlisting and selection of appropriate climate models for the UIB under two RCPs: RCP 4.5 and RCP 8.5, based on: (a) range of projected mean changes, (b) range of projected extreme changes, and (c) skill in reproducing the past climate. Furthermore, because of higher uncertainties in climate projection for high mountainous regions like the UIB, a wider range of future GCM climate projections is considered by using all possible extreme future scenarios (wet-warm, wet-cold, dry-warm, dry-cold). Based on this two-fold procedure, a limited number of climate models is pre-selected, from of which the final selection is done by assigning ranks to the weighted score for each of the mentioned selection criteria. The dynamically downscaled climate projections from the Coordinated Regional Downscaling Experiment (CORDEX) available for the top-ranked GCMs are further statistically downscaled (bias-corrected) over the UIB. The downscaled projections up to the year 2100 indicate temperature increases ranging between 2.3 °C and 9.0 °C and precipitation changes that range from a slight annual increase of 2.2% under the drier scenarios to as high as 15.9% in the wet scenarios. Moreover, for all scenarios, future precipitation will be more extreme, as the probability of wet days will decrease, while, at the same time, precipitation intensities will increase. The spatial distribution of the downscaled predictors across the UIB also shows similar patterns for all scenarios, with a distinct precipitation decrease over the south-eastern parts of the basin, but an increase in the northeastern parts. These two features are particularly intense for the "Dry-Warm" and the "Median" scenarios over the late 21st century.

Keywords: GCM; RCM; CMIP5; CORDEX; climate change; climate model selection; upper Indus basin

---

1. Introduction

Future climate projections provided by general circulation models (GCMs) can serve as the basic input for climate change impact studies on water resources. As the outputs from these general circulation models (GCMs) have only coarse spatial resolution, and so are often not suitable as direct input to distributed or semi-distributed hydrologic models, they have to be downscaled in most cases to appropriate (higher) resolutions. Such a downscaling can be done either through applying statistical downscaling or through dynamical downscaling via use of a regional climate model (RCM) embedded in a larger GCM.

Despite the availability of a large number of GCM outputs in the CMIP5 archive, and the on-going improvements in their process representations, issues of large uncertainties with regard to the future climate are not yet avoidable. The inherent uncertainties, along with other factors such as time limitations, human resource availability, or computational constraints, make it imperative to sort out the most appropriate individual GCM or small ensemble of GCMs suitable for downscaling and subsequent use in the assessment of climate change impacts.

This aforementioned selection of GCMs is not simple or straightforward, as there can be nearly an unlimited number of criteria and approaches through which climate models can be evaluated for their skill and suitability for specific purposes and regions. In most cases, though, the selection can be based either on a single criterion or a whole set of criteria. One approach may be to consider the total change projected by the GCMs, in the means and/or extremes of a climate variable and its location on the overall spectrum of the future projected by all GCMs. Another approach may place more emphasis on the success of GCMs in simulating past climate for either the means, extremes, or seasonality [1,2] of the study region. Additionally, there may be approaches based on some combination of the aforementioned approaches. The first approach, which considers all the possible projected futures (stretching from warm and wet to cold and dry, or opting for the middle path of all possible futures) is becoming more relevant, especially in regions such as the Hindu Kush Himalayas (HKH) and UIB, where GCMs/RCMs have been reported to struggle in simulating the past climate [3–6]. As no individual model can be separated out as superior in simulating the past climate in the HKH region, it is therefore important to consider the full range of possible projected futures when focusing on assessments of climate change impacts.

The criteria to be used for selecting the most appropriate model runs are also defined based on their intended purpose or the region. Both of these factors are important, as a different intended uses my require consideration of assessment based on totally different skills or variables, while the importance of a specific selection criteria may differ for different locations and topographically contrasting areas. Additionally, as not all the available models may be equally good for specific locations, regions or topographies, the need for the assessment of the ability of climate models to reproduce important processes in the study region is vital and essential.

In the current study, we consider a combination of these approaches to shortlist climate model runs, along with utilizing new and improved data for the past climate in the UIB [7] for assessment of model skill in simulating the seasonal cycles in the region. The main aim of the study was to select a set of GCM simulations that can represent the full spectrum of the future climate, as projected by the entire pool of climate models, in term of both means and extremes, and which can be subsequently used as climate forcing for hydrological modelling to assess a wider range of possible climate change hydrological impacts, especially for the expected changes in water yield, annual cycle, high and low flows, and floods.

The specific objectives of the study included:

1. To devise a procedure for the identification/filtering of a limited number of climate model runs that can represent the full spectrum of future climate as projected by the entire pool of climate models, in term of both means and extremes;
2. To devise procedures/methodologies for evaluating the skills of climate models in simulating the annual climatic cycle of the recent past;
3. To select suitable climate model runs using the devised methodologies, based on their skills in simulating past climate, as well as on their ability to represent specific parts of the full spectrum of climate model projections; and
4. To downscale and/or bias correct the selected GCMs (or GCM-RCM chains, using the selected GCMs as boundary conditions) through appropriate methods.

2. Study Area and Data Used

2.1. Study Area

In the current study, the climate change model selection procedure was carried out for the UIB, which is spread over the Hindu-Kush, Karakorum and Himalayan ranges, and feeds the largest canal system in the world (Figure 1). This river basin is very important due to two main reasons: first, the irrigated agriculture of Pakistan overwhelmingly depends on the inputs from this river basin; and second, the region is probably a climate change hot-spot [8,9], with an extremely uncertain future hydro-climatology. The future scenario data from the selected models are intended to be used, after downscaling and bias correction, as input to the SWAT hydrological model [10] for quantifying possible climate change impacts on the hydrological dynamics of the basin.

Climatic variables are usually strongly influenced by topographic altitude. Thus, the northern valley floors of the UIB are arid and warm, with an annual precipitation of only 100–200 mm. These totals increase to 600 mm at 4400 m altitude, and glaciological studies suggest annual accumulation rates of 1500–2000 mm at height of 5500 m [11]. The UIB draws more than 50% of its water from melting of seasonal and permanent snow cover in the Himalaya, Karakoram and the Hindu Kush (HKH) mountains [5,12–15]. A rise in temperature in the UIB will, therefore, result in elevated melt rates with huge impacts on the timing and magnitude of the generated flows. This will not only lead to a higher average stream flow, but also to an increase in the occurrence and magnitude of extremes, especially during high-precipitation events [16]. There is also the possibility that the peak flows may shift to earlier months or other seasons, with a rise in temperature [5] in the UIB.

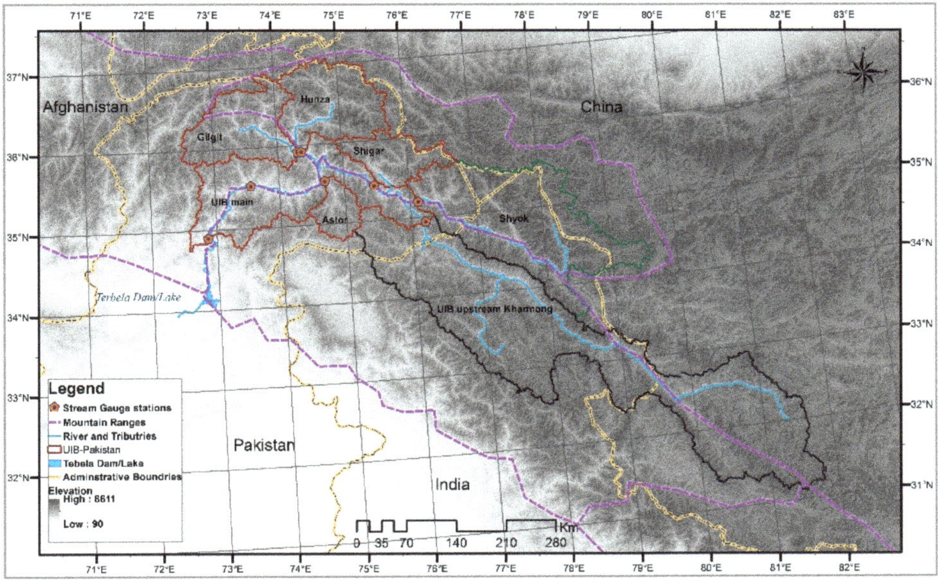

Figure 1. Upper Indus Basin (UIB): Main catchments, meteorological stations, streams and tributaries.

All these facts make UIB a very sensitive region to possible climate change, and even, according to some [17], a climate-change "hotspot". However, despite the necessity of intensified investigations on different aspects of climate change and its possible implications, the task is hindered by the harshness of the environment and the unavailability of representative data. The climatic data available in the UIB lacks suitable coverage, since the in situ meteorological observations in the UIB are sparse and mostly taken at valley stations. Furthermore, the complex orography of the UIB region also affects the amounts, spatial patterns and seasonality of the precipitation. Therefore, neither the sparsely observed station data and gridded data products based on them, nor the sensors-based data, fully represent the precipitation regime of the region [6].

2.2. Data Used

2.2.1. GCM Outputs

In the IPCC 5th assessment report, four representative concentration pathways (RCPs) are normally used as a basis for future climate modelling: one very high baseline emission scenario (RCP8.5), two medium stabilization scenarios (RCP4.5 and RCP6) and one mitigated scenario (RCP2.6) (Table 1).

Table 1. Representative concentration pathways (RCPs), their radiative forcing, emissions (CO_2 equivalent and growth rate %) and temperature increase.

RCP	Radiative Forcing	CO_2 Equiv. (ppm)	Temperature Increase (°C)	Pathway	CO_2 Growth Rate (%)
RCP8.5	8.5 Wm^{-2} in 2100	1370	4.9	Rising	≈2.5
RCP6.0	6 Wm^{-2} post 2100	850	3.0	Stabilization without overshoot	≈1
RCP4.5	4.5 Wm^{-2} post 2100	650	2.4	Stabilization without overshoot	≈1.5
RCP2.6 (RCP3PD)	3 Wm^{-2} before 2100, declining to 2.6 Wm^{-2} by 2100	490	1.5	Peak and decline	≈1.6

Source: [18–20]

The current study intended to include emission scenarios, covering a wider range of Radiative forcing and future temperature anomaly, while remaining close to the reality and considering RCP's showing minimum differences with the 2005 onwards actual observed CO_2 emission trend and growth rates. Keeping these prerequisites in mind, out of the four options: RCP2.6 was not considered in the current selection as it seemed to be the least likely [18,21] and the mitigation effort implied by this RCP, is unfeasible in the current circumstances [22,23], because it needs a sustained global CO_2 mitigation rate of around 3% per year, not a likely prospect, at least in the near future [20].

Out of the remaining three RCPs, the high baseline emission scenario (RCP8.5) and one medium-stabilization scenario (RCP4.5) were selected for the current study. The RCP8.5 was included because it covers the higher end of radiative forcing, as well as the temperature change, and it is also in line with the observed trend of around 3% in the average annual CO_2 emission growth rates for 2005–2012 [20,23].

For the medium-stabilization scenarios, both RCP4.5 and RCP6 are equally acceptable, but due to time constraints, and because RCP4.5 shows a better match (≈1.5%) of the trends of the average annual CO_2 emission growth rates for the period of 2005–2012 than RCP6 (≈1.0%) [20,23], RCP4.5 was picked along RCP8.5 for the GCM selection procedure.

Additionally, in the current study, only the available GCM runs for the ensemble member *r1p1i1* in the CMIP5 repository [24] are included in the initial list. This is done so as to keep open the possibility of using dynamically downscaled projections (driven by the selected GCMs as boundary conditions) by Regional Climate Models (RCMs), which in most cases have utilized boundary conditions from the ensemble member *r1p1i1* of the GCMs.

In the current study, a total number of 42 available model runs (ensemble member *r1p1i1*) are evaluated for RCP4.5 and of 39 for RCP8.5.

2.2.2. Extremes Indices

For the assessment of model runs for extremes, the ETCCDI extremes indices are utilized. The annual extremes of the daily CMIP5 data were acquired from the ETCCDI extremes indices archive [25,26], provided at the "Canadian Centre for Climate Modelling and Analysis". This data was indirectly obtained and downloaded through the "KNMI Climate Explorer", which is a web-based research tool to investigate climate and climate change.

2.2.3. Observed Data

The climate station network in the UIB has historically been comprised of only a few low-altitude, valley-based stations. Although the number of in situ observational points has increased since the mid-nineties, with the installations of a few higher altitude automatic weather stations, the coverage is still very thin, and the data is often not very representative, especially for different elevation zones. Similarly, while most of the weather stations have become operational after the mid-nineties, long-term data is a rare commodity and is only available at limited locations.

Similarly, owing to the complex orography of the UIB region and to the co-action of different hydro-climatic regimes, neither the sparse observed station data or the gridded data products based on them, nor the sensor-based climatic datasets fully represent the precipitation regime of the region [6,16,27,28]. Several studies have pointed out that precipitation and other climatic variables in the HKH region exhibit large changes over short distances and considerable vertical gradients [11,29–34].

In the absence of long-term climate data with acceptable representation of the UIB climate, most climate-change studies have relied on either the very thin climatic observation network records or the gridded datasets based on them. In all these cases, either the data have acceptable quality, but shorter duration, or they have huge biases, especially, in the case of precipitation in regions with higher altitudes. These biases are further amplified when this data is used as a reference for bias correction or downscaling of climate projections, making the results questionable.

In the current study, therefore, a new long-term climate dataset was prepared (Figure 2). The work related to this new long-term gridded data product [7] is not included in this paper, but we utilized this new dataset instead of the readily available global or regional gridded historical climate datasets, for bias correction, downscaling and assessment of the reliability of climate models for the simulation of the past climate in the region.

These gridded precipitation and temperature data are derived, based on all the available in situ observations available in the UIB, through reconstruction for the periods before the mid-nineties, interpolation and correction for the orography and elevation-induced effects guided by available data for runoff, actual evapotranspiration and glacier mass-balance [7].

2.2.4. RCM Outputs

Five CORDEX-SA experiments (Table 6), including IPSL-CM5A-MR_RCA4, MPI-ESM-LR_RCA4, NorESM1-M_RCA4, Can ESM2_RegCM4-4, and GFDL-ESM2M_RCA4, were downscaled and bias corrected. These five GCMs have been dynamically downscaled by CORDEX, using two different RCMs (RCA4 and RegCM4). Their RCM outputs are at considerably finer scale (0.44°) then the source GCMs.

3. Methods

3.1. Selection and Shortlisting of GCMs/RCMs

The full spectrum of GCM projections is wide, with large uncertainties attached [35–37], and it cascades to even a larger spectrum when downscaled or translated into possible impacts. Furthermore, the available future projections differ vastly from each other and may range from very wet to drier or very warm to colder future climates, so that the models can be categorized as representing either

Warm-Wet, Warm-Dry, Cold-Wet and Cold-Dry corners of the full spectrum, in addition to the projections which are around the median tendency of future model projections.

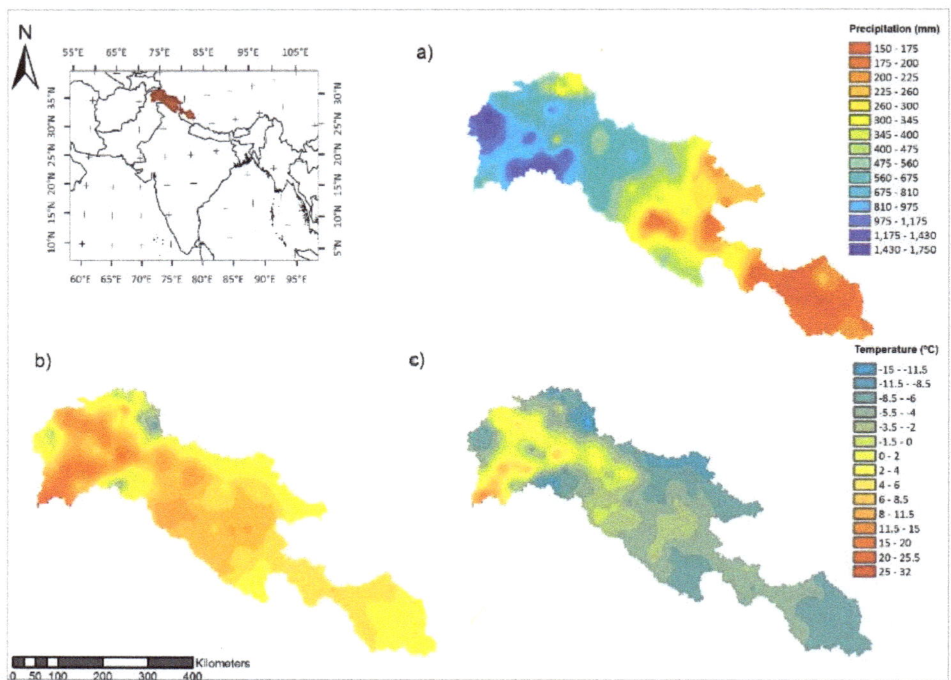

Figure 2. Reference climate data: (**a**) Mean annual precipitation (mm), (**b**) mean temperature- maximum (°C), and (**c**) mean temperature- minimum (°C).

These issues have led to diverse views on how to select or use these climate model projections, or even whether these climate models or their downscaled outputs should explicitly be used at all, or should only be indirectly used, instead, as guides to generate a range of plausible scenarios more suited for targeted impact studies and practical adaptation planning [38].

In mountainous regions, such as the Upper Indus Basin (UIB), the issue of how to proceed with the climate change impact studies becomes more complicated, because not only may the uncertainties shown by the climate models for these regions be even greater [4,39], but also because of the lower margin for error, as the lives and livelihood of millions of people depend purely on the water resources generated in these basins.

The usual approach of selecting results of a certain model or group of models or opting for a scenario with the mean trend of future projections may not be practical, as the full range of possible future climatic conditions needs to be covered in order to assess the full range of expected impacts required for climate adaptation needs.

As mentioned earlier, the selection of GCMs can be done following different approaches and may be based on a single criterion or a set of criteria. These approaches may include criteria such as: the total amount of change in the mean and/or an extreme of a projected climate variable; the success of a GCM in simulating the past climate for means or extremes; or maybe the skill in presenting the same pattern of tele-connections that drive the climate of the study region, and so on.

The current study adopted a combination of some of these approaches and applied a step-wise shortlisting of climate models based on a range of projected change in the (a) mean, (b) extremes, and (c) skill in reproducing the past climate. As the aim was to arrive at a limited number of models that can represent not only all the possible futures as projected by the entire pool of climate models, but also changes in climatic extremes, so that the selected model runs can provide representation of the full spectrum of future climate projections by GCMs in terms of change in mean, as well as extremes. In other words, for each selected RCP, we intended to filter and select five climate model runs, each representing one the four corners of the spectrum or the median tendencies.

3.1.1. Shortlisting Based on Changes in the Means

As a first step, the total number of available model runs (ensemble member *r1p1i1*) for RCP4.5 (42) and RCP8.5, (39) were evaluated and shortlisted based on the change presented by them, in terms of the mean annual precipitation sum (ΔP) and the mean air temperature (ΔT), averaged across the UIB, between the simulated reference period historical data (1976–2005) and the late 21st-century projected data (2071–2100). The calculations were done using the web-based application "Climate Explorer" managed by the Royal Netherlands Meteorological Institute (KNMI) (http://climexp.knmi.nl).

As our intention was to identify fewer model runs that best represent the four corners of the full spectrum, as well as the central and middle tendencies, we first determined the 10th, 50th and 90th percentile values of ΔP and ΔT for the entire ensemble considered for each RCP, to explore the extent of the full spectrum of the projected changes in temperature and precipitation under that RCP. This was followed by determining the four (4) closest projections to each of the corners, as well as the center of the spectrum. The total number of shortlisted model runs for each of the two RCPs then amounted to 20.

Details of the different parts of the full spectrum considered during this study are as follows:

- the Dry-Cold corner, represented by the 10th percentile ΔP as well as 10th percentile value of ΔT;
- the Dry-Warm corner, represented by the 10th percentile ΔP but the 90th percentile value of ΔT;
- the Wet-Cold corner, represented by the 90th percentile ΔP and the 10th percentile value of ΔT;
- the Wet-Warm corner, represented by the 90th percentile values for both ΔP as well as ΔT; and finally
- the median projected future climate, represented by the 50th percentile values of both ΔP and ΔT

The identification of the closest model runs to any corner point was done according to the procedure suggested by [19]. It should be noted that 10th and 90th percentiles were selected as the central points of the corners, rather than the maximum or minimum values, in order to avoid selection of any outlier projections.

3.1.2. Ranking Based on Changes in Climate Extremes

To ascertain that preference will be given to those climate model runs that represent the full range of projected change in extremes, all 20 shortlisted model runs for each of RCP4.5 and RCP8.5, were further scrutinized and ranked based on their projected changes in climatic extremes. To that end, the ETCCDI indices [25] (Table 2) were used to evaluate changes in climatic extremes for air temperature, as well as precipitation. For the former, changes in the extremes were ranked and evaluated based on two indices—the warm spell duration index (WSDI), and the cold spell duration index (CSDI)—while for the latter, consecutive dry days (CDD) and the precipitation due to extremely wet days (R99pTOT) were considered.

To keep the work manageable, we only analyzed four indices in total, two indices to represent changes in precipitation extremes and two for changes in temperature extremes. Furthermore, as the intended use of the selected climate model ensemble was to force the hydrological model for assessing climate change impacts on both flows and extremes, we chose the four most obvious indicators of precipitation and temperature extremes.

The R99pTOT (precipitation due to extremely wet days (>99th percentile)) and CDD (consecutive dry days: maximum length of dry spell (P < 1 mm)) are appropriate indicators for precipitation extremes and suitable for assessment of associated hydrological extremes. R99pTOT is an important indicator for wet spells in terms of their length and magnitude, which are both key influencing factors in shaping extreme hydrological events (floods and high flows). Similarly, CDD is an important indicator for dry spells that can provide a good opportunity for assessment of the associated low flow episodes. The two temperature-related extreme indices used, i.e., WSDI (count of days in a span of at least 6 days where TX > 90th percentile) and CSDI (count of days in a span of at least 6 days where TN < 10th percentile), seemed best suited for their effects on evapotranspiration and cryospheric processes. The snow and glacier melt/accumulation, as well as evapotranspiration dynamics, are very important in the highly glacierized study area.

The changes in these indices, averaged over the UIB and over 30 years, between the reference period (1976–2005) and the late 21st century projections (2071–2100), were calculated using the database available at the ETCCDI extremes indices archive (http://climexp.knmi.nl), constructed by [26,40].

Only the relevant index for the air temperature or for the precipitation was considered for each of the previously selected group of models (a set of four) initially shortlisted models for each corner or the center), so that for the models in the Wet-Warm corner, only the R99pTOT index for precipitation and the WSDI index for temperature were considered, because they were the only relevant indices, as R99pTOT indicates extreme precipitation events, while WSDI indicates warm spells (Table 2). The other two indices, i.e., CDD and CSDI, were not considered in this case; however, they were the only indices considered for models in the Dry-Cold corner. For each corner, the relevant indices were given scores based on the ratio of the extreme index to the mean of that index, for all four models in a corner. For example, in the Wet-Warm corner, the % change in R99pTOT for a single model is divided by the mean of the % change in R99pTOT for all four models in that corner. The same procedure was applied for WSDI and, finally, both scores were averaged to obtain a final score.

Table 2. List of ETCCDI extreme indices used during the GCM selection procedure.

Climate Variable	ETCCDI Index	Description of the ETCCDI Index
Precipitation	R99pTOT	Precipitation due to extremely wet days (>99th percentile)
	CDD	Consecutive dry days: maximum length of dry spell (P < 1 mm)
Air Temperature	WSDI	Warm spell duration index: count of days in a span of at least 6 days where TX > 90th percentile
	CSDI	Cold spell duration index: count of days in a span of at least 6 days where TN < 10th percentile

For each of the extreme indices, a weighted rank/skill score (Sk_{EI}) was calculated, with the highest value among the group getting the highest weighted rank/skill score of 1, and the others getting a rank according to their difference from this highest value, i.e.,

$$Sk_{EI} = 1 - \frac{EI_h - EI_t}{EI_h} \tag{1}$$

where Sk is the weighted rank for the specific extreme index EI, h denotes the highest index value in a group, and t denotes the target index to be ranked.

Similarly, in the case of the change in means, i.e., ΔT (°C) and ΔP (%), the ranking (Sk_m) was done based on the difference ΔT (°C) or ΔP (%) shown by each member with the percentile value relevant to that group,

$$Sk_m = 1 - \frac{(\Delta\,T\;or\;\Delta\,P)_{10,\,50\;or\;90^{th}\,percentile} - (\Delta\,T\;or\;\Delta\,P)_{target}}{(\Delta\,T\;or\;\Delta\,P)_{10,\,50\;or\;90^{th}\,percentile}} \tag{2}$$

3.1.3. Ranking Based on Skill in Reproducing the Reference Climate

The models were also evaluated with respect to their skill at simulating the past climate during the reference period (1976–2005). The selected climate model simulations were compared to the reference temperature and the precipitation gridded dataset [7] and were assigned skill scores. We did not use the same method for assigning skill score to temperature and precipitation. For assessing the performance of models in simulating past temperature, the method we applied was adopted from Perkins et al. [41]. In this method, the skill score for temperature is calculated based on the identification of similarities between PDFs of modelled data and the observed reference data. A metric is generated to calculate the cumulative minimum value of each binned value for the two distributions, which represent the common area between two PDFs. This skill score (Sk_{Tmp}) can be expressed as follows:

$$Sk_{Tmp} = \sum_{1}^{n} minimum \left(Z_{CM}, Z_{Obs} \right) \tag{3}$$

where n is the number of bins used to calculate the PDF, Z_{CM} is the frequency of values in a given bin from the model while Z_{CM} is the frequency of values in a given bin from the observed data. This skill score is 1, when there is a perfect match between simulated and the observed data, while a score of 0 means no similarities at all.

The number of bins used in this study to generate the PDFs was 50.

In the case of precipitation, the skill score is calculated by a method proposed by [42] as the product of five skill functions, each assessing similarities between modelled and observed data, while covering different aspects of precipitation behavior. These five skill score functions for a particular model j are listed below:

$$f_{1j} = 1 - \left(\frac{\left| A_{CMj} - A_{Obs} \right|}{2 \cdot A_{Obs}} \right)^{0.5} \tag{4}$$

$$f_{2j} = 1 - \left(\frac{\left| A^{+}_{CMj} - A^{+}_{Obs} \right|}{2 \cdot A^{+}_{Obs}} \right)^{0.5} \tag{5}$$

$$f_{3j} = 1 - \left(\frac{\left| A^{-}_{CMj} - A^{-}_{Obs} \right|}{2 \cdot A^{-}_{Obs}} \right)^{0.5} \tag{6}$$

$$f_{4j} = 1 - \left(\frac{\left| \overline{P_{CMj}} - \overline{P_{Obs}} \right|}{2 \cdot \overline{P_{Obs}}} \right)^{0.5} \tag{7}$$

$$f_{5j} = 1 - \left(\frac{\left| \sigma_{CMj} - \sigma_{Obs} \right|}{2 \cdot \sigma_{Obs}} \right)^{0.5} \tag{8}$$

where A_{CMj} and A_{Obs} are the areas below the simulated (climate model j) and the observed precipitation cumulative density function (PDF) curves, respectively, and A+ and A− are the fractional areas over (+) and under (−) the 50th percentile. P denotes the average annual precipitation over UIB and σ is the standard deviation of the probability distribution function.

Each of the above factors is intended to cover different aspects of probability distribution characteristics of the climate models, so that the distribution as a whole is taken into account through the mean and the total area (Equations (4) and (7)), the smaller and higher precipitation amounts are accounted for, through the 50th-percentile limit (Equations (5) and (6)), while the shape of the distribution is defined through the variance (Equation (8)).

These five factors are multiplied together to yield a single final skill score (*Sk$_{Prec}$*) for precipitation estimated by each model *j*:

$$Sk_{Prec} = f_{1j} \cdot f_{2j} \cdot f_{3j} \cdot f_{4j} \cdot f_{5j} \tag{9}$$

As a final step, all the rankings/scores, based on the changes in the means and in the extremes, as well as the skill scores for reproducing reference temperature and precipitation, are multiplied together to get the final overall skill or rank as follows:

$$Final\ Skill\ Score = Sk_{EI_1} \cdot Sk_{EI_2} \cdot Sk_{\Delta\ T} \cdot Sk_{\Delta\ P} \cdot Sk_{Temp} \cdot Sk_{Prec} \tag{10}$$

Under this skill score, a higher value indicates better performance, while a lower value indicates otherwise. These skill scores can be further translated to a simple ranking of 1 to 4 for each group of climate models.

The climate model selection procedure adopted in this study is in line with the approach and methods suggested by [21,41,42], although with certain modifications in the evaluation criteria. For assessing the performance of models in simulating past temperature, the method applied is adopted from [41], while in the case of precipitation, guidance is taken from [42]. A major difference from [21] in assessing model performances in simulating past climate is the use of a new long-term climate data set and an additional evaluation step for assessing model runs for their skill in reproducing the annual cycle of precipitation and temperature as well.

3.2. Downscaling and Bias Correction

After the shortlisting and ranking of the GCMs, the next step was to address the two primary issues inhibiting impact studies: firstly, the coarse spatial scales represented by the GCM may not be as fine as required by regional- and local-scale environmental modelling or impact studies; and secondly, the GCM raw outputs, or their downscaled versions, are deemed to contain systematic errors (bias) of certain magnitude, relative to the observational data, and therefore need post-processing by correcting it with and towards observations prior their its use in environmental modelling or impact studies.

The downscaling can be done either by applying statistical downscaling methods or through dynamical downscaling via application of a regional climate model (RCM). As dynamical downscaling was too demanding in terms of time and computational resource requirements, we decided only to explore whether any dynamically downscaled RCM projections were available for the already shortlisted and ranked GCMs. The Coordinated Regional Downscaling Experiment (CORDEX) has generated fine-scale climate projections for different regions of the world, of which the CORDEX-South Asia experiments cover the UIB region. We found that CORDEX-RCM model projections were available for four of the selected GCMs at the 1st rank and one GCM at the 2nd rank. These RCM projections provide dynamically downscaled data at a resolution of ~50 km for all of our selected GCMs. The data for the relevant GCM–RCM combinations were downloaded, but needed further downscaling, as the scale was still not fine enough, and also needed to undergo bias correction before further use in hydrological modelling.

This downscaling and bias correction was achieved by the Distribution Mapping method (DM) [43], which was selected out of five different bias correction methods for the precipitation climate variable. These methods included: (1) Linear Scaling (LS); (2) Local Intensity Scaling (LIS); (3) Power Transformation (PT); (4) Distribution Mapping (DM); and (5) Distribution Mapping followed by Intensity and frequency Scaling (DM-IS).

For the temperature, the selection was made after evaluating the performance of the following three bias correction methods: (1) Linear Scaling (LS); (2) Variance Scaling (VS); and (3) Distribution Mapping (DM). Further details of these methods can be found in [43].

The calibration and validation statistics, along with brief explanations, are provided as appendices (Supplementary Materials: Appendix A, Tables A1 and A2).

4. Results

4.1. Selection of Climate Models

4.1.1. Shortlisting of Models: Changes in Climatic Means

The results of the initial shortlisting of the GCM model runs are given in Figures 3 and 4. In this step, only those GCM runs were retained which showed minimal difference with the 10th, 50th and 90th percentile values of ΔT (°C) and ΔP (%), so that, for each RCP, we were left with sets of 4 GCM runs at each corner and 4 in the middle, while the remaining model runs were not processed any further. In this way, a total of 20 model runs were selected for each RCP.

It is worth mentioning that the range of projections for ΔT and ΔP for the RCP8.5 model pool was much larger than for the RCP4.5 model pool. For the latter, more extreme RCP, ΔP ranges from -5.42% to 19.56%, and ΔT ranges from 1.26 °C to 5.41 °C; while for the former (RCP4.5), these ranges are much higher, with ΔP ranging between -12.01% and 35.12% and ΔT between 1.48 °C and 8.57 °C.

The shortlisted GCM runs were also ranked according to their differences with the 10th, 50th or 90th percentile values in the respective corner or center. This ranking was intended for use in the final selection step, so that those model runs which show closest representation of the group of models or type of scenarios (Warm-Wet, Warm-Dry, Cold-Wet, Cold-Dry or the Median) get preference during the final selection.

It should be noted that the term Cold used in the "Wet-Cold" and "Dry-Cold" scenarios does not mean that the future temperatures will be colder than those of the reference period, but rather indicates that the warming will be less than that of the Warm scenarios. Similarly, the term Dry in the scenarios "Dry-Cold"and "Dry-Warm" is also only indicative of its comparative position relative to other climate models.

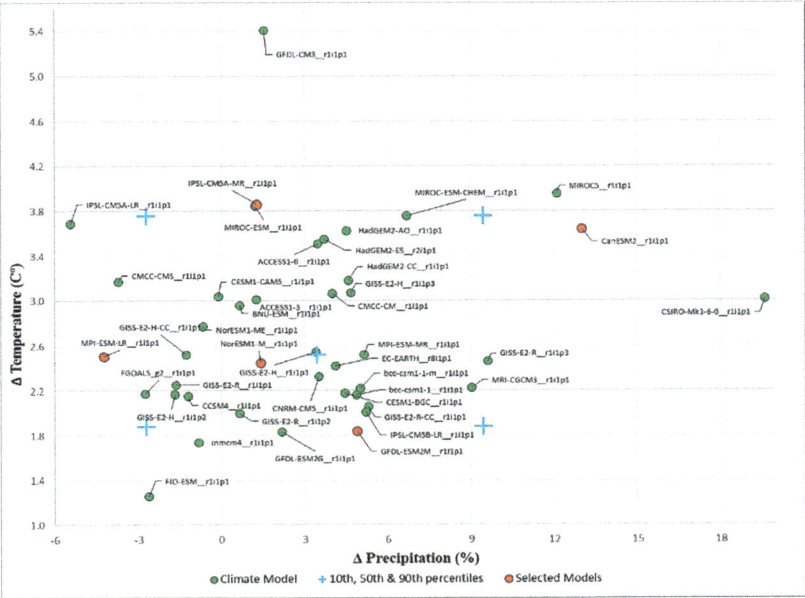

Figure 3. Projected changes in mean air temperature (ΔT) and annual precipitation sum (ΔP) between 2071 and 2100 and 1971 and 2000 for all included RCP4.5 GCM runs. Blue crosses indicate the 10th, 50th and 90th percentile values for ΔT and ΔP. The model runs shortlisted during this step are indicated in red color.

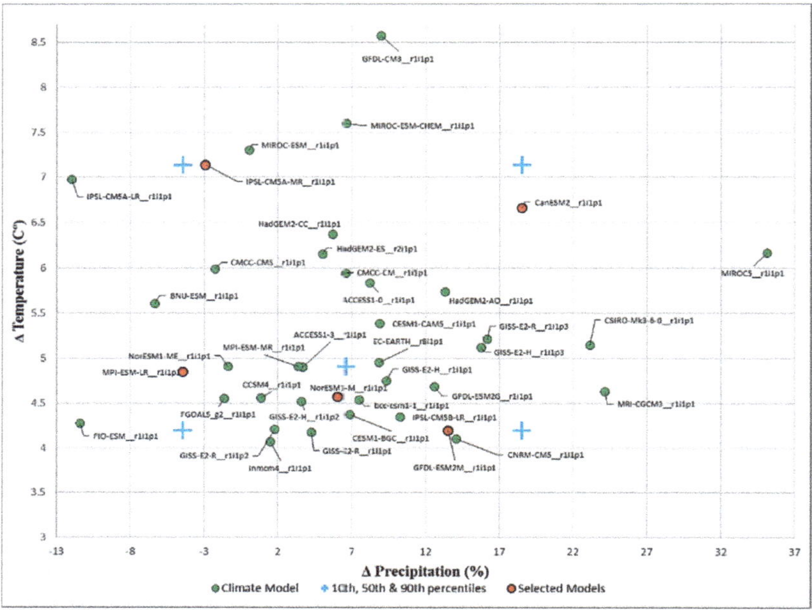

Figure 4. Similar to Figure 3, but for RCP8.5 GCM runs.

4.1.2. Ranking Based on Changes in Climatic Extremes

The 20 shortlisted model runs for each RCP were further scrutinized based on their projected changes in climatic extremes. The details of the projected changes in selected extreme indices are given in Table 3. The darker colors indicate the higher values, while the lighter indicates lower values. These indices were given a weighted rank/score based on their difference from the highest value in the group of four model runs in a corner.

Similar to the rank assigned based on changes in the means, this ranking was also intended for use in the final selection step, so that the model runs, which show the largest changes in the extreme indices for each of the corner: Warm-Wet, Warm-Dry, Cold-Wet or Cold-Dry, get preference during the final selection. Unlike the four corners, evaluation based on the extreme indices was not carried out for the central or the mean scenario.

The ranking and scores for means and extreme indices, as well as the skill scores for simulating reference climate, are presented in Tables 4 and 5 for RCP4.5 and RCP8.5, respectively.

In most cases, the model run with the highest or the lowest changes in mean precipitation or temperature coincide with the highest change in relevant extreme index as well.

The index Δ R99pTOT (%) was evaluated to represent the "Wet" scenarios, while the Δ CDD (%) represented the "Dry" scenarios. Similarly, Δ WSDI (%) was considered for the "Warm" scenarios, while Δ CSDI (%) was considered for the "Cold" scenarios. In this way, a set of two (2) indices out of the four (4) were evaluated for each of the scenarios: Warm-Wet, Warm-Dry, Cold-Wet, and Cold-Dry.

Table 3. Percentage change in ETCCDI indices (R99pTOT, CDD, WSDI, and CSDI) along with changes in mean precipitation (ΔP) and temperature (ΔT), for all corners/scenarios (Warm-Wet, Warm-Dry, Cold-Wet and Cold-Dry) and both RCPs (RCP4.5 and RCP8.5).

Projection	Model	Δ R99pTOT (%)	Δ CDD (%)	Δ WSDI (%)	Δ CSDI (%)	ΔP (%) (%)	ΔT (°C)
				RCP 4.5			
	CanESM2	29.0	−7.2	814	−96.2	13.0	3.6
Wet-Warm	HadGEM2-ES	28.6	12.5	1002	−98.7	3.7	3.6
	MIROC5	76.4	−8.8	938	−96.3	12.1	4.0
	MIROC-ESM-CHEM	19.8	2.2	611	−89.9	6.7	3.8
	bcc-csm1-1-m	45.3	−1.0	298	−87.6	5.0	2.2
Wet-Cold	GFDL-ESM2M	42.4	−4.9	202	−61.6	4.9	1.8
	IPSL-CM5B-LR	32.2	−11.7	293	−81.6	5.2	2.0
	MRI-CGCM3	59.6	−7.5	471	−89.8	9.0	2.2
	ACCESS1-0	46.4	0.9	656	−92.1	3.47	3.5
Dry-Warm	CMCC-CMS	61.9	7.1	454	−89.8	−3.35	3.6
	IPSL-CM5A-MR	54.5	12.0	604	−90.2	1.28	3.9
	MIROC-ESM	26.8	1.8	718	−97.0	2.41	4.2
	CCSM4	4.8	−0.8	323	−92.0	4.54	2.4
Dry-Cold	GFDL-ESM2G	16.2	−0.1	373	−70.9	2.14	2.2
	inmcm4	2.0	4.3	216	−48.9	−5.29	4.2
	MPI-ESM-LR	42.3	17.7	406	−89.1	−5.76	2.8
				RCP 8.5			
	CanESM2	101.7	−12.3	1181	−97.3	18.5	6.7
Wet-Warm	GFDL-CM3	9.7	−5.0	1426	−100.0	9.0	8.6
	MIROC5	257.2	−13.4	1640	−98.5	35.1	6.2
	MIROC-ESM-CHEM	28.5	14.5	1314	−100.0	6.6	7.6
	GFDL-ESM2G	95.9	−1.0	668	−99.0	12.6	4.7
Wet-Cold	GFDL-ESM2M	72.9	−3.1	1696	−95.5	13.5	4.2
	CNRM-CM5	68.6	−3.5	638	−96.1	14.1	4.1
	MRI-CGCM3	195.5	−12.6	1309	−98.4	24.1	4.6
	IPSL-CM5A-LR	94.5	23.3	1022	−97.6	−12.01	7.0
Dry-Warm	IPSL-CM5A-MR	194.6	9.3	1358	−99.1	−2.95	7.1
	MIROC-ESM	9.5	4.4	1521	−100.0	0.06	7.3
	CMCC-CMS	143.9	18.8	985	−99.9	−2.26	6.0
	MPI-ESM-LR	136.0	29.1	1067	−98.2	−4.49	5.2
Dry-Cold	CCSM4	48.3	7.0	871	−99.5	0.86	4.6
	inmcm4	61.3	4.7	849	−85.9	1.48	4.1
	NorESM1-M	107.1	3.5	1010	−98.6	6.01	4.6

4.1.3. Ranking Based on Skill in Reproducing the Reference Climate

After checking the model runs for their projected changes in means and extreme indices, they were finally evaluated for their skill at reproducing the reference precipitation and temperature data.

The ranking for past performance utilized a new set of reference precipitation and temperature data [7], averaged over the UIB. The skill scores were calculated following the procedure of Section 4.1.3, and are presented in columns g and h in Tables 4 and 5. For most scenarios, the same models performed better than the others for both RCPs in simulating past climate.

After allocating the skill score based on the past performance, the final skill scores and ranks were calculated by multiplying all the relevant skill scores allocated to each model run. The final ranks were allocated to each scenario, with the highest rank allotted to the model run with highest final skill score, and so on.

It is interesting to note that for the 4 scenarios, Warm-Dry, Cold-Wet, Cold-Dry and Median, for both RCPs, the same GCMs get the highest skill scores and ranks. The only exception is the Warm-Wet scenario, where different models top the ranking. In this scenario, for RCP4.5, the GCM "MIROC5" is in the top rank, followed by "CanESM2", while for RCP8.5, the ranking of these two GCMs is reversed.

Table 4. Weighted ranks for all shortlisted RCP4.5 GCM runs based on change in means (e and f), change in extremes (a, b, c and d) and their skill scores for simulating reference precipitation and air temperature(g and h).

Projection	Climate Model	a Weighted Rank Δ R99pTOT (%)	b Weighted Rank Δ CDD (%)	c Weighted Rank Δ WSDI (%)	d Weighted Rank Δ CSDI (%)	e Weighted Rank Δ T (°C)	f Weighted Rank Δ P (%)	g Skill Score for Temperature (Sk_{Tmp})	h Skill Score for Precipitation (Sk_{Perc})	Final Skill Score (a*b*c*d*e*f*g*h*10)	Final Rank
Wet-Warm	CanESM2	0.38		0.85		0.97	0.63	0.79	0.36	0.57	2
	HadGEM2-ES	0.37		1.00		0.94	0.39	0.73	0.29	0.29	3
	MIROC5	1.00		0.94		0.95	0.72	0.81	0.38	1.93	1
	MIROC-ESM-CHEM	0.26		0.61		1.00	0.71	0.71	0.25	0.20	4
Wet-Cold	bcc-csm1-1-m	0.74			0.93	0.77	0.53	0.71	0.40	0.80	3
	GFDL-ESM2M	0.71			0.75	0.97	0.52	0.79	0.41	0.88	2
	IPSL-CM5B-LR	0.54			1.00	0.94	0.55	0.71	0.14	0.28	4
	MRI-CGCM3	1.00			0.90	0.82	0.96	0.78	0.35	1.91	1
Dry-Warm	ACCESS1-0		0.07	0.91		0.93	0.78	0.77	0.35	0.14	2
	CMCC-CMS		0.59	0.75		0.97	0.81	0.70	0.31	0.75	4
	IPSL-CM5A-MR		1.00	1.00		0.97	0.59	0.79	0.33	1.52	1
	MIROC-ESM		0.15	0.81		0.87	0.92	0.74	0.22	0.16	3
Dry-Cold	CCSM4		0.05		1.00	0.64	0.34	0.79	0.26	0.02	3
	GFDL-ESM2G		0.00		0.77	0.58	0.56	0.75	0.32	0.00	4
	inmcm4		0.24		0.53	0.89	0.76	0.66	0.35	0.20	2
	MPI-ESM-LR		0.98		0.97	0.75	0.72	0.75	0.32	1.23	1
Mean	NorESM1-M					0.94	0.58	0.79	0.43	1.83	1
	bcc-csm1-1-m					0.76	0.70	0.76	0.44	1.76	2
	GFDL-ESM2G					0.87	0.85	0.75	0.32	1.77	2
	CMCC-CMS					0.56	0.20	0.70	0.31	0.24	4

Table 5. Similar to Table 4, but for RCP8.5.

Projection	Climate Model	a — Weighted Rank Δ R99pTOT (%)	b — Weighted Rank Δ CDD (%)	c — Weighted Rank Δ WSDI (%)	d — Weighted Rank Δ CSDI (%)	e — Weighted Rank Δ T (°C)	f — Weighted Rank Δ P (%)	g — Skill Score for Temperature (Sk_{Tmp})	h — Skill Score for Precipitation (Sk_{Perc})	Final Skill Score (a*b*c*d*e*f*g*h*10)	Final Rank
Wet-Warm	CanESM2	0.40		0.72		0.93	1.00	0.79	0.36	0.76	1
	GFDL-CM3	0.04		0.87		0.80	0.48	0.71	0.39	0.04	3
	MIROC5	1.00		1.00		0.86	0.10	0.81	0.38	0.27	2
	MIROC-ESM-CHEM	0.11		0.80		0.94	0.36	0.71	0.25	0.05	3
Wet-Cold	GFDL-ESM2G	0.49			0.97	0.89	0.68	0.75	0.32	0.69	4
	GFDL-ESM2M	0.50			1.00	1.00	0.73	0.79	0.34	0.97	2
	CNRM-CM5	0.35			1.00	0.98	0.76	0.81	0.41	0.87	3
	MRI-CGCM3	1.00			0.98	0.90	0.70	0.78	0.35	1.65	1
Dry-Warm	IPSL-CM5A-LR		1.00	0.67		0.98	0.41	0.79	0.33	0.71	2
	IPSL-CM5A-MR		0.40	0.89		1.00	0.84	0.78	0.34	0.79	1
	MIROC-ESM		0.19	1.00		0.98	0.60	0.74	0.22	0.18	4
	CMCC-CMS		0.81	0.65		0.84	0.78	0.70	0.31	0.74	2
Dry-Cold	MPI-ESM-LR		1.00		0.86	0.73	0.96	0.75	0.31	1.42	1
	NorESM1-M		0.12		0.85	0.64	0.01	0.79	0.50	0.00	2
	CCSM4		0.24		0.84	0.64	0.48	0.79	0.26	0.13	4
	inmcm4		0.16		1.00	0.57	0.42	0.66	0.34	0.09	2
Mean	NorESM1-ME					0.93	0.91	0.79	0.50	3.34	1
	GFDL-ESM2G					0.95	0.09	0.75	0.32	0.21	2
	CCSM4_r1i1p1					0.93	0.13	0.79	0.26	0.25	2
	bcc-csm1-1					0.92	0.86	0.77	0.28	1.70	4

4.1.4. Limitations of the Model Selection Procedure

In the previous section, the step-wise shortlisting of the various climate models was based on the range of projected change in (a) mean, (b) extremes, and (c) skill in reproducing the past climate. Although the main aim of this approach was to combine the strengths of two different methodologies, i.e., the selection of the GCMs based on the properties of the full range of projections and the selection procedures based on past performance, certain limitations are unavoidable and need to be discussed.

First of all, the analysis considered only models selected based on the changes in the means and only the ensemble member *r1p1i1*, resulting in a reduced number of GCM runs for evaluation and possibly a smaller range of climatic extremes. This may also have led to possibly screening out models which may have had better past performance.

Similarly, another issue is concerned with the scale at which the method was applied. During the shortlisting step, and also during the evaluation of extreme indices, the projected changes or ETCCDI extremes indices were averaged over the entire UIB, which has the possibility of decreasing the spatial variation in projected changes.

Additionally, the weighting of different skill scores in this study also differed for similar work, such as [21]. In our study, the final skill score was a combination of scores allocated for change in mean, change in extremes, and performance in reproducing past climate. This may have reduced the chances of selecting the climate model with the best past performance, but increased the chance of a better spread of scenarios over the entire range, while still taking past performance as a key factor in selections. Our model selection approach also assumes that all the evaluated model runs are independent of each other, which may not be the case, as some models use the same forcing and validation data or may share similar model codes [44,45].

Despite these limitations, the adopted approach made it possible for us to identify a limited number of model runs, representative of the full range of future projected means and extremes while giving due preference to models which perform better at simulating the reference climate.

4.2. Bias Correction and Downscaling of Future Climate Scenarios

The future climate projections of the selected climate models needed to be downscaled and corrected for biases, before further use in hydrological model simulation. Therefore, as a first option, all the dynamically downscaled climate projections available for UIB were checked for whether any Regional Climate Model (RCM) projections were available that have dynamically downscaled projections for the already shortlisted GCM at first or second position in the ranking. We found that, for both RCPs, the outputs of at least three (3) CORDEX-SA experiments were based on the GCMs ranked 1st in our study (IPSL-CM5A-MR_RCA4, MPI-ESM-LR_RCA4 and NorESM1-M_RCA4). The GSM CanESM2, which is at 2nd rank for RCP4.5 and at 1st for RCP8.5, has dynamically downscaled projections under CORDEX-SA experiments, CanESM2_RegCM4-4. The output of one CORDEX-SA experiment (GFDL-ESM2M_RCA4) was based on GFDL-ESM2M, which is ranked at 2nd position for both the RCPs in our study (Tables 4 and 5)

It was decided to utilize the available dynamically downscaled data for our selected GCMs, ranked at 1st or 2nd positions. Therefore, five CORDEX-SA experiments (Table 6), including IPSL-CM5A-MR_RCA4, MPI-ESM-LR_RCA4, NorESM1-M_RCA4, Can ESM2_RegCM4-4, and GFDL-ESM2M_RCA4 were selected for further processing and bias correction. These five GCMs were dynamically downscaled by CORDEX using two different RCMs (RCA4 and RegCM4). Their RCM outputs are at a considerably finer scale (0.44°) than the source GCMs.

Table 6. List of CORDEX South Asia experiments for RCP8.5 and their CMIP5-GCM forcing.

Nr	Scenario	Experiment		Driving AOGCM	RCM	RCM Description
		Name	Short Form			
1	Wet-Warm	CanESM2_RegCM4-4	CAN	CCCma-CanESM2 (1st)	RegCM4	Abdus Salam International Centre for Theoretical Physics (ICTP) Regional Climatic Model version 4 (RegCM4; [46])
2	Wet-Cold	GFDL-ESM2M_RCA4	GFDL	NOAA-GFDL-GFDL-ESM2M (2nd)	RCA4	Rossby Centre regional atmospheric model version 4 (RCA4; [47])
3	Mean	NorESM1-M_RCA4	NOR	Nor-ESM1-M (1st)		
4	Dry-Cold	MPI-ESM-LR_RCA4	MPI	MPI-ESM-LR (1st)		
5	Dry-Warm	IPSL-CM5A-MR_RCA4	IPSL	IPSL-CM5A-MR (1st)		

4.3. Projected Changes in Temperature and Precipitation

The five (5) selected (CORDEX-SA) RCM outputs were further bias-corrected using the "distribution mapping technique" [43] for RCP4.5 and RCP8.5 for two sets of durations, i.e., mid-century (2041–2070) and end-century (2071–2100). Major properties of the downscaled projections are given in Table 7.

Table 7. Future precipitation and temperature projections from 5 GCM models, 2 RCPs and 2 periods.

| Model | Duration | Precipitation (mm) | | | | | | | | Temperature (°C) | | | | | |
| | | RCP 4.5 Values and (change %) | | | | RCP 8.5 Values and (change %) | | | | RCP 4.5 Values and change | | | RCP 8.5 Values and change | | |
		PCP (mm) (Av-An)	90th percentile (mm)	Probability-Wet Days (days)	Intensity-Wet Days (mm)	PCP (mm) (Av-An)	90th percentile (mm)	Probability-Wet Days (days)	Intensity-Wet Days (mm)	TMP ($C°$) (mean)	90th Percentile ($C°$)	10th Percentile ($C°$)	TMP ($C°$) (mean)	90th Percentile ($C°$)	10th Percentile ($C°$)
IPSL-CM5A-MR_RCA4	41–70	539 (2.9%)	19.7 (36.8%)	107.4 (−7.8%)	5.0 (13.2%)	532 (1.7%)	19.5 (35.4%)	106.9 (−8.3%)	4.9 (11.8%)	5.52 (4.12)	17.6 (3.85)	−6.1 (4.93)	6.3 (4.91)	18.4 (4.66)	−5.3 (5.70)
	71–00	557 (6.2%)	20.7 (42.1%)	107.4 (−7.8%)	5.2 (17%)	502 (−4.2%)	20.5 (18%)	94.4 (−1.9%)	5.4 (22%)	7.7 (6.33)	13.1 (5.34)	−3.6 (7.44)	10.4 (9.0)	21.9 (8.17)	−1.2 (9.82)
MPI-ESM-LR_RCA4	41–70	536 (2.3%)	18.6 (29.1%)	110.9 (−4.8%)	4.6 (5.2%)	535 (2.0%)	18.6 (29.1%)	108.5 (−6.9%)	4.7 (7.3%)	4.0 (2.64)	15.2 (2.44)	−8.1 (2.85)	4.5 (3.08)	16.7 (2.99)	−7.7 (3.27)
	71–00	537 (2.4%)	18.7 (41.1%)	109.1 (−6.4%)	4.7 (7.7%)	559 (6.7%)	20.3 (41.1%)	106.1 (−8.9%)	5.1 (15.5%)	5.5 (4.11)	17.4 (3.67)	−6.5 (4.48)	7.3 (5.86)	19.2 (5.51)	−4.9 (6.09)
NorESM1-M_RCA4	41–70	536 (2.4%)	20.3 (46.1%)	109.0 (−6.4%)	4.9 (10.7%)	555 (6.0%)	21.1 (46.1%)	111.3 (−4.5%)	4.9 (12.3%)	3.8 (2.36)	15.8 (3.03)	−8.0 (3.02)	4.3 (2.92)	17.9 (4.2)	−7.5 (3.53)
	71–00	537 (2.5%)	20.3 (54.4%)	109.0 (−6.4%)	4.9 (12%)	548 (4.6%)	22.2 (54.4%)	107.0 (−8.2%)	5.2 (17.5%)	4.9 (3.50)	17.7 (3.99)	−6.76 (4.23)	6.6 (5.23)	19.9 (6.16)	−5.1 (5.93)
GFDL-ESM2M_RCA4	41–70	540 (3.1%)	17.9 (42.6%)	111.9 (−4%)	4.7 (7%)	578 (10.4%)	20.5 (42.6%)	114.9 (−1.4%)	4.9 (11.1%)	3.8 (2.41)	15.0 (2.31)	−7.8 (3.22)	4.1 (2.73)	16.2 (2.43)	−7.0 (4.02)
	71–00	536 (2.2%)	19.4 (52.8%)	112.8 (−3.2%)	4.7 (5.7%)	612 (16.8%)	22.0 (52.8%)	114.7 (−1.5%)	5.2 (18.6%)	5.1 (3.70)	17.14 (3.42)	−6.4 (4.57)	6.6 (5.22)	18.8 (5.03)	−4.8 (6.17)
CanESM2_RegCM4-4	41–70	560 (6.9%)	21.1 (43.8%)	119.6 (2.7%)	4.7 (6.4%)	557 (6.3%)	20.7 (43.8%)	115.6 (−0.8%)	4.6 (5.5%)	4.5 (3.14)	16.8 (3.08)	−7.8 (3.20)	4.9 (3.51)	16.9 (3.16)	−7.2 (3.75)
	71–00	607 (15.9%)	23.2 (51.6%)	117.2 (0.6%)	5.05 (14.8%)	590 (12.5%)	21.8 (51.6%)	114.9 (−1.3%)	5.0 (13.2%)	5.6 (4.24)	17.5 (3.8)	−6.5 (4.47)	7.4 (6.03)	20.0 (6.24)	−5.1 (5.89)
Observed	1976–2005	524	14.4	116.5	4.4	524.1	14.4	116.5	4.4	1.4	13.7	−11.0	1.4	13.7	−11.0

The downscaled projections show changes in temperature ranging from 2.3 °C to 6.33 °C for RCP4.5 and of 2.92 °C to 9.0 °C for RCP8.5. The downscaled and bias-corrected precipitation ranges from a minor increase of 2.2% for the drier scenarios to as high as 15.9% for the wet scenarios. Thus, both temperature and precipitation show increases, as do the extremes, since the probabilities of the wet days are projected to decrease, while the precipitation intensities are projected to increase unanimously by both RCPs.

The spatial distribution of the projected future changes for precipitation and temperature across the UIB also show certain distinct trends. Thus, the precipitation (Figure 5) over the mid-century (2041–2070), as well as the late century (2071–2100), reveals for all scenarios a remarkable decrease in the southeastern parts of the basin, but an increase in the northeastern parts. This decrease/increase is particularly intense for the "Dry-Warm" and the "Median" scenarios over the late 21st century.

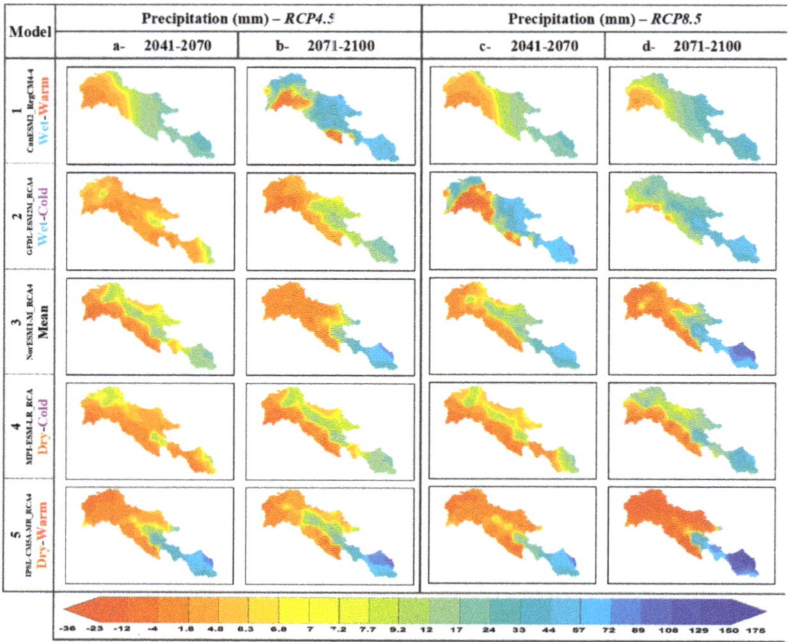

Figure 5. Spatial distribution of projected precipitation change across the UIB over the mid (2014–2070) and the late- (2071–2100) 21st century for 5 models and 2 RCPs. The figure is arranged in a tabular form where the 1st and 2nd column represent projected change in precipitation for RCP 4.5, for the mid-century (2014–2070) and the late-century (2071–2100), respectively, while the 3rd and 4th columns show the projected change in the mid-century and the late-century precipitation for RCP 8.5, respectively. The rows represent the climate models used.

The spatial distribution of the projected changes for in temperature (Figure 6) also shows similarities across all scenarios, with the northern and northwestern parts of the basin exhibiting higher increases, while the eastern and southern parts experience a comparatively smaller temperature increase.

For RCP8.5, the projected temperature changes appear to be very high over the late 21st century and this occurs under all scenarios, especially for the "Warm" scenarios, with an almost uniform spread across the whole UIB. The projected temperature changes range for all RCPs and the two 20th-century periods from a minimum increase of 3.76 °C (NorESM1-M_RCA4, RCP4.5, Period: 2041–2070) to a maximum increase as high as 10.4 °C (IPSL-CM5A-MR_RCA4, RCP8.5 and period: 2071–2100).

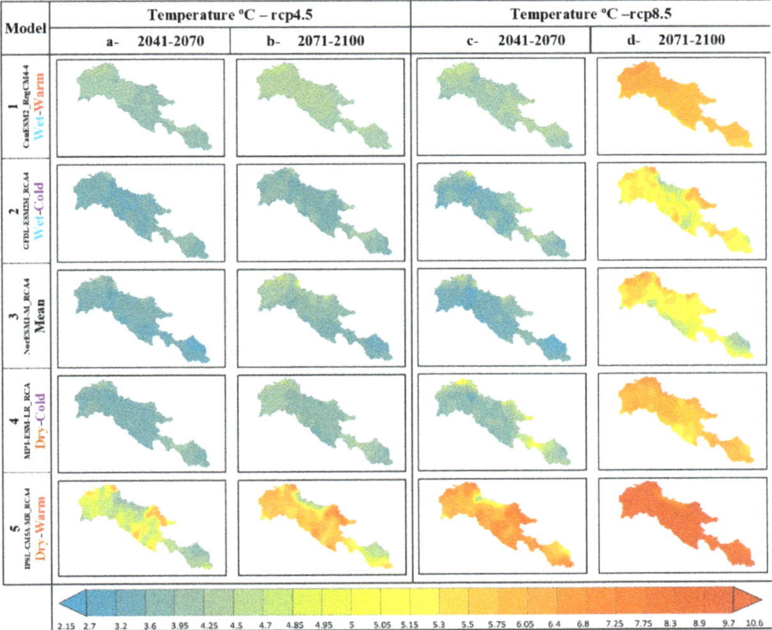

Figure 6. Similar to Figure 5, but for temperature changes.

4.4. Limitations of the Downscaling and Bias Correction Approach Adopted

The downscaling and bias correction approaches described in the previous sections, although adopted as the best available option considering the availability of time, resources and data, are still not without limitations.

Firstly, we opted for using GCM-RCM chains for the already shortlisted climate model runs where the RCM add further detail to global climate simulations and may provide regional- to local-scale information, but their outputs are still subject to inherited or new systematic errors and may therefore require a bias correction or further downscaling to a higher resolution, along with the fact that they may also produce quite a different response in terms of the temperature and precipitation change than the forcing GCMs. In fact, many authors recognized this affliction of GCM-RCM model chains with systematic errors (biases) [48–55], but still the representation of explicit atmospheric and surface processes and the level of spatial details provided by the RCMs [56] make them a better option for regions with high topographic variability, The selection procedure may in future, though, directly include the RCM runs instead of GCM, when an appropriate amount of RCM runs are available.

The systematic errors (biases) in GCM/RCM outputs make them unsuitable for certain uses, and application of different bias-correction methods has increasingly become a standard procedure, especially in climate change impact studies [57].

Although these bias correction approaches improve the agreement of climate model output with observations, and therefore narrows the uncertainty range of predictions and simulations, they do so without a sound physical basis [57]. At the same time, a growing number of authors are showing reservations on the use of bias correction methods for a variety of reasons. Some argue that the main assumption behind bias correction approaches is the stationarity of the correction parameters, which is not realistic and may not be the case, especially under climate change [57,58]. While others believe that as bias correction cannot overcome major climate model errors, inexperienced application might result in ill-informed adaptation decisions [59]. Despite being critical of bias correction methods, many of these authors e.g., [57,59,60] acknowledge the need for some type of bias correction and recommend different precautions to avoid any pitfalls.

5. Conclusions

It is essential to have representative future climate projections of appropriate quality for climate change impact studies, especially in the water resource sector. Despite the availability of an increasing number of GCM outputs in the CMIP5 archive and the on-going improvements in their process representations, issues of large uncertainties in their future climate predictions cannot be avoided. This situation, along with other factors, such as time, human resources or computational constraints, make it imperative to sort out the most appropriate individual GCM or small ensemble of GCMs for a more reliable assessment of climate change impacts.

The approach presented in the present study seeks the most suitable set of climate model runs, while considering not only the full ranges of projected changes in terms of means and extremes by different climate models, but also their skills in simulating the past climate in a reference period.

This selection procedure was applied for future climate projections over the Upper Indus Basin for two representative concentration pathways (RCPs), the RCP 4.5 and RCP 8.5. All available model runs for the r1p1i1 ensemble member of each GCM in the CMIP5 repository were included in the initial list. The total number of model runs available for RCP4.5 was 42, and 39 for RCP8.5.

Based on the huge uncertainties reported in the GCM runs for the UIB, all possible extreme future scenarios (Wet-Warm, Wet-Cold, Dry-Warm, Dry-Cold) were considered, in addition to the selection of GCMs representing the mean future climate change, with respect to both changes in the projected means and the extremes. This procedure made it possible to arrive at a limited number of climate models, from which the final selection was performed by assigning ranks based on the weighted score for each of the mentioned selection criteria.

Finally, the precipitation and temperature time series of the selected GCM model runs were bias corrected and further downscaled to the scale of the reference data by means of a distribution mapping technique. The ensembles of the selected GCM runs for RCP4.5 and RCP8.5 scenarios show that the uncertainty of future climate in the study region is very large for the raw data, as well as their downscaled versions.

The downscaled projections indicate increases of temperature ranging between 2.3 °C and 9.0 °C and changes in precipitation that range from a slight annual increase of 2.2% under the drier scenarios, to as high as 15.9% for the wet scenarios. Thus, for both temperature and precipitation, the future projections under all scenarios and both RCP's only show increases in the mean annual values, with no negative trend. Moreover, for all scenarios, the future precipitation is projected to be more extreme, as the probability of wet days decreases, while at the same time, the precipitation intensities will increase.

The spatial distribution of the downscaled predictors, namely, the precipitation, also shows distinct patterns across the UIB, such that this variable shows for all time periods/scenarios considered a distinct decrease in the southeastern parts, but an increase in the northeastern parts of the basin. This decrease/increase is particularly intense for the "Dry-Warm" and the "Median" scenarios over the late 21st century.

Overall, the future climate of the UIB region remains very uncertain, which justifies the selection procedure proposed here to arrive at a wider range of possible climate scenarios that can then be further utilized and translated into a wider spectrum of climate change impact scenarios.

Supplementary Materials: The following are available online at http://www.mdpi.com/2225-1154/6/4/89/s1.

Author Contributions: Conceptualization, A.J.K.; Formal analysis, A.J.K.; Investigation, A.J.K.; Methodology, A.J.K.; Resources, M.K.; Supervision, M.K.; Validation, A.J.K.; Writing—original draft, A.J.K.; Writing—review & editing, M.K.

Funding: This research received no external funding.

Acknowledgments: We acknowledge provision of data by the following sources:

- The data sets for changes in mean/extremes were acquired from "KNMI Climate Explorer", web application to analysis climate data statistically. It is part of the WMO Regional Climate Centre at KNMI, Netherland;
- CORDEX-South Asia experiment, RCMs data, hosted at Centre for Climate Change Research (CCCR) at the Indian Institute of Tropical Meteorology (IITM).

Conflicts of Interest: The authors declare no conflict of interest.

References

1. Biemans, H.; Speelman, L.H.; Ludwig, F.; Moors, E.J.; Wiltshire, A.J.; Kumar, P.; Gerten, D.; Kabat, P. Future water resources for food production in five South Asian river basins and potential for adaptation—A modeling study. *Sci. Total Environ.* **2013**, *468*, S117–S131. [CrossRef] [PubMed]
2. Pierce, D.W.; Barnett, T.P.; Santer, B.D.; Gleckler, P.J. Selecting global climate models for regional climate change studies. *Proc. Natl. Acad. Sci. USA* **2009**, *106*, 8441–8446. [CrossRef] [PubMed]
3. Turner, A.G.; Annamalai, H. Climate change and the South Asian summer monsoon. *Nat. Clim. Chang.* **2012**, *2*, 587–595. [CrossRef]
4. Mishra, V. Climatic uncertainty in Himalayan water towers. *J. Geophys. Res. Atmos.* **2015**, *120*, 2689–2705. [CrossRef]
5. Lutz, A.F.; Immerzeel, W.W.; Kraaijenbrink, P.D.A.; Shrestha, A.B.; Bierkens, M.F.P. Climate Change Impacts on the Upper Indus Hydrology: Sources, Shifts and Extremes. *PLoS ONE* **2016**, *11*, e0165630. [CrossRef] [PubMed]
6. Palazzi, E.; Hardenberg, J.; von Provenzale, A. Precipitation in the Hindu-Kush Karakoram Himalaya: Observations and future scenarios. *J. Geophys. Res. Atmos.* **2013**, *118*, 85–100. [CrossRef]
7. Khan, A.J.; Koch, M. Correction and Informed Regionalization of Precipitation Data in a High Mountainous Region (Upper Indus Basin) and Its Effect on SWAT-Modelled Discharge. *Water* **2018**, *10*, 1557. [CrossRef]
8. De-Souza, K.; Kituyi, E.; Harvey, B.; Leone, M.; Murali, K.S.; Ford, J.D. Vulnerability to climate change in three hot spots in Africa and Asia: Key issues for policy-relevant adaptation and resilience-building research. *Reg. Environ. Chang.* **2015**, *15*, 747–753. [CrossRef]
9. Nepal, S.; Shrestha, A.B. Impact of climate change on the hydrological regime of the Indus, Ganges and Brahmaputra river basins: A review of the literature. *Int. J. Water Resour. Dev.* **2015**, *31*, 201–218. [CrossRef]
10. Arnold, J.G.; Srinivasan, R.; Muttiah, R.S.; Williams, J.R. Large area hydrologic modeling and assessment part I: Model development. *J. Am. Water Resour. Assoc.* **1998**, *34*, 73–89. [CrossRef]
11. Wake, C.P. Glaciochemical Investigations as a Tool for Determining the Spatial and Seasonal Variation of Snow Accumulation in the Central Karakoram, Northern Pakistan. *Ann. Glaciol.* **1989**, *13*, 279–284. [CrossRef]
12. Tahir, A.A.; Chevallier, P.; Arnaud, Y.; Ahmad, B. Snow cover dynamics and hydrological regime of the Hunza River basin, Karakoram Range, Northern Pakistan. *Hydrol. Earth Syst. Sci.* **2011**, *15*, 2275–2290. [CrossRef]
13. Ali, K.F.; de Boer, D.H. Spatial patterns and variation of suspended sediment yield in the upper Indus River basin, northern Pakistan. *J. Hydrol.* **2007**, *334*, 368–387. [CrossRef]
14. Archer, D. Contrasting hydrological regimes in the upper Indus Basin. *J. Hydrol.* **2003**, *274*, 198–210. [CrossRef]
15. Immerzeel, W.W.; van Beek, L.P.H.; Bierkens, M.F.P. Climate Change Will Affect the Asian Water Towers. *Science* **2010**, *328*, 1382–1385. [CrossRef] [PubMed]
16. Wijngaard, R.R.; Lutz, A.F.; Nepal, S.; Khanal, S.; Pradhananga, S.; Shrestha, A.B.; Immerzeel, W.W. Future changes in hydro-climatic extremes in the Upper Indus, Ganges, and Brahmaputra River basins. *PLoS ONE* **2017**, *12*, e0190224. [CrossRef] [PubMed]
17. Kilroy, G. A review of the biophysical impacts of climate change in three hotspot regions in Africa and Asia. *Reg. Environ. Chang.* **2015**, *15*, 771–782. [CrossRef]
18. Van Vuuren, D.P.; Edmonds, J.; Kainuma, M.; Riahi, K.; Thomson, A.; Hibbard, K.; Hurtt, G.C.; Kram, T.; Krey, V.; Lamarque, J.-F.; et al. The representative concentration pathways: An overview. *Clim. Chang.* **2011**, *109*, 5–31. [CrossRef]

19. Wayne, G.P. The Beginner's Guide to Representative Concentration Pathways. Available online: https://www.skepticalscience.com/docs/RCP_Guide.pdf (accessed on 1 January 2018).

20. Peters, G.P.; Andrew, R.M.; Boden, T.; Canadell, J.G.; Ciais, P.; Le Quéré, C.; Marland, G.; Raupach, M.R.; Wilson, C. The challenge to keep global warming below 2 °C. *Nat. Clim. Chang.* **2013**, *3*, 4–6. [CrossRef]

21. Lutz, A.F.; ter Maat, H.W.; Biemans, H.; Shrestha, A.B.; Wester, P.; Immerzeel, W.W. Selecting representative climate models for climate change impact studies: An advanced envelope-based selection approach. *Int. J. Climatol.* **2016**, *36*, 3988–4005. [CrossRef]

22. Mora, C.; Frazier, A.G.; Longman, R.J.; Dacks, R.S.; Walton, M.M.; Tong, E.J.; Sanchez, J.J.; Kaiser, L.R.; Stender, Y.O.; Anderson, J.M.; et al. The projected timing of climate departure from recent variability. *Nature* **2013**, *502*, 183–187. [CrossRef] [PubMed]

23. Sanford, T.; Frumhoff, P.C.; Luers, A.; Gulledge, J. The climate policy narrative for a dangerously warming world. *Nat. Clim. Chang.* **2014**, *4*, 164–166. [CrossRef]

24. Taylor, K.E.; Stouffer, R.J.; Meehl, G.A. An Overview of CMIP5 and the Experiment Design. *Bull. Am. Meteorol. Soc.* **2012**, *93*, 485–498. [CrossRef]

25. Peterson, T.C. Climate change indices. *WMO Bull.* **2005**, *54*, 83–86.

26. Sillmann, J.; Kharin, V.V.; Zhang, X.; Zwiers, F.W.; Bronaugh, D. Climate extremes indices in the CMIP5 multimodel ensemble: Part 1. Model evaluation in the present climate. *J. Geophys. Res. Atmos.* **2013**, *118*, 1716–1733. [CrossRef]

27. Yatagai, A.; Kamiguchi, K.; Arakawa, O.; Hamada, A.; Yasutomi, N.; Kitoh, A. APHRODITE: Constructing a Long-Term Daily Gridded Precipitation Dataset for Asia Based on a Dense Network of Rain Gauges. *Bull. Am. Meteorol. Soc.* **2012**, *93*, 1401–1415. [CrossRef]

28. Palazzi, E.; Filippi, L.; von Hardenberg, J. Insights into elevation-dependent warming in the Tibetan Plateau-Himalayas from CMIP5 model simulations. *Clim. Dyn.* **2017**, *48*, 3991–4008. [CrossRef]

29. Dahri, Z.H.; Ludwig, F.; Moors, E.; Ahmad, B.; Khan, A.; Kabat, P. An appraisal of precipitation distribution in the high-altitude catchments of the Indus basin. *Sci. Total Environ.* **2016**, *548*, 289–306. [CrossRef] [PubMed]

30. Pang, H.; Hou, S.; Kaspari, S.; Mayewski, P.A. Influence of regional precipitation patterns on stable isotopes in ice cores from the central Himalayas. *Cryosphere* **2014**, *8*, 289–301. [CrossRef]

31. Hewitt, K. Glacier Change, Concentration, and Elevation Effects in the Karakoram Himalaya, Upper Indus Basin. *Mt. Res. Dev.* **2011**, *31*, 188–200. [CrossRef]

32. Winiger, M.; Gumpert, M.; Yamout, H. Karakorum-Hindukush-western Himalaya: Assessing high-altitude water resources. *Hydrol. Process.* **2005**, *19*, 2329–2338. [CrossRef]

33. Weiers, S. *Zur Klimatologie des NW-Karakorum und Angrenzender Gebiete. Statistische Analysen Unter Einbeziehung von Wettersatellitenbildern und Eines Geographischen Informationssystems (GIS)*; Kommission bei F. Dümmler: Bonn, Germany, 1995.

34. Dhar, O.N.; Rakhecha, P.R. The Effect of Elevation on Monsoon Rainfall Distribution in the Central Himalayas. In *Monsoon Dynamics*; Lighthill, M.J., Pearce, R.P., Eds.; Cambridge University Press: New York, NY, USA, 1981; pp. 253–260.

35. Wilby, R.L.; Dawson, C.W.; Murphy, C.; O'Connor, P.; Hawkins, E. The Statistical DownScaling Model-Decision Centric (SDSM-DC): Conceptual basis and applications. *Clim. Res.* **2014**, *61*, 259–276. [CrossRef]

36. Pielke, R.A.; Wilby, R.L. Regional climate downscaling: What's the point? *Eos Trans. Am. Geophys. Union* **2012**, *93*, 52–53. [CrossRef]

37. Stakhiv, E.Z. Pragmatic Approaches for Water Management Under Climate Change Uncertainty1. *J. Am. Water Resour. Assoc.* **2011**, *47*, 1183–1196. [CrossRef]

38. Wilby, R.L.; Dessai, S. Robust adaptation to climate change. *Weather* **2010**, *65*, 180–185. [CrossRef]

39. Sanjay, J.; Krishnan, R.; Shrestha, A.B.; Rajbhandari, R.; Ren, G.-Y. Downscaled climate change projections for the Hindu Kush Himalayan region using CORDEX South Asia regional climate models. *Adv. Clim. Chang. Res.* **2017**, *8*, 185–198. [CrossRef]

40. Sillmann, J.; Kharin, V.V.; Zwiers, F.W.; Zhang, X.; Bronaugh, D. Climate extremes indices in the CMIP5 multimodel ensemble: Part 2. Future climate projections. *J. Geophys. Res. Atmos.* **2013**, *118*, 2473–2493. [CrossRef]

41. Perkins, S.E.; Pitman, A.J.; Holbrook, N.J.; McAneney, J. Evaluation of the AR4 Climate Models' Simulated Daily Maximum Temperature, Minimum Temperature, and Precipitation over Australia Using Probability Density Functions. *J. Clim.* **2007**, *20*, 4356–4376. [CrossRef]
42. Sánchez, E.; Romera, R.; Gaertner, M.A.; Gallardo, C.; Castro, M. A weighting proposal for an ensemble of regional climate models over Europe driven by 1961–2000 ERA40 based on monthly precipitation probability density functions. *Atmos. Sci. Lett.* **2009**, *10*, 241–248. [CrossRef]
43. Teutschbein, C.; Seibert, J. Bias correction of regional climate model simulations for hydrological climate-change impact studies: Review and evaluation of different methods. *J. Hydrol.* **2012**, *456*, 12–29. [CrossRef]
44. Knutti, R.; Masson, D.; Gettelman, A. Climate model genealogy: Generation CMIP5 and how we got there. Geophys. *Res. Lett.* **2013**, *40*, 1194–1199. [CrossRef]
45. Jun, M.; Knutti, R.; Nychka, D.W. Local eigenvalue analysis of CMIP3 climate model errors. *Tellus A Dyn. Meteorol. Oceanogr.* **2008**, *60*, 992–1000. [CrossRef]
46. Giorgi, F.; Coppola, E.; Solmon, F.; Mariotti, L.; Sylla, M.B.; Bi, X.; Elguindi, N.; Diro, G.T.; Nair, V.; Giuliani, G.; et al. RegCM4: Model description and preliminary tests over multiple CORDEX domains. *Clim. Res.* **2012**, *52*, 7–29. [CrossRef]
47. Samuelsson, P.; Jones, C.G.; Will´En, U.; Ullerstig, A.; Gollvik, S.; Hansson, U.; Jansson, E.; Kjellstro, M.C.; Nikulin, G.; Wyser, K. The Rossby Centre Regional Climate model RCA3: Model description and performance. *Tellus A Dyn. Meteorol. Oceanogr.* **2011**, *63*, 4–23. [CrossRef]
48. Wilby, R.L.; Hay, L.E.; Gutowski, W.J.; Arritt, R.W.; Takle, E.S.; Pan, Z.T.; Leavesley, G.H.; Clark, M.P. Hydrological responses to dynamically and statistically downscaled climate model output. *Geophys. Res. Lett.* **2000**, *27*, 1199–1202. [CrossRef]
49. Wood, A.W.; Leung, L.R.; Sridhar, V.; Lettenmaier, D.P. Hydrologic implications of dynamical and statistical approaches to downscaling climate model outputs. *Clim. Chang.* **2004**, *62*, 189–216. [CrossRef]
50. Randall, D.A.; Wood, R.A.; Bony, S.; Colman, R.; Fichefet, T.; Fyfe, J.; Kattsov, V.; Pitman, A.; Shukla, J.; Srinivasan, J.; et al. Climate Models and Their Evaluation. In *Climate Change 2007: The Physical Science Basis*; Contribution of Working Group I to the Fourth Assessment Report of the Intergovernmental Panel on Climate Change; Cambridge University Press: Cambridge, UK; New York, NY, USA, 2007.
51. Piani, C.; Weedon, G.P.; Best, M.; Gomes, S.M.; Viterbo, P.; Hagemann, S.; Haerter, J.O. Statistical bias correction of global simulated daily precipitation and temperature for the application of hydrological models. *J. Hydrol.* **2010**, *395*, 199–215. [CrossRef]
52. Hagemann, S.; Chen, C.; Haerter, J.O.; Heinke, J.; Gerten, D.; Piani, C. Impact of a sta25 tistical bias correction on the projected hydrological changes obtained from three GCMs and two hydrology models. *J. Hydrometeorol.* **2016**, *12*, 556–578. [CrossRef]
53. Chen, C.; Haerter, J.O.; Hagemann, S.; Piani, C. On the contribution of statistical bias correction to the uncertainty in the projected hydrological cycle, Geophys. *Res. Lett.* **2011**, *38*, L20403. [CrossRef]
54. Rojas, R.; Feyen, L.; Dosio, A.; Bavera, D. Improving pan-European hydrological simulation of extreme events through statistical bias correction of RCM-driven climate simulations. *Hydrol. Earth Syst. Sci.* **2011**, *15*, 2599–2620. [CrossRef]
55. Haddeland, I.; Heinke, J.; Voß, F.; Eisner, S.; Chen, C.; Hagemann, S.; Ludwig, F. Effects of climate model radiation, humidity and wind estimates on hydrological simulations. *Hydrol. Earth Syst. Sci.* **2012**, *16*, 305–318. [CrossRef]
56. Trzaska, S.; Schnarr, E.; A Review of Downscaling Methods for Climate Change Projections. United States Agency for International Development (USAID) under the African and Latin American Resilience to Climate Change (ARCC) Project by Tetra Tech ARD, Burlington, Vermont, USA. 2014, pp. 1–42. Available online: http://www.ciesin.org/documents/Downscaling_CLEARED_000.pdf (accessed on 1 January 2018).
57. Ehret, U.; Zehe, E.; Wulfmeyer, V.; Warrach-Sagi, K.; Liebert, J. "HESS Opinions" Should we apply bias correction to global and regional climate model data? *Hydrol. Earth Syst. Sci.* **2012**, *16*, 3391–3404. [CrossRef]
58. Vannitsem, S. Bias correction and post-processing under climate change. *Nonlinear Process. Geophys.* **2011**, *18*, 911–924. [CrossRef]

59. Maraun, D.; Shepherd, T.G.; Widmann, M.; Zappa, G.; Walton, D.; Gutiérrez, J.M.; Hagemann, S.; Richter, I.; Soares, P.M.; Hall, A.; Mearns, L.O. Towards process-informed bias correction of climate change simulations. *Nat. Clim. Chang.* **2017**, *7*, 764. [CrossRef]
60. Maraun, D. Bias correcting climate change simulations-a critical review. *Curr. Clim. Chang. Rep.* **2016**, *2*, 211–220. [CrossRef]

 *climate*

Article

Estimating the Impact of Artificially Injected Stratospheric Aerosols on the Global Mean Surface Temperature in the 21th Century

Sergei A. Soldatenko

St. Petersburg Institute for Informatics and Automation of the Russian Academy of Sciences, No. 39, 14-th Line, St. Petersburg 199178, Russia; soldatenko@iias.spb.su; Tel.: +7-931-354-0598

Received: 28 September 2018; Accepted: 26 October 2018; Published: 28 October 2018

Abstract: In this paper, we apply the optimal control theory to obtain the analytic solutions of the two-component globally averaged energy balance model in order to estimate the influence of solar radiation management (SRM) operations on the global mean surface temperature in the 21st century. It is assumed that SRM is executed via injection of sulfur aerosols into the stratosphere to limit the global temperature increase in the year 2100 by 1.5 °C and keeping global temperature over the specified period (2020–2100) within 2 °C as required by the Paris climate agreement. The radiative forcing produced by the rise in the atmospheric concentrations of greenhouse gases is defined by the Representative Concentration Pathways and the 1pctCO$_2$ (1% per year CO$_2$ increase) scenario. The goal of SRM is formulated in terms of extremal problem, which entails finding a control function (the albedo of aerosol layer) that minimizes the amount of aerosols injected into the upper atmosphere to satisfy the Paris climate target. For each climate change scenario, the optimal albedo of the aerosol layer and the corresponding global mean surface temperature changes were obtained. In addition, the aerosol emission rates required to create an aerosol cloud with optimal optical properties were calculated.

Keywords: climate change; optimal control; geoengineering; climate manipulation

---

1. Introduction

Climate change is among the most significant threats to human civilization in the 21st century and beyond [1]. The Paris Climate Accord proposed to hold average temperature increase "to well below 2 °C above pre-industrial levels" and to pursue efforts to keep warming "below 1.5 °C above pre-industrial levels" [2]. To reach these goals, eight countries have already presented long-term low-emission strategies, which aims to reduce greenhouse gas emissions; several countries are currently in the process of preparing such strategies [3]. Meanwhile, the World Meteorological Organization's (WMO) "Statement of the State of the Global Climate in 2017" released in January 2018 said, "The global mean temperature in 2017 was approximately 1.1 °C above the pre-industrial era" [4]. There is high confidence that planetary warming will continue throughout the 21st century even if we immediately stopped emitting greenhouse gases into the atmosphere (e.g., References [5–9]). Some resent studies (e.g., References [10–13]) suggest that geoengineering technologies can serve as a supplementary measure to stabilize climate as "in the absence of external cooling influence" [14], it is hard to achieve the Paris Agreement climate goals.

Solar radiation management (SRM) by injection of sulfur aerosols into the stratosphere [15,16] is one of the most feasible and promising solutions for inducing negative radiative forcing (RF) from aerosols in order to at least partially compensate the positive RF from atmospheric greenhouse gases. The current state of understanding of climate engineering technologies, including SRM, has been discussed in References [17–26]. Over the years, climate models have played a key role in exploring geoengineering techniques and predicting and quantifying their potential effects on Earth's climate (e.g., References [27–36]). Due to the uncertainties inherent in climate models that could not be sufficiently reduced over the last decade [37], the resulting range of possible outcomes of hypothetical geoengineering efforts remains quite vague. To handle the climate response uncertainties, some studies (e.g., References [38–45]) have suggested modeling the Earth's climate as a control system with feedbacks, which allows planning scenarios for geoengineering using the so-called "design model". This formulation makes it possible to design the control law and calculate the amount of SRM forcing as a function of time needed to offset the rise in global mean surface temperature due to human-caused positive RF. Meanwhile, exploring Earth's global climate as controlled dynamical system, we can approach geoengineering from the perspective of optimal control theory [46–49]. Within the optimal control framework, the goal of geoengineering can be formulated in terms of extremal problem, which involves finding control functions and the corresponding climate system trajectory that minimize or maximize a certain objective functional (also referred to as performance measure or index) subject to various constraints (e.g., References [50,51]. If $\mathbf{x}$ is the state vector of climate system and $\mathbf{u}$ is the vector of control variables, then the abstract extremal problem can be formulated as follows:

$$\mathcal{J}(\mathbf{x}, \mathbf{u}) \rightarrow extr, \ \ \mathcal{F}(\mathbf{x}, \mathbf{u}) = 0, \ \ (\mathbf{x}, \ \mathbf{u}) \in \mathcal{M} \subset \mathcal{X} \times \mathcal{U} \tag{1}$$

The statement of this problem includes a set $\mathcal{X} \times \mathcal{U}$ on which the (real) functional $\mathcal{J}(\mathbf{x}, \mathbf{u})$ is defined and constraints imposed on state and control variables given by the model of control object $\mathcal{F}(\mathbf{x}, \mathbf{u}) = 0$ (dynamic constraints) and by the subset $\mathcal{M}$ in $\mathcal{X} \times \mathcal{U}$. The solution to the extremal problem (Equation (1)) is the optimal process $(\mathbf{x}^*, \mathbf{u}^*)$. Thus, by solving the optimal control problem (OCP), we can obtain the mathematically rigorous control law and the corresponding system's trajectory that are relevant for the specified performance measure $\mathcal{J}(\mathbf{x}, \mathbf{u})$.

This paper deals with a simple mathematical model for controlling the global mean surface temperature T_{sfc} in the 21st century by the injection of sulfur aerosols into the stratosphere to limit the global temperature increase in the year 2100 by 1.5 °C above pre-industrial level and keeping global temperature over the period of 2020–2100 within 2 °C as required by the Paris climate agreement. The objective is to minimize resources (the total mass of aerosols) required to achieve the desired final state of the climate system. In the model, the positive RF produced by the rise in the atmospheric concentrations of greenhouse gases is specified in accordance with the Representative Concentration Pathways [52] and the 1pctCO$_2$ (1% per year CO$_2$ increase) scenario.

The mathematical statement of OCP is the collection of the following key elements: objective function defined to judge the effectiveness of control process, mathematical model of the controlled object, equality and inequality constraints to be satisfied by state and control variables, and boundary and initial conditions (if any) for state variables. To imitate the behavior of the climate system, we applied a two-component energy balance model [53–55] in which the global mean surface temperature anomaly (perturbation) represents the variable that interests us the most, and the albedo of the global aerosol layer is designated as the control variable. We derived analytical expressions for both the optimal albedo of the global aerosol layer and the corresponding change in the global mean surface temperature.

The results of illustrative calculations are presented for the period 2020–2100. For each climate change scenario, the optimal albedo of the aerosol layer—and therefore the aerosol emission rates—as well as the associated global mean surface temperature changes were found.

We need to emphasize that the main reason for using such a model is that similar two-layer models have been considered and analyzed in a number of papers considering the response to forced climate change. For example, Geoffroy et al. [56,57] obtained and discussed the general analytical solutions of the two-layer model for different hypothetical climate forcing scenarios and suggested the approach of calibrating the model parameters to imitate the time response of coupled general circulation models (CGCMs) from CMIP5 to radiative forcing. Gregory et al. [58] analyzed the two-layer model and discussed the transient climate response, the global mean surface air temperature change under two scenarios: one with a step forcing (the abrupt 4xCO$_2$ experiment) and one with the 1pctCO$_2$ scenario. Despite the fact that the two-layer model is one of the simplest tools to mimic climate dynamics under external radiative forcing, it was able to simulate the evolution of average global surface temperature over time in response to both abrupt and time-dependent forcing with reasonable accuracy (e.g., References [56,59]).

In the two-layer model, climate control is carried out via changing Earth's planetary albedo by injection of sulfur aerosols into the stratosphere ("albedo modification"). Sulfur aerosols increase the amount of sunlight that is scattered back to space, thereby reducing the amount of sunlight absorbed by Earth. Inherently, the planetary albedo is an average of the local albedo, averaged over the entire globe. The local albedo, in turn, is a highly variable dimensionless parameter that depends on a number of local factors, such as the composition of the atmosphere and in particular the presence of aerosols, the cloud amount and properties [60], the sea ice cover [61], the land use [62–64], the snow cover [65], etc. A typical value of Earth's planetary albedo is about 0.3 [66].

As change in the albedo of our planet is a powerful driver of climate (indeed, a 1% change in the Earth's planetary albedo generates the radiative effect of 3.42 Wm^{-2}, which is commensurate with radiative forcing due to a doubling of CO$_2$ concentrations in the atmosphere), scientists have proposed "albedo modification" as a powerful tool to deal with global warming (e.g., References [16,20,23,29,31,33,67]).

2. Materials and Methods

2.1. The Model of Control Object

The control object is Earth's climate system. To simulate the climate system dynamics under the influence of external radiative forcing, we have applied the mathematical model consisting of two subsystems: One is the upper layer subsystem, which combines the atmosphere, the land surface, and the upper ocean; the other is the lower layer subsystem, which represents the deep ocean [53–55]. The state of each subsystem is characterized by the corresponding temperature perturbation (anomaly) with respect to initial climate "equilibrium" state. Denoting temperature anomalies for upper and lower subsystems by T and T_D, respectively, the equations that govern these perturbations can be written as follows:

$$C_U \frac{dT}{dt} = -\lambda T - \gamma(T - T_D) + \Delta R_{CO_2} + (1 - \alpha_0)\Delta R_A \tag{2}$$

$$C_D \frac{dT_D}{dt} = \gamma(T - T_D) \tag{3}$$

Here, C_U and C_D are the effective heat capacities of the upper and lower models, respectively (note that $C_U \ll C_D$); λ is a climate radiative feedback parameter; γ is a coupling strength parameter that describes the rate of heat loss by the upper layer; ΔR_{CO_2} is the radiative forcing caused by global increase in the atmospheric CO$_2$ concentration; ΔR_A is the negative radiative forcing generated by the artificial aerosols at the top of the atmosphere; and α_0 is Earth's planetary albedo. We will assume that the temperature anomaly T is identified with the global mean surface temperature change T_{sfc} [53,54].

Despite its simplicity, this model imitates climate changes under external radiative forcing with reasonable accuracy [56–59]. We have chosen values of 7.34 W yr m^{-2} K^{-1}, 105.5 W yr m^{-2} K^{-1}, 1.13 W m^{-2} K^{-1}, and 0.7 W m^{-2} K^{-1} for parameters C_U, C_D, λ, and γ, respectively. These values are taken in accordance with values consistent with the CMIP5 multimodel mean under climate change derived in Reference [56].

For convenience sake, we have rewritten the model Equations (2) and (3) as follows:

$$\frac{dT}{dt} = -aT + bT_D + \frac{\Delta R_{CO_2}}{C_U} + \frac{(1 - \alpha_0)\Delta R_A}{C_U} \tag{4}$$

$$\frac{dT_D}{dt} = pT - pT_D \tag{5}$$

where

$$a = \frac{\lambda + \gamma}{C_U}, \quad b = \frac{\gamma}{C_U}, \quad p = \frac{\gamma}{C_D}. \tag{6}$$

Ordinary Differential Equations (4) and (5) represent a mathematical model of the object to be controlled.

2.2. Parameterization of the Aerosols' Radiative Effect

The climate control is assumed to be executed through the injection of nonabsorptive sulfate aerosols into the stratosphere. Injected aerosol particles scatter shortwave solar radiation back to the outer space and consequently change the radiative balance of our planet, increasing Earth's planetary albedo and therefore causing the negative RF at the top of the atmosphere [68–71]:

$$\Delta R_A = -\alpha_A Q_0 \tag{7}$$

Here, α_A is the instant albedo of the global aerosol layer; Q_0 is the global average incoming solar radiation on the top of the atmosphere defined as $Q_0 = I_0/4$, where $I_0 = 1368$ W m^2 is a solar constant [72,73]. Thus, to estimate the radiative effect of stratospheric aerosol, we need to calculate the albedo α_A, which is considered as the control variable. However, in reality, we have the ability to manipulate the emission rate of aerosols injected into the stratosphere E_A. To determine E_A from the known α_A, the mass balance equation is used:

$$\frac{dM_A}{dt} = E_A - \frac{M_A}{\tau_A} \tag{8}$$

where τ_A is the residence time of stratospheric aerosol particles; M_A is the global mass of the stratospheric aerosols, which is linearly related to the albedo α_A [69]:

$$\alpha_A = M_A(\beta_A k_A / Q_0 S_e) \tag{9}$$

where the coefficient $\beta_A = 24$ W m^{-2} [70,71]; $k_A = 7.6$ m^{2}g^{-1} is the mass extinction coefficient [69]; S_e is Earth's area determined as $S_e = 4\pi R_e^2$, where $R_e = 6371$ km is Earth's radius.

In geoengineering, sulfate aerosol particles are not directly injected into the stratosphere but can be formed from gaseous precursors, such as sulfur dioxide SO_2, hydrogen sulfide H_2S, carbonyl sulfide OCS, or dimethyl sulfide (DMS), which then convert into aerosols. We will express the emission rate of aerosol precursors as well as the mass of sulfate aerosols in units of sulfur, denoting them by E_S (in Tg S yr^{-1}) and M_S (in Tg S), respectively. Assuming that 1 Tg of sulfur injected into the stratosphere forms approximately 4 Tg of aerosol particles [74], we obtain that $E_S \approx E_A/4$ and $M_S \approx M_A/4$. As the relationship between M_A and α_A is linear, the following predictive equation for α_A can be derived from Equation (8):

$$\frac{d\alpha_A}{dt} = \chi^{-1} E_S - \frac{\alpha_A}{\tau_A} \tag{10}$$

where $\chi = Q_0 S_e / (4\beta_A k_A) \approx 2.39 \times 10^2$ Tg S.

Thus, solving the OCP, we can find the optimal control law $\alpha_A^*(t)$ and then calculate the optimal aerosol emission rate $E_S^*(t)$ using Equation (10).

2.3. Parameterization of the Anthropogenic Radiative Forcing

In energy balance models, simple empirical expressions are generally used to calculate radiative forcing due to the increase in atmospheric greenhouse gases. For example, the radiative forcing caused by a perturbation of the atmospheric burden of CO_2 can be parameterized as a function of CO_2 only [75,76]: $\Delta R_{CO_2} = \kappa \times \ln\left[C_{CO_2}(t)/C_{CO_2}^{(0)}\right]$, where κ (W m^{-2}) is the empirical coefficient; $C_{CO_2}(t)$ is the CO_2 concentration at time t; and $C_{CO_2}^{(0)}$ is the reference CO_2 concentration level. A typical value for the parameter κ is near 5.35 W m^2 [75,76]. In our model, we have taken the total global mean anthropogenic and natural radiative forcing ΔR_N as prescribed by the different scenarios and approximated by a linear function of time:

$$\Delta R_N = \eta t \tag{11}$$

where η is the annual rate of forcing (see Table 1).

Table 1. Annual radiative forcing rate η.

Scenario	RCP8.5	1pctCO$_2$	RCP6	RCP4.5
η (W m^{-2} yr^{-1})	7.14$\times$ 10^{-2}	5.29$\times$ 10^{-2}	3.814$\times$ 10^{-2}	2.17$\times$ 10^{-2}

2.4. Optimal Control Problem Formulation

We let [t_0, t_f] be a finite and fixed time interval. The OCP is defined as follows:

We find the control function $\alpha_A(t)$ generating the corresponding temperature anomalies $T(t)$ and $T_D(t)$ that minimizes the objective function:

$$\mathcal{J} = \frac{1}{2} \int_{t_0}^{t_f} \alpha_A^2(t) dt \tag{12}$$

subject to the dynamics (4) and (5) and given initial $T(t_0) = 0$ *and* $T_D(t_0) = 0$, *as well as final (terminal)* $T\left(t_f\right) = T^f$ *conditions.*

In this formulation, the terminal condition T^f is interpreted as a target change in the global mean surface temperature at $t = t_f$, and the performance index (Equation (12)) characterizes the aerosol consumption for SRM operations (recall that α_A and M_A are linearly dependent functions). Thus, we wish to minimize the mass of aerosols required to reach the target surface temperature change at the final time. The global mean deep ocean temperature anomaly at the final time t_f is not defined because changes in the global mean surface temperature are of primary concern, while changes in the deep ocean temperature are only of secondary concern. The total amount of aerosols annually emitted to the stratosphere can be limited by the available technical equipment. In this case, the minimization problem (Equation (12)) should be considered within the framework of control-constrained OCP. The set of admissible controls is given formally by $\alpha_A \in [0, U]$, where U is the maximum value of technically feasible and affordable albedo α_A.

2.5. Method for Solving the Optimal Control Problem

Before proceeding further, we rewrite the model Equations (4) and (5) by replacing ΔR_{CO_2} with ηt (11) and ΔR_A with $-\alpha_A Q_0$ (7):

$$\frac{dT}{dt} = -aT + bT_D + ct - q\alpha_A \tag{13}$$

$$\frac{dT_D}{dt} = pT - pT_D \tag{14}$$

where $c = \eta/C_U$ and $q = (1 - \alpha_0)Q_0/C_U$.

We solve the formulated OCP using the Pontryagin's maximum principle (PMP) [47]. The Hamiltonian function for the problem (12) is defined as follows:

$$H = -\frac{1}{2}\alpha_A^2 + \psi_1(-aT + bT_D + ct - q\alpha_A) + \psi_2(pT - pT_D) \tag{15}$$

where ψ_1 and ψ_2 are time-varying Lagrange multipliers, also known as costate or adjoint variables, which satisfy the adjoint system:

$$\frac{d\psi_1}{dt} = -\frac{\partial H}{\partial T} = a\psi_1 - p\psi_2 \tag{16}$$

$$\frac{d\psi_2}{dt} = -\frac{\partial H}{\partial T_D} = -b\psi_1 + p\psi_2 \tag{17}$$

The PMP states that the optimal control $\alpha_A^*(t) \in [0, U]$ is one that would maximize the Hamiltonian (Equation (13)) at each fixed time $t \in \left[t_0, t_f\right]$:

$$\alpha_A^* = \underset{\alpha_A \in [0,U]}{\arg\,\max} H(\alpha_A) \tag{18}$$

Therefore, to find the optimal control α_A^*, we must maximize H with respect to α_A, where the control belongs to the admissible control region $\alpha_A \in [0, U]$. The Hamiltonian maximization condition is as follows:

$$\frac{\partial H}{\partial \alpha_A} = -\alpha_A - q\psi_1 = 0 \tag{19}$$

Thus, to find the optimal control and the corresponding climate system's trajectory, we need to solve the set of four ordinary Differential Equations (13), (14), (16), and (17) in four unknowns T, T_D, ψ_1, and ψ_2 with given initial and terminal conditions. As the variable T_D is not defined at t_f, the following transversality condition for costate variable ψ_2 applies: $\psi_2\left(t_f\right) = 0$ [48,49]. The analytic expressions derived for the control variable α_A and temperature anomalies T and T_D can be written as follows:

$$\alpha_A(t) = -C_1 q \left[v_{11} e^{\lambda_1 t} + e^{(\lambda_1 - \lambda_2) t_f} v_{21} e^{\lambda_2 t} \right] \tag{20}$$

$$T(t) = C_1 \alpha_1 \left(e^{\lambda_1 t} - e^{\lambda_2 t} \right) + C_3 e^{-\lambda_1 t} + C_4 e^{-\lambda_2 t} + + w_2 t + w_1 \tag{21}$$

$$T_D(t) = C_3 \frac{a - \lambda_1}{b} e^{-\lambda_1 t} + C_4 \frac{a - \lambda_2}{b} e^{-\lambda_2 t} + \tag{22}$$

$$+ C_1 \left[\frac{\alpha_1 (a + \alpha_1) - q^2 v_{11}}{b} e^{\lambda_1 t} - \frac{\alpha_2 (a + \alpha_2) - q^2 v_{21} e^{(\lambda_1 - \lambda_2) t_f}}{b} e^{\lambda_2 t} \right] +$$

$$+ \frac{a w_2 - c}{b} t + \frac{a w_1 + w_2}{b}$$

where C_1, C_3, and C_4 are arbitrary integration constants (note that the integration constant $C_2 = -C_1 e^{(\lambda_1 - \lambda_2) t_f}$); λ_1 and λ_2 are the eigenvalues of the coefficient matrix of the adjoint system, Equations (14) and (15); v_{11} and v_{21} are the components of the corresponding eigenvectors.

$$\alpha_1 = \frac{q^2 v_{11} (\lambda_1 + p)}{\lambda_1^2 + \lambda_1 (a + p) + (ap - pb)}$$

$$\alpha_2 = \frac{q^2 v_{21} (\lambda_2 + p) e^{(\lambda_1 - \lambda_2) t_f}}{\lambda_1^2 + \lambda_1 (a + p) + (ap - pb)}$$

$$w_1 = \frac{c[(ap - pb) - p(a + p)]}{(ap - pb)^2}$$

$$w_2 = \frac{pc}{ap - pb}$$

The constants of integration are determined by applying the boundary conditions.

If we consider climate engineering as a state-constrained OCP with constraints on the state variables, then additional necessary conditions for optimality, known as the complementary slackness conditions, should be specified [77]. In this study, we express the OCP with the following state constraint:

$$T(t) \le C_T \qquad \forall t \in \left[t_0, \, t_f\right] \tag{23}$$

where C_T is the threshold parameter whose value should be set. The meaning of the condition (Equation (23)) known as a path constraint is that throughout the geoengineering project, the global mean surface temperature change should not exceed a certain value C_T, which is determined a priori. We should highlight that state constraints add a great deal of complexity to the OCP [77,78].

3. Results and Discussion

In the calculations, we took calendar years 2020 and 2100 as the initial t_0 and the final (terminal) t_f time, respectively, which meant that we were examining the climate control problem on the finite time interval 2020–2100. To formulate the boundary conditions and impose a constraint on change in the global mean surface temperature T_{sfc}, we assumed the following:

- The temperature anomalies T and T_D were calculated relative to 2020, i.e., the boundary conditions for T and T_D at $t = t_0$ were $T_{2020} = 0$ and $T_{D,\,2020} = 0$, respectively, where the numerical subscript referred to the year 2020.
- By 2020, T_{sfc} would exceed the pre-industrial level by 1.1 °C, i.e., $\Delta T_{2020} = 1.1$.
- By 2100, T_{sfc} would exceed the pre-industrial level by 1.5 °C, i.e., $\Delta T_{2100} = 1.5$.
- For the 2020 to 2100 period, the rise in T_{sfc} should not exceed 2 °C above the pre-industrial level.

Then, the permissible increase in the temperature anomaly T_{2100} by year 2100 relative to 2020 would be $T_{2100} = \Delta T_{2100} - \Delta T_{2020} = 0.4$. This value represents the boundary condition for T at $t = t_f$. The threshold parameter, which defines a path constraint (Equation (23)), is $C_T = 2 - \Delta T_{2020} = 0.9$.

Changes in both global mean surface temperature and deep ocean temperature calculated for different climate change scenarios in the absence of climate engineering interventions are illustrated in Figure 1. The corresponding temperature changes in the year 2100 are shown in Table 2. According to Reference [79], without additional measures to reduce GHG emissions (RCP8.5 scenario), increases in global mean surface temperatures are expected to be between 3.7 and 4.8 °C by the year 2100 versus pre-industrial levels (this range is based on median climate response). As seen in Table 2, by year 2100, the model outlined here projects globally averaged surface temperature increases of 4.26, 3.44, and 2.80 °C for the RCP8.5, 1pctCO$_2$, and RCP6.0 scenarios, respectively (relative to pre-industrial period). Thus, geoengineering can be regarded as one of supplementary measures needed to achieve the climate targets of the Paris Agreement.

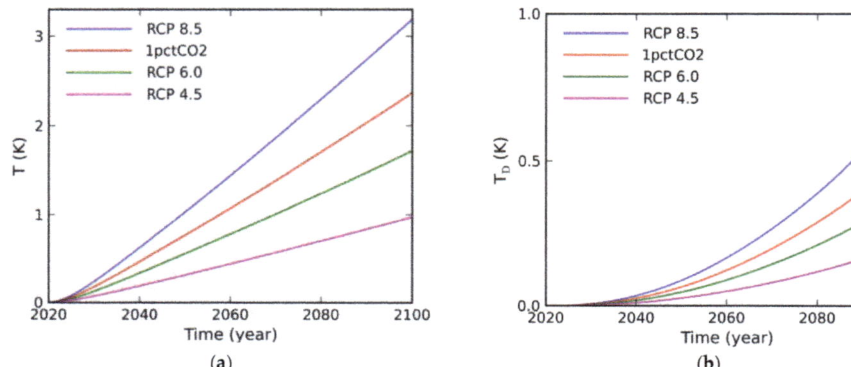

Figure 1. Changes in (**a**) global mean surface temperature and (**b**) deep ocean temperature calculated for different climate change scenarios in the absence of climate engineering interventions.

Table 2. Calculated temperature changes T and T_D from 2020 to 2100 (changes relative to the pre-industrial level are shown in brackets).

Scenario	RCP8.5	1pctCO$_2$	RCP6.0	RCP4.5
T (K)	3.16 (4.26)	2.34 (3.44)	1.70 (2.80)	0.96 (2.06)

We considered results of calculations for the RCP8.5 (the worst-case) scenario in more detail. Figure 2 shows (a) the optimal albedo of the global stratospheric aerosol layer, (b) the corresponding surface temperature anomaly, (c) the mass of the global aerosol layer, and (d) the optimal emission rate of aerosol particles calculated for RCP8.5 pathway with and without constraint on the global mean surface temperature increase. In the absence of state constraint, the optimal albedo α_A^* and, accordingly, the optimal emission rate of aerosol particles E_S^* would increase exponentially. This optimal aerosols emission rate ensures that the target temperature anomaly $T_{2100} = 0.4$ is satisfied. However, within the given time interval 2020–2100, a temperature rise would exceed the set point C_T, i.e., $T(t) > C_T$ (the "overshooting" phenomenon [80]). The maximum increases in global mean surface temperature for different climate change scenarios are presented in Table 3. The use of the constraint (Equation (21)) allows us to avoid overshoot; however, compared to the unconstrained case, keeping the increase in global mean surface temperature below the target constrained level C_T would require additional amount of aerosols (see Table 4). For example, for the RCP8.5 scenario, the total mass of aerosol particles injected in the stratosphere from the year 2020 to 2100 is about 73.6 Tg S, which is about 2 times larger than $M_{S,\,tot}^*$, calculated by solving an unconstrained OCP.

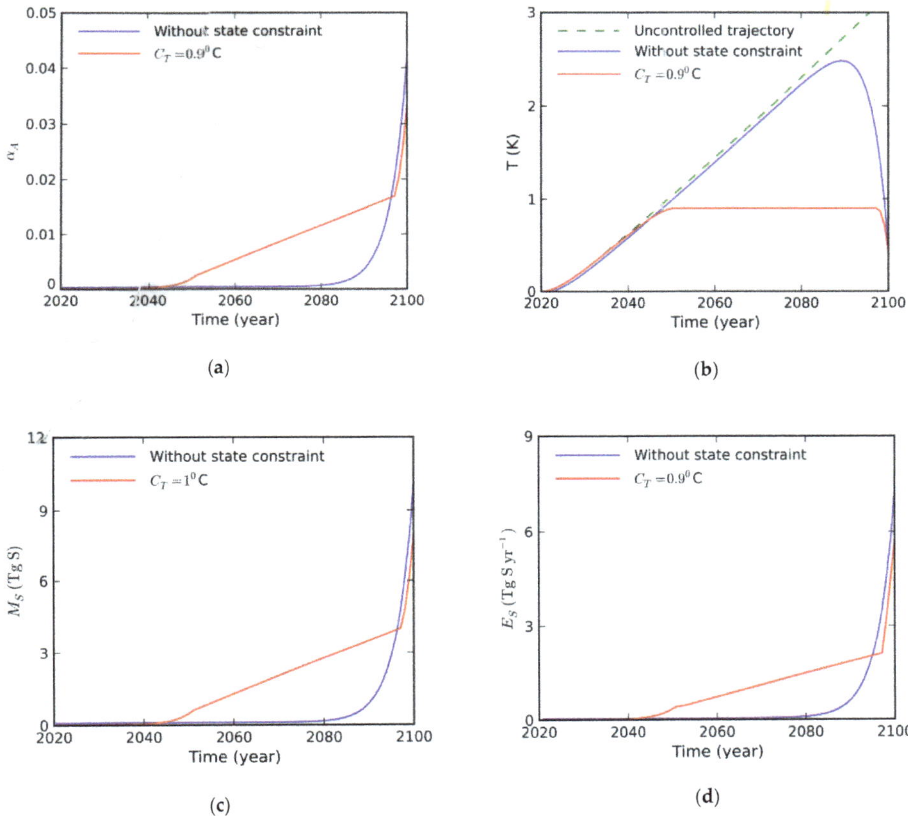

Figure 2. Results for the RCP8.5 pathway: (**a**) optimal albedo of aerosol layer α_A^*; (**b**) the corresponding temperature anomaly T^*; (**c**) total mass of aerosols M_S^*; and (**d**) the optimal emission rate E_S^*.

Table 3. Maximum global mean surface temperature anomaly T calculated without state constraint.

Scenario	RCP8.5	1pctCO$_2$	RCP6	RCP4.5
T (K)	2.48	1.84	1.34	0.78

Table 4. The mass of aerosols $M_{S,\ tot}$ (Tg S) injected into the stratosphere from 2020 to 2100.

Scenario	RCP8.5	1pctCO$_2$	RCP6	RCP4.5
$M_{S,tot}$ without state constraint	36.5	25.6	17.0	7.7
$M_{S,tot}$ with state constraint	73.6	44.46	23.3	-

Results obtained for 1pctCO$_2$, RCP6.0, and RCP4.5 scenarios are represented in Figures 3–5, respectively. These figures show that the overshooting phenomenon is also observed for the 1pctCO$_2$ and RCP6.0 scenarios. The only exception is the RCP4.5 scenario.

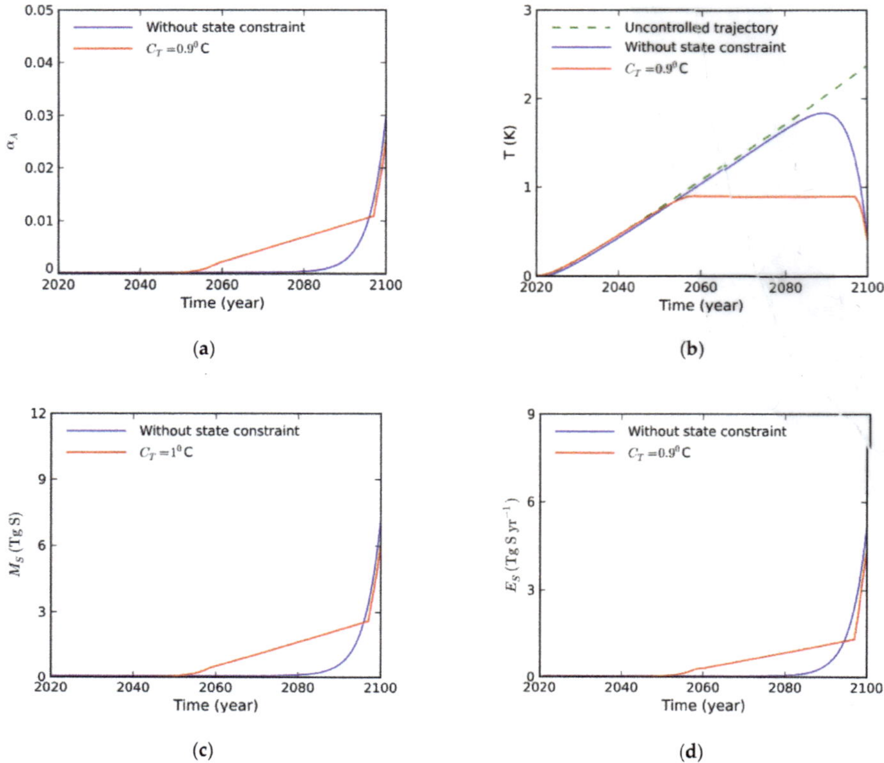

Figure 3. Results for the 1pctCO$_2$ scenario: (**a**) optimal albedo of aerosol layer α_A^*; (**b**) the corresponding temperature anomaly T^*; (**c**) total aerosol mass M_S^*; and (**d**) the optimal emission rate E_S^*.

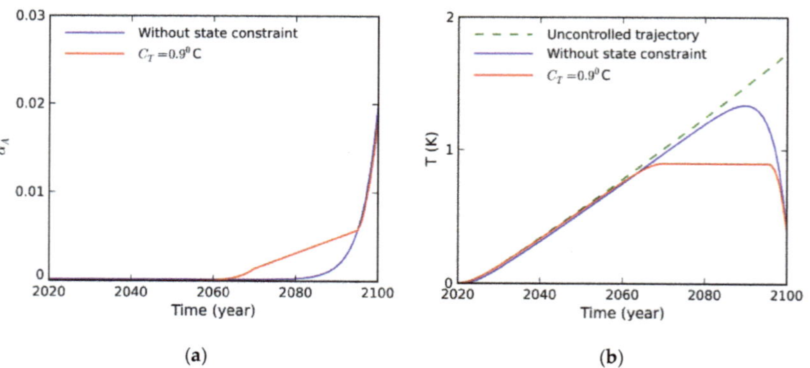

Figure 4. *Cont.*

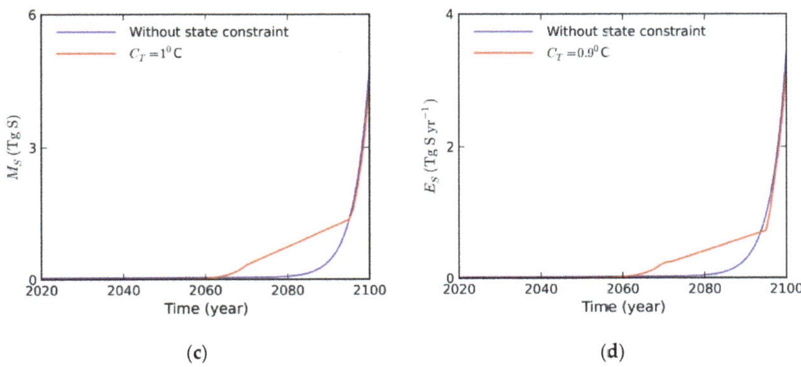

Figure 4. Results for the RCP6.0 pathway: (**a**) optimal albedo of aerosol layer α_A^*; (**b**) the corresponding temperature anomaly T^*; (**c**) total mass of aerosols M_S^*; and (**d**) the optimal emission rate E_S^*.

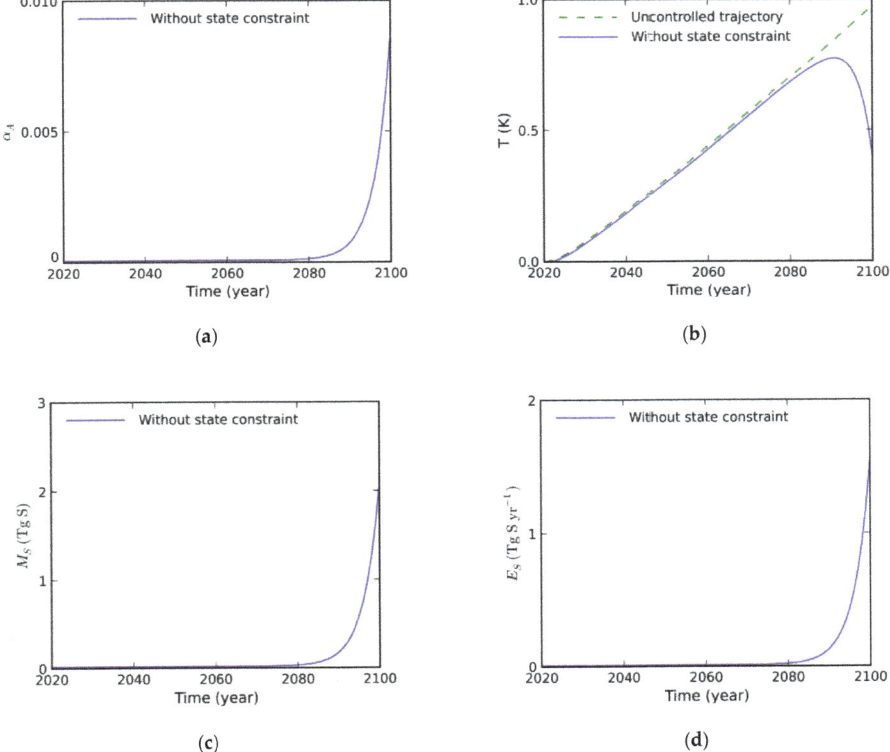

Figure 5. Results for the RCP4.5 pathway: (**a**) optimal albedo of aerosol layer α_A^*; (**b**) the corresponding temperature anomaly T^*; (**c**) total mass of aerosols M_S^*; and (**d**) the optimal emission rate E_S^*.

If the optimal control problem is considered with control variable constraint $\alpha_A(t) \leq U$, then the target value for the temperature anomaly in the year 2100 may not necessarily be achieved (this depends on the value of the constraint U and the scenario in question). Assuming, for example, that $U = 0.02$, then the corresponding instant mass of aerosols is estimated to be 4.8 Tg S. In such a case, for the RCP8.5 scenario, the calculated temperature anomaly in the year 2100 relative to 2020 would exceed the target value by 0.3 °C, which is equivalent to exceeding the pre-industrial level by 1.8 °C. It needs to be recalled that constraint on the control variable is associated with a possible limitation on resources required to implement the project, namely, the amount of aerosols available to the project executors.

We emphasize that the results of calculations discussed above are for illustration purposes only. The primary outcome presented in this paper is the optimal control-based approach that can be used to design projects targeting purposeful manipulation of climate and weather.

4. Concluding Remarks

The use of fine aerosol particles, artificially injected into the stratosphere, is considered to be one of the most effective and feasible measures to counter global warming in the 21st century and beyond. Computer simulation using mathematical climate models of various degrees of sophistication and complexity is the most popular and reliable technique for exploring and estimating the effectiveness of stratospheric aerosol climate engineering and climate and weather manipulation. Numerical simulation of climate engineering requires the design of fairly realistic scenarios for aerosol injections. This paper introduced the optimal-control-based method for designing climate engineering scenarios. Considering Earth's climate as controlled dynamical system, we proposed to approach geoengineering from the standpoint of the optimal control theory, thereby formulating the goal of geoengineering projects in terms of extremal problem. The capability to apply this technique was illustrated using the two-layer energy balance model in which the global mean surface temperature anomaly and the deep ocean temperature perturbation were the state variables, and the emission rate of aerosol precursors was the control variable. Solutions to the unconstrained as well as state and control constrained problems were obtained on the basis of classical Pontryagin's maximum principle. The proposed method will provide additional useful insights for the development of optimal climate manipulation strategies to counter global warming in the 21st century.

Funding: This research received no external funding.

Acknowledgments: The author would like to thank two anonymous reviewers for their helpful comments.

Conflicts of Interest: The author declares no conflict of interest.

References

1. Stocker, T.F.; Qin, D.; Planner, G.; Tignor, M.S.; Allen, K.; Boschumg, J.; Alexander, N.; Yu, X.; Vincent, B.; Pauline, M.M. *Climate Change 2013: The Physical Science Basis. Contribution of Working Group I to the Fifth Assessment Report of the Intergovernmental Panel on Climate Change*; Cambridge University Press: Cambridge, UK, 2013.
2. Paris Agreement. Available online: https://unfccc.int/sites/default/files/english_paris_agreement.pdf (accessed on 16 August 2018).
3. New Climate Institute. Available online: https://unfccc.int/process/the-paris-agreement/long-term-strategies (accessed on 16 August 2018).
4. World Meteorological Organization. *WMO Statement on the State of the Global Climate in 2017*; WMO-No. 1212; World Meteorological Organization: Geneva, Switzerland, 2018.
5. Rodelj, J.; den Elzen, M.; Höhne, N.; Fransen, T.; Fekete, H.; Winkler, H.; Schaeffer, R.; Sha, F.; Riahi, K.; Meinshausen, M. Paris Agreement climate proposals need a boost to keep warming well below 2 °C. *Nature* **2016**, *534*, 631–639. [CrossRef]
6. Brown, P.; Caldeira, K. Greater future global warming inferred from Earth's recent energy budget. *Nature* **2017**, *552*, 45–50. [CrossRef] [PubMed]
7. Raftery, A.E.; Zimmer, A.; Frierson, D.M.W.; Startz, R.; Liu, P. Less than 2 °C warming by 2100 unlikely. *Nat. Clim. Chang.* **2017**, *7*, 637–641. [CrossRef] [PubMed]
8. Kong, Y.; Wang, C.-H. Responses and changes in the permafrost and snow water equivalent in the Northern Hemisphere under a scenario of 1.5 °C warming. *Adv. Clim. Chang. Res.* **2017**, *8*, 235–244. [CrossRef]
9. Jacob, D.; Kotova, L.; Teichmann, C.; Stefan, P.S.; Vautard, R.; Chantal, D.; Aristeidis, G.; KoutroulisManolis, G.; GrillakisIoannis, K.T.; Andrea, D. Climate impacts in Europe under +1.5 °C global warming. *Earth Future* **2018**, *6*, 264–285. [CrossRef]

10. Tanaka, K.; O'Neill, B.C. The Paris Agreement zero-emissions goal is not always consistent with the 1.5 °C and 2 °C temperature targets. *Nat. Clim. Chang.* **2018**, *8*, 319–324. [CrossRef]

11. Keith, D.; Mac-Martin, D. A temporary, moderate and responsive scenario for solar geoengineering. *Nat. Clim. Chang.* **2015**, *5*, 201–206. [CrossRef]

12. Chen, Y.; Xin, Y. Implications of geoengineering under the 1.5 °C target: Analysis and policy suggestions. *Adv. Clim. Chang. Res.* **2017**, *8*, 123–129. [CrossRef]

13. MacMartin, D.G.; Ricke, K.L.; Keith, D.W. Solar geoengineering as part of an overall strategy for meeting the 1.5 °C Paris target. *Phil. Trans. R. Soc.* **2018**, *376*. [CrossRef] [PubMed]

14. Henley, B.; King, A. Trajectories toward the 1.5 °C Paris target: Modulation by the Interdecadal Pacific Oscillation. *Geophys. Res. Lett.* **2017**, *44*, 4256–4262. [CrossRef]

15. Budyko, M.I. *Climate and Life*; Academic Press: New York, USA, 1974.

16. Crutzen, P.J. Albedo enhancement by stratospheric sulfur injections: A contribution to resolve a policy dilemma? *Clim. Chang.* **2006**, *77*, 211–220. [CrossRef]

17. Bellamy, R.; Chilvers, J.; Vaughan, N.E.; Lenton, T.M. A review of climate geoengineering appraisals. *WIREs Clim. Chang.* **2012**, *3*, 597–615. [CrossRef]

18. Shepherd, J.G. Geoengineering the climate: An overview and update. *Philos. Trans. R. Soc. A* **2012**, *370*, 4166–4175. [CrossRef] [PubMed]

19. Zhang, Z.; Moore, J.C.; Huisingh, D.; Zhao, Y. Review of geoengineering approaches to mitigating climate change. *J. Clean. Prod.* **2015**, *103*, 898–907. [CrossRef]

20. Irvine, P.J.; Kravitz, B.; Lawrence, M.G.; Muri, H. An overview of the Earth system science of solar geoengineering. *WIREs Clim. Chang.* **2016**, *7*, 815–833. [CrossRef]

21. Irvine, P.J.; Kravitz, B.; Lawrence, M.G.; Gerten, D.; Caminade, C.; Gosling, S.N.; Hendy, E.J.; Kassie, B.T.; Kissling, W.D.; Muri, H.; et al. Towards a comprehensive climate impact assessment of solar engineering. *Earth Future* **2017**, *5*, 93–106. [CrossRef]

22. Visioni, D.; Pitari, G.; Aquila, V. Sulfate geoengineering: A review of the factors controlling the needed injection of sulfur dioxide. *Atmos. Chem. Phys.* **2017**, *17*, 3879–3889. [CrossRef]

23. Caldeira, K.; Bala, G. Reflecting on 50 years of geoengineering research. *Earth Future* **2017**, *5*, 1–17. [CrossRef]

24. Boettcher, M.; Schäfer, S. Reflecting upon 10 years of geoengineering research. *Earth Future* **2017**, *5*, 266–277. [CrossRef]

25. Keith, D.W. Geoengineering the Climate: History and Prospect. *Annu. Rev. Energy Environ.* **2000**, *25*, 245–284. [CrossRef]

26. Robock, A.; Marquardt, A.; Kravitz, B.; Stenchikov, G. Benefits, risks, and costs of stratospheric geoengineering. *Geophys. Res. Lett.* **2009**, *36*, L19703. [CrossRef]

27. Robock, A.; Kravitz, B.; Boucher, O. Standardizing experiments in geoengineering. *Eos Trans. Am. Geophys. Union* **2011**, *92*, 197. [CrossRef]

28. Schmidt, H.; Alterskjær, K.; Karam, B.D.; Boucher, O.; Jones, A.; Kristjánsson, J.E.; Niemeier, U.; Schulz, M.; Aaheim, A.; Benduhn, F.; et al. Solar irradiance reduction to counteract radiative forcing from a quadrupling CO_2: Climate responses simulated by four earth system models. *Earth Syst. Dyn.* **2012**, *3*, 63–78. [CrossRef]

29. Kravitz, B.; Robock, A.; Boucher, O.; Schmidt, H.; Taylor, K.E.; Stenchikov, G.; Schulz, M. The Geoengineering Model Intercomparison Project (GeoMIP). *Atmos. Sci. Lett.* **2011**, *12*, 162–167. [CrossRef]

30. Kravitz, B.; Robock, A.; Haywood, J.M. Progress in climate model simulations of geoengineering. *Eos Trans. Am. Geophys. Union* **2012**, *93*, 340. [CrossRef]

31. Kravitz, B.; Caldeira, K.; Boucher, O.; Robock, A.; Rasch, P.J.; Alterskjær, K.; Karam, D.B.; Cole, J.N.S.; Curry, C.L.; Haywood, J.M.; et al. Climate model response from the Geoengineering Model Intercomparison Project (GeoMIP). *J. Geophys. Res.* **2013**, *118*, 8320–8332. [CrossRef]

32. Kravitz, B.; Robock, A.; Irvine, P. Robust results from climate model simulations of geoengineering. *Eos Trans. Am. Geophys. Union* **2013**, *94*, 292. [CrossRef]

33. Kravitz, B.; Robock, A.; Tilmes, S.; Boucher, O.; English, J.M.; Irvine, P.J.; Jones, A.; Lawrence, M.G.; MacCracken, M.; Muri, H.; et al. The Geoengineering Model Intercomparison Project Phase 6 (GeoMIP6): Simulation design and preliminary results. *Geosci. Model Dev.* **2015**, *8*, 2279–2292. [CrossRef]

34. MacMartin, D.G.; Keith, D.W.; Kravitz, B.; Caldeira, K. Management of trade-offs in geoengineering through optimal choice of non-uniform radiative forcing. *Nat. Clim. Chang.* **2013**, *3*, 365–368. [CrossRef]

35. Kalidindi, S.; Bala, G.; Modak, A.; Caldeira, K. Modeling of solar radiation management: A comparison of simulations using reduced solar constant and stratospheric sulfate aerosols. *Clim. Dyn.* **2015**, *44*, 2909–2925. [CrossRef]

36. Crook, J.A.; Jackson, L.S.; Osprey, S.M.; Forster, P.M. A comparison of temperature and precipitation responses to different Earth radiation management schemes. *J. Geophys. Res.* **2015**, *120*, 9352–9373. [CrossRef]

37. Qian, Y.; Jackson, C.; Giorgi, F.; Booth, B.; Duan, Q.; Forest, C.; Higdon, D.; Hou, Z.J.; Huerta, G. Uncertainty quantification in climate modelling and prediction. *Bull. Am. Meteorol. Soc.* **2016**, *97*, 821–824. [CrossRef]

38. MacMartin, D.G.; Kravitz, B.; Keith, D.W.; Jarvis, A. Dynamics of the coupled human-climate system resulting from closed-loop control of solar geoengineering. *Clim. Dyn.* **2014**, *43*, 243–258. [CrossRef]

39. Jarvis, A.J.; Young, P.C.; Leedal, D.T.; Chotai, A. A robust sequential CO_2 emissions strategy based on optimal control of atmospheric CO_2 concentrations. *Clim. Chang.* **2008**, *86*, 357–373. [CrossRef]

40. Jarvis, A.J.; Leedal, D.T.; Taylor, C.J.; Young, P.C. Stabilizing global mean surface temperature: A feedback control perspective. *Environ. Model. Softw.* **2009**, *24*, 665–674. [CrossRef]

41. Ban-Weiss, G.A.; Caldeira, K. Geoengineering as an optimization problem. *Environ. Res. Lett.* **2010**, *5*, 034009. [CrossRef]

42. Jarvis, A.; Leedal, D. The geoengineering model intercomparison project (GeoMIP): A control perspective. *Atmos. Sci. Lett.* **2012**, *13*, 157–163. [CrossRef]

43. Kravitz, B.; MacMartin, D.G.; Leedal, D.T.; Rasch, P.J.; Jarvis, A.J. Explicit feedback and the management of uncertainty in meeting climate objectives with solar geoengineering. *Environ. Res. Lett.* **2014**, *9*, 044006. [CrossRef]

44. Dykema, J.A.; Keith, D.W.; Anderson, J.G.; Weisenstein, D. Stratospheric controlled perturbation experiment: A small-scale experiment to improve understanding of the risks of solar geoengineering. *Philos. Trans. A Math. Phys. Eng. Sci.* **2014**, *372*. [CrossRef] [PubMed]

45. Weller, S.R.; Schultz, B.P. Geoengineering via solar radiation management as a feedback control problem: Controller design for disturbance rejection. In Proceedings of the 4th Australian Control Conference (AUCC), Canberra, Australia, 17–18 November 2014.

46. Bellman, R. *Dynamical Programming*; Princeton University Press: Princeton, NJ, USA, 1957.

47. Pontryagin, L.S.; Bolryanskii, V.G.; Gamktelidze, R.V.; Mishchenko, E.F. *The Mathematical Theory of Optimal Processes*; Wiley: New York, NY, USA, 1962.

48. Kirk, D. *Optimal Control Theory: An Introduction*; Prentice Hall: Englewood Cliffs, NJ, USA, 1970.

49. Sontag, E.D. *Mathematical Control Theory: Deterministic Finite Dimensional Systems*; Springer: New York, NY, USA, 1990.

50. Yusupov, R.M. *An Introduction to Geophysical Cybernetics and Environmental Monitoring*; St. Petersburg State University Press: St. Petersburg, Russia, 1998.

51. Soldatenko, S.A. Weather and climate manipulation as an optimal control for adaptive dynamical systems. *Complexity* **2017**, 1–12. [CrossRef]

52. Meinshausen, M.; Smith, S.J.; Calvin, K.; Daniel, J.S.; Kainuma, M.L.T.; Lamarque, J.-F.; Matsumoto, K.; Montzka, S.A.; Raper, S.C.B.; Riahi, K.; et al. The RCP greenhouse gas concentrations and their extensions from 1765 to 2300. *Clim. Chang.* **2011**, *109*, 213–241. [CrossRef]

53. Gregory, J.M.; Mitchell, J.F.B. The climate response to CO_2 of the Hadley Centre coupled AOGCM with and without flux adjustment. *Geophys. Res. Lett.* **1997**, *24*, 1943–1946. [CrossRef]

54. Gregory, J.M. Vertical heat transports in the ocean and their effect on time-dependent climate change. *Clim. Dyn.* **2000**, *16*, 501–515. [CrossRef]

55. Held, I.M.; Winton, M.; Takahashi, K.; Delworth, T.; Zeng, F.; Vallis, G.K. Probing the fast and slow components of global warming by returning abruptly to preindustrial forcing. *J. Clim.* **2010**, *23*, 2418–2427. [CrossRef]

56. Geoffroy, O.; Saint-Martin, D.; Olivié, D.J.L.; Voldoire, A.; Bellon, G.; Tytéca, S. Transient climate response in a two-layer energy-balance model. Part I: Analytical solution and parameter calibration using CMIP5 AOGCM experiments. *J. Clim.* **2012**, *26*, 1841–1857. [CrossRef]

57. Geoffroy, O.; Saint-Martin, D.; Bellon, G.; Voldoire, A.; Olivié, D.J.L.; Tytéca, S. Transient climate response in a two-layer energy-balance model. Part II: Representation of the efficacy of deep-ocean heat uptake and validation for CMIP5 AOGCMs. *J. Clim.* **2013**, *26*, 1859–1876. [CrossRef]

58. Gregory, J.M.; Andrews, T.; Good, P. The inconstancy of the transient climate response parameter under increasing CO_2. *Philos. Trans. R. Soc. A* **2015**, *373*, 20140417. [CrossRef] [PubMed]

59. Gregory, J.M.; Andrews, T. Variation in climate sensitivity and feedback parameters during the historical period. *Geophys. Res. Lett.* **2016**, *43*, 3911–3920. [CrossRef]

60. Farmer, G.T.; Cook, J. Earth's Albedo, Radiative Forcing and Climate Change. In *Climate Change Science: A Modern Synthesis*; Springer: Dordrecht, The Netherlands, 2018; pp. 217–229.

61. Pistone, K.; Eisenman, I.; Ramanathan, V. Observational determination of albedo decrease caused by vanishing Arctic sea ice. *Proc. Natl. Acad. Sci. USA* **2014**, *111*, 3322–3326. [CrossRef] [PubMed]

62. Calabrò, E.; Magazù, S. Correlation between Increases of the Annual Global Solar Radiation and the Ground Albedo Solar Radiation due to Desertification—A Possible Factor Contributing to Climatic Change. *Climate* **2016**, *4*, 64. [CrossRef]

63. Rutherford, W.A.; Painter, T.H.; Ferrenberg, S.; Belnap, J.; Okin, G.S.; Flagg, C.; Reed, S.C. Albedo feedbacks to future climate via climate change impacts on dryland biocrusts. *Sci. Rep.* **2017**, *7*, 44188. [CrossRef] [PubMed]

64. Kreidenweis, U.; Humpenöder, F.; Stevanović, M.; Bodirsky, B.L.; Kriegler, E.; Lotze-Campen, H.; Popp, A. Afforestation to mitigate climate change: Impacts on food prices under consideration of albedo effects. *Environ. Res. Lett.* **2016**, *11*, 085001. [CrossRef]

65. Fassnacht, S.R.; Cherry, M.L.; Venable, N.B.H.; Saavedra, F. Snow and albedo climate change impacts across the United States Northern Great Plains. *Cryosphere* **2016**, *10*, 329–339. [CrossRef]

66. Cohen, S.; Stanhill, G. Widespread Surface Solar Radiation Changes and Their Effects: Dimming and Brightening. In *Climate Change. Observed Impacts on Planet Earth*, 2nd ed.; Letcher, T.M., Ed.; Elsevier: Amsterdam, The Netherlands, 2016; pp. 491–511.

67. Kravitz, B.; Rasch, P.J.; Wang, H.; Robock, A.; Gabriel, C.; Boucher, O.; Cole, J.N.S.; Haywood, J.; Ji, D.; Jones, A.; et al. The climate effects of increasing ocean albedo: An idealized representation of solar geoengineering. *Atmos. Chem. Phys.* **2018**, *18*, 13097–13113. [CrossRef]

68. Bluth, G.J.S.; Doiron, S.D.; Schnetzler, C.C.; Krueger, A.J.; Walter, L.S. Global tracking of the SO_2 clouds from the June, 1991 Mount-Pinatubo eruptions. *Geophys. Res. Lett.* **1992**, *19*, 151–154. [CrossRef]

69. Eliseev, A.V.; Chernokulsky, A.V.; Karpenko, A.A.; Mokhov, I.I. Global warming mitigation by sulfur loading in the stratosphere: Dependence of required emissions on allowable residual warming rate. *Theor. Appl. Climatol.* **2010**, *101*, 67–81. [CrossRef]

70. Hansen, J.; Sato, M.; Ruedy, R.; Nazarenko, L.; Lacis, A.; Sato, R.; Ruedy, L.; Nazarenko, A.; Lacis, G.A.; Schmidt, G.; et al. Efficacy of climate forcing. *J. Geophys. Res.* **2005**, *110*, D18104. [CrossRef]

71. Lenton, T.M.; Vaughan, N.E. The radiative forcing potential of different climate geoengineering options. *Atmos. Chem. Phys.* **2009**, *9*, 5539–5561. [CrossRef]

72. McGuffie, K.; Henderson-Sellers, A. *A Climate Modelling Primer*, 3rd ed.; Wiley: New York, NY, USA, 2005.

73. Karper, H.; Engler, H. *Mathematics and Climate*; SIAM: Philadelphia, PA, USA, 2013.

74. Rasch, P.J.; Tilmes, S.; Turco, R.; Robock, A.; Oman, L.; Chen, C.-C.; Stenchikov, G.L.; Garcia, R.R. An overview of geoengineering of climate using stratospheric sulphate aerosols. *Philos. Trans. R. Soc. A* **2008**, *366*, 4007–4037. [CrossRef] [PubMed]

75. Myhre, G.; Highwood, E.J.; Shine, K.P.; Stordal, F. New estimates of radiative forcing due to well mixed greenhouse gases. *Geophys. Res. Lett.* **1998**, *25*, 2715–2718. [CrossRef]

76. Myhre, G.D.; Shindell, F.-M.; Bréon, W.; Collins, J.; Fuglestvedt, J.; Huang, D.; Koch, J.-F.; Lamarque, D.; Lee, B.; Mendoza, T.; et al. Anthropogenic and Natural Radiative Forcing Supplementary Material. In *Climate Change 2013: The Physical Science Basis. Contribution of Working Group I to the Fifth Assessment Report of the Intergovernmental Panel on Climate Change*; Stocker, T.F., Qin, D., Plattner, G.-K., Tignor, M., Allen, S.K., Boschung, J., Nauels, A., Xia, Y., Bex, V., Midgley, P.M., Eds.; World Meteorological Organization: Geneva, Switzerland, 2013.

77. Bryson, A.E.; Ho, Y.-C. *Appled Optimal Control: Optimization, Estimation, and Control*; Wiley: New York, NY, USA, 1975.

78. Sethi, S.P.; Thompson, G.L. *Optimal Control Theory: Application to Management Science and Economics*; Springer: New York, NY, USA, 2000.

79. Intergovernmental Panel on Climate Change (IPCC). Summary for Policymakers. In *Climate Change 2014: Mitigation of Climate Change. Contribution of Working Group III to the Fifth Assessment Report of the Intergovernmental Panel on Climate Change*; Edenhofer, O.R., Pichs-Madruga, Y., Sokona, E., Farahani, S., Kadner, K., Seyboth, A., Adler, I., Baum, S., Brunner, P., Eickemeier, B., et al., Eds.; Cambridge University Press: Cambridge, UK; New York, NY, USA, 2014.
80. Ricker, K.L.; Millar, R.J.; MacMartin, D.G. Constraints on global temperature target overshoot. *Sci. Rep.* **2017**, *7*, 14743. [CrossRef] [PubMed]

Article

A Proposal to Evaluate Drought Characteristics Using Multiple Climate Models for Multiple Timescales

Nguyen Tien Thanh

Department of Hydro-meteorological Modeling and Forecasting, Thuyloi University, 175 Tay Son, Dong Da, Hanoi, Vietnam; thanhnt@tlu.edu.vn

Received: 23 August 2018; Accepted: 18 September 2018; Published: 20 September 2018

Abstract: This study presents a method to investigate meteorological drought characteristics using multiple climate models for multiple timescales under two representative concentration pathway (RCP) scenarios, RCP4.5 and RCP8.5, during 2021–2050. The methods of delta change factor, unequal weights, standardized precipitation index, Mann–Kendall and Sen's slope are proposed and applied with the main purpose of reducing uncertainty in climate projections and detection of the projection trends in meteorological drought. Climate simulations of three regional climate models driven by four global climate models are used to estimate weights for each run on the basic of rank sum. The reliability is then assessed by comparing a weighted ensemble climate output with observations during 1989–2008. Timescales of 1, 3, 6, 9, 12, and 24 months are considered to calculate the standardized precipitation index, taking the Vu Gia-Thu Bon (VG-TB) as a pilot basin. The results show efficient precipitation simulations using unequal weights. In the same timescales, the occurrence of moderately wet events is smaller than that of moderately dry events under the RCP4.5 scenario during 2021–2050. Events classified as "extremely wet", "extremely dry", "very wet" and "severely dry" are expected to rarely occur under the RCP8.5 scenario.

Keywords: multiple climate models; standardized precipitation index (SPI); droughts; weights; Vu Gia-Thu Bon

1. Introduction

Drought is a natural hazard related to a deficiency of precipitation for an extended period that results in water shortage for some activities or for some economic sectors [1]. The meanings of "drought" depend on different perspectives of stakeholders from farmers to meteorologists [2]. Commonly, according to the studies of droughts [3–5], the concept of drought is clustered into four types consisting of meteorological, agricultural, hydrological and socioeconomic types. The Intergovernmental Panel on Climate Change (IPCC) Fourth Assessment Report [6] emphasizes that the world indeed has become more drought-prone during the past 25 years. Drought-affected areas will likely increase in frequency and severity, with implications for sustainable development (e.g., agriculture and forestry production or land degradation). Observed changes in characteristics of droughts (i.e., more intense and longer duration droughts) are widely documented for a variety of regional and ocean basin scales since the 1970s with the emphasis on tropics and substropics [6]. In comparison to the Medieval Climate Anomaly (950–1250), more megadroughts appeared in monsoon Asia and wetter conditions became dominant in arid Central Asia and the South American monsoon region during the Little Ice Age (1450–1850) [7]. Over a global scale, it is observed that the intensity and/or duration of droughts likely increase in the Mediterranean and West Africa and decrease in central North America and north-west Australia [6]. Increased drying is directly linked to higher temperatures and decreased precipitation. It is noteworthy that the palaeoclimate records show that droughts prolonging with a scale of decades or longer have been very likely a repetitive feature of the climate in several regions over the last 2000 years [6].Overall, these studies show that several extreme droughts occured in the last millennium. Moreover, it poses questions about the uncertainty of calculated results as well as how the projected droughts express their footprint on a local scale under different scenarios of climate change. There is an urgent need to find away to evaluate the droughts. This plays a central role in drought management strategies, and social responses to manage the risks and mitigate the drought impacts.

To evaluate the impacts of droughts on different fields (e.g., water shortages, concomitant shortages, crop growth), many drought indices have been developed and applied under timescales of short- or long-term. These indices widely vary from simple indices (e.g., percentage of normal precipitation) to sophisticated indices (e.g., Palmer drought severity index). Obviously, no drought index is suitable for all circumstances of a specific region. Some indices can be better suited than others for certain regional applications. Svoboda and Brian [8] showed some disadvantages of the standardized precipitation index (SPI) such as an assumption of distribution that can bias the results, particularly when examining short-duration events. In this study, however, the SPI index is still selected to calculate the drought-related components for numerous reasons: (1) the SPI index has been suggested and highlighted by the World Meteorological Organization [5], and nowadays many national meteorological services and drought monitoring centers use it (e.g., in the US [9]; in Europea [10]; and in Canada [11]); (2) the SPI is a powerful, flexible indicator based on robust underlying probability functions and it has high spatial coherence. Moreover, it is simple to calculate and needs only the required input parameter of precipitation. More importantly, Salehnia et al. [12] showed that eight precipitation-based drought indices, namely SPI, PNI (percent of normal index), DI (deciles index), EDI (effective drought index), CZI (China-Z index), MCZI (modified CZI), RAI (rainfall anomaly index), and ZSI (Z-score index) have similar trends; (3) lots of publications illustrated that precipitation data alone could explain most of the variability of drought (e.g., [13]). Furthermore, some indices (i.e., SPI, the standardized precipitation evapotranspiration index (SPEI: [14]), the Palmer drought severity index (PDSI: [15]); the self-calibrated PDSI [16] and the reconnaissance drought index (RDI: [17])) have been tested on a global scale by Spinoni et al. [18]. They showed that these indices have more difficulties than the SPI with possible error values such as a heat wave for a meteorological drought for the SPEI index or unrealistic extreme values for the RDI index; (4) the SPI index is designed to quantify the precipitation deficit for multiple timescales, reflecting the impact of drought on the water availability. For example, shorter SPI timescales (from 1 to 6 months) mainly indicate the drought index for agriculture practices like soil moisture conditions, whereas longer SPI timescales (from 12 to 48 months) indicate the drought index for hydrology like groundwater, streamflow and reservoir [5]. In this study, multiple timescales of the SPI index (1-, 3-, 6-, 9-, 12- and 24-month SPI) are considered using multiple climate models. The reason for this is that Vietnam is an agricultural country and has nearly 7000 reservoirs over the whole country. VG-TB has 6 reservoirs including the A Vuong, DakMi 4, Song Tranh 2, Song Bung 4, Song Bung 4A and Song Bung 5 reservoirs.

For climate projections, strictly speaking, the understanding of nature and its representation in climate models is mostly incomplete with sources of uncertainty (e.g., emissions of greenhouse gases, parameterization schemes of convective cloud and land surface, grid systems, map projections and climatic forcing factors). Thus, to be more reliable and accurate for climate projections, a combination of multiple climate models is widely applied. This pragmatic approach has received much attention over the availability of numerical weather and climate forecasts from institutions and centers of weather and climate research. To date, the metadata has been constructed such as the Ensemble-Based Predictions of Climate Changes and Their Impacts (ENSEMBLES) project [19] and phase 3, phase 5 and phase 6 of the Coupled Model Intercomparison Project (CMIP3, CMIP5 and CMIP6) using multiple climate models. Usually, the simplest method of constructing a multiple climate model is used with "equal weighting" in which weights equal $1/M$, where M is the number of models. In a more sophisticated manner, the approach of "unequal weighting" or "optimum weighting" is voted. Weigel et al. [1] discussed two ways of "equal weighting" and "unequal weighting". The study showed that "equal weighting" from multiple models perform better than the single models. Furthermore, the projection errors can be further eliminated by using "unequal weighting". To get this, however, single model skills and relative contributions of the joint model error are required. More importantly, Timothy et al. [20] used a statistical test to define whether an ensembles of multi models with "unequal weighting" is significantly better than without "unequal weighting". The study showed that a value for the relatively small global fraction is illustrated with the method of "unequal weighting". Sanderson et al. [18] suggested a weighting scheme to eliminate some aspects of model codependency in the ensemble. Also, a weighting strategy for an ensemble of CMIP5 was presented by Sanderson et al. [21] in the fourth National Climate Assessment. In general, these studies use a distance metric of models to observations and the distance metric of a pair of models i and j, and a relationship to convert those into a weight. The equal weighting is, remarkably, often used to develop the global ensemble scenarios as a safer and more transparent way to combine models [1], but unequal weights can be better for some areas of the global [20]. In most cases of existing ensemble members, the application of any kind of weighting to ensemble variance is mostly discarded, but only considered the weighted mean [1,22,23]. This can ignore the intermodel relationships and unexpected values, as extreme weather variables can potentially be more sensitive to changes in the variance [24]. In addition, ensemble members typically come from the same model.

In the context of changing climate, studies in drought events at scales of region and basin are valuable for understanding their evolution and impacts on a wide range of fields (e.g., agriculture, socio-economic, environment and natural resources) that occur over certain areas. In this sense, the present study aims to evaluate the wet and drought events in the 21stcentury using multiple climate models for multiple timescales. The precipitation projections from multiple regional climate models driven by multiple global climate models are separately corrected using the method of delta change factor. In addition, an unequal weight method is proposed in the expectation of a better performance of multiple climate projections of precipitation at basin scale in Vietnam as a case study. Weights for each climate simulation are calculated on the basic of rank sum metric of each climate simulation. The rank sum is defined from statistical indices of each climate simulation in comparison with observation. In comparison with the existing methods mentioned above, this approach can measure not only the absolute performance of each model, but the performance compared with the other models in the ensemble with its ranks. The methods of the non-parametric Mann–Kendall (MK) test [25,26] and Sen's slope [27] are then applied to detect the projection trends in meteorological drought for multiple timescales at a significance level of 0.05. The reason for this is that the MK test is widely applied [28–30] with advantages of a rank correlation without any request of a particular distribution of data and not affected by the data errors and outliers.

2. Materials and Methodology

2.1. Description of the Case Study Area: Vu Gia-Thu Bon Basin

A plot basin, Vu Gia-Thu Bon (VG-TB), is selected in this study. It is located in central Vietnam, elongating from 16°55′ through 14°55′ and from 107°15′ through 108°24′ and covers a total of area of approximately 12,577 km². The VG-TB basin is surrounded by two main provincial administrative territories Quang Nam and Da Nang. The basin is characterized by a steep topography and the altitude ranging from 0 m at the coast to 2567 m in elevation above sea level (m.a.s.l) in the west (Figure 1).

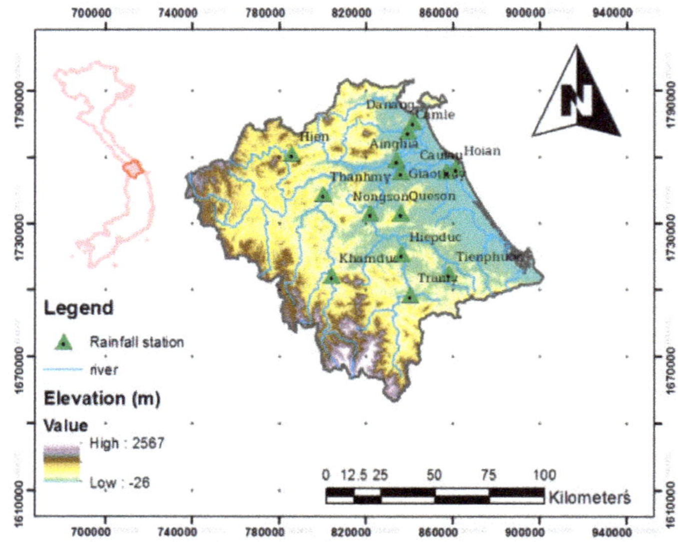

Figure 1. Network of hydro-meteorological stations at Vu Gia-Thu Bon (VG-TB) basin.

2.2. Data

2.2.1. Observational Data

The monthly precipitation records are obtained from the Vietnam HydroMeteorological Data Center of the Ministry of National Resources and Environment of Vietnam (MONRE). They are aggregated from the daily series of data. There are two national rain gauge stations (i.e., Danang and Tramy). Other stations including Ainghia, Camle, Giaothuy, Caulau, Hien, Hiepduc, Hoian, Khamduc, Nongson, Queson, Thanhmy and Tienphuoc are popular rain gauge stations which operate manually on the base of volunteers. The location of these stations is displayed on Figure 1. The data is available from 1986–2015.

2.2.2. Gridded Data

The precipitation products are from different assembliese of regional models: (1) The Regional Climate Model version 4 (RegCM4), developed by the International Centre for Theoretical Physics (ICTP). The dynamical structure of RegCM4 firstly developed at the National Center of Atmosphere Research (NCAR) and Penn State University (PSU) for a hydrostatic version of the Meso-scale Model (MM5). A detailed description of RegCM4 can be found in Giorgi et al. [31]. The model output of HadGEM2-AO produced by the National Institute of Meteorological Research (NIMR)/Korea Meteorological Administration (KMA) are used as an initial and boundary conditions, referred to REG/HadGEM. Details of HadGEM2-AO are given by Collins et al. [32]; (2) The model of SNU-MM5(Seoul National University Meso-scale Model version 5) [33] is based on a hydrostatic version of the Meso-scale Model and the community land model version 3 (CLM3). The future climatic projections are produced with the HadGEM2-AO, referred to SNU/HadGEM; (3) A regional spectral model, which is also known as Regional Model Program (RMP) of the Global/Regional Integrated Model System (GRIMs) [34] is used in this study. The dynamic frame of RMP is rooted in the National Center for Environmental Prediction (NCEP) RSM. More detailed information about the GRIMs-RMP is provided by Hong et al. [34]. This model is driven by the HadGEM2-AO, referred to RSM/HadGEM; (4) The RegCM4 is forced by the model of Max Planck Institute for Meteorology Earth System Model MR (MPI-ESM-MR), which has an ocean horizontal resolution of $0.4° \times 0.4°$ and atmosphere horizontal resolution of $1.9° \times 1.9°$. It is written under a short symbol of REG/MPI; (5) The RegCM4 is forced by the model of Institut Pierre Simon Laplace CM5A-LR (IPSL-CM5A-LR), which is the low-resolution version of the IPSL-CM5A Earth system model. It has a horizontal resolution of 1.875° × 3.75° with 39 vertical level for the atmosphere and about 2° (with a meridional increased resolution of 0.5° near the equator) and with 31 vertical levels for the ocean. In this study, it implies to REG/IPSL; (6) The RegCM4 is forced by the model of Irish Centre for High-End Computing European community Earth-System (ICHEC-EC-EARTH), which is a new Earth system model on the basic of the operational seasonal forecast system of the European Centre for Medium-Range Weather Forecasts (ECMWF). This case is written with a short symbol of the REG/ICHEC. More importantly, all simulations and projections of climatic are run under two IPCC RCP4.5/8.5 scenarios. The RCP4.5 is a stabilization scenario where total radiative forcing is stabilized before 2100 by employment of a range of technologies and strategies for reducing greenhouse gas emissions. Meanwhile, the RCP8.5 is characterized by increasing greenhouse gas emissions over time representative for scenarios in the literature leading to high greenhouse gas concentration. Table 1 lists the name of the models and the number of runs of historical control and RCPs as well as the considered periods. A total of 18 climatic simulations and projections are considered in this study as shown in Table 1.

Table 1. Climate models and number of runs.

	Historical (1986–2005)	RCP4.5 (2021–2050)	RCP8.5 (2021–2050)	Spatial Resolution	Temporal Resolution
RegCM4 forced by MPI-ESM-MR (REG/MPI)	1	1	1	20 km	Monthly
RegCM4 forced by IPSL-CM5A-LR (REG /IPSL)	1	1	1	20 km	Monthly
RegCM4 forced by ICHEC-EC-EARTH (REG/ICHEC)	1	1	1	20 km	Monthly
RegCM4 forced by HadGEM2-AO (REG/HadGEM)	1	1	1	20 km	Monthly
SNU-MM5 forced by HadGEM2-AO (SNU/HadGEM)	1	1	1	20 km	Monthly
RSM forced by HadGEM2-AO (RSM/HadGEM)	1	1	1	20 km	Monthly

2.3. Methods

2.3.1. Delta Change Factor

The raw climate simulations, especially for precipitation time series, are highly biased as mentioned in many studies [6,35,36]. Thus, an additional post-processing step (e.g., bias correction of climatic variables) is a standard procedure for related climate change studies. In this study, the delta change method is adopted due to its simple and common use as described in Olsso et al. [37], Lenderink et al. [38], Teutschbein and Seibert [36] and Maraun [35]. This approach does not adjust the output of climate models, but uses observations and the change signal of regional climate models forced by global climate models to generate future data. Also, this approach is to avoid considerable variability in the day-to-day change signals and changes in extremes are linearly scaled with changes in the mean. The core of this method is that the historical observations are transformed into future projections using monthly average correction factors that derived from the regional climate models forced by global climate model outputs for the baseline and future climate. It is expressed by the equation:

$$P_{OBS.f} = \frac{\overline{P_{GCM.f}}}{\overline{P_{GCM.b}}} \cdot P_{OBS.b} \tag{1}$$

where $\overline{P_{GCM.f}}$ is the monthly precipitation from the future climate. $\overline{P_{GCM.b}}$ is the monthly precipitation from the baseline climate. $P_{OBS.f}$ is future projections and $P_{OBS.b}$ is historical observations.

With this approach, the correlation structure of downscaled future data in spatio-temporal terms are physically reasonable because it reflects observed conditions. The drawback of this approach, however, is that the future and baseline scenarios just differ in terms of their means and intensity, but other statistics of the data (e.g., skewness or structure of dry and wet days) are mostly unchanged. Also, the sample is limited to the length of the observed record. It should be noted that for a near-term future (2021–2050), the changes in the dry/wet days are probably trivial. Sun et al. [39] used the lag 1 autocorrelation to investigate the stationary land-based gridded annual precipitation (1940–2009). The results showed a stationary annual precipitation over an area of about 80% of the global land surface. Wilks and Wilby [40] show that calculating the autocorrelation of non-zero precipitation amounts is essential at time step of hourly (or sub-hourly) rather than a daily time step. Meanwhile, the autocorrelation between successive nonzero precipitation amounts is usually of little practical importance and quite small. At a time step of monthly precipitation, thus, a stationary assumption in the temporal correlation is made in this study. In other words, rescaling the precipitation time series observed during the baseline could lead to not realistic results when future scenarios that preserve the observed autocorrelation in time are not considered.

2.3.2. Unequal Weights

In this study, a method is suggested to estimate the unequal weights based on the rank sum from each climate simulation run for the analytical hierarchy process. It is called unequal weights. The rank sum is calculated using the statistical indices. The statistical indices are Nash–Sutcliffe efficiency with logarithmic value (ln(Nash)), root-mean-square error (RMSE) and Nash–Sutcliffe efficiency (Nash) for the studied domain. They are basically quantified by measuring the difference between the observed data and the outputs of regional climate models forced by lateral and surface boundary conditions from the European Centre for Medium-Range Weather Forecasts at the monthly scale. The Nash–Sutcliffe efficiency with logarithmic values ln(Nash) is selected because it can be added to expect a better quantification of the performance in different conditions (maximum and minimum values) [41]. The other remaining value is widely applied in hydrometeorological fields. The formulations used to compute the goodness-of-fit statistical indicators for each climate simulation run are presented as follows:

$$\text{RMSE} = \sqrt{\frac{1}{n}\sum_{i=1}^{n}(F_i - O_i)^2} \tag{2}$$

$$\ln(\text{Nash}) = 1 - \frac{\sum_{1}^{n}(\ln O_i - \ln F_i)^2}{\sum_{1}^{n}(\ln O_i - \ln \overline{O})^2} \tag{3}$$

$$\text{Nash} = 1 - \frac{\sum_{1}^{n}(O_i - F_i)^2}{\sum_{1}^{n}(O_i - \overline{O})^2} \tag{4}$$

where n denotes number of months, F_i and O_i represent simulated and observed monthly data, respectively. The method of unequal weights is briefly expressed with the following steps with an assumption of N climate simulations:

(1) Calculation of the statistical indices on the basic of the historical observations and climate simulations from regional climate models forced by the reanalysis data of the European Centre

for Medium-Range Weather Forecasts during 1989–2008. Each climate simulation receives a rank from 1 to N depending on the levels of perfect score for each statistical index, starting with the best as 1 and the worst is N. As an example, if the RMSE index of the ith climate model has the best score (the perfect score of RMSE is zero), the received rank is 1. Then, an ensemble rank order (r) as an integer number is calculated from the average of the ranks they span for each climate simulation.

(2) Calculation of rank sum for each climate simulation by N-r + 1 with N is the number of climate simulations.

(3) Establishing a reciprocal matrix between sets of models $a_{ij} = 1/a_{ji}$ with i,j ranging from 1 corresponding to the best climate simulation which has the largest rank sum to N and $a_{ij} = 1$ as i = j. a_{ij} is determined by the difference of rank sum between sets of models plus 1.

(4) Estimation of weights matrix $w_{ij} = a_{ij}/\sum_1^N a_{ij}$ (i,j = 1, N)

(5) Estimation of each weight for each climate simulation $w_i = \sum_{j=1}^N w_{ji}$ (i = 1, N) with $\sum_{i=1}^N w_i = 1$.

2.3.3. Standardized Precipitation Index (SPI)

In this study, the SPI is constructed for multiple timescales ranging from 1 month to 24 months. The calculation of the SPI is separately applied for each month on the basis of the Gamma distribution. The reason for this is that the Gamma distribution is the most frequently used and fits well with daily precipitation in different studies across Vietnam [42]. The alpha and beta parameters of the Gamma probability density function are estimated for each station and for the required multiple timescales. They are used to calculate the cumulative probability distribution of accumulated precipitation. The maximum likelihood solutions [43] and Thom's study [44] are applied to optimally estimate alpha (α) and beta (β). It is especially noteworthy that the reference period adopted to compute the best-fit parameters for the gamma distribution spans 30 years (1986–2015).

A probability transformation is then applied to transform monthly precipitation to a standard normal distribution with a zero mean and standard deviation of one to yield SPI values by preserving probabilities [45]. Figure 2 shows the fitness of the SPI data [46].

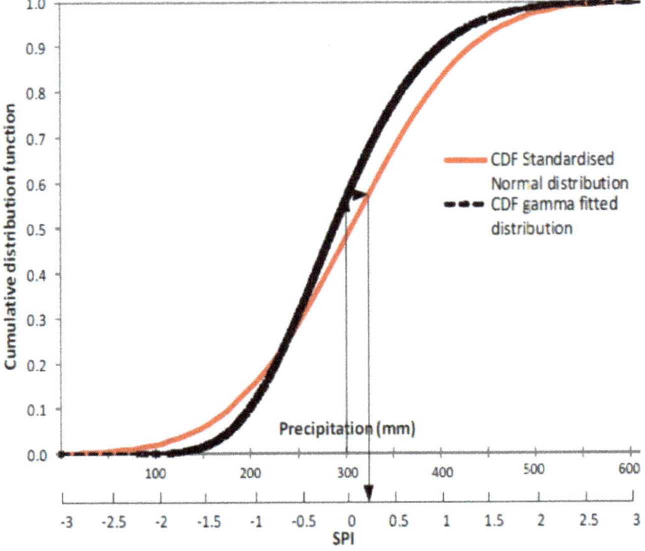

Figure 2. The probability transformation from fitted gamma distribution to the standard normal distribution.

2.3.4. Non-Parametric Mann–Kendall Test

The non-parametric Mann–Kendall (MK) test [25,26] is widely applied to detect the possible trends in many countries such as in China [47], in Serbia [48] in Brazil [49], in Canada [50]. In this study, the MK test statistic, S, is applied and briefly represented by:

$$S = \sum_{k}^{n-1} \sum_{j=k+1}^{n} sign(x_j - x_k) \tag{5}$$

where n is the number of data points; x_j and x_k are the data values in time series j and k respectively, and

$$sign(x_j - x_k) = \begin{cases} +1 & \text{if } x_j - x_k > 0 \\ 0 & \text{if } x_j - x_k = 0 \\ -1 & \text{if } x_j - x_k < 0 \end{cases} \tag{6}$$

In this test, the null hypothesis (H_o) assumes that there is no trend in meteorological droughts over time; the alternative hypothesis (H_1) assumes that there is an upward or downward trend over time. The mathematical equations for calculating Var(S) and standardized test statistics Z are presented in previous studies [25,26,47,51,52]. An upward, downward, or no trend will be assessed at α significance level of 0.05. The computed probability is greater than the specific significance level α (H_o is rejected); the increasing trend responds to a positive value of Z and a negative value of Z indicates a decreasing trend. There is no trend if the computed probability is less than the level of significance (H_o is accepted). At the α significance level of 0.05, the null hypothesis of no trend is rejected if $|Z_{MK}| > 1.96$.

2.3.5. The Sen's Slope Estimator

In order to get the magnitude of a consistent trend (Q), the Sen's non-parametric method [19] is used. It is estimated by the slope of all data pairs (N = n(n − 1)/2) as the following formula:

$$Q_i = \frac{x_j - x_k}{j - k} \text{ for } i = 1, N \tag{7}$$

where Q is slope between data points x_j and x_k, x_j and x_k are the data values at time j and k (j > k) respectively, j is time after time k. The Sen's is computed by the median slope as:

$$Q_{med} = \begin{cases} Q_{[\frac{N+1}{2}]} & \text{if N is odd} \\ \frac{Q_{[\frac{N}{2}]} + Q_{[\frac{N+2}{2}]}}{2} & \text{if N is even} \end{cases} \tag{8}$$

where N is the number of calculated slopes.

The confidence limits for the nonparameter slope estimator are estimated by the well-known studies [27,29,53,54].

3. Results and Discussions

3.1. Calculation of Weights for Each Climate Model

As the first step, the delta change factor is applied in this study. Then, the unequal weights are added to define weights for each climate simulations as mentioned in Table 1 over the whole VG-TB. Table 2 shows the statistical indices for these climate simulations. For RMSE index, the perfect score is zero with a wide range ($0 \leq$ RMSE $< \infty$). While the perfect score of Nash and ln(Nash) indices is 1 with a wide span ($-\infty \leq$ Nash < 1; $-\infty \leq$ ln(Nash) < 1). Based on these score, ranks (i.e., ensemble rank, rank sum) are calculated and presented in Table 3. It is observed that REG/IPSL gives the best score of ln(Nash), RMSE and Nash. This means that REG/IPSL simulation is well matched to observations.

As opposed to this, RSM/HadGEM gives the worst score of ln(Nash), RMSE and Nash. Although REG/IPSL is the best, a multiple climate model is always highlighted in the climate simulations and projections due to reducing uncertainty [7]. The reason for this is that the uncertainty comes from lots of factors such as convective cloud parameterization, greenhouse gases scenarios or land surface parameterization. The best rank sum belongs to the model REG/IPSL, followed by REG/ICHEC. In contrary to this, the worst rank sum belongs to the model RSM/HadGEM as shown in Table 3.

Table 2. Statistical indices for multiple climate models.

	ln(Nash)	RMSE	Nash
REG/ICHEC	0.935	170.67	0.523
REG/IPSL	0.941	157.44	0.594
REG/MPI	0.937	171.90	0.516
REG/HadGEM	0.891	256.72	−0.079
SNU/HadGEM	0.913	252.75	−0.046
RSM/HadGEM	0.880	278.93	−0.274

Table 3. Ranks of climate models.

	ln(Nash)	RMSE	Nash	Average	Ensemble Rank	Rank Sum
REG/ICHEC	3	2	2	2.3	2	5
REG/IPSL	1	1	1	1.0	1	6
REG/MPI	2	3	3	2.7	3	4
REG/HadGEM	5	5	4	4.7	5	2
SNU/HadGEM	4	4	5	4.3	4	3
RSM/HadGEM	6	6	6	6.0	6	1

A reciprocal matrix between sets of models is created based on the rank sum of climate models. As shown in Table 4, it is illustrated that REG/IPSL is more important than REG/ICHEC and much more important than REG/MPI. Meanwhile, REG/ICHEC is more important than REG/MPI and much important than SNU/HadGEM. The least importance within all considered climate simulations is RSM/HadGEM, followed by REG/HadGEM.

Table 4. A reciprocal matrix between sets of models.

	REG/IPSL	REG/ICHEC	REG/MPI	SNU/HadGEM	REG/HadGEM	RSM/HadGEM	Total
REG/IPSL	1	2	3	4	5	6	21
REG/ICHEC	1/2	1	2	3	4	5	15.5
REG/MPI	1/3	1/2	1	2	3	4	10.83
SNU/HadGEM	1/4	1/3	1/2	1	2	3	7.08
REG/HadGEM	1/5	1/4	1/3	1/2	1	2	4.28
RSM/HadGEM	1/6	1/5	1/4	1/3	1/2	1	2.45
Total							**61.15**

After establishing a weights matrix, weights are estimated for each climate model based on the 4th and 5th steps in Section 2.3.2. Weights are 0.344 (REG/IPSL), 0.254 (REG/ICHEC), 0.177 (REG/MPI), 0.116 (SNU/HadGEM), 0.07 (REG/HadGEM) and 0.038 (RSM/HadGEM). The results from this method are against the results from the method of equal weights. Both of them are compared with observations using the statistical indices of mean absolute error (MAE) and RMSE. The pair of equal weights and observations gives results of 99.3 for MAE and 164 for RMSE, meanwhile a pair of unequal weights and observations gives results of 95.4 and 155 for MAE and RMSE, respectively.

The perfect score of MAE and RMSE is zero. In other words, the minimum of the RMSE and MAE is obtained when simulated time series perfectly matches observed time series. As mentioned in the delta change factor approach, an assumption of the delta factor is equal to one (i.e., there is no variation) for each climate runs, and the condition of minimum of the RMSE and MAE is obtained.

Climate models with convective precipitation schemes, however, often tend to produce higher precipitation [6,55] compared with the observed data. Moreover, incomplete understanding of natural and its representation is presented within the climate models. Thus, the delta factor assumed by one is unreal, especially in Vietnam where the rain regime is strongly dominated by convective processes. The calculated statistics of MAE and RMSE indicate a better implementation of unequal weights. Climate simulations are closely fit to observations with the unequal weights method as presented in Figure 3.

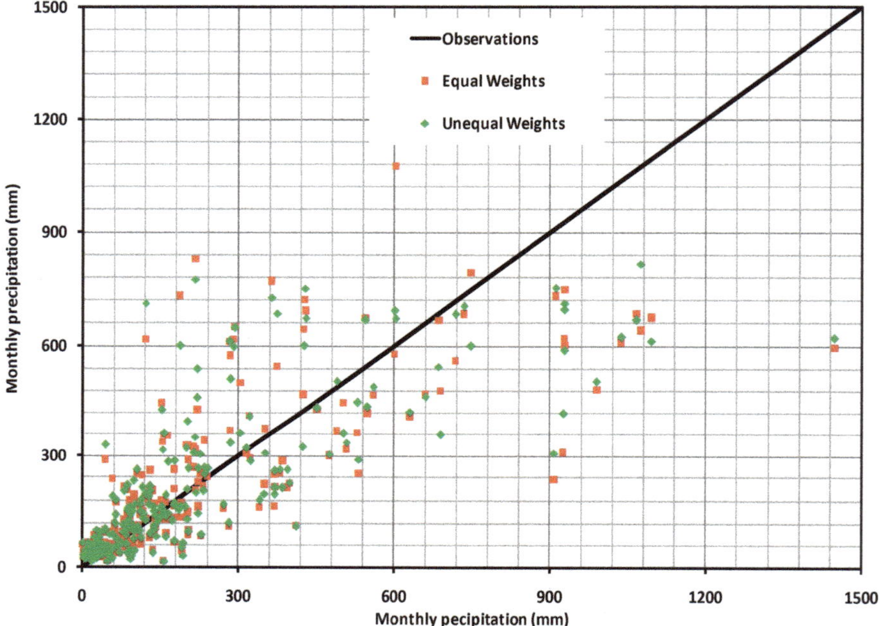

Figure 3. Scatter plot ofthe precipitation simulations with equal and unequal weights minus observations over the whole VG-TB during 1989–2008 (mm/month).

3.2. Calculation of SPI

The weights estimated from Section 3.1 are used to calculate the SPI for 14 stations over the whole VG-TB. It is noted that the weights are applied for both scenarios of RCP4.5 and RCP8.5 during 2021–2050 with an assumption of no change in future. The SPI is designed to project the precipitation deficit for multiple timescales with a weighted ensemble of multiple climate models in future. SPI values are computed for the timescales of 1- (SPI-1), 3- (SPI-3), 6- (SPI-6), 9- (SPI-9), 12- (SPI-12) and 24- (SPI-24) month for 14 stations. For instance, the temporal variation of drought for the timescales of 1 to 24months under RCP4.5 (Figure 4) and RCP8.5 (Figure 5) at Danang station from 2021 to 2050 is displayed.

According to the classification of SPI values [56], generally, there are no events classified as "extremely wet" and "extremely dry" for the Danang station under both scenarios of RCP4.5 and RCP8.5 during 2021–2050 for considered multiple timescales. Under these scenarios, events classified as "near normal" hold more than 90% out of considered months for all timescales of 14 stations. In particular, events classified as "extremely wet" and "extremely dry" are not found out for the all stations under both scenarios of RCP4.5 and RCP8.5 during 2021–2050 for SPI-1, SPI-3, SPI-6, SPI-9, SPI-12 and SPI-24 indices. More importantly, under RCP4.5 scenario, events classified as "very wet" and "severely dry" are lower than 0.6% (mainly in stations located in the east of basin as Caulau) (see Supplementary Material). Meanwhile there is a little change (less than 2.8%) in events classified as "very wet", "moderately wet", "moderately dry" and "severely dry" for SPI-1 for Ainghia, Camle, Giaothuy, Danang and Hien stations under the RCP8.5 scenario. Under this scenario, the remaining stations have no change in drought and wet events for all considered multiple time scales (see Supplementary Material). This indicates only little changes in drought and wet events in the eastern domain of basin under RCP8.5 for considered time scales. "Moderately wet"events for the SPI-1 range from 1.4% (Hien, Hoian and Khamduc stations) to 2.8% (Danang station) under the RCP4.5 scenario. Hence, an emphasis of events classified as "moderately wet" and "moderately dry" for multiple timescales over the whole basin under the RCP4.5 scenario during 2021–2050 is documented in Table 5.

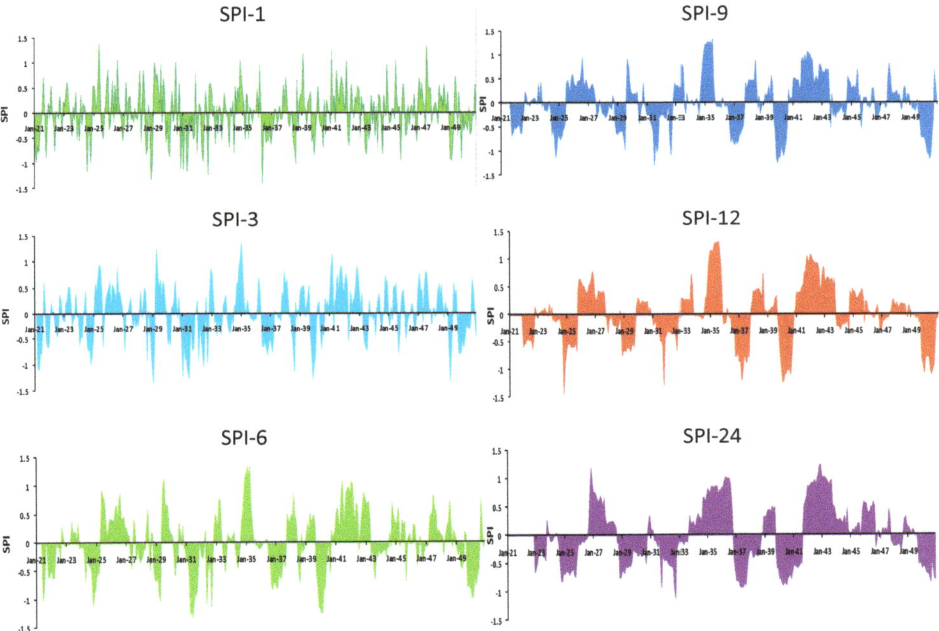

Figure 4. Standardized precipitation index (SPI)for multiple timescales under RCP4.5 scenario (2021–2050) for Danang station.

As shown in Table 5, the maximum of moderately dry event is 6.5% for SPI-24, followed by 6.3% and 6.0% for SPI-12 and SPI-9, respectively. The median of moderately dry is projected to get the maximum value of 4.2% for SPI-24, followed by 4.0% and 3.7% for SPI-9 and SPI-12, respectively. Importantly, the maximum of moderately wet event is 5.9% for SPI-24, followed by 3.7% and 3.4% for SPI-12 and SPI-9, respectively. It is noteworthy that the values reported from Table 5 and Tables S1–S6 are assigned by the percentages of drought and wet events under considered multiple timescales. In other words, these values are defined as the percentage of number of months suffering wet and drought events among all months during a 30-year period under considered scenarios. Under the RCP4.5 scenario, in general, timescales are longer, intensity of moderately dry is stronger. In the same timescale, occurrence of moderately wet events is smaller than that of moderately dry under RCP4.5 scenario during 2021–2050.

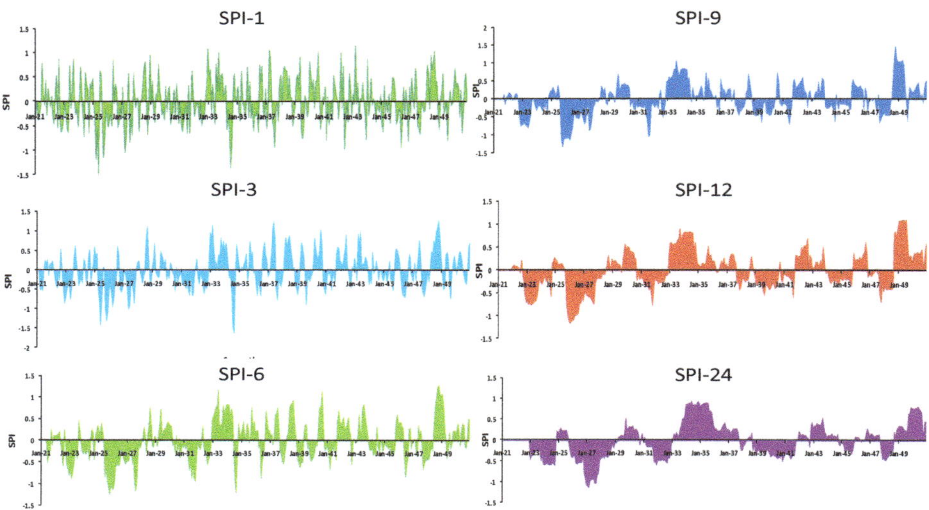

Figure 5. SPI for multiple timescales under RCP8.5 scenario (2021–2050) for Danang station.

Table 5. Percentages of drought and wet events under considered multiple timescales under RCP4.5 scenario over basin during 2021–2050 (%).

	Classification	Min	25%	Median	75%	Max
SPI-1	Moderatelywet	1.4	1.5	2.1	2.2	2.8
	Moderately dry	1.4	1.7	2.1	2.2	3.1
SPI-3	Moderatelywet	0.8	1.2	1.4	1.9	2.2
	Moderately dry	1.1	2.6	3.1	3.4	3.9
SPI-6	Moderately wet	0.6	1.1	1.3	1.9	2.5
	Moderately dry	1.4	3.7	3.8	4.2	5.1
SPI-9	Moderately wet	0.3	0.3	1.3	2.4	3.4
	Moderately dry	1.7	3.3	4.0	4.5	6.0
SPI-12	Moderately wet	0.0	0.3	0.7	1.8	3.7
	Moderately dry	0.9	3.2	3.7	4.4	6.3
SPI-24	Moderately wet	0.3	1.8	2.4	3.5	5.9
	Moderately dry	0.0	0.4	4.2	5.5	6.5

3.3. Projection Trends in SPI

Projection trends in SPI for multiple timescales are considered with an emphasis of drought events under the RCP4.5 scenario. The purpose of this is to detect any trends or changes in drought events. In addition, as clarified in the previous section, just little, even no wet and drought events of extremely wet, extremely dry, very wet and severely dry are found out under the RCP8.5 scenarios. During 2021–2050, out of 14 stations, only five show significant upward trends in November for the SPI-1 index as presented in Figure 6b. There is a similar of upward trends for SPI-3 and SPI-6 in November. Importantly, no trends are detected in November for SPI-9, SPI-12 and SPI-24. Besides that, it is noteworthy that a poor signal of upward trends in SPI-9, SPI-12 and SPI-24 indices is illustrated for months in a year. It indicates an importance of short-term, plans more than long-term plans in related aspects such as water resources planning.

In February, insignificant downward trends are mostly observed over the whole basin (Figure 6a). In the months of March, May, July, October and December, more than half of total considered stations are defined with no trends in SPI-1 index. No trends in SPI-3 index are calculated for almost all the months in a year, except for January, April and November. Downward trends in SPI-1 (in the months of February, April and June), SPI-3 (in April) and SPI-6 (in the months of July and August) indices are clearly observed. More importantly, the intensity of drought events in the north and east of the basin is generally stronger than that in other areas.

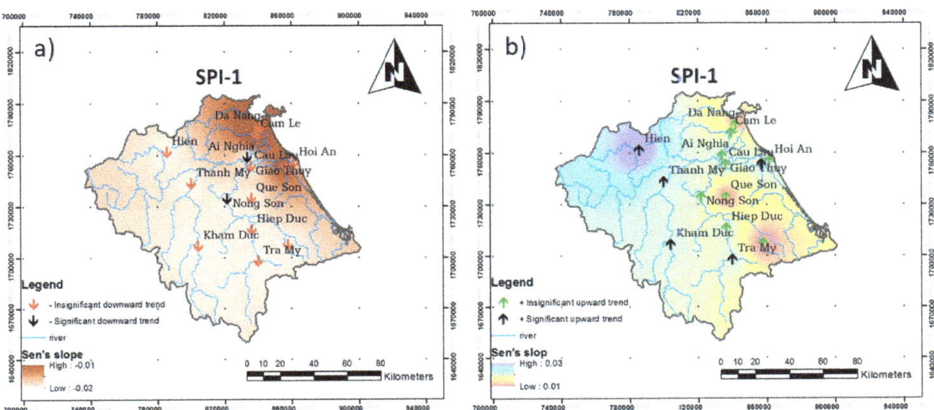

Figure 6. Spatial distribution of Sen's slope in February (**a**) and in November (**b**) for SPI-1 index.

4. Conclusions

This study presents a proposed approach to evaluate drought characteristics using multiple climate models for multiple timescales under a context of global warming. The method of unequal weights is proposed to reduce uncertainty of climate projections. The Mann–Kendall and Sen's slope methods are then applied to find the trends in drought characteristics for multiple timescales of 1, 3, 6, 9, 12, and 24 months with six climate projections. The unequal weights are proposed on the basic of rank sum of each climate model. Vu Gia-Thu Bon basin located in central Vietnam is selected to implement this study as a pilot basin. The major findings of the present study can be exposed as an impulse of the precipitation simulations using unequal weights for multiple climate models. Under the scenarios of RCP4.5 and RCP8.5, there are no events classified as "extremely wet" and "extremely dry" for the all stations during 2021–2050 for SPI-1, SPI-3, SPI-6, SPI-9, SPI-12 and SPI-24 indices. A higher magnitude of the drought conditions in the north and east of the basin compared toother areas is found out. Under the RCP4.5 scenario, more importantly, timescales are longer, and the intensity of moderately dry is stronger. Moreover, the occurrence of moderately wet events is smaller than that of moderately dry under the RCP4.5 scenario during 2021–2050 in the same timescales. Under the RCP8.5 scenario, events classified as "extremely wet", "extremely dry", "very wet" and "severely dry" are expected to rarely occur.

Supplementary Materials: The following are available online at http://www.mdpi.com/2225-1154/6/4/79/s1.

Funding: This research received no external funding.

Conflicts of Interest: The author declares no conflicts of interest.

References

1. Weigel, A.P.; Knutti, R.; Liniger, M.A.; Appenzeller, C. Risks of model weighting in multimodel climate projections. *J. Clim.* **2010**, *23*, 4175–4191. [CrossRef]
2. Moreland, J. *Drought: Us Geological Survey Water Fact Sheet*; Open-File Report; U.S. Geological Survey: Reston, VA, USA, 1993; pp. 93–642.

3. Dracup, J.A.; Lee, K.S.; Paulson, E.G., Jr. On the definition of droughts. *Water Resour. Res.* **1980**, *16*, 297–302. [CrossRef]

4. Wilhite, D.A.; Glantz, M.H. Understanding: The drought phenomenon: The role of definitions. *Water Int.* **1985**, *10*, 111–120. [CrossRef]

5. WMO. *Standardized Precipitation Index—User Guide*; WMO-No. 1090; WMO: Geneva, Switzerland, 2012.

6. IPCC. *Climate Change 2007—The Physical Science Basis: Working Group I Contribution to the Fourth Assessment Report of the IPCC*; Cambridge University Press: Cambridge, UK, 2007; Volume 4.

7. IPCC. *Climate Change 2013: The Physical Science Basis: Working Group I Contribution to the Fifth Assessment Report of the Intergovernmental Panel on Climate Change*; Cambridge University Press: Cambridge, UK, 2014.

8. Svoboda, M.; Fuchs, B. *Handbook of Drought Indicators and Indices*; WMO-No. 1173; WMO: Geneva, Switzerland, 2016.

9. Svoboda, M.; LeComte, D.; Hayes, M.; Heim, R.; Gleason, K.; Angel, J.; Rippey, B.; Tinker, R.; Palecki, M.; Stooksbury, D. The drought monitor. *Bull. Am. Meteorol. Soc.* **2002**, *83*, 1181–1190. [CrossRef]

10. Vogt, J.; Barbosa, P.; Hofer, B.; Magni, D.; Jager, A.; Singleton, A.; Horion, S.; Sepulcre, G.; Micale, F.; Sokolova, E. *Developing a European Drought Observatory for Monitoring, Assessing and Forecasting Droughts across the European Continent*; AGU Fall Meeting Abstracts; Copernicus Publications: Göttingen, Germany, 2011.

11. Anctil, F.; Larouche, W.; Viau, A.; Parent, L.-E. Exploration of the standardized precipitation index with regional analysis. *Can. J. Soil Sci.* **2002**, *82*, 115–125. [CrossRef]

12. Salehnia, N.; Alizadeh, A.; Sanaeinejad, H.; Bannayan, M.; Zarrin, A.; Hoogenboom, G. Estimation of meteorological drought indices based on agmerra precipitation data and station-observed precipitation data. *J. Arid Land* **2017**, *9*, 797–809. [CrossRef]

13. Ntale, H.K.; Gan, T.Y. Drought indices and their application to East Africa. *Int. J. Climatol.* **2003**, *23*, 1335–1357. [CrossRef]

14. Vicente-Serrano, S.M.; Beguería, S.; López-Moreno, J.I. A multiscalar drought index sensitive to global warming: The standardized precipitation evapotranspiration index. *J. Clim.* **2010**, *23*, 1696–1718. [CrossRef]

15. Palmer, W.C. *Meteorological Drought*; Research Paper No. 45; US Department of Commerce Weather Bureau: Washington, DC, USA, 1965; p. 59.

16. 1Wells, N.; Goddard, S.; Hayes, M.J. A self-calibrating palmer drought severity index. *J. Clim.* **2004**, *17*, 2335–2351. [CrossRef]

17. Tsakiris, G.; Pangalou, D.; Vangelis, H. Regional drought assessment based on the reconnaissance drought index (rdi). *Water Resour. Manag.* **2007**, *21*, 821–833. [CrossRef]

18. Spinoni, J.; Naumann, G.; Carrao, H.; Barbosa, P.; Vogt, J. World drought frequency, duration, and severity for 1951–2010. *Int. J. Climatol.* **2014**, *34*, 2792–2804. [CrossRef]

19. Weisheimer, A.; Doblas-Reyes, F.; Palmer, T.; Alessandri, A.; Arribas, A.; Déqué, M.; Keenlyside, N.; MacVean, M.; Navarra, A.; Rogel, P. Ensembles: A new multi-model ensemble for seasonal-to-annual predictions—Skill and progress beyond demeter in forecasting tropical pacific ssts. *Geophys. Res. Lett.* **2009**, *36*. [CrossRef]

20. DelSole, T.; Yang, X.; Tippett, M.K. Is unequal weighting significantly better than equal weighting for multi-model forecasting? *Q. J. R. Meteorol. Soc.* **2013**, *139*, 176–183. [CrossRef]

21. Sanderson, B.M.; Wehner, M.; Knutti, R. Skill and independence weighting for multi-model assessments. *Geosci. Model Dev.* **2017**. [CrossRef]

22. Knutti, R.; Abramowitz, G.; Collins, M.; Eyering, V.; Gleckler, P.J.; Hewitson, B.; Mearns, L.O. Good Practice Guidance Paper on Assessing and Combining Multi Model Climate Projections. In Proceedings of the IPCC Expert Meeting on Assessing and Combining Multi Model Climate Projections, Boulder, CO, USA, 28–27 January 2010; p. 1.

23. Krishnamurti, T.N.; Kishtawal, C.; Zhang, Z.; LaRow, T.; Bachiochi, D.; Williford, E.; Gadgil, S.; Surendran, S. Multimodel ensemble forecasts for weather and seasonal climate. *J. Clim.* **2000**, *13*, 4196–4216. [CrossRef]

24. Taylor, K.E.; Stouffer, R.J.; Meehl, G.A. An overview of cmip5 and the experiment design. *Bull. Am. Meteorol. Soc.* **2012**, *93*, 485–498. [CrossRef]

25. Kendall, M.G. *Rank Correlation Methods*; Hafner Publishing Co.: Oxford, UK, 1955.

26. Mann, H.B. Nonparametric tests against trend. *Econ. J. Econ. Soc.* **1945**, 245–259. [CrossRef]

27. Sen, P.K. Estimates of the regression coefficient based on kendall's tau. *J. Am. Stat. Assoc.* **1968**, *63*, 1379–1389. [CrossRef]

28. Douglas, E.; Vogel, R.; Kroll, C. Trends in floods and low flows in the united states: Impact of spatial correlation. *J. Hydrol.* **2000**, *240*, 90–105. [CrossRef]

29. Helsel, D.R.; Hirsch, R.M. *Statistical Methods in Water Resources*; Elsevier: New York, NY, USA, 1992; Volume 49.

30. Hirsch, R.M.; Slack, J.R.; Smith, R.A. Techniques of trend analysis for monthly water quality data. *Water Resour. Res.* **1982**, *18*, 107–121. [CrossRef]

31. Giorgi, F.; Coppola, E.; Solmon, F.; Mariotti, L.; Sylla, M.; Bi, X.; Elguindi, N.; Diro, G.; Nair, V.; Giuliani, G. Regcm4: Model description and preliminary tests over multiple cordex domains. *Clim. Res.* **2012**, *52*, 7–29. [CrossRef]

32. Collins, W.; Bellouin, N.; Doutriaux-Boucher, M.; Gedney, N.; Halloran, P.; Hinton, T.; Hughes, J.; Jones, C.; Joshi, M.; Liddicoat, S. Development and evaluation of an earth-system model—Hadgem2. *Geosci. Model Dev.* **2011**, *4*, 1051–1075. [CrossRef]

33. Lee, D.-K.; Cha, D.-H.; Kang, H.-S. Regional climate simulation of the 1993 summer flood over East Asia. *J. Meteorol. Soc. Jpn. Ser. II* **2004**, *82*, 1735–1753. [CrossRef]

34. Hong, S.-Y.; Park, H.; Cheong, H.-B.; Kim, J.-E.E.; Koo, M.-S.; Jang, J.; Ham, S.; Hwang, S.-O.; Park, B.-K.; Chang, E.-C. The global/regional integrated model system (grims). *Asia-Pac. J. Atmos. Sci.* **2013**, *49*, 219–243. [CrossRef]

35. Maraun, D. Bias correcting climate change simulations—A critical review. *Curr. Clim. Chang. Rep.* **2016**, *2*, 211–220. [CrossRef]

36. Teutschbein, C.; Seibert, J. Is bias correction of regional climate model (rcm) simulations possible for non-stationary conditions? *Hydrol. Earth Syst. Sci.* **2013**, *17*, 5061–5077. [CrossRef]

37. Olsson, J.; Berggren, K.; Olofsson, M.; Viklander, M. Applying climate model precipitation scenarios for urban hydrological assessment: A case study in Kalmar City, Sweden. *Atmos. Res.* **2009**, *92*, 364–375. [CrossRef]

38. Lenderink, G.; Buishand, A.; Deursen, W.V. Estimates of future discharges of the river rhine using two scenario methodologies: Direct versus delta approach. *Hydrol. Earth Syst. Sci.* **2007**, *11*, 1145–1159. [CrossRef]

39. Sun, F.; Roderick, M.L.; Farquhar, G.D. Rainfall statistics, stationarity, and climate change. *Proc. Natl. Acad. Sci. USA* **2018**, *115*, 2305–2310. [CrossRef] [PubMed]

40. Wilks, D.S.; Wilby, R.L. The weather generation game: A review of stochastic weather models. *Prog. Phys. Geogr.* **1999**, *23*, 329–357. [CrossRef]

41. Krause, P.; Boyle, D.; Bäse, F. Comparison of different efficiency criteria for hydrological model assessment. *Adv. Geosci.* **2005**, *5*, 89–97. [CrossRef]

42. Nguyen, T.T.; Remo, L.D.A. Projected changes of precipitation idf curves for short duration under climate change in central Vietnam. *Hydrology* **2018**, *5*, 33.

43. Choi, S.C.; Wette, R. Maximum likelihood estimation of the parameters of the gamma distribution and their bias. *Technometrics* **1969**, *11*, 683–690. [CrossRef]

44. Thom, H.C. A note on the gamma distribution. *Mon. Weather Rev.* **1958**, *86*, 117–122. [CrossRef]

45. Edwards, D.C. *Characteristics of 20th Century Drought in the United States at Multiple Time Scales*; Air Force Inst of Tech: Wright-Patterson AFB, OH, USA, 1997.

46. Azam, M.; Maeng, S.; Kim, H.; Lee, S.; Lee, J. Spatial and temporal trend analysis of precipitation and drought in South Korea. *Water* **2018**, *10*, 765. [CrossRef]

47. Yue, S.; Wang, C. The mann-kendall test modified by effective sample size to detect trend in serially correlated hydrological series. *Water Resour. Manag.* **2004**, *18*, 201–218. [CrossRef]

48. Gocic, M.; Trajkovic, S. Analysis of changes in meteorological variables using mann-kendall and sen's slope estimator statistical tests in Serbia. *Glob. Planet. Chang.* **2013**, *100*, 172–182. [CrossRef]

49. Blain, G.C. Removing the influence of the serial correlation on the mann-kendall test. *Revista Brasileira de Meteorologia* **2013**, *29*. [CrossRef]

50. Yue, S.; Pilon, P.; Phinney, B.; Cavadias, G. The influence of autocorrelation on the ability to detect trend in hydrological series. *Hydrol. Process.* **2002**, *16*, 1807–1829. [CrossRef]

51. Burn, D.H.; Elnur, M.A.H. Detection of hydrologic trends and variability. *J. Hydrol.* **2002**, *255*, 107–122. [CrossRef]

52. Yue, S.; Pilon, P.; Cavadias, G. Power of the mann-kendall and spearman's rho tests for detecting monotonic trends in hydrological series. *J. Hydrol.* **2002**, *259*, 254–271. [CrossRef]

53. Gilbert, R.O. *Statistical Methods for Environmental Pollution Monitoring*; John Wiley & Sons: Hoboken, NJ, USA, 1987.

54. Theil, H. A rank-invariant method of linear and polynomial regression analysis. In *Henri Theil's Contributions to Economics and Econometrics*; Springer: Berlin, Germany, 1992; pp. 345–381.

55. Grell, G.A.; Dudhia, J.; Stauffer, D.R. A Description of the Fifth-Generation Penn State/Ncar Mesoscale Model (mm5). 1994. Available online: http://danida.vnu.edu.vn/cpis/files/Books/MM5%20Discription%20-%201995.pdf (accessed on 23 August 2018).

56. McKee, T.B.; Doesken, N.J.; Kleist, J. The relationship of drought frequency and duration to time scales. In Proceedings of the 8th Conference on Applied Climatology, Anaheim, CA, USA, 17–22 January 1993; American Meteorological Society: Boston, MA, USA, 1993; pp. 179–183.

Article

Spatial and Temporal Rainfall Variability over the Mountainous Central Pindus (Greece)

Stefanos Stefanidis * and Dimitrios Stathis

Faculty of Forestry and Natural Environment, Aristotle University of Thessaloniki, Laboratory of Mountainous Water Management and Control, 54124 Thessaloniki, Greece; dstatis@for.auth.gr
* Correspondence: ststefanid@gmail.com; Tel.: +30-2310-992712

Received: 19 July 2018; Accepted: 5 September 2018; Published: 6 September 2018

Abstract: In this study, the authors evaluated the spatial and temporal variability of rainfall over the central Pindus mountain range. To accomplish this, long-term (1961–2016) monthly rainfall data from nine rain gauges were collected and analyzed. Seasonal and annual rainfall data were subjected to Mann–Kendall tests to assess the possible upward or downward statistically significant trends and to change-point analyses to detect whether a change in the rainfall time series mean had taken place. Additionally, Sen's slope method was used to estimate the trend magnitude, whereas multiple regression models were developed to determine the relationship between rainfall and geomorphological factors. The results showed decreasing trends in annual, winter, and spring rainfalls and increasing trends in autumn and summer rainfalls, both not statistically significant, for most stations. Rainfall non-stationarity started to occur in the middle of the 1960s for the annual, autumn, spring, and summer rainfalls and in the early 1970s for the winter rainfall in most of the stations. In addition, the average magnitude trend per decade is approximately −1.9%, −3.2%, +0.7%, +0.2%, and +2.4% for annual, winter, autumn, spring, and summer rainfalls, respectively. The multiple regression model can explain 62.2% of the spatial variability in annual rainfall, 58.9% of variability in winter, 75.9% of variability in autumn, 55.1% of variability in spring, and 32.2% of variability in summer. Moreover, rainfall spatial distribution maps were produced using the ordinary kriging method, through GIS software, representing the major rainfall range within the mountainous catchment of the study area.

Keywords: rainfall; trend analysis; Mann–Kendall; kriging interpolation

---

1. Introduction

Rainfall is the most important meteorological and climatological parameter for natural ecosystems and human life on earth, as it affects the enrichment of lakes and underground aquifers, river flow regime, and many natural hazards (floods, drought, landslides, etc.). Accurate knowledge of the spatial and seasonal variations of long-term rainfall time series is required for rural and forest development and planning, sustainable development, as well as infrastructure work scheduling.

The Mediterranean basin has a wide range of climatic conditions [1]. In particular, the rainfall regime in Greece presents a highly irregular behavior, both on spatial and temporal scales, namely in rainfall amount and rainfall distribution [1,2]. It is well accepted that the main physical and physico-geographical factors controlling the spatial distribution of rainfall over Greece are the following: the atmospheric circulation, the mountains in the west and east, the Mediterranean Sea-surface temperature distribution, the dehumidification of the air masses crossing the Aegean Sea, and land and sea interactions [3]. Furthermore, the highest rainfall totals for western Greece were found to be related to the atmospheric circulation associated with the Mediterranean Sea-surface temperature distribution and the complex topography of the region, as imposed by the orography of the Pindus Mountains in northwestern and central Greece and the mountains of Olympus and Crete [4].

Mountainous areas are of great interest, because runoff is generated and supplies lowlands (through catchments) with water. Moreover, the plain areas receive the eroded material deposited by mountainous catchments, due to intense rainfall. Variability is considered particularly higher in a mountainous environment, because the rainfall pattern is influenced by complex terrain conditions [5–7]. The assessment of climate variability is a common issue that should be treated by hydrologists; in particular, the total rainfall in an ungauged site over an area (e.g., catchment) should be evaluated. However, hydrologists face a crucial challenge when it comes to mountainous terrains, since data from only a few meteorological stations are usually available.

To overcome the lack of rainfall data, interpolation methods have been developed over the last few years for rainfall modeling and mapping. These methods are based on the similarity and the topological relationship between nearby sample points and on the value of the variable to be measured [8]. Interpolation can be achieved using simple methods (splines, inverse distance weighting, Thiessen polygons, etc.) or advanced geostatistical methods (e.g., kriging). Geostatistical interpolation has become the most appropriate downscale technique in applied climatology and for areas with complex terrain, since it is based on the spatial variability of the variables of interest and allows the quantification of the estimation uncertainty [9–11].

In recent decades, the interest in climate variability and climate change has augmented. Climate change has emerged as a key issue facing environmental and economic aspects, as it affects floods [12], soil erosion [13], drought phenomena [14], agriculture [15], tourism [16], groundwater aquifers [17], and forest fires [18].

According to IPCC reports [19], the Mediterranean basin is expected to become warmer and drier due to the anthropogenic increase of greenhouse gas (GHG) emissions (CO_2, CH_4, N_2O, and F-gases), until the end of the 21st century [20,21]. Moreover, in Mediterranean regions, future warming is expected to be greater than the global mean, accompanied by a significant decrease in rainfall [22]. Based on the above, researchers are orienting their work to investigate trends in rainfall conditions [23–28] and to estimate future rainfalls [29,30] within Greece. Research results highlighted the decreasing trend of rainfalls recorded from long-term time series analysis, whereas this reduction is expected to be higher in the future, based on regional climate models (RCMs) that have been proposed. Even though much research has been conducted in Greece on trend analysis and spatial mapping of rainfall, only limited research efforts concern mountainous areas with consideration given to long-term time series using a dense network of stations. The identification and recording of seasonal trends can improve water resources management through the selection of appropriate management practices.

The main object of this study was to detect annual and seasonal variation and trends in rainfall time series based on data from rain gauge stations located in mountainous areas. Furthermore, variation and uncertainty in the small-scale rainfall interpolation in mountainous catchments were also evaluated.

2. Materials and Methods

2.1. Study Area

The study was conducted over the central Pindus mountain range, in Central Greece. The area is considered highly important from a hydrological point of view because it is located in the mountainous area of two hydrological basins (Pinios and Acheloos), which supply Central Greece with water, and where many hydropower dams have been constructed. For this purpose, a dense network of meteorological stations (compared to other regions in Greece) has been established, in mountainous terrain. The characteristics of the meteorological stations used in this study are given in Table 1.

Table 1. Meteorological stations in the study area.

A/A	Meteorological Station	Coordinates		Altitude (m)	Period (years)
		Longitude(°)	Latitude(°)		
1	Agiofylo	21.34	39.52	581	1961–2016
2	Chrysomilia	21.3	39.36	940	1961–2016
3	Elati	21.32	39.51	900	1961–2016
4	Katafyto	21.28	39.38	980	1961–2016
5	Malakasi	21.17	39.47	849	1961–2016
6	Mesochora	21.20	39.26	849	1961–2016
7	Pertouli	21.28	39.33	1180	1961–2016
8	Polyneri	21.22	39.34	801	1961–2016
9	Stournareika	21.29	39.28	761	1961–2016

Observations of monthly rainfall totals for a period of 55 years of rainfall (1961–2016) were used from all nine stations of the wider region (see Figure 1). These stations are equipped with pluviometer and Hellmann-type rain gauges (Fuess Meteorologische Instrumente KG, Königs Wusterhausen, Germany) with a precision of 0.1 mm. The data series are complete, that is, they have no missing values. Moreover, the instruments and observing practices were common among all stations used, and they remained the same during this study's research. The double mass method and two parametric statistical tests (Student's *t*-test and chi-squared test) were applied to adjust any heterogeneity of the rainfall data, and the details regarding these methods can be obtained from the WMO [31]. The latter tests demonstrated that the precipitation data were indeed homogeneous and ready to be entered into the subsequent procedures of the study.

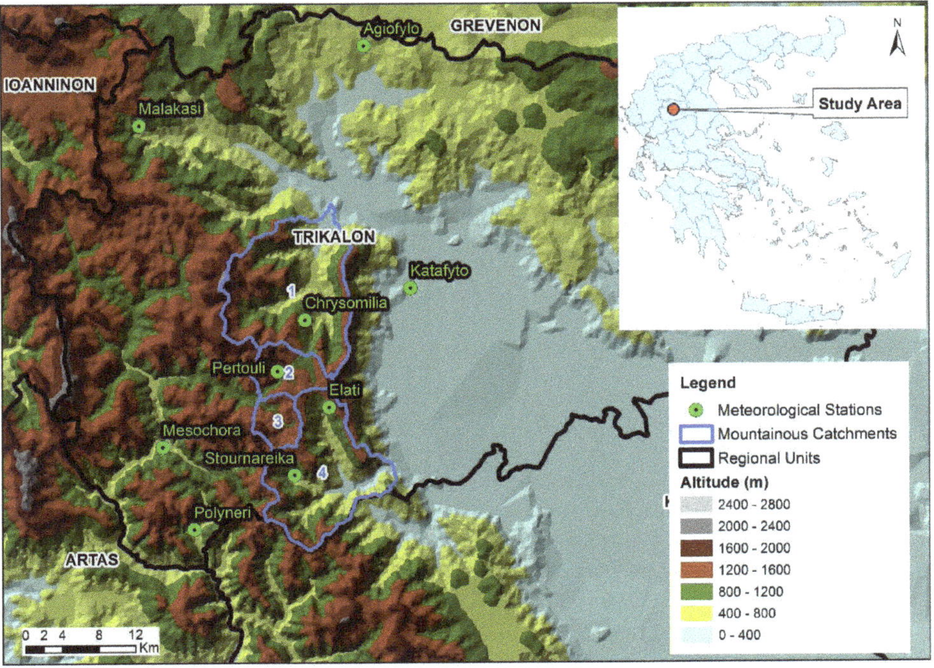

Figure 1. The study area and location of stations.

These stations are operated by the Ministry of Environment & Energy (Agiofylo, Chrysomilia, Elati, Katafyto, and Malakasi), the Public Power Corporation (Mesochora, Polyneri, and Stournareika) and the University Forest Administration and Management Fund (Pertouli).

The study area is an area of increased importance, because it is located in the mountainous area of two hydrological basins (Pinios and Acheloos), which supply Central Greece with fresh water. The mountainous catchments examined within this study are (1) Klinovitikos, (2) Aspropotamos, (3) Korpos, and (4) Portaikos, as showed in Figure 1. Additionally, the basic morphometrical and hydrographical characteristics are given in Table 2.

Table 2. Morphometrical and hydrographical characteristics of the mountainous catchments.

	Catchment Name			Klinovitikos	Aspropotamos	Korpos	Portaikos
	Code			(1)	(2)	(3)	(4)
A/A	Morphometrical Characteristics	Symbol	Units				
1	Area	F	km^2	171.1	36.3	23.7	136.4
2	Perimeter	U	km	61.7	28.1	20.1	60
3	Minimum elevation	H_{min}	m	320	1020	1020	240
4	Maximum elevation	H_{max}	m	2204	2074	1721	1862
5	Mean elevation	H_{med}	m	1112	1420	1397	963
6	Mean catchment slope	J_λ	%	48.4	41.6	39.3	52.93
	Hydrographic Characteristics						
7	Density of hydrographic network	D	km/km^2	2.86	3.13	3.36	2.69
8	Main stream length	L	km	20.2	12.4	8.3	16.9
9	Main stream slope	J_κ	%	6.5	5.6	9.6	7.8

The study area is characterized as mountainous, whereas the relief is rather intense. Regarding geology, the main rocks are flysch and limestones, quite vulnerable to landslides and weathering phenomena. The forest cover is high and distributed to the mountainous catchment as follows: (1) Klinovitikos, 66%; (2) Aspropotamos, 73%; (3) Korpos, 72%; and (4) Portaikos, 44%. The dominant forest species in the study area are *Abies borisii-regis*, *Quercus frainetto*, *Quercus petraea*, *Pinus nigra*, and *Fagus Sylvatica*. Moreover, the study region is of great environmental importance, belonging to the European nature conservation network Natura 2000 according to the criteria of Directive 92/43/EEC.

2.2. Trend Analysis

Time series of annual and seasonal rainfall were subjected to the Mann–Kendall test to detect possible trends over the period of 1961–2016. It is the most widely used test for trend analysis in climatological time series [32].

The Mann–Kendall test is a non-parametric statistical test to detect the presence of a monotonic increasing or decreasing trend within a time series [33,34]. The advantage of the non-parametric tests over the parametric tests is that they are robust and more suitable for non-normally distributed data with missing and extreme values, frequently encountered in environmental time series [35].

The Mann–Kendall test statistics S is calculated as:

$$S = \sum_{k=1}^{n-1} \sum_{j=k+1}^{n} sign(x_j - x_k), \tag{1}$$

where n is the number of data points, x_i and x_k are the data values in the time series j and k $(j > k)$, respectively, and sign $(x_j - x_i)$ is the sign function as follows:

$$sign(x_j - x_k) = \begin{cases} 1 \ if \ (x_j - x_k) > 0 \\ 0 \ if \ (x_j - x_k) = 0 \\ -1 \ if \ (x_j - x_k) < 0 \end{cases} . \tag{2}$$

The variance is computed as:

$$VAR(S) = \frac{n(n-1)(2n+5) - \sum_{i=1}^{P} t_i(t_i - 1)(2t_i + 5)}{18}, \tag{3}$$

where n is the number of data points, P is the number of tied groups, the summary sign (P) indicates the summation over all tied groups, and t_i is the number of data values in the Pth group. In case of no tied groups, this summary process can be ignored. A tied group is a set of sample data having the same value. In the case where the sample size $n > 30$, the standard normal test statistic Z is estimated by:

$$Z = \begin{cases} \frac{S-1}{\sqrt{VAR(S)}} \ if \ S > 0 \\ 0 \ if \ S = 0 \\ \frac{S+1}{\sqrt{VAR(S)}} \ if \ S < 0 \end{cases} . \tag{4}$$

Positive values of Z indicate increasing trends, whereas negative Z values indicate decreasing trends. Trend testing is done at a specific significance level. When $|Z| > Z_{1-a/2}$, the null hypothesis is rejected and a significant trend exists in the time series. The value of $Z_{1-a/2}$ is obtained from the standard normal distribution table. In this study, the significance level $\varepsilon = 0.05$ was used. At the 5% significance level, the null hypothesis of no trend is rejected if $|Z| > 1.96$.

Furthermore, a change-point analysis approach was applied, using the Change-Point Analyzer (CPA) [36]. This method iteratively uses a combination of cumulative sum charts (CUSUM) and bootstrapping to detect whether a change in the mean of the rainfall time series has taken place. A sudden change in the direction of the CUSUM indicates a sudden shift or change in the average. Additionally, trend magnitudes were computed by employing the Theil–Sen approach (TSA) [37,38], which is based on slope β, often referred to as Sen's slope [38]. It is preferable to linear regression, because it limits the influence of outliers on the slope [39].

2.3. Spatial Mapping of Rainfall

Initially, the relationship between altitude and rainfall height (mm) for both an annual and a seasonal basis was evaluated using different types of trendlines (linear, logarithmic, polynomial, power, and exponential). Moreover, the spatial distribution was determined, applying multiple regression equation and taking into account not only the altitude but also the longitude and latitude. The multiple linear regression equation has the following form:

$$P = a + b_1 X_1 + b_2 X_2 + b_3 X_3, \tag{5}$$

where P represents the rainfall (mm), a is constant, $b_1 \dots b_3$ are coefficients obtained for each independent variable, X_1 is longitude (°), X_2 is latitude (°), and X_3 is altitude (m).

Furthermore, the geostatistical interpolation method of ordinary kriging (spherical variogram) was employed, using the ArcGIS 10.2 software. At this point, it should be noted that geostatistical methods are more valid for increasing sample size. To this end, automatic points were generated in a 1 km × 1 km grid resolution within the catchments, using the Fishnet command of the ArcGIS 10.2 software's Data Management toolbar. Therefore, rainfall height was calculated for each point and all seasons, based on the multiple regression equation described above and the calculation of the individual variables for each point.

Finally, cross-validation was performed, in order to compare results of rainfall spatial interpolation derived from ordinary kriging with other spatial interpolation methods, for example, inverse distance weighting (IDW), radial basis function (RBF), and universal kriging (UK), and a combination of variograms (spherical, exponential), using the Geostatistical Wizard tool of ArcGIS [40].

Cross-validation is any of various similar model validation techniques for assessing how the results of a statistical analysis will generalize to an independent dataset. It is mainly used in settings where the goal is prediction, and where one wants to estimate how accurately a predictive model will perform in practice. In a prediction problem, a model is usually given a dataset of known data on which training is run (training dataset), and a dataset of unknown data (or first seen data) against which the model is tested. The goal of cross-validation is to test the model's ability to predict new data that were not used in estimating it, in order to flag problems like overfitting and to give an insight on how the model will generalize to an independent dataset.

The root mean square error (RMSE) and mean error were used as evaluation indexes in this case study. The mathematical description of these indexes is given below:

$$RMSE = \sqrt{\frac{1}{n}\sum_{i=1}^{n}(y_i - x_i)^2} \, , \tag{6}$$

$$MAE = \frac{1}{n}\sum_{i=1}^{n}(y_i - x_i), \tag{7}$$

where n is the number of observations, and x_i and y_i are the observed and interpolated rainfall values, respectively, for $i = 1, 2 \ldots$ n. The RMSE is considered one of the most reliable indexes because it depicts the deviation from the truth rather than the mean value, as in the case of standard deviation. The RMSE gives the weighted variations (residuals) in errors between the estimated and observed values, whereas mean error measures the weighted average magnitude of the errors. Mean error is the most natural and unambiguous measure of average error magnitude [41,42]. RMSE, on the other hand, is one of the most widely used error measures [43].

3. Results

The rainfall pattern in the case study area demonstrated certain particularities and varied greatly in both space and time, in line with the main characteristics of the climate type in the Mediterranean basin. The seasonal distribution of rainfall based on the examined meteorological stations' data is shown in Figure 2. As depicted, 35% of the annual rainfall occurs during winter, 32% in autumn, 24% in spring, and only 9% in summer.

The results of the Mann–Kendall statistics test indicated that most of the meteorological stations (around 67%) recorded a downward trend in annual rainfall, which could be considered as statistically significant for the Katafyto station. In addition, decreasing trends of rainfall time series were recorded in winter and autumn for most of the stations. During spring, half of the stations revealed a decreasing trend, whereas the other half revealed an increasing one. Finally, summer was the only time season when rainfall trends were recorded increasing in most of the stations.

Detailed results of the application of the Mann–Kendall test are given in Table 3. The upward arrow ↑ indicates an increasing trend, whereas the downward arrow ↓ indicates a decreasing one. Furthermore, the light grey cell color shows that the trend is not statistically significant (for a significance level of a = 0.05), whereas the dark grey color shows a statistically significant trend. In addition, the number within the parenthesis indicates the time of occurrence for changes in the mean of the rainfall time series.

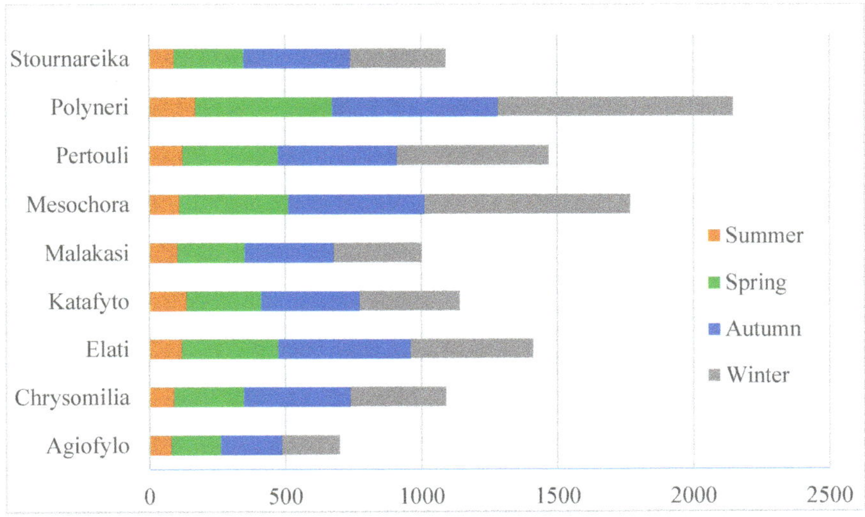

Figure 2. Seasonal distribution of rainfall.

Table 3. Trend detection using the Mann–Kendall test and the change-point analysis.

a/a	Meteorological Station	Period				
		Annual	Winter	Autumn	Spring	Summer
1	Agiofylo	↑ (86)	↑ (66)	↑ (96)	↑ (62)	↑ (62)
2	Chrysomilia	↓ (67)	↓ (89)	↑ (62)	↑ (64)	↑ (68)
3	Elati	↓ (65)	↓ (88)	↓ (63)	↓ (96)	↑ (62)
4	Katafyto	↓ (75)	↓ (01)	↓ (63)	↓ (64)	↑ (63)
5	Malakasi	↓ (65)	↓ (74)	↓ (63)	↓ (64)	↑ (62)
6	Mesochora	↑ (67)	↓ (72)	↑ (73)	↑ (71)	↑ (68)
7	Pertouli	↓ (65)	↓ (71)	↑ (62)	↓ (62)	↓ (71)
8	Polyneri	↑ (64)	↓ (70)	↑ (62)	↑ (64)	↑ (63)
9	Stournareika	↓ (64)	↓ (70)	↑ (63)	↑ (63)	↓ (70)

As shown in Table 3, the rainfall non-stationarity starts to occur in the middle of 1960s for the annual, autumn, spring, and summer rainfalls and the early 1970s for the winter rainfall in most of the stations.

Moreover, Sen's slope was used to compute the trend magnitude per decade, which ranged from approximately −5.3% to +1.5% (average −1.9%) in annual rainfalls, from −14.5% to +7.5% (average −3.2%) in winter, from −4.7% to +5.5% (average +0.7%) in autumn, from −4.2% to +3.6% (average +0.2%) in spring, and from −0.3% to +4.8% (average +2.4%) in summer. Detailed results for each station are given in the next table (see Table 4).

The investigation of the relationships between the variation of rainfall and altitude showed that the derived coefficients of determination are rather low. This indicates that only a small percentage (19–25%) of rainfall variation in the study area was due to the change in altitude. Furthermore, it was observed that the power trendline performed the best fit in all cases. In the equations below (see Table 5), y is the rainfall (mm) and x is altitude (m).

Table 4. Trend magnitude (%) per decade using Sen's slope method.

a/a	Meteorological Station	Trend Magnitude (% per Decade)				
		Annual	**Winter**	**Autumn**	**Spring**	**Summer**
1	Agiofylo	−5.3%	7.5%	5.3%	2.9%	4.6%
2	Chrysomilia	−0.4%	−2.1%	0.9%	2.3%	1.3%
3	Elati	−4.3%	−2.2%	−0.2%	−4.2%	0.1%
4	Katafyto	−5.1%	−14.5%	−4.7%	−0.3%	9.7%
5	Malakasi	−0.7%	−3.7%	−1.2%	−1.5%	4.8%
6	Mesochora	1.5%	−1.9%	5.5%	3.6%	0.9%
7	Pertouli	−3.1%	−5.5%	0.3%	−3.8%	−0.3%
8	Polyneri	1.4%	−3.1%	0.1%	1.5%	0.8%
9	Stournareika	−1.1%	−2.9%	0.4%	0.9%	−0.3%

Table 5. Relationship between rainfall and altitude.

	Trend Equation	R^2
Annual	$y = 6.14x^{0.8}$	0.21
Winter	$y = 0.60x^{0.9}$	0.18
Autumn	$y = 2.63x^{0.6}$	0.25
Spring	$y = 2.94x^{0.7}$	0.19
Summer	$y = 2.57x^{0.6}$	0.21

For that reason, the spatial variability of both seasonal and annual rainfall was assessed using a multiple regression analysis. All the necessary factors affecting rainfalls were included into the multiple regression procedure, including longitude, latitude, and altitude. The coefficient obtained for each factor, based on a regression analysis, is given in Table 6.

Table 6. Coefficients of multiple regression models and statistics.

	Coefficients				
	Annual	**Winter**	**Autumn**	**Spring**	**Summer**
a	77,660	35,750	22,056	16,815	2963
b_1	−0.018	−0.010	−0.003	−0.004	−0.001
b_2	−0.016	−0.007	−0.005	−0.003	−0.001
b_3	0.21	0.06	0.07	0.05	0.03
Adjusted R^2	0.62	0.59	0.76	0.55	0.32

It is noteworthy that the multiple regression model can explain 62.2% of the spatial variability of the annual rainfall, 58.9% of variability in winter, 75.9% of variability in autumn, 55.1% of variability in spring, and 32.2% of variability in summer. In order to evaluate the statistical significance of the examined factors, p-values were estimated (see Table 7). In cases where the significant level of the examined factor is less than 95% ($p > 0.05$), the factor should be eliminated from the model and the multi-linear regression must be performed again. In this study, the p-values for all factors were less than 0.05, which means a strong presumption against null hypothesis.

Table 7. Output p-values of the examined coefficients.

	p-Values				
	Annual	**Winter**	**Autumn**	**Spring**	**Summer**
a	0.021	0.031	0.005	0.030	0.028
b_1	0.046	0.038	0.026	0.028	0.046
b_2	0.028	0.041	0.007	0.041	0.031
b_3	0.048	0.049	0.039	0.044	0.049

Furthermore, regarding the results of cross validation amongst different spatial interpolation methods it was revealed that better results were achieved by Ordinary Kriging combined with spherical semivariogram (Table 8).

Table 8. Cross-validation results from the interpolation of annual and seasonal rainfall.

	Annual	Winter	Autumn	Spring	Summer
Inverse Distance Weighting					
Mean Error	78.52	33.82	24.57	16.98	3.04
RMSE	315.04	148.19	95.99	63.67	24.79
Radial Basis Function					
Mean Error	38.78	16.87	12.28	8.08	1.42
RMSE	282.96	128.60	83.62	53.87	20.38
Ordinary Kriging (Spherical)					
Mean Error	16.58	7.32	4.58	3.35	0.38
RMSE	254.24	117.13	76.57	48.92	20.22
Ordinary Kriging (Exponential)					
Mean Error	22.02	9.55	6.93	4.68	0.43
RMSE	268.34	123.06	79.76	51.44	22.86
Universal Kriging (Spherical)					
Mean Error	16.62	8.11	−4.76	−8.85	−2.41
RMSE	255.70	119.20	77.80	73.26	25.88
Universal Kriging (Exponential)					
Mean Error	−49.77	−27.67	−12.50	4.52	−5.42
RMSE	384.76	192.78	110.46	52.60	29.54

The semivariogram/covariance cloud tool shows the empirical semivariogram and covariance values for all pairs of locations within a dataset and plots them as a function of the distance that separates the two locations. It can be used to examine the local characteristics of spatial autocorrelation within a dataset and look for local outliers. The selection of lag size has an important effect on the semivariogram. If the lag size is too large, the short-range autocorrelation may be masked, whereas if the lag size is too small, there may be many empty bins. A rule of thumb is to multiply the lag size times the number of lags, which should be about half of the largest distance among all points.

Important characteristics of the semivariogram are also the nugget and the partial sill. The nugget is a parameter of covariance or semivariogram model that represents independent error, and a microscale variation at spatial scales that are too fine to detect. As for the partial sill, it is a parameter that represents the variance of a spatially autocorrelated process without any nugget effect. These parameters for the spherical variogram that was used in this study are given in the following table (see Table 9).

Regarding the semivariograms (see Figure 3), it can be assumed that the phenomenon to estimate is smooth (i.e., rainfall values change gradually with the distance). The semivariogram represents the continuity structure quite well also. Additionally, the semivariogram diagrams showed that the samples did not show autocorrelation in any direction.

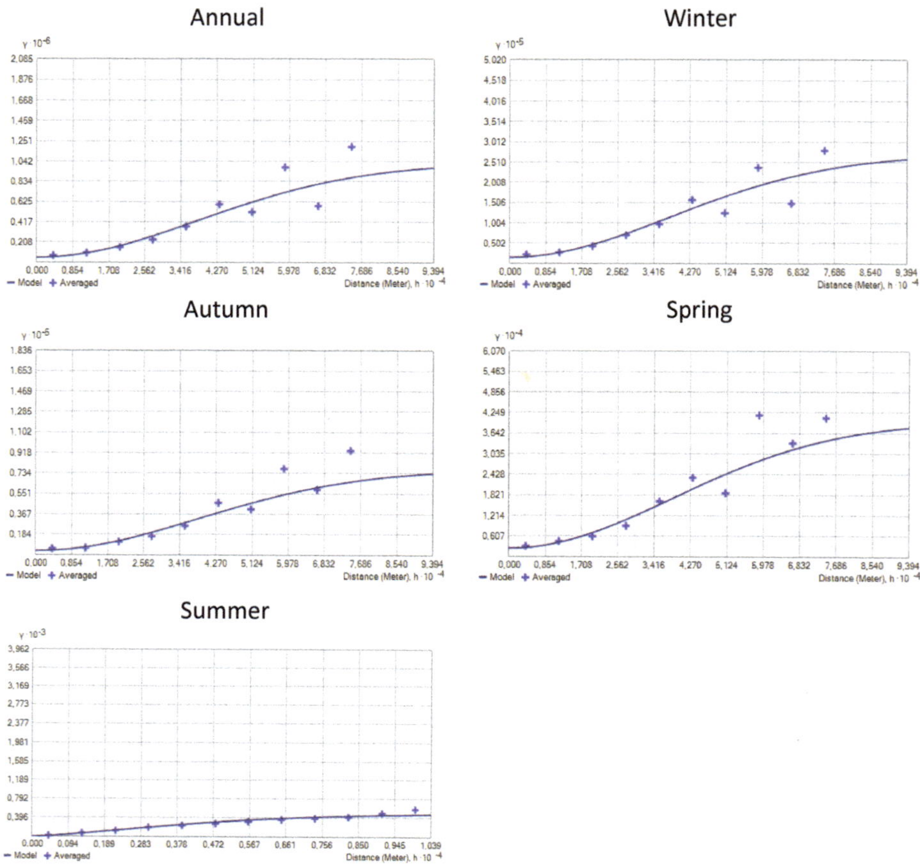

Figure 3. Spherical semivariograms of ordinary kriging interpolation models for annual and seasonal rainfalls.

Table 9. Variogram Statistics.

Parameter	Annual	Winter	Autumn	Spring	Summer
Nugget	0	0	0	0	283.01
Partial Sill	678185.10	7828.59	93943.10	27528.71	493.07
Lag size	7828.59	7828.59	7828.59	7828.59	2100.93

The rainfall spatial distribution maps over the mountainous catchment of the study area were produced using ordinary kriging and are given in the following figure (see Figure 4).

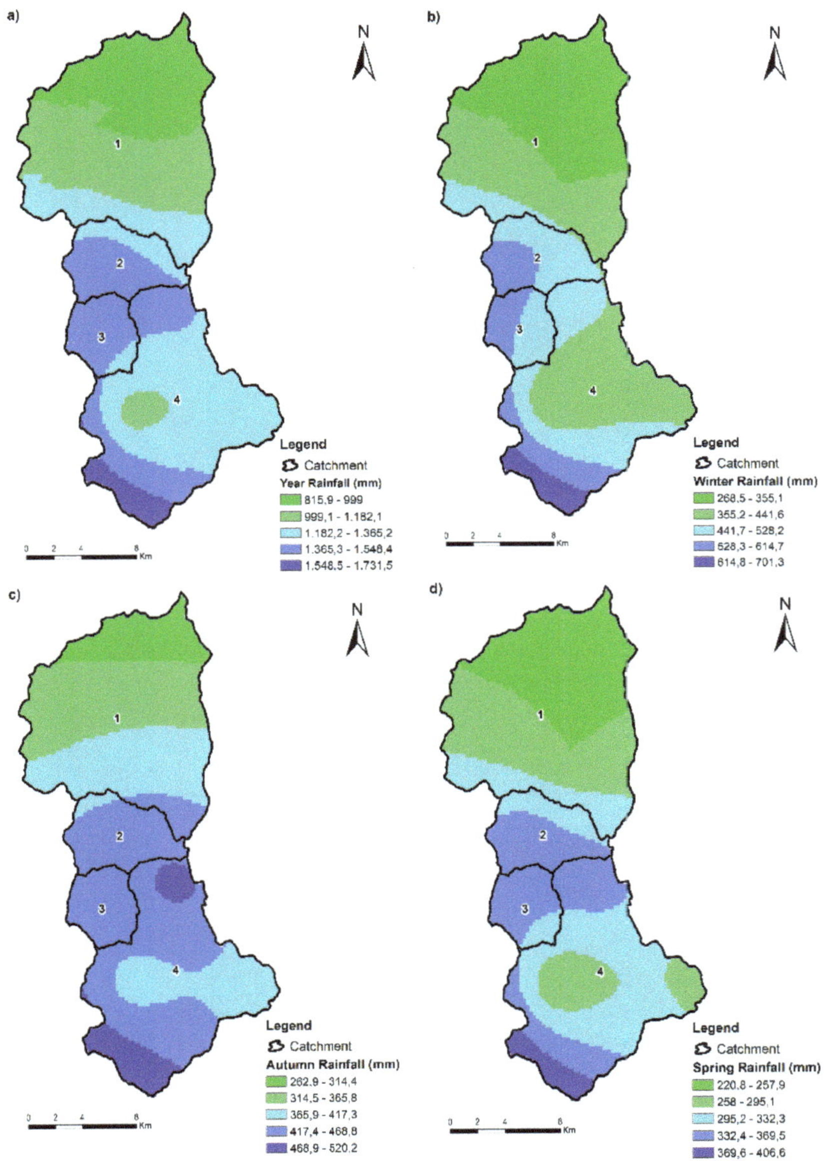

Figure 4. *Cont.*

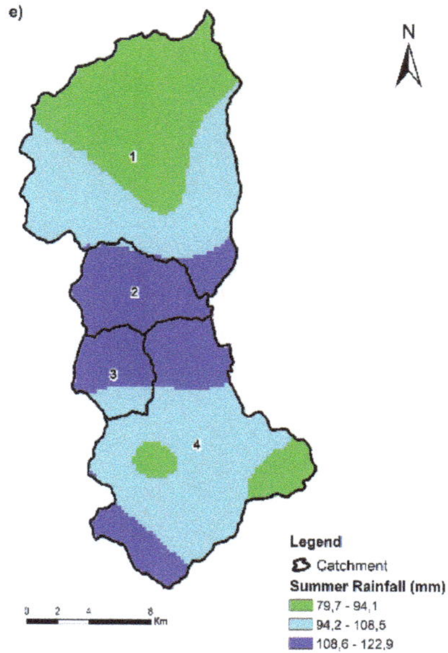

Figure 4. Spatial distribution of (**a**) annual, (**b**) winter, (**c**) autumn, (**d**) spring and (**e**) summer rainfall over the study area.

The major range of rainfall conditions within the mountainous catchment of the study area is shown in Figure 2. The rainfall range (mm) in each catchment and every season is given in Table 10.

Table 10. Annual and seasonal rainfall (mm) ranges in the catchments of the study area.

a/a	Catchment	Annual		Winter		Autumn		Spring		Summer	
		Min	Max	Min	Max	Min	Max	Min	Max	Min	Max
1	Klinovitikos	815.9	1336.7	268.5	494.6	262.9	449.1	220.8	328.6	79.7	115.1
2	Aspropotamos	1279.1	1483.2	434.7	562.7	406.5	463.9	308.4	351.9	106.5	121.6
3	Korpos	1281.2	1530.4	437.2	592.0	428.9	456.4	308.2	358.9	101.9	117.7
4	Portaikos	1100.2	1731.5	355.7	703.0	381.6	520.2	261.1	406.6	88.1	122.9

4. Conclusions

Rainfall variability is crucial for rational water resource management especially in Mediterranean countries, such as Greece, with rainfalls presenting temporal and spatial variation. In this study, monthly rainfall data from nine meteorological stations in the central Pindus mountain range were collected and analyzed for the period of 1961–2016. The conclusions reached are summarized below:

- Rainfall is characterized by great seasonal variability. Of the whole year's rainfall, 35% falls during winter, 32% during autumn, 24% during spring, and 9% during summer. Previous studies [44–46] have shown that there is a high degree of correlation between seasonal rainfall amounts and seasonal rainy days and the corresponding frequency of cyclonic circulation types at 500 hPa.

- Regarding the results of the Mann–Kendall test, it is highlighted that at most of the examined meteorological stations, decreasing trends were recorded on an annual basis; winter and spring rainfalls showed a decreasing trend, as well. As for the autumn and summer rainfalls, increasing trends were recorded in most stations. The above-mentioned trends are not statistically significant, except for the annual rainfall decreasing trend at the Katafyto station and the winter rainfall decreasing trends at the Katafyto and Pertouli stations. In addition, it was found that rainfall non-stationarity starts to occur in the middle of the 1960s for the annual, autumn, spring, and summer rainfalls and the early 1970s for the winter rainfall in most of the stations. Finally, the average trend magnitude per decade, using Sen's slope method, was −1.9% for the annual rainfall, −3.2% for the winter rainfall, +0.7% for the autumn rainfall, +0.2% for the spring rainfall, and +2.4 for the summer rainfall. The observed downward trends in rainfall in Greece was linked mainly to a rising trend in the hemispheric circulation modes of the North Atlantic Oscillation Index and its connection with the Mediterranean Oscillation Index [47]. In addition, the link between precipitation variability in Greece and the Mediterranean pressure oscillation is very reasonable from a physical point of view.

- The ordinary kriging method gives better results in spatial rainfall mapping in the study area in comparison with other spatial interpolation methods. The spatial variability in rainfall is extremely high. The relationship between geomorphological factors (longitude, latitude, and altitude) and rainfall was obtained and proposed using a multiple regression technique. The results indicated that the developed regression models could better explain the variability of rainfall in autumn and winter, rather than in spring and summer. Moreover, the spatial distribution maps obtained using ordinary kriging through the GIS software revealed a wide range of rainfalls (for all seasons) through the catchments, whereas many different zones of rainfalls could be recognized. These maps are easy to produce for every area of interest and are reliable and useful for various stakeholders.

- According the rainfall pattern of the study area, there is a lack of water during summer months and great differences of rainfall amounts between the mountainous areas and the lowlands. The need for the rational management of mountainous catchments is now a necessity. Appropriate silvicultural treatments must be applied in order to achieve an all-aged forest stand structure, which can increase water production from the mountainous catchments [48,49]. Additionally, stream regulation using check dams could have a positive effect on water availability. After siltation waters have infiltrated through the deposits, they act as ideal artificial aquifers. To accomplish this goal, appropriate pipelines to the lower body of the check dams must be constructed. The retained water is supplied with piping and by natural flow to the areas that have a demand. This method of water conservation provides clean and filtered water and avoids evaporation losses [50].

Author Contributions: S.S. analyzed the GIS data, investigated the spatial distribution of rainfall, and wrote the paper, whereas D.S. collected the meteorological data and applied the Mann–Kendall statistical test.

Funding: This research received no external funding.

Conflicts of Interest: The authors declare no conflict of interest.

References

1. Maheras, P.; Anagnostopoulou, C. Circulation types and their influence on the interannual variability and precipitation changes in Greece. In *Mediterranean Climate*; Springer: Berlin/Heidelberg, Germany, 2003; pp. 215–239.
2. Tolika, K.; Maheras, P.; Anagnostopoulou, C. The exceptionally wet year of 2014 over Greece: A statistical and synoptical-atmospheric analysis over the region of Thessaloniki. *Theor. Appl. Climatol.* **2017**, *132*, 809–821. [CrossRef]
3. Xoplaki, E.; Luterbacher, J.; Burkard, R.; Patrikas, I.; Maheras, P. Connection between the large-scale 500 hPa geopotential height fields and precipitation over Greece during wintertime. *Clim. Res.* **2000**, *14*, 129–146. [CrossRef]
4. Metaxas, D.A. Evidence on the importance of diabatic heating as a divergence factor in the Mediterranean. *Arch. Meteorol. Geophys. Bioklimatol.* **1978**, *27*, 69–80. [CrossRef]
5. Stathis, D. The role of topography in the distribution of precipitation over mountainous Greece as reveal by principal component analysis. *J. Meteorol.* **1999**, *24*, 50–56.

6. Buytaert, W.; Celleri, R.; Willems, P.; De Bievre, B.; Wyseure, G. Spatial and temporal rainfall variability in mountainous areas: A case study from the south Ecuadorian Andes. *J. Hydrol.* **2006**, *329*, 413–421. [CrossRef]

7. Stathis, D.; Myronidis, D. Principal component analysis of precipitation in Thessaly region (Central Greece). *Glob. NEST J.* **2009**, *11*, 467–476.

8. Maris, F.; Kitikidou, K.; Angelidis, P.; Potouridis, S. Kriging interpolation method for estimation of continuous spatial distribution of precipitation in Cyprus. *Br. J. Appl. Sci. Technol.* **2013**, *3*, 1286–1300. [CrossRef]

9. De Amorim Borges, P.; Franke, J.; da Anunciação, Y.M.T.; Weiss, H.; Bernhofer, C. Comparison of spatial interpolation methods for the estimation of precipitation distribution in Distrito Federal, Brazil. *Theor. Appl. Climatol.* **2016**, *123*, 335–348. [CrossRef]

10. Paparrizos, S.; Maris, F.; Matzarakis, A. A downscaling technique for climatological data in areas with complex topography and limited data. *Int. J. Eng. Res. Dev.* **2016**, *12*, 17–23.

11. Syed, M.A.; Al Amin, M. Geospatial modeling for investigating spatial pattern and change trend of temperature and rainfall. *Climate* **2016**, *4*, 21. [CrossRef]

12. Arnell, N.W.; Goslin, S.N. The impacts of climate change on river flood risk at the global scale. *Clim. Chang.* **2016**, *134*, 387–401. [CrossRef]

13. Simonneaux, V.; Cheggour, A.; Deschamps, C.; Mouillot, F.; Cerdan, O.; Le Bissonnais, Y. Land use and climate change effects on soil erosion in a semi-arid mountainous watershed (High Atlas, Morocco). *J. Arid Environ.* **2015**, *122*, 64–75. [CrossRef]

14. Anagnostopoulou, C. Drought episodes over Greece as simulated by dynamical and statistical downscaling approaches. *Theor. Appl. Climatol.* **2017**, *129*, 587–605. [CrossRef]

15. Blanc, E.; Reilly, J. Approaches to assessing climate change impacts on agriculture: An overview of the debate. *Rev. Environ. Econ. Policy* **2017**, *11*, 247–257. [CrossRef]

16. Michailidou, A.V.; Vlachokostas, C.; Moussiopoulos, N. Interactions between climate change and the tourism sector: Multiple-criteria decision analysis to assess mitigation and adaptation options in tourism areas. *Tour. Manag.* **2016**, *55*, 1–12. [CrossRef]

17. Theodossiou, N. Assessing the Impacts of Climate Change on the Sustainability of Groundwater Aquifers. Application in Moudania Aquifer in N. Greece. *Environ. Process.* **2016**, *3*, 1045–1061. [CrossRef]

18. Kalabokidis, K.; Palaiologou, P.; Gerasopoulos, E.; Giannakopoulos, C.; Kostopoulou, E.; Zerefos, C. Effect of climate change projections on forest fire behavior and values-at-risk in Southwestern Greece. *Forests* **2015**, *6*, 2214–2240. [CrossRef]

19. Intergovernmental Panel on Climate Change (IPCC). *Climate Change 2013: The Physical Science Basis. Contribution of Working Group I to the Fifth Assessment Report of the Intergovernmental Panel on Climate Change*; Cambridge University Press: Cambridge, UK, 2013.

20. Cox, P.M.; Betts, R.A.; Jones, C.D.; Spall, S.A.; Totterdell, I.J. Acceleration of global warming due to carbon-cycle feedbacks in a coupled climate model. *Nature* **2000**, *408*, 184–187. [CrossRef] [PubMed]

21. Solomon, S.; Plattner, G.K.; Knutti, R.; Friedlingstein, P. Irreversible climate change due to carbon dioxide emissions. *Proc. Natl. Acad. Sci. USA* **2009**, *106*, 1704–1709. [CrossRef] [PubMed]

22. Giorgi, F.; Lionello, P. Climate change projections for the Mediterranean region. *Glob. Planet. Chang.* **2008**, *63*, 90–104. [CrossRef]

23. Sahsamanoglou, H.; Makrogiannis, T.; Hatzianastasiou, N.; Rammos, N. Long term change of precipitation over the Balkan Peninsula. In *Eastern Europe and Global Change*; European Commission: Brussels, Belgium, 1997; pp. 111–124.

24. Maheras, P.; Tolika, K.; Anagnostopoulou, C.; Vafiadis, M.; Patrikas, I.; Flocas, H. On the relationships between circulation types and changes in rainfall variability in Greece. *Int. J. Climatol.* **2004**, *24*, 1695–1712. [CrossRef]

25. Nastos, P.T.; Zerefos, C.S. Spatial and temporal variability of consecutive dry and wet days in Greece. *Atmos. Res.* **2009**, *94*, 616–628. [CrossRef]

26. Mavromatis, T.; Stathis, D. Response of the water balance in Greece to temperature and precipitation trends. *Theor. Appl. Climatol.* **2011**, *104*, 13–24. [CrossRef]

27. Stefanidis, S.; Chatzichristaki, C. Response of soil erosion in a mountainous catchment to temperature and precipitation trends. *Carpathian J. Earth Environ. Sci.* **2017**, *12*, 35–39.

28. Markonis, Y.; Batelis, S.C.; Dimakos, Y.; Moschou, E.; Koutsoyiannis, D. Temporal and spatial variability of rainfall over Greece. *Theor. Appl. Climatol.* **2017**, *130*, 217–232. [CrossRef]

29. Tolika, K.; Zanis, P.; Anagnostopoulou, C. Regional climate change scenarios for Greece: Future temperature and precipitation projections from ensembles of RCMs. *Glob. NEST J.* **2012**, *14*, 407–421.

30. Paparrizos, S.; Maris, F.; Matzarakis, A. Integrated analysis of present and future responses of precipitation over selected Greek areas with different climate conditions. *Atmos. Res.* **2016**, *169*, 199–208. [CrossRef]

31. World Meteorological Organization (WMO). *Guidelines on the Quality Control of Surface Climatological Data*; WCP-85; WMO: Geneva, Switzerland, 1986.

32. Goossens, C.; Berger, A. Annual and seasonal climatic variations over the northern hemisphere and Europe during the last century. *Ann. Geophys.* **1986**, *4*, 385–400.

33. Mann, B. Nonparametric tests against trend. *Econom. J. Econom. Soc.* **1945**, *13*, 245–259. [CrossRef]

34. Kendall, M. *Multivariate Analysis*; Charles Griffin: London, UK, 1975.

35. Sicard, P.; Dalstein-Richier, L. Health and vitality assessment of two common pine species in the context of climate change in southern Europe. *Environ. Res.* **2015**, *137*, 235–245. [CrossRef] [PubMed]

36. Taylor, W. *Change-Point Analyzer 2.0 Shareware Program*; Taylor Enterprises: Libertyville, IL, USA, 2000; Available online: http://www.variation.com/cpa/ (accessed on 19 July 2018).

37. Thiel, H. A rank-invariant method of linear and polynomial regression analysis, part 3. In *Advanced Studies in Theoretical and Applied Econometrics*; Springer: Berlin, Germany, 1992; pp. 345–381.

38. Sen, P.K. Estimates of the regression coefficients based on Kendall's tau. *J. Am. Stat. Assoc.* **1968**, *63*, 1379–1389. [CrossRef]

39. Hirsch, R.M.; Slack, J.R.; Smith, R.A. Techniques of trend analysis for monthly water quality data. *Water Resour. Res.* **1982**, *18*, 107–121. [CrossRef]

40. Goovaerts, P. Geostatistical approaches for incorporating elevation into the spatial interpolation of rainfall. *J. Hydrol.* **2000**, *228*, 113–129. [CrossRef]

41. Willmott, J.; Matsuura, K. Advantages of the mean absolute error (MAE) over the root mean square error (RMSE) in assessing average model performance. *Clim. Res.* **2015**, *30*, 79–82. [CrossRef]

42. Chai, T.; Draxler, R. Root mean square error (RMSE) or mean absolute error (MAE)?—Arguments against avoiding RMSE in the literature. *Geosci. Model Dev.* **2014**, *7*, 1247–1250. [CrossRef]

43. Mavromatis, T. Spatial resolution effects on crop yield forecasts: An application to rainfed wheat yield in north Greece with CERES-Wheat. *Agric. Syst.* **2016**, *143*, 38–48. [CrossRef]

44. Norrant-Romand, C.; Douguédroit, A. Significant rainfall decreases and variations of the atmospheric circulation in the Mediterranean (1950–2000). *Reg. Environ. Chang.* **2014**, *14*, 1725–1741. [CrossRef]

45. Casado, M.J.; Pastor, M.A. Circulation types and winter precipitation in Spain. *Int. J. Climatol.* **2016**, *36*, 2727–2742. [CrossRef]

46. Putniković, S.; Tošić, I. Relationship between atmospheric circulation weather types and seasonal precipitation in Serbia. *Meteorol. Atmos. Phys.* 2017. [CrossRef]

47. Feidas, H.; Noulopoulos, C.; Makrogiannis, T.; Bora-Senta, E. Trend analysis of precipitation time series in Greece and their relationship with circulation using surface and satellite data. *Theor. Appl. Climatol.* **2007**, *87*, 155–177. [CrossRef]

48. Kotoulas, D. Hydrological influences of forests in torrent catchment regions of Greece. *Schweiz. Z. Forstwes.* **1980**, *131*, 617–626.

49. Gatzojannis, S.; Stefanidis, P.; Kalabokidis, K. An inventory and evaluation methodology for non-Tiber functions of forests. *Mitt. Abt. Forstl. Biom.* **2001**, *1*, 3–49.

50. Kotoulas, D. The possibilities of increase water supply in Greece. *Sci. Ann. Dep. For. Nat. Environ.* **1992**, *LE*, 141–149. (In Greek)

Article

Intercomparison of Univariate and Joint Bias Correction Methods in Changing Climate From a Hydrological Perspective

Olle Räty [1,*], Jouni Räisänen [1], Thomas Bosshard [2] and Chantal Donnelly [2]

[1] Institute for Atmospheric and Earth System Research (INAR), University of Helsinki, 00100 Helsinki, Finland; jouni.raisanen@helsinki.fi

[2] Swedish Meteorological and Hydrological Institute (SMHI), 60176 Norrköping, Sweden; thomas.bosshard@smhi.se (T.B.); chantal.donnelly@bom.gov.au (C.D.)

* Correspondence: olle.raty@helsinki.fi; Tel.: +358-504-160-500

Received: 26 February 2018; Accepted: 23 April 2018; Published: 26 April 2018

Abstract: In this paper, the ability of two joint bias correction algorithms to adjust biases in daily mean temperature and precipitation is compared against two univariate quantile mapping methods when constructing projections from years 1981–2010 to early (2011–2040) and late (2061–2090) 21st century periods. Using both climate model simulations and the corresponding hydrological model simulations as proxies for the future in a pseudo-reality framework, these methods are inter-compared in a cross-validation manner in order to assess to what extent the more sophisticated methods have added value, particularly from the hydrological modeling perspective. By design, bi-variate bias correction methods improve the inter-variable relationships in the baseline period. Cross-validation results show, however, that both in the early and late 21st century conditions the additional benefit of using bi-variate bias correction methods is not obvious, as univariate methods have a comparable performance. From the evaluated hydrological variables, the added value is most clearly seen in the simulated snow water equivalent. Although not having the best performance in adjusting the temperature and precipitation distributions, quantile mapping applied as a delta change method performs well from the hydrological modeling point of view, particularly in the early 21st century conditions. This suggests that retaining the observed correlation structures of temperature and precipitation might in some cases be sufficient for simulating future hydrological climate change impacts.

Keywords: regional climate modeling; hydrological modeling; bias correction; multivariate; pseudo reality

---

1. Introduction

In recent years, bias adjustment has become the de facto standard for preprocessing global (GCM) and regional (RCM) climate model simulations for climate change impact studies, hydrological modeling being no exception. The use is driven by practical needs. Due to systematic errors in climate model simulations with respect to the observed climate, GCM and RCM output usually cannot be directly used in impact modeling, as impact models require unbiased, high-resolution information as their input. This is because of non-linear and threshold processes within impact models. For example, a cold bias in forcing data to a hydrological model could lead to an impact result indicating no change in snow depths if the cold bias kept temperatures below 0 degrees.

Numerous methods belonging to so-called model output statistics (MOS) have been developed to adjust biases in temperature and precipitation data from climate models. These range from simple scaling of time-mean climate to more sophisticated methods addressing biases in the daily variability. This group also covers the widely used quantile mapping techniques of which there are a number of different variations [1–8]. Studies have illustrated that bias correction methods are able to reduce biases in climate model output [6] and also to provide noticeable improvements to hydrological simulations in the present-day climate [9,10]. However, most of these methods are restricted to the independent adjustment of biases in the marginal aspects of GCM-RCM simulations and do not take biases in inter-variable correlation structures into account. For example, studies have indicated that highest precipitation intensities co-occur with high surface temperatures in winter, as indicated by Clausius–Clapeyron relation, while mostly negative relationships between temperature and precipitation have been observed in summer in Europe [11,12]. In case a GCM-RCM has difficulties to reasonably capture such relationships, a bias correction method that does not explicitly take inter-variable correlations into account might not be sufficient for certain applications such as hydrological modeling.

To address this issue, different types of bi- and multivariate bias correction algorithms have been recently proposed [13–17]. These studies have given evidence that jointly bias correcting multiple variables improves the multivariate aspects of bias corrected model simulations when compared against the observed climate, and might outperform their univariate counterparts in further applications, such as in the calculation of Canadian Forest Fire Weather Index [17]. However, most of the intercomparison studies have concentrated on evaluating the relative performance of bias correction methods in the present-day climate, which does not inform on their ability to predict climate variables in changing climatic conditions. In other words, it is not known how well the adjustment of inter-variable correlations, which is inherently constrained by biases in the present-day, is able to capture the inter-variable correlations in the future climate. This information is crucial for reliably assessing potential climate change impacts, particularly as concerns have been expressed on the shortcomings and potentially unjustified use of bias correction in non-stationary conditions [18,19].

Due to the lack of an observational basis, surrogate data emulating the future observations has been proposed to be used as proxy data (hereafter referred to as pseudo-realities) to assess the ability of bias adjustment to improve projections in a changing climate. The pseudo-reality approach has been used in recent studies [20–23] and was also considered in the European Concerted Research Action ES1102 VALUE (Validating and Integrating Downscaling Methods for Climate Change Research) framework as an important, although not sufficient step, when evaluating bias adjustment method performance [24].

Most of the pseudo-reality studies have concentrated on the analysis of the application of bias adjustment directly to climate model output, which does not give direct information on their usability to construct future projections for the purposes of hydrological climate change impact studies. One of the first attempts to extend the pseudo-reality approach to hydrological simulations was made by Velázquez et al. [25], whose study evaluated implications of non-stationarity to bias correction for future conditions and how they affect the estimation of future changes in river discharges. Their results showed that although monthly mean river discharges were improved in some cases after bias correcting the hydrological model input, biases still remained in the results. In their study the pseudo-reality approach was applied without taking pseudo-reality biases in the present-day climate into account. If a hydrological model is sensitive to absolute biases in climate model outputs, the hydrological model behavior and its response to the projected changes might be unrealistic, which would hamper the evaluation of bias adjustment methods in the pseudo-reality framework. Furthermore, the study used only two GCM-RCM combinations and one bias correction method that did not take inter-variable correlations directly into account in the bias correction step. From the hydrological modeling perspective, a physically plausible description of co-variations of temperature and precipitation might be important to reasonably describe the surface fluxes such as evapotranspiration and processes affecting water stored in soil and snow pack, which together regulate the river discharge generation. It is also important to assess the performance of the hydrological modeling of low frequency impacts (e.g., high and low flows) as these impacts are often of interest to users, but could be subject to different biases than mean flows.

Here, we extend the study of Velázquez et al. [25] to assess the relative performance of four bias adjustment methods from the hydrological impact modeling perspective. More specifically, the aim is to address the following questions:

1. How does the relative performance of bias adjustment methods vary when assessing them from the perspective of the impacts of climate change on different hydrological variables rather than from climate modeling perspective?
2. What is the added value of bias correcting inter-variable relationships between daily mean temperature and precipitation in comparison to the adjustment of their marginal distributions only?

We use five GCM-RCMs produced in the European branch of the Coordinated Regional Climate Downscaling Experiment (EURO-CORDEX) initiative [26]. In addition to the separate adjustment of temperature and precipitation distributions, two methods [14,17] which take biases in their inter-variable relationships into account are compared against univariate quantile mapping to assess the extent to which hydrological simulations benefit from the additional correction of temperature and precipitation correlations in the changing climate. We perform cross-validation tests in a pseudo-reality framework broadly similar to the one used in Velázquez et al. [25] and extend the analysis by taking into account GCM-RCM biases in the calibration data in order to bring the hydrological simulations closer to the observations (i.e., by bias adjusting the pseudo-reality data).

The paper is structured as follows. In the next section we introduce the GCM-RCM simulations used in this study together with the hydrological model used to conduct the hydrological simulations. In addition, the bias adjustment methods, the pseudo-reality framework and the cross-validation statistics used to assess the relative method performance are also discussed in Section 2. The results are shown in Section 3 and the discussion together with conclusions are presented in Section 4.

2. Materials and Methods

2.1. Reference Data

WATCH Forcing Data based on ERA-Interim reanalysis (WFDEI) is used as the reference climatology in real-world illustrations as well as in hydrological modeling exercises for both daily mean temperature and precipitation [27]. In this data set the monthly means of daily mean temperature and precipitation have been adjusted for biases in relation to the large scale gridded observations produced by the Climatic Research unit (CRU) and the Global Precipitation Climatology Centre (GPCC) with subsequent elevation corrections applied to both variables. Rust et al. [28] discuss some of the known issues in WFDEI such as spurious jumps between individual months caused by the bias adjustment procedure. These introduced discontinuities might affect the hydrological simulations, for example, by triggering snow melt unrealistically. However, the effect of the spurious jumps is less severe in the European region than in the tropics and cold regions [28], which facilitates the use of WFDEI in this study.

2.2. GCM-RCM Data

Five high-resolution GCM-RCM simulations (Table 1) produced by the European branch of the Coordinated Regional Climate Downscaling Experiment (EURO-CORDEX) initiative were downloaded from the Earth System Grid Federation (ESGF) database [29]. The cross-validation within a pseudo-reality framework (see Section 2.5) requires all projections to be forced by the same representative concentration pathways (RCP). Here, we chose RCP4.5 as a mid-range emission scenario, which corresponds to end-of-21st century radiative forcing of 4.5 Wm^{-2} [30]. Each of the five model simulations has a different GCM and RCM component. One should note that the models were not selected based on their performance in the present-day conditions but to ensure that the ensemble members are independent from each other to the extent possible.

The model simulations were re-gridded to the regular $0.125° \times 0.125°$ EURO-CORDEX (EUR-11i) grid using nearest neighbor interpolation before bias adjusting them. The grid box closest to each sub-basin was then used to drive the hydrological model. The difference in the spatial resolution to WFDEI was not taken into account, as most of the results are based on comparisons between the GCM-RCM simulations (see Section 3). For both calibration and validation of the bias adjustment methods, we selected three 32-year periods including a 2-year spin-up period for the hydrological model runs. After excluding the spin-up period, years 1981–2010 were used as the baseline period, while years 2011–2040 and 2061–2090 were used to validate the methods in early and late 21st century conditions.

Table 1. List of GCM-RCM simulations selected for this study. The first column shows the abbreviation used in the text, the second and third columns show the GCM and RCM part of each model, respectively and the last column shows the name of the providing institution.

Abbreviation	GCM Component	RCM Component	Institution
CNRM-A	CNRM-CM5	ALADIN	CNRM
CCLM-MPI	MPI-ESM-LR	CCLM4.8.17	CLM Community
RACMO-EC	EC-EARTH	RACMO22E	KNMI
RCA4-H	HadGEM	RCA4	SMHI
WRF-I	IPSL-CMA5-MR	WRF	IPSL-INERIS

2.3. Hydrological Simulations

Hydrological impacts are simulated using sub-models extracted from the European scale hydrological model E-HYPE v2.5 [31]. E-HYPE is an application of the Hydrological Predictions for the Environment (HYPE) model developed in the Swedish Meteorological and Hydrological Institute [32]. The model is process-based, semi-distributed and designed for hydrological modeling at different spatial scales also in ungauged regions. The source code for HYPE (v4.10) is available at [33]. The model calibration and evaluation details together with a list of used topographical and land-use data sets can be found from [31]. E-HYPE was chosen due to its large spatial coverage, distributed nature and also to see how a hydrological model commonly used in impact assessments [34] responds to the bias adjustment step. Because the model is distributed (over sub-basins of median size 215 km^2), spatial variability in biases can also be assessed. Hydrological simulations were first conducted using the full 32-year periods. The first two years served to spin-up the hydrological model and have not been included in further analysis.

To sample the varying hydrological conditions in the European region, the model was run with bias adjusted daily mean temperature and precipitation within four sub-models selected from the catchments shown in Figure 1. These catchments have predominantly natural flow conditions which the hydrological model was capable of capturing well in the present-day climate. The northernmost model domain is located in the upper parts of Tornio river catchment, where water stored in snow pack and variations in it strongly regulate the annual cycle of surface hydrology, leading to peak river discharges in late spring and early summer. Two domains with maritime mild climatic conditions, which cover parts of Trent and Ems catchments, are less affected by snow processes and river discharges reach their maximum values in winter months. The southernmost study region is located in the Sava tributary of the Danube river and is characterized by a mixture of Alpine and Mediterranean climates. The seasonal cycle of river discharge has two distinctive peaks here, the first one caused by snow melt in Alpine regions in spring and the latter one by heavy rainfalls in autumn.

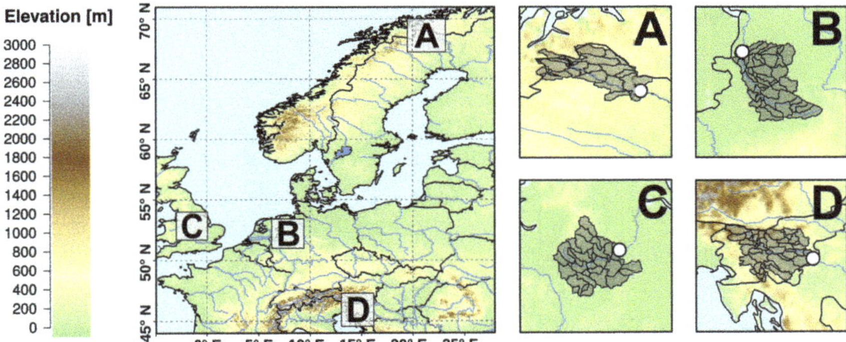

Figure 1. Geographical locations of the four sub-models selected for the hydrological simulation tests. The sub-models cover parts of (**A**) Tornio, (**B**) Ems, (**C**) Trent and (**D**) Sava river catchments. White dots denote the locations of the discharge gauging stations for which the reference period statistics are shown in Table 2.

Table 2. Statistics for daily time-series of simulated river discharge calculated against the observed discharge at the mouth of each hydrological sub-model (cf. Figure 1). The first column shows the catchment name, the second column Nash-Sutcliffe efficiency coefficient and the third column the relative volume error.

Sub-Model	NSE	RE (%)
Tornio	0.78	−17.0
Trent	0.66	−3.0
Ems	0.83	3.1
Sava	0.52	6.0

The ability of HYPE to simulate river flows in the reference period (1981–2010), when WFDEI is used directly as forcing, is briefly illustrated in Table 2, which shows the Nash-Sutcliffe efficiency coefficient (NSE) and relative volume error (RE) in simulated river discharge for the four gauging stations located at the outlets of the selected sub-models. The NSE values vary from 0.83 in Ems to 0.52 in the Sava region. These values are reasonable considering that E-HYPE has been calibrated uniformly for all of Europe to optimize predictions in ungauged regions. The RE values range from −17.0% to 6.0% with largest deviations seen in the Tornio sub-model, where the model tends to underestimate river discharge volume, particularly during the spring season. These differences are at least partially explained by the limitations of the WFDEI data set; the representation of daily precipitation variability is not sufficient in regions with large topographical variations, and subject to gauge undercatch for which the corrections are particularly uncertain in windy, snow dominated regions. Also temperature discontinuities might have a role in explaining the differences to the observations. In addition, the inherent limitations in the HYPE formulation and parameterisation likely explain part of this discrepancy (as would any other hydrological model).

2.4. Model Output Statistics

Non-parametric quantile mapping applied both in the delta change (M1) and bias correction (M2) mode is used to benchmark the ability of joint bias correction methods to adjust temperature and precipitation for biases in the GCM-RCM simulations (see Table 3). In the following, the formulation is shown for the bias correction form of quantile mapping. For given simulated values of daily mean temperature or precipitation in the scenario period (s_i), the projected values p_i are obtained by transforming s_i according to

$$p_i = F_o^{-1}\left(F_c\left(s_i\right)\right). \tag{1}$$

here F_c denotes the cumulative distribution function of the baseline period simulation and F_o^{-1} its inverse estimated from the observations. Formulation for the delta change form is simply obtained by switching the indexes for the observations (o) and the future period simulation (s).

Before the transformation shown in Equation (1) is applied, the quantile-quantile relationship between F_c and F_o is smoothed by replacing individual quantiles with a running average taken over a specified quantile range using the approach and numerical values described in Räty [21] and Räty et al. [22]. If the future simulated values are outside the baseline period observations and model simulation, the quantile relationship is extrapolated assuming a constant, additive, relationship above the highest and below the lowest quantile for daily mean temperature. For precipitation, relative values are used. For further implementation details, the reader is referred to Räty [21] and Räty et al. [22].

Table 3. List of bias adjustment methods used in this study together with a short description.

Name	Description	References
M1	Univariate delta change: quantile mapping with smoothing	Räisänen and Räty [21], Räty et al. [22]
M2	Univariate bias correction: quantile mapping with smoothing	Räisänen and Räty [21], Räty et al. [22]
M3	Bi-variate bias correction: copula-based, precipitation conditioned on temperature	Li et al. [14], Gennaretti et al. [35]
M4	Bi-variate bias correction: full 2-dimensional distribution using the N-pdft algorithm	Pitié et al. [36], Cannon [17]

To take biases in the co-variations of daily mean temperature and precipitation into account, two bi-variate bias correction methods were implemented and compared with their univariate counterparts. In the first one (M3), the dependence structure is modeled separately from the marginal (i.e., unconditional) distributions of temperature and precipitation using a copula-based approach as described in Li et al. [14]. The implementation of this method is based on the properties of copula described by Sklar's theorem [37], which states that, given two random variables X and Y such as daily mean temperature and precipitation, their joint cumulative distribution $(H(x,y))$ can be constructed as

$$H(x,y) = C(F(x), G(y)), \tag{2}$$

where $C()$ denotes the cumulative copula distribution and $F(x)$ and $G(y)$ are the cumulative marginal distributions for X and Y. Here, it is assumed that the dependence structure of daily mean temperature and precipitation can be reasonably modeled using a Gaussian copula as in Li et al. [14]. The Gaussian copula was chosen due to its relatively simple formulation and its ability to model both positive and negative correlations. Although several other parametric copulas are available for modeling the dependence structure, testing them is beyond the scope of this paper. Marginal distributions were modeled with parametric distributions in a similar manner as in Yang et al. [2] and Li et al. [14]. Temperature is assumed to follow Gaussian distribution, while gamma distribution is used to model precipitation above a wet-day threshold, here defined as 0.1 mmd^{-1}. To improve the performance of M3 at daily temporal scales, separate temperature distributions were fitted both on dry and wet days following a pre-adjustment of the fraction of wet days in model simulations to the observed one [2,35]. In M3, bias correction of the wet-day part of the joint distribution needs to be applied conditionally on either temperature or precipitation, thus offering two approaches to correct the joint distribution. Based on tests with these two alternatives, we decided to bias correct precipitation conditionally on temperature due to the slightly better overall performance of this option (not shown). The correlation parameter values obtained from fitting Gaussian copula to the baseline period simulations are illustrated in the supplementary material (Figure S1). For further details, the reader is referred to an R package available in GitHub [38], which contains implementations of methods M1–M3 written by the authors.

The second bi-variate bias correction method (M4) was recently proposed by Cannon [17] based on the N-pdft algorithm designed by Pitié et al. [36]. In this method the full 2-dimensional distribution structure is adjusted iteratively by reducing the adjustment of the 2-dimensional distribution to a series of 1-dimensional bias corrections of the marginal distributions. First, both temperature and precipitation distributions are normalized and randomly rotated to a new orthogonal coordinate system. Second, quantile mapping is applied to the rotated distributions. The adjusted distributions are then rotated back to the original coordinate system before repeating the described sequence. After several successive iterations it can be shown that the joint distribution converges to the target distribution [36]. As discussed by Cannon [17], the algorithm constructs the joint distribution at each iteration step as a linear combination of the bias corrected marginal distributions, which allows modifying the dependence structure. In the original article quantile delta mapping [4,8] was used to adjust the marginal distributions along the iteration cycles. Here, we use the same quantile mapping algorithm as in M2 instead of quantile delta mapping when bias correcting the marginal distributions of temperature and precipitation. Doing this, the bias corrected temperature and precipitation distributions are identical in M2 and M4. No smoothing was applied to the marginal distributions in the rotation step, as this would had contracted the underlying joint distribution of observations and the control period simulation. The algorithm was terminated after 50 iterations, which should be sufficient for the algorithm to converge to the target distribution, as illustrated by Cannon [17]. The implementation of M4 was based on the R package [39] available in the CRAN repository [40].

To take biases in the annual cycle into account, daily mean temperature and precipitation time series were adjusted on a monthly basis at each sub-model domain. As sampling errors are likely to affect the estimation of simulated changes (M1) and biases (M2–M4) in the GCM-RCM distributions, the effect of increased sample size on method performance was addressed by using both one- and two-month time windows when estimating model biases and simulated changes from GCM-RCM simulations. Using an even larger time window could in principle reduce the sampling noise [41], although with the expense of possibly introducing systematic biases to the future results. Tests with longer time windows did not show significant changes in the results, although a more systematic comparison would be needed to fully assess the potential benefits of reducing sampling noise.

To illustrate how each method represents the dependence structure in the calibration period, Figure 2 shows differences in the empirical copula density for the wet-day values of temperature and precipitation in comparison to the WFDEI copula in winter months of years 1981–2010 in the Tornio sub-model. The copula density has been estimated as a 2-dimensional histogram of the normalized ranks for both variables (see a more detailed description in Section 2.6). The density values can be interpreted as the ratio of the joint probability density to the case, where both variables were independent from each other. For example, values larger than one suggest a larger-than expected joint probability density in this part of the two-dimensional space. Differences in the empirical copula density roughly denote the difference in the strength of dependence for particular cumulative probability values of both temperature and precipitation. The panel for the reference data shows that temperature and precipitation are positively correlated in this example (the highest values slope from bottom-left to top-right). By design, M1 takes the inter-variable relationships directly from the reference. In contrast to the delta change mode, M2 inherits the multivariate dependence structure from the uncorrected GCM-RCM simulation (M0), with some modifications to it due to changes in the fraction of wet days [42]. Therefore, these methods can be thought to give the "limits" in which the multivariate methods can operate on adjusting the inter-variable correlations. Although the dependence structure of temperature and precipitation can be reasonably modeled using Gaussian copula on monthly scales, this might not be as feasible at daily scales, at least in cold climates. From Figure 2, it is immediately seen that although the Pearson correlation coefficient is well captured by M3, the overall copula structure shows noticeable deviations from the target distribution. In this particular case, M3 tends to overestimate the strength of the co-occurrence of low precipitation intensities and medium temperature values. Although the behavior of M3 strongly depends on the selected GCM-RCM, season and location, this example highlights the importance of evaluating the full multivariate dependence structure to reveal such issues. In contrast to M3, M4 performs very well in capturing the WFDEI dependence structure and differences in the copula density field are small in most parts of the distribution.

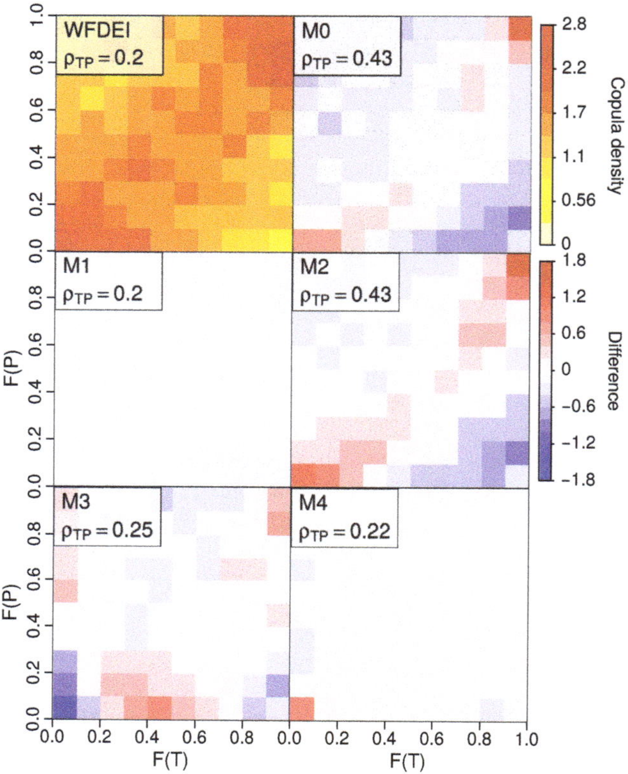

Figure 2. Empirical copula density of wet-day ($P > 0.1$ mmd^{-1}) precipitation and temperature in winter months (December-January-February) estimated from the reference data (WFDEI) in Tornio river catchment separately for each sub-basin and then averaged over the whole domain. In addition, differences in the estimated densities, when compared against WFDEI, are shown for CNRM-A GCM-RCM without bias adjustments (M0) and after applying each of the four methods (M1–M4) in the baseline period (1981–2010). The sub-model-averaged Pearson correlation coefficient is also shown on the top-left corner of each panel.

As another example, Figure 3 shows how the four methods capture the hydrological conditions in the Tornio sub-model, when adjusting the GCM-RCM simulations against WFDEI in the baseline period (1981–2010). For simplicity, a one-month time window has been used when estimating the simulated changes and biases in the GCM-RCM simulations. As expected, M1 has essentially a perfect correspondence with the observed hydrological conditions. The remaining biases in temperature and precipitation are very similar for M2 and M4 but not identical as small differences arise from the re-shuffling of daily values over the full 32-year period in M4, which is an inherent property of this and many other multivariate bias correction methods. Furthermore, M3 shows a very similar pattern for temperature, while the differences to M2 and M4 are more visible in the remaining precipitation biases. The differences between the methods are also visible in the annual monthly mean cycle of different aspects of the simulated surface hydrology. The differences to WFDEI are largest for M3 in most cases, which is expected, as the (potentially sub-optimal) parametric marginal distributions used in M3 match the GCM-RCM-simulated temperature and particularly precipitation less accurately with WFDEI than the non-parametric versions used in M2 and M4. Apart from M1, M4 has generally the smallest differences to WFDEI, particularly in total runoff and snow water equivalent, although the remaining evapotranspiration biases are similar for M2 and M4.

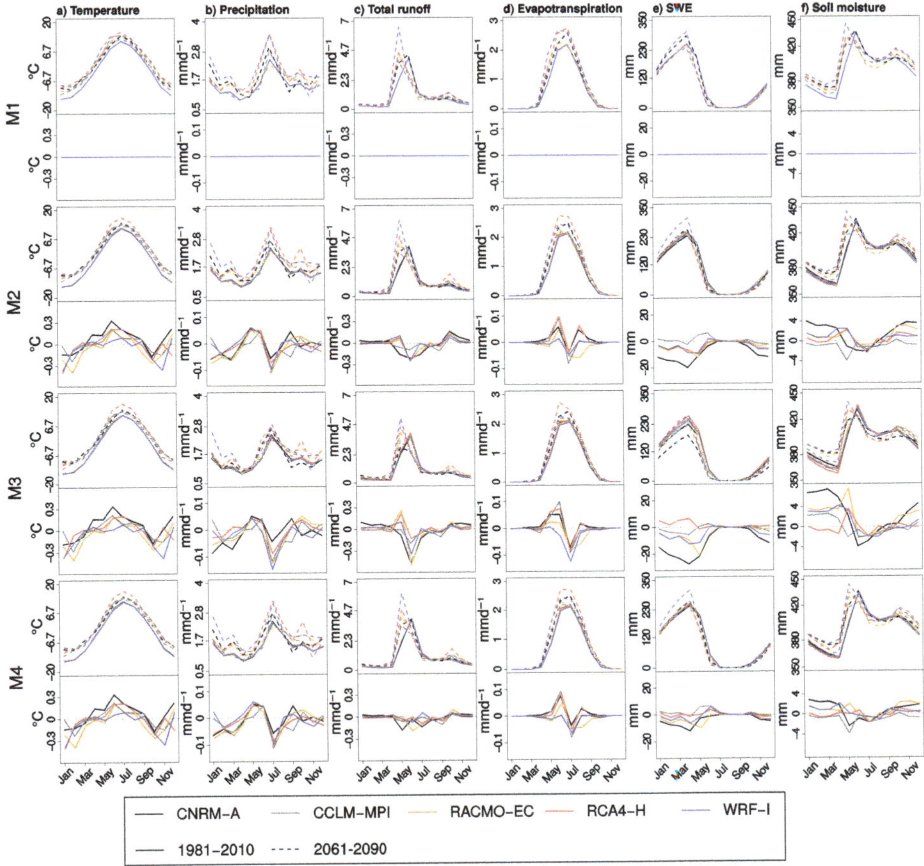

Figure 3. A real-world example showing the annual cycle of (**a**) daily mean temperature, (**b**) precipitation, (**c**) total runoff, (**d**) evapotranspiration, (**e**) snow water equivalent and (**f**) soil moisture in the Tornio river catchment in years (solid lines) 1981–2010 and (dashed lines) 2061–2090 separately for each of the GCM-RCMs when adjusted against WFDEI using the four (M1–M4) bias correction and delta change methods. To readily illustrate differences in comparison to WFDEI, the remaining biases in 1981–2010 are shown below the annual cycle panels separately for each method and variable.

2.5. Pseudo-Reality Framework

To make inferences on the potential future performance of the selected univariate and multivariate methods, intermodel cross-validation is performed using the so-called pseudo-reality approach (Figure 4). In the first stage, one GCM-RCM at a time is used as the verifying model (i.e., pseudo-reality) against which the rest of the models are adjusted using the selected methods. The bias adjusted simulations are then compared against the pseudo-reality GCM-RCM with a set of performance measures. The resulting cross-validation statistics are then averaged over all pseudo-realities to obtain an overall view of the bias correction performance in changing climatic conditions. The same framework is applied to the hydrological simulations to see to what extent the relative performance of the selected MOS methods differs when inspected from the hydrological modeling point of view. To this end, the bias adjusted temperature and precipitation time-series are used as input to the E-HYPE sub models, which are then run to simulate the future hydrological conditions in the selected catchments. Hydrological simulations are cross-validated in a similar manner as the GCM-RCM simulations using complementary performance measures.

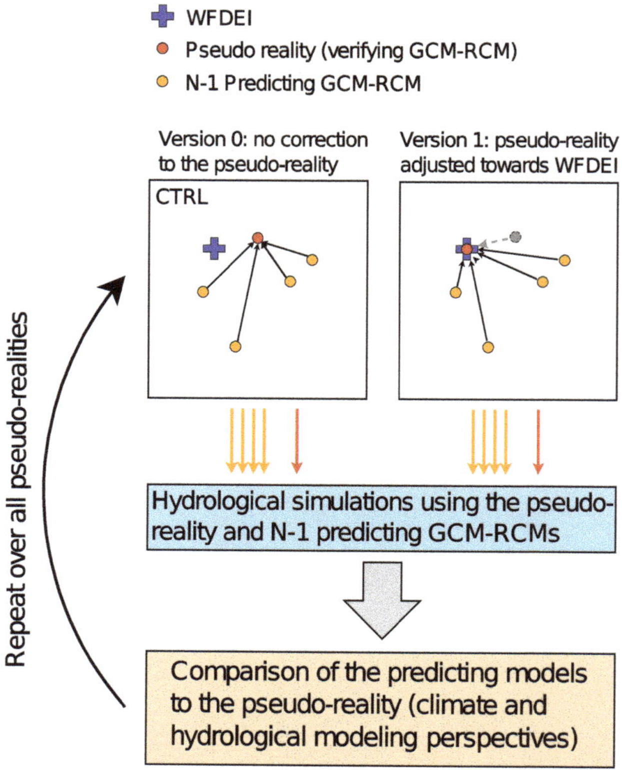

Figure 4. An illustration of the pseudo-reality framework procedures in the baseline period, applied both from climate modeling and hydrological modeling perspectives.

In order to improve the applicability of the pseudo-reality approach to hydrological simulations, two ways to construct pseudo-realities were tested (Figure 4): (a) raw GCM-RCM simulations were used as pseudo-realities without taking biases in relation to observations (i.e., WFDEI) into account [25]; (b) the annual cycle of the GCM-RCM acting as pseudo-reality was adjusted to biases in comparison to WFDEI by simply removing the mean bias at each day of the annual cycle using a 30-day sliding window. Daily adjustments were applied instead of monthly ones in order to avoid additional jumps in the annual cycle of the pseudo-reality time series. This shift in the mean values obviously alters the bias between pseudo-reality and the verifying models but leaves the changes in this relatively untouched (see Figure S2 in supplementary material). The motivation for the second approach is apparent: biases in relation to the observed climate are substantial in some of the selected GCM-RCMs, which leads to unrealistic hydrological model behavior both in the pseudo-reality runs and the verifying hydrological simulations. For example, substantial cold biases at high altitude regions in Sava sub-model and during winter in Tornio sub-model cause unrealistic volumes of snow to accumulate throughout the simulation periods. We argue that without this additional bias adjustment step the use of GCM-RCMs as pseudo-realities when cross-validating bias adjustment methods from hydrological modeling perspective might not be reasonable due to unrealistic shifts in hydrological regimes. One should note that although the intention is to keep the daily variability in the pseudo-reality time series untouched, the multiplicative scaling applied to daily precipitation slightly modifies the spread of precipitation distributions both in the baseline and scenario periods. This also slightly changes the daily variability of hydrological simulations accordingly.

2.6. Metrics for GCM-RCM Simulations

To assess the general similarity between the empirical cumulative probability distributions of the predicting models F_{pred} and the GCM-RCM acting as pseudo-reality F_{ver}, 2-sample Cramér–von Misés (CM) statistic [43] was calculated according to

$$\text{CM} = A\left\langle \frac{mn}{(m+n)^2} \left\{ \sum_{i=1}^{m} \left[\hat{F}_{\text{pred}}(x_i) - F_{\text{ver}}(x_i) \right]^2 + \sum_{j=1}^{n} \left[\hat{F}_{\text{pred}}(y_j) - F_{\text{ver}}(y_j) \right]^2 \right\} \right\rangle, \tag{3}$$

where $(\hat{\ })$ denotes the pooled sample of the four predicting GCM-RCM simulations, while m and n are the numbers of values within the pooled sample (x) and in pseudo-reality (y), respectively. The actual calculations were made for binned data using bin widths of $1\,^\circ\text{C}$ and $1\,\text{mm}\,\text{d}^{-1}$ and the same number of bins with identical bin boundaries for both predicting GCM-RCMs and the pseudo-reality GCM-RCM. $A\langle\rangle$ indicates an average over 12 months and the area of a sub-model. CM measures the similarity of two empirical distributions in probability space and puts more weight on discrepancies in the tails of the cumulative distributions than the widely used Kolmogorov–Smirnov statistic, which measures the maximum distance between the cumulative probability distributions. Comparison with these statistics did not reveal significant differences, and the results are shown only for CM.

The second statistic, mean absolute error (MAE), was calculated over quantiles i ($i \in [1, ..., 100]$) of the predicting and verifying (i.e., pseudo-reality) model distributions following

$$\text{MAE} = A\langle |\hat{F}_{\text{pred}}^{-1}(i) - F_{\text{ver}}^{-1}(i)| \rangle, \tag{4}$$

where $A\langle\rangle$ encompasses averaging over the distribution quantiles in addition to temporal and spatial averaging. The analysis was also repeated using the mean squared error, but the results did not show substantial differences to MAE. Thus, the relative method performance is illustrated in terms of MAE in the remainder of the paper.

Two statistics measuring errors in inter-variable correlations were calculated. First, to assess to what extent the linear correlation is modified by different methods, MAE in the Pearson correlation coefficient was calculated between the average correlation coefficient of the four verifying models and pseudo-reality, averaged in a similar manner as in Equation (3). Secondly, to evaluate the remaining errors in the full dependence structure, the empirical copula density was approximated from the pseudo-observations (u, v), estimated for the ith temperature (x) and precipitation (y) value as $u = \text{rank}(x_i)/(n+1)$ and $v = \text{rank}(y_i)/(n+1)$, where n is the number of values for both variables. These values were binned 2-dimensionally and normalized such that the histogram approximately corresponds to the copula density. The 2-dimensional binning was done at 0.1 interval. MAE between empirical copula densities of the predicting GCM-RCMs and pseudo-reality was then calculated according to

$$\text{MAE}_{\text{c}} = \frac{1}{n} \sum_{i=1}^{n} |\hat{c}(i)_{\text{pred}} - c(i)_{\text{ver}}|. \tag{5}$$

In Equation (5), $\hat{c}_{\text{pred}}$ denotes the empirical copula density averaged over the four predicting models, c_{ver} the copula density calculated for pseudo-reality and n is the number of bins used to estimate the copula density. In the following, the subscript c is dropped for brevity. MAE based on kernel density estimates were also tested but the resulting statistics depended substantially on the used kernel method and the kernel width and thus, were not considered further in this study. To reduce the effect of sampling noise to the results, temperature-precipitation pairs were pooled over the area of each sub-model and season before estimating the empirical copula densities. Identical values were handled using the same approach as in Gennaretti et al. [35]: ranks were first given randomly to identical values before estimating the empirical copula density. This was repeated 10 times and the final copula density was calculated as the average of the randomly ranked estimates. Despite being a simple and not a proper goodness-of-fit measure, this statistic readily illustrates how well each method is capable to adjust the full dependence structure. Gennaretti et al. [35] briefly pointed out that the measured performance depended on whether dry days were included when estimating the empirical copula density. While the focus is here on the copula density including the full time-series, the results for the wet-day copula can be found from the supplementary material (Figure S3).

2.7. Metrics for Hydrological Simulations

An additional set of cross-validation statistics was calculated for the hydrological indexes. First, quantile distributions of river discharge Q (i.e., flow-duration curves) were estimated at the outflow sub-basin of each of the sub-models. The average of the four predicting distributions was then compared against the pseudo-reality distribution using a logarithmic accuracy ratio (LAR10) defined as

$$\text{LAR10} = A\left\langle \left| \log_{10} \left(\frac{\hat{F}_{\text{pred}}^{-1}(i)}{F_{\text{ver}}^{-1}(i)} \right) \right| \right\rangle, \tag{6}$$

where $A\langle\rangle$ has the same meaning as in Equation (4). The statistic is symmetric in the sense that the same value is assigned for under- and overestimation of the same relative magnitude [17]. This alleviates the issue of most other relative accuracy measures penalizing overestimation more strongly than underestimation. In addition to distribution-averaged statistics, LAR10 was also inspected individually for the 5th (Q_5) and 99th (Q_{99}) percentile of the flow duration curve to see how the relative performance of the selected MOS methods varies in the tails of the monthly flow duration curves.

The analysis of river discharges is complemented by evaluating a set of individual flux and storage elements, which affect the overall water balance and river discharge generation. To this end, MAE was calculated for the monthly mean values of total runoff (R), evapotranspiration (E), soil moisture (S) and snow water equivalent (SWE) in a similar manner as for daily mean temperature and precipitation. MAE in the mean annual maximum SWE (SWE_{max}) was also calculated to evaluate how method differences are reflected in the simulation of highest snow pack depths. This allows us to make inferences on how the remaining errors in daily mean temperature and precipitation and in their inter-variable correlations affect the different hydrological elements, as well as to gain insights on the relative performance of the selected bias adjustment methods in terms of hydrological modeling results.

3. Results

3.1. Distribution-Averaged Statistics for Daily Mean Temperature and Precipitation

We first inspect the overall performance of the four methods from the GCM-RCM perspective. Figure 5 illustrates the distribution-averaged cross-validation statistics in the two scenario periods (see Figure S4 for the statistics in years 1981–2010). When first concentrating on the results shown for the years 2061–2090, it is seen that bias correction methods M2–M4 slightly outperform M1 in adjusting the temperature distribution in terms of both CM and MAE. On the other hand, both CM and MAE of M3 are very close to M2 and M4, which indicates that temperature can be reasonably modeled using a normal distribution. Although the general picture is mostly similar for precipitation, CM and MAE give a partially contrasting picture about the relative method performance. The CM values are smallest and almost identical for methods M2–M4, while the relative performance of M3 is slightly worse than M2 and M4 in terms of MAE. This might be partially explained by how the fraction of dry days is explicitly taken into account by M3 in its corrections, while using a gamma distribution to model the precipitation distribution might not capture biases in it as efficiently as non-parametric quantile mapping. Using a two-month time window generally reduces errors in both temperature and precipitation distributions, which is in line with the results of Räisänen and Räty [21] and Räty et al. [22]. As expected, all methods have substantially smaller errors in the marginal distributions of temperature and precipitation in comparison to the uncorrected model simulations (M0) in both periods.

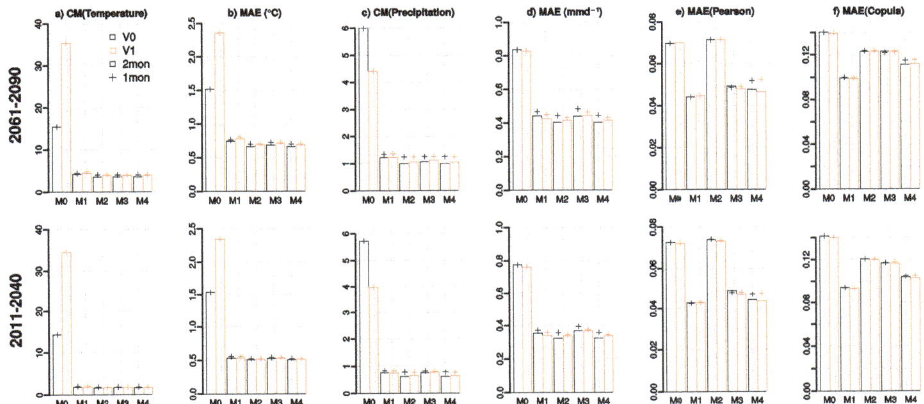

Figure 5. Cross-validated CM and MAE for (**a**,**b**) daily mean temperature and (**c**,**d**) daily precipitation distribution in years 2011–2040 (**bottom**) and 2061–2090 (**top**). Also shown are the MAE in (**e**) the Pearson correlation coefficient and (**f**) the empirical copula density. Black color denotes the cross-validation statistics for the pseudo-reality approach without additional adjustments (V0), while the results for the approach where pseudo-realities have been adjusted to biases in relation to WFDEI are shown in red (V1). Furthermore, crosses (bars) indicate the results for the one-month (two-month) time window used to estimate simulated changes or model biases, shown for both V0 and V1. Note that the differences between the one- and two-month time windows are typically small, as indicated by the small differences between the bars and crosses.

The MAE for the Pearson correlation coefficient and the empirical copula density, when calculated over the full monthly time-series of temperature and precipitation, is also shown in Figure 5. The results for the Pearson correlation coefficient show that, although M3 and M4 improve the results in comparison to method M2, M1 performs slightly better in capturing the linear correlation between temperature and precipitation than the other methods. Moreover, M4 seems to be susceptible to the effect of noise, as M3 has a somewhat smaller MAE when the one-month time window is used. The situation is slightly different when the MAE in the empirical copula density is considered. While M1 has again the best performance out of all methods, M2 has now MAE values which are closer to the bi-variate methods. The modest improvement obtained with M4 in comparison to M1 is again at least partially related to the small sample size, as indicated by the reduction in the MAE values for the two-month time window. Yet, this highlights the difficulty to robustly estimate biases in inter-variable correlations in a changing climate. As M1 has a superior performance in terms of both of the two measures regardless of the period considered, this suggests that the inter-variable correlations do not change substantially among the selected models and within the studied regions.

The bottom row shows the cross-validation statistics for the near-term scenario period (2011–2040). As expected, the remaining errors are generally smaller for all methods in this period. The marginal distributions of both temperature and precipitation are slightly better captured by method M1 in comparison to other methods, while the relative performance of other methods does not show marked differences between the two periods. Furthermore, the MAE in the Pearson correlation coefficient and the copula density indicate a slightly improved performance for M3 and M4 in comparison to M1, although M1 still has the smallest MAE in all cases.

In qualitative terms, the cross-validation statistics are similar for temperature and precipitation regardless of the pseudo-reality approach. By far, the largest differences are shown by method M0 for which the cross-validation statistics calculated for temperature deteriorate when correcting the pseudo-reality GCM-RCM toward WFDEI (V1), while the opposite happens for the precipitation statistics. For temperature, the larger MAE in V1 is explained by the systematic cold bias within the GCM-RCM ensemble. However, for methods M1-M4 the results are mostly similar between the two pseudo-reality approaches, although the cross-validation statistics for the temperature and precipitation distributions tend to be slightly worse for the two-month time window after pseudo-realities have been adjusted against WFDEI (V1). This suggests that, from the climate modeling perspective, the additional adjustment step does not substantially modify the cross-validation statistics apart from the uncorrected model simulations, backing up its use in the hydrological modeling step.

While not the specific target of this study, it should be mentioned that an inherent property of M4 is that in order to obtain correct ranks for each temperature and precipitation pair, both time series need to be temporally re-ordered. This is to a lesser extent an issue in M3, in which only the temporal sequence of precipitation is potentially modified. As the temporal re-ordering might affect the hydrological simulations, a modified version of M4 was tested. First, the time series of uncorrected temperature and precipitation were divided into dry and wet days in a similar manner as in M3. Next, M2 was applied separately on wet-day and dry-day distributions to retain the improved statistics for them, as obtained with M4. Finally, the N-pdft algorithm was applied only on wet-day distributions of temperature and precipitation. Tests with the modified algorithm showed, however, that although the cross-validated MAE of both correlation measures decreased slightly, changes in the cross-validation statistics for hydrological variables in comparison to the original method varied non-systematically depending on the season, region and variable considered (not shown) and, thus, did not offer systematic improvements in comparison to the original algorithm.

3.2. Cross-Validation Statistics for Hydrological Simulations

3.2.1. Spatially Distributed Variables

How are the differences in the ability of the four methods to adjust the joint distribution of temperature and precipitation reflected in the hydrological simulations? Figure 6 shows the average cross-validation statistics for the five hydrological components (R, E, S, SWE and SWE_{max}) in the two scenario periods (see Figure S5 for the statistics in years 1981–2010). For clarity, the results are shown only for the two-month time window in the remainder of the paper. When looking at the MAE in monthly mean runoff (R) in the late 21st century period, it is seen that M4 outperforms the other methods in pseudo-reality approach V0, although the differences are small in comparison to M1. In pseudo-reality approach V1, however, MAE is practically identical for M1–M2 and M4. M3 has a somewhat worse performance than the other methods, which is likely caused by the larger remaining errors in the precipitation distribution than for the other methods. When evapotranspiration (E) as simulated by HYPE is considered, bi-variate methods M3 and M4 have the smallest MAE in monthly mean values and M3 actually has the smallest MAE in pseudo-reality approach V0. In addition, the results for E illustrate a side-effect of the additional pseudo-reality adjustment (V1): the MAE in E is systematically larger in V1 for all methods, as the systematic underestimation of temperature in V0 likely leads to too weak evapotranspiration in comparison to real-world hydrological simulations.

We next take a look at the cross-validation statistics for the two storage variables. The MAE of soil moisture (S) is almost identical for all methods, which indicates that it is relatively insensitive to the adjustment of daily-scale inter-variable correlations. The small differences between the four methods tend to follow those seen in the MAE calculated over the precipitation distribution, as the

MAE is smallest and almost identical for methods M4 and for M2. The last two panels in Figure 6 show the MAE of the monthly mean SWE and SWE$_{max}$. These results illustrate the main benefit of pseudo-reality approach V1. As predicted, the adjustment of pseudo-reality GCM-RCMs reduces biases in snow variables, although with the expense of increased MAE for E, as discussed before. This also causes differences in the relative ranking of the correction methods between the two pseudo-reality approaches (V0 and V1); M4 performs slightly worse in relation to M1 in reality approach V0, whereas the opposite is seen after adjusting the pseudo-reality GCM-RCMs towards WFDEI (V1). Overall, these results indicate that the simulation of most hydrological aspects is only marginally improved by joint bias correction and that the accurate adjustment of marginal distributions plays a more important role, at least when only temperature and precipitation are used as input in a hydrological model, such as HYPE.

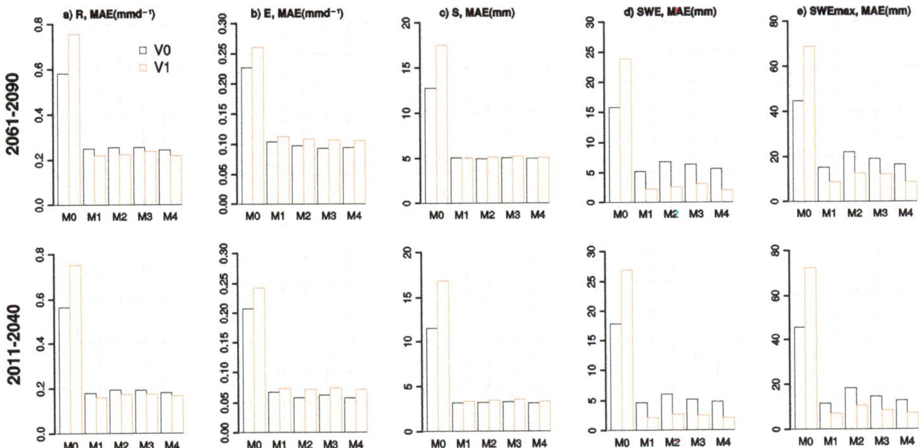

Figure 6. Similar to Figure 5 but for the cross-validated MAE of monthly mean (**a**) total runoff, (**b**) evapotranspiration, (**c**) soil moisture, (**d**) snow water equivalent and (**e**) the mean annual maximum of snow water equivalent in years (**bottom**) 2011–2040 and (**top**) 2061–2090.

The cross-validation statistics for the near-term scenario period are in line with the corresponding statistics of temperature and precipitation, with generally smaller errors in all studied hydrological aspects than in the later scenario period. The relatively better performance of M1, when adjusting the joint distribution of temperature and precipitation at that time is to some extent reflected in the hydrological simulations (bottom of row Figure 6), as R, E and S are all better captured by M1 in the near-future period. In contrast to the later scenario period, M2 and M4 have smaller MAE values in monthly mean evapotranspiration than M3. The cross-validation statistics of monthly mean SWE and SWE$_{max}$ show the largest differences between bias adjustment methods also in this period, indicating that method choice is most important for this variable from the studied hydrological aspects.

3.2.2. Evaluation of Future River Discharges

The analysis is complemented by illustrating the cross-validated LAR10 for Q$_5$, Q$_{99}$ as well as for the distribution-averaged LAR10 in the two scenario periods (Figure 7). The absolute values of LAR10 vary to some extent between the two pseudo-reality approaches. For example, LAR10 of Q$_5$ is systematically smaller in V1 for all methods (apart from M0) in both periods, while the opposite is seen in the Q$_{99}$ in the early 21st century period. Furthermore, the performance of all methods is extremely consistent when the distribution-averaged LAR10 is considered. Methods M2 and M4 have a marginally smaller LAR10 than M1 and M3, while in the earlier scenario period method M1 performs equally well or even better than M2 and M4. Also the best performing method depends on the pseudo-reality approach when low flows (Q_5) are considered. In V0, method M2 somewhat outperforms the other methods in both periods, while in V1 method M3 has a slightly better performance in comparison to the other methods. On the other hand, M1 has the largest LAR10 values in both periods, which is probably related to the larger errors in temperature and evapotranspiration accordingly. The simulation of Q_{99} seems to marginally benefit from the adjustment of inter-variable correlations, as M4 has the smallest LAR10 among the four methods, particularly in years 2011–2040. Again, the LAR10 is larger for M3 than for the other methods, most likely due to the combination of the aforementioned issues.

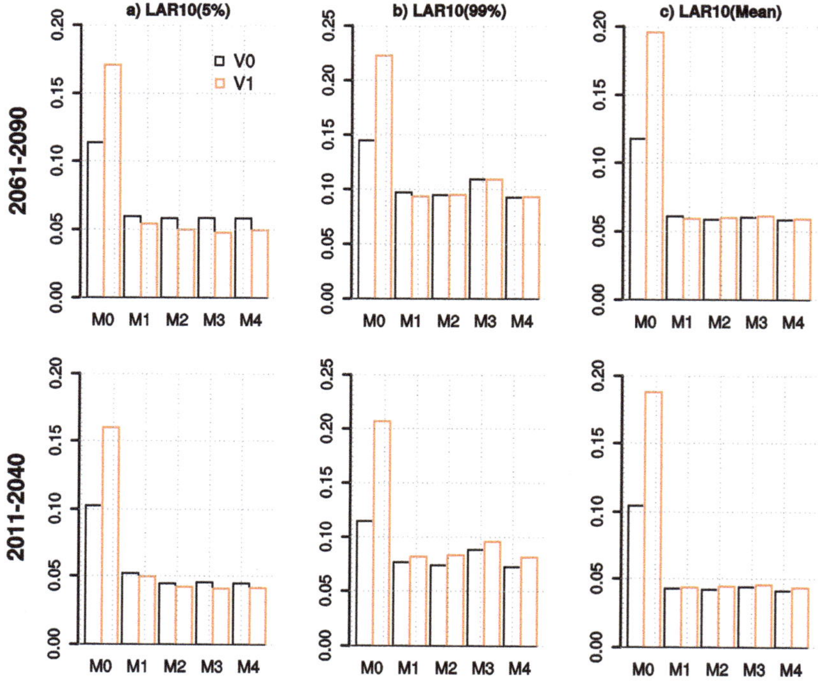

Figure 7. Similar to Figure 5 but for the cross-validated LAR10 in (**a**) the 5th and (**b**) 99th percentile of flow duration curves shown together with (**c**) the distribution-averaged LAR10 in years 2011–2040 (**bottom**) and 2061–2090 (**top**).

3.3. Temporal and Spatial Variations in the Cross-Validation Statistics

Seasonal and spatial variations in both the relative performance of the studied methods and contributions of different hydrological processes are reflected in the cross-validation statistics for the hydrological simulations. To better infer the reasons for these variations, Figures 8 and 9 show matrices of cross-validated MAE for different aspects of the joint distribution of temperature and precipitation and the distributed hydrological variables in winter and summer seasons in the four sub-models (see Figures S6 and S7 for the same statistics in spring and autumn seasons). In absolute terms the largest MAE values for temperature are seen in Tornio in both seasons, while for precipitation the MAE is generally largest in Sava. The pattern is less clear for the correlation measures, which show larger variations between the four regions. The numeric values in the panels denote the percentage difference to M1. These values indicate that methods M2–M4 perform better than M1 in most cases in winter, while in summer the temperature is relatively well captured by M1. On the other hand, the statistics for both correlation measures indicate that M2 has systematically poorer performance than the other methods and the relative difference is particularly large in Tornio and Ems. The most striking feature in Figure 8 is the consistently better performance of M1 in capturing the empirical copula density when compared to the other methods, regardless of region or season; apart from M4, which has a relatively similar performance in winter, M1 outperforms the other methods by a large margin.

When the cross-validation statistics for temperature and precipitation are compared against the statistics for the hydrological variables, it is evident that the relative differences between the methods are mostly explained by their capability to adjust the marginal distributions of temperature and precipitation, particularly in those regions and seasons, where snow processes play a less important role in the hydrological cycle. Backing up the previous conclusions, the added value of the adjustment of temperature and precipitation dependence structure, as indicated by differences between methods M2 and M4, is most visible in Tornio and Sava sub-models, where M4 systematically improves the simulation of SWE. The link between the improved simulation of SWE and improvements in total runoff and soil moisture is apparent, as M4 has a smaller MAE than M2 in both variables. Tornio and Sava also show the largest differences between pseudo-reality approaches V0 and V1, as SWE has substantial errors remaining even in summer in Tornio, when V0 is used.

Results from the previous section suggest that quantile mapping applied as a delta change method (M1) has a relatively robust performance from a hydrological modeling perspective. On sub-model scale this is only partially true, as M2 and M4 tend to have better performance in the northern sub-models, while M1 performs particularly well in Sava catchment. However, the relative differences are small in many cases also in other regions, which suggests that the delta change approach might be a good alternative for bias correction. From individual methods, M3 has largest variations in its relative performance between the four sub-models and two seasons. These variations seem not to be solely due to issues with the marginal distributions, and in Tornio, for example, failures to capture the full dependence structure in winter (cf. Figure 2) might deteriorate the statistics for M3.

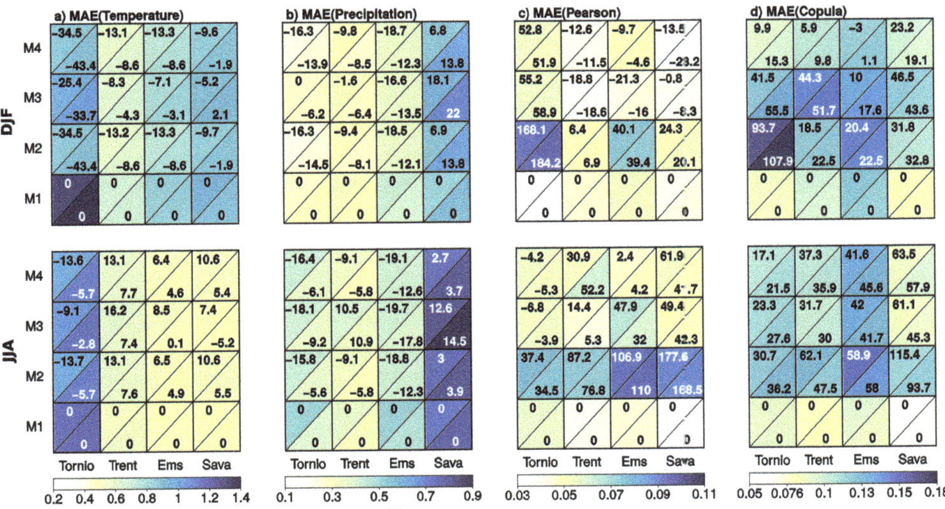

Figure 8. Panels showing the cross-validated MAE (colors) of (**a**) daily mean temperature, (**b**) daily precipitation, (**c**) Pearson correlation coefficient and (**d**) empirical copula density separately for each method (panel rows) at each hydrological sub-domain (panel columns) in years 2061–2090, when two-month time window has been used to estimate simulated changes or model biases. Values for the pseudo-reality approach V0 (V1) are plotted in the upper (lower) triangle of each cell and are shown separately for (**top**) winter and (**bottom**) summer months. In addition, percentage differences to M1 are shown as numeric values for each element.

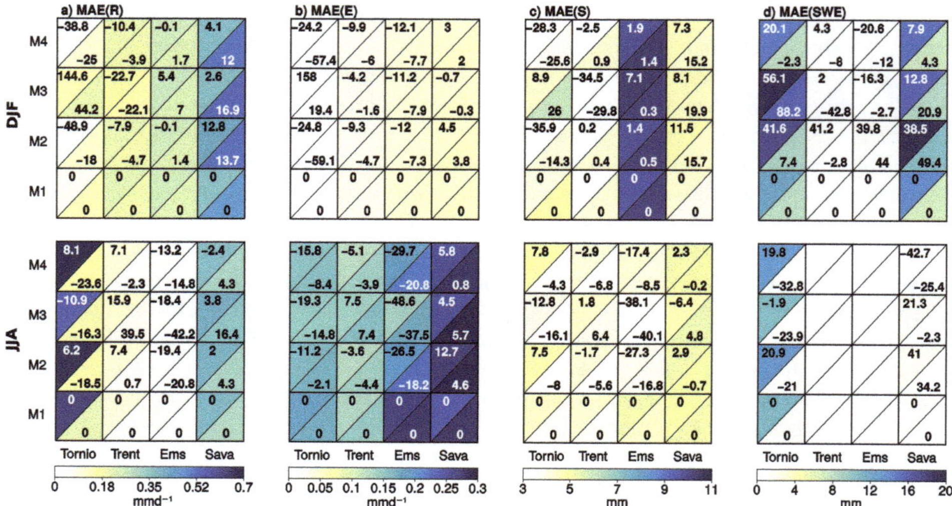

Figure 9. As in Figure 8 but shown for monthly mean (**a**) total runoff, (**b**) evapotranspiration, (**c**) soil moisture and (**d**) snow water equivalent.

4. Discussion and Conclusions

This paper presents the results from a study in which two joint bias corrections methods applied in the bi-variate mode are compared against quantile mapping applied both traditionally and in the delta change mode using five EURO-CORDEX GCM-RCM simulations as a proxy for the future climate. The evaluation is two-fold: first, cross-validation is performed to obtain quantitative estimates for the relative method performance in the early and late 21st century conditions, when applied to construct future projections of the joint distribution of daily mean temperature and precipitation; second, these projections were fed to a hydrological model to assess, whether or not the bi-variate adjustments improve future hydrological simulations in comparison to univariate quantile mapping.

The main results of these exercises are summarized as follows:

- By design, joint bias correction brings the inter-variable correlations closer to the observed one in the baseline period. In particular, the iterative N-pdft algorithm (M4) reproduces the full dependence structure (as measured by the MAE in the empirical copula density) well in comparison to univariate quantile mapping (M2). The adjustment of inter-variable correlation in M3 might fail in certain situations, as the method tends to have larger remaining biases in the copula structure in winter conditions in HYPE Tornio sub-model.
- Cross-validation statistics in years 2011–2040 and 2061–2090 indicate that although the correlation structure is improved in terms of Pearson correlation, the benefit of bi-variate methods is less clear when the full dependence structure is considered. Part of the modest improvement is likely explained by the limited sample size, which might lead to over-fitting to the present-day climatic conditions. On the other hand, quantile mapping applied in the delta change mode (M1) often performs better than the other methods, which indicates that retaining present-day correlation structures of temperature and precipitation might be sufficient also in future projections.
- The results suggest that the pseudo-reality approach is potentially useful for evaluating the relative performance of bias adjustment methods from hydrological modeling perspective in the future climate. However, to improve the validity of the conclusions in these types of studies, the implementation of pseudo-reality framework needs to be designed on case-by-case basis, for example by first bias adjusting the pseudo-reality GCM-RCMs to avoid unrealistic shifts in hydrological regimes.

- For the hydrological variables, the bi-variate approaches offered no substantial advantage over the univariate methods with M4 often having similar performance to M2. Only marginal improvements in comparison to methods M1 and M2 are seen in the cross-validation statistics for high flows and for the monthly mean and annual maximum snow water equivalent in Tornio and Sava. Although quantile mapping applied as a delta change method (M1) has slightly poorer performance in projecting marginal distributions of temperature and precipitation than quantile mapping-based bias correction (M2) and its bi-variate version (M4), the cross-validation statistics indicate that it has a relatively good ability to capture the future hydrological conditions. Nevertheless, for the hydrological variables studied (apart from snow), there were only small differences in cross-validation statistics between the tested methods, indicating that care should be taken when selecting MOS methods for particular purposes and (ideally) several methods should be used in parallel. Overall, the results highlight the difficulty to illustrate the added value of more complex methods, when applying them in producing projections for daily mean temperature and precipitation.

The main shortcoming of this study is the limited number of GCM-RCM simulations available for cross-validation tests, and optimally a larger set of model simulations should be used. Furthermore, the response to different bias correction and delta change algorithms is likely dependent on the hydrological model and the used parameterisations. For example, earlier studies have shown e.g., [44] that projections for evapotranspiration based on parameterising potential evapotranspiration using only temperature are not suitable for all climatic conditions. Furthermore, snow processes were parameterized in the HYPE simulations using a simple degree-day algorithm, which does not take solar radiation and other meteorological factors into account. More complex parameterizations, which require multiple variables as input, should be evaluated in further studies. If bias correction of a higher dimensional joint distribution were required, more sophisticated bias correction methods could, at least in principle, provide larger improvements in comparison to univariate methods, depending on the available data for robust calibration. This has been demonstrated by Cannon [17], who showed that M4 performs very well when adjusting a higher dimensional distribution for Canadian Forest Fire Weather Index calculations in the present-day climate. Moreover, different implementations of M3 could also be studied in future research. Most importantly, Gaussian copula is unlikely to be the optimal choice for describing the temperature and precipitation dependence structure in some cases and the use of other copulas should be further explored.

The presented framework allows to make some inferences about the ability of bias correction and delta change methods in constructing projections for future climate and their applicability from hydrological modeling perspective, an information of major interest to the impact modeling community. However, these tests are not sufficient alone to determine whether a particular method is suitable for climate change assessments. Additional tests such as those implemented in the VALUE framework [24] should be conducted to obtain a complete picture of benefits and limitations of bias correcting inter-variable correlations. As discussed above, the used approach does not easily allow to evaluate the potential benefits/adverse effects of the modification of temporal sequencing to the hydrological model results, which would require temporally synchronized model simulations and reference data. Furthermore, the effect of errors in the spatial representation of climate model output caused (e.g.,) by differing topography should be studied comprehensively to see how sensitive future hydrological simulations are to the correct representation of spatial fields.

Supplementary Materials: Figures S1–S7 are available online at http://www.mdpi.com/2225-1154/6/2/33/s1.

Author Contributions: C.D. selected and provided the hydrological sub-models and O.R. ran the hydrological simulations. O.R. and J.R. implemented the bias adjustment methods. All authors contributed to the planning of the study, the analysis of the results and writing the manuscript.

Acknowledgments: The study is supported by the Academy funded Center of Excellence (project No. 307331). Olle Räty is funded by the Vilho, Lauri and Yrjö Väisälä Foundation and by NordForsk through project number 74456 "Statistical Analysis of Climate Projections" (eSACP). Further funding was received from the projects HazardSupport and AQUACLEW. HazardSupport is financed by the Swedish Civil Contingencies Agency, MSB (grant No. 2015-3631). AQUACLEW is part of ERA4CS, an ERA-NET initiated by JPI Climate and funded by FORMAS (SE), DLR (DE), BMWFW (AT), IFD (DK), MINECO (ES) and ANR (FR) with co-funding by the European Union (Grant 690462). We would also like to thank the climate modeling groups participating the European branch of Coordinated Regional Climate Downscaling Experiment (EURO-CORDEX) for producing and making their model output publicly available.

Conflicts of Interest: The authors declare no conflict of interest.

References

1. Déqué, M. Frequency of precipitation and temperature extremes over France in an anthropogenic scenario: Model results and statistical correction according to observed values. *Glob. Planet. Chang.* **2007**, *57*, 16–26. [CrossRef]

2. Yang, W.; Andréasson, J.; Phil Graham, L.; Olsson, J.; Rosberg, J.; Wetterhall, F. Distribution-based scaling to improve usability of regional climate model projections for hydrological climate change impacts studies. *Hydrol. Res.* **2010**, *41*, 211–229. [CrossRef]

3. Olsson, J.; Berggren, K.; Olofsson, M.; Viklander, M. Applying climate model precipitation scenarios for urban hydrological assessment: A case study in Kalmar City, Sweden. *Atmos. Res.* **2009**, *92*, 364–375. [CrossRef]

4. Li, H.; Sheffield, J.; Wood, E.F. Bias correction of monthly precipitation and temperature fields from Intergovernmental Panel on Climate Change AR4 models using equidistant quantile matching. *J. Geophys. Res. Atmos.* **2010**, *115*. [CrossRef]

5. Maraun, D.; Wetterhall, F.; Ireson, A.M.; Chandler, R.E.; Kendon, E.J.; Widmann, M.; Brienen, S.; Rust, H.W.; Sauter, T.; Themeßl, M.; et al. Precipitation downscaling under climate change: Recent developments to bridge the gap between dynamical models and the end user. *Rev. Geophys.* **2010**, *48*, RG3003. [CrossRef]

6. Themeßl, M.J.; Gobiet, A.; Leuprecht, A. Empirical-statistical downscaling and error correction of daily precipitation from regional climate models. *Int. J. Climatol.* **2011**, *31*, 1530–1544. [CrossRef]

7. Hempel, S.; Frieler, K.; Warszawski, L.; Schewe, J.; Piontek, F. A trend-preserving bias correction –the ISI-MIP approach. *Earth Syst. Dyn.* **2013**, *4*, 219–236. [CrossRef]

8. Cannon, A.J.; Sobie, S.R.; Murdock, T.Q. Bias Correction of GCM Precipitation by Quantile Mapping: How Well Do Methods Preserve Changes in Quantiles and Extremes? *J. Clim.* **2015**, *28*, 6938–6959. [CrossRef]

9. Teutschbein, C.; Seibert, J. Bias correction of regional climate model simulations for hydrological climate-change impact studies: Review and evaluation of different methods. *J. Hydrol.* **2012**, *456–457*, 12–29. [CrossRef]

10. Chen, J.; Brissette, F.P.; Chaumont, D.; Braun, M. Finding appropriate bias correction methods in downscaling precipitation for hydrologic impact studies over North America. *Water Resour. Res.* **2013**, *49*, 4187–4205. [CrossRef]

11. Trenberth, K.E.; Shea, D.J. Relationships between precipitation and surface temperature. *Geophys. Res. Lett.* **2005**, *32*, L14703. [CrossRef]

12. Berg, P.; Haerter, J.O.; Thejll, P.; Piani, C.; Hagemann, S.; Christensen, J.H. Seasonal characteristics of the relationship between daily precipitation intensity and surface temperature. *J. Geophys. Res. Atmos.* **2009**, *114*. [CrossRef]

13. Piani, C.; Haerter, J.O. Two dimensional bias correction of temperature and precipitation copulas in climate models. *Geophys. Res. Lett.* **2012**, *39*. [CrossRef]

14. Li, C.; Sinha, E.; Horton, D.E.; Diffenbaugh, N.S.; Michalak, A.M. Joint bias correction of temperature and precipitation in climate model simulations. *J. Geophys. Res. Atmos.* **2014**, *119*, 153–162. [CrossRef]

15. Mehrotra, R.; Sharma, A. A Multivariate Quantile-Matching Bias Correction Approach with Auto- and Cross-Dependence across Multiple Time Scales: Implications for Downscaling. *J. Clim.* **2016**, *29*, 3519–3539. [CrossRef]

16. Cannon, A.J. Multivariate Bias Correction of Climate Model Output: Matching Marginal Distributions and Intervariable Dependence Structure. *J. Clim.* **2016**, *29*, 7045–7064. [CrossRef]

17. Cannon, A.J. Multivariate quantile mapping bias correction: An N-dimensional probability density function transform for climate model simulations of multiple variables. *Clim. Dyn.* **2017**, *50*, 31–49. [CrossRef]

18. Ehret, U.; Zehe, E.; Wulfmeyer, V.; Warrach-Sagi, K.; Liebert, J. HESS Opinions "Should we apply bias correction to global and regional climate model data?". *Hydrol. Earth Syst. Sci.* **2012**, *16*, 3391–3404. [CrossRef]

19. Maraun, D. Bias Correcting Climate Change Simulations—A Critical Review. *Curr. Clim. Chang. Rep.* **2016**, *2*, 211–220. [CrossRef]

20. Maraun, D. Nonstationarities of regional climate model biases in European seasonal mean temperature and precipitation sums. *Geophys. Res. Lett.* **2012**, *39*, L06706. [CrossRef]

21. Räisänen, J.; Räty, O. Projections of daily mean temperature variability in the future: cross-validation tests with ENSEMBLES regional climate simulations. *Clim. Dyn.* **2013**, *41*, 1553–1568. [CrossRef]

22. Räty, O.; Räisänen, J.; Ylhäisi, J.S. Evaluation of delta change and bias correction methods for future daily precipitation: intermodel cross-validation using ENSEMBLES simulations. *Clim. Dyn.* **2014**, *42*, 2287–2303. [CrossRef]

23. Van Schaeybroeck, B.; Vannitsem, S. Assessment of calibration assumptions under strong climate changes. *Geophys. Res. Lett.* **2016**, *43*, 1314–1322. [CrossRef]

24. Maraun, D.; Widmann, M.; Gutiérrez, J.M.; Kotlarski, S.; Chandler, R.E.; Hertig, E.; Wibig, J.; Huth, R.; Wilcke, R.A. VALUE: A framework to validate downscaling approaches for climate change studies. *Earth's Future* **2015**, *3*, 1–14. [CrossRef]

25. Velázquez, J.A.; Troin, M.; Caya, D.; Brissette, F. Evaluating the Time-Invariance Hypothesis of Climate Model Bias Correction: Implications for Hydrological Impact Studies. *J. Hydrometeorol.* **2015**, *16*, 2013–2026. [CrossRef]

26. Jacob, D.; Petersen, J.; Eggert, B.; Alias, A.; Christensen, O.B.; Bouwer, L.M.; Braun, A.; Colette, A.; Déqué, M.; Georgievski, G.; et al. EURO-CORDEX: New high-resolution climate change projections for European impact research. *Reg. Environ. Chang.* **2014**, *14*, 563–578. [CrossRef]

27. Weedon, G.P.; Balsamo, G.; Bellouin, N.; Gomes, S.; Best, M.J.; Viterbo, P. The WFDEI meteorological forcing data set: WATCH Forcing Data methodology applied to ERA-Interim reanalysis data. *Water Resour. Res.* **2014**, *50*, 7505–7514. [CrossRef]

28. Rust, H.W.; Kruschke, T.; Dobler, A.; Fischer, M.; Ulbrich, U. Discontinuous Daily Temperatures in the WATCH Forcing Datasets. *J. Hydrometeorol.* **2015**, *16*, 465–472. [CrossRef]

29. Federated ESGF-CoG Nodes. Available online: https://esgf.llnl.gov/nodes.html (accessed on 27 April 2016).

30. Moss, R.H.; Edmonds, J.A.; Hibbard, K.A.; Manning, M.R.; Rose, S.K.; van Vuuren, D.P.; Carter, T.R.; Emori, S.; Kainuma, M.; Kram, T.; et al. The next generation of scenarios for climate change research and assessment. *Nature* **2010**, *463*, 747–756. [CrossRef] [PubMed]

31. Donnelly, C.; Andersson, J.C.; Arheimer, B. Using flow signatures and catchment similarities to evaluate the E-HYPE multi-basin model across Europe. *Hydrolog. Sci. J.* **2016**, *61*, 255–273. [CrossRef]

32. Lindström, G.; Pers, C.; Rosberg, J.; Strömqvist, J.; Arheimer, B. Development and testing of the HYPE (Hydrological Predictions for the Environment) water quality model for different spatial scales. *Hydrol. Res.* **2010**, *41*, 295–319. [CrossRef]

33. HYPE Open Source Code. Available online: http://hypecode.smhi.se/ (accessed on 13 December 2016).

34. Donnelly, C.; Greuell, W.; Andersson, J.; Gerten, D.; Pisacane, G.; Roudier, P.; Ludwig, F. Impacts of climate change on European hydrology at 1.5, 2 and 3 degrees mean global warming above preindustrial level. *Clim. Chang.* **2017**, *143*, 13–26. [CrossRef]

35. Gennaretti, F.; Sangelantoni, L.; Grenier, P. Toward daily climate scenarios for Canadian Arctic coastal zones with more realistic temperature-precipitation interdependence. *J. Geophys Res. Atmos.* **2015**, *120*, 862–877. [CrossRef]

36. Pitié, F.; Kokaram, A.C.; Dahyot, R. Automated colour grading using colour distribution transfer. *Comput. Vis. Image Underst.* **2007**, *107*, 123–137. [CrossRef]

37. Sklar, A. Fonctions de répartition à n dimensions et leurs marges. *Publ. Inst. Stat. Univ. Paris* **1959**, *8*, 229–231.

38. BCUH: University of Helsinki bias adjustment tools. Available online: https://github.com/RatyO/BCUH (accessed on 9 April 2018).

39. R Core Team. *R: A Language and Environment for Statistical Computing*; R Foundation for Statistical Computing: Vienna, Austria, 2017.

40. Cannon, A. MBC: Multivariate Bias Correction of Climate Model Outputs. Available online: https://CRAN. R-project.org/package=MBC (accessed on 20 April 2017).
41. Reiter, P.; Gutjahr, O.; Schefczyk, L.; Heinemann, G.; Casper, M. Does applying quantile mapping to subsamples improve the bias correction of daily precipitation? *Int. J. Climatol.* **2017**, *38*, 1623–1633. [CrossRef]
42. Rajczak, J.; Kotlarski, S.; Schär, C. Does Quantile Mapping of Simulated Precipitation Correct for Biases in Transition Probabilities and Spell Lengths? *J. Clim.* **2016**, *29*, 1605–1615. [CrossRef]
43. Anderson, T.W. On the distribution of the two-sample Cramer-von Mises criterion. *Ann. Math. Stat.* **1962**, *33*, 1148–1159. [CrossRef]
44. Hagemann, S.; Chen, C.; Haerter, J.O.; Heinke, J.; Gerten, D.; Piani, C. Impact of a Statistical Bias Correction on the Projected Hydrological Changes Obtained from Three GCMs and Two Hydrology Models. *J. Hydrometeorol.* **2011**, *12*, 556–578. [CrossRef]

Article

Projected Changes in Precipitation, Temperature, and Drought across California's Hydrologic Regions in the 21st Century

Minxue He *, Andrew Schwarz, Elissa Lynn and Michael Anderson

California Department of Water Resources, 1416 9th Street, Sacramento, CA 95814, USA;
Andrew.Schwarz@water.ca.gov (A.S.); Elissa.Lynn@water.ca.gov (E.L.);
Michael.L.Anderson@water.ca.gov (M.A.)
* Correspondence: kevin.he@water.ca.gov; Tel.: +1-916-651-9634

Received: 21 March 2018; Accepted: 20 April 2018; Published: 23 April 2018

Abstract: This study investigated potential changes in future precipitation, temperature, and drought across 10 hydrologic regions in California. The latest climate model projections on these variables through 2099 representing the current state of the climate science were applied for this purpose. Changes were explored in terms of differences from a historical baseline as well as the changing trend. The results indicate that warming is expected across all regions in all temperature projections, particularly in late-century. There is no such consensus on precipitation, with projections mostly ranging from −25% to +50% different from the historical baseline. There is no statistically significant increasing or decreasing trend in historical precipitation and in the majority of the projections on precipitation. However, on average, precipitation is expected to increase slightly for most regions. The increases in late-century are expected to be more pronounced than the increases in mid-century. The study also shows that warming in summer and fall is more significant than warming in winter and spring. The study further illustrates that, compared to wet regions, dry regions are projected to become more arid. The inland eastern regions are expecting higher increases in temperature than other regions. Particularly, the coolest region, North Lahontan, tends to have the highest increases in both minimum and maximum temperature and a significant amount of increase in wet season precipitation, indicative of increasing flood risks in this region. Overall, these findings are meaningful from both scientific and practical perspectives. From a scientific perspective, these findings provide useful information that can be utilized to improve the current flood and water supply forecasting models or develop new predictive models. From a practical perspective, these findings can help decision-makers in making different adaptive strategies for different regions to address adverse impacts posed by those potential changes.

Keywords: California; hydrologic regions; warming; drought

---

1. Introduction

Understanding hydroclimatic changes and trends is of important scientific and practical significance for water resources management [1,2]. In particular, this understanding helps: (1) characterize the behavior of hydroclimatic variables (e.g., precipitation and temperature) as well as extreme events (e.g., droughts); (2) inform the development and enhancement of predictive tools to forecast future occurrence of these events; and (3) develop mitigation and adaptation plans to minimize the adverse impacts of unavoidable changes. This is particularly critical in arid and semi-arid areas including the State of California.

As the home to more than 37 million people [3] and a top-ten economy in the world, California's growth has been largely dependent on its ability to manage limited water resources [4]. In California, most of the precipitation falls in the northern half of the state, while the majority of the demand comes from the southern half where most of the population and farmlands are located. In addition, available water for supply in the state mostly comes during the wet season (November to April) as most precipitation falls in this period, while the demand is typically the highest in the dry season (late spring and summer) [5]. Furthermore, the state is prone to hydroclimatic extremes [1], with the most recent examples being the record-setting 2012–2015 drought and flooding in 2017. In the face of the geographically and temporally uneven distribution of water resources, the state traditionally relies on statewide and regional water storage and transfer projects, including the State Water Project (SWP) and the Central Valley Project (CVP), to redistribute water to meet multiple and often competing water management objectives [6]. However, the system was designed using hydroclimatic data of the first half of the 20th century. Since then, significant changes have been observed and reported, including increasing temperature, declining mountain snowpack, earlier snowmelt and streamflow peaking, higher percentage of precipitation falling as rainfall rather than snowfall, and increasing sea level, among others [7–17]. Those changes would likely amplify and accelerate in the future as the state's hydroclimate continues to change in a changing climate. In addition, as the population and economy continue to grow, natural hazards including extreme flooding and drought events pose a greater risk [18,19]. Those factors collectively make reliable water supply and drought and flood management in the state unprecedentedly challenging [20].

In light of their importance, many studies have focused on characterizing potential future hydroclimatic events in California [21–30]. These studies mostly used climate model projections from the Coupled Model Intercomparison Project Phase 3 (CMIP3) [31], which were produced more than a decade ago and do not represent the latest climate science. There are a few exceptions [21,24,25] that employed the latest climate model projections from the Coupled Model Intercomparison Project Phase 5 (CMIP5) [32]. However, these studies generally focused on spatial scales not directly relevant to water resources management practices. For instance, Sun et al. [24] selected mountainous areas in Southern California as their study focus. In addition, the linear regression approach was generally used in trend assessment in those studies. The results of this method are largely affected by the starting and ending values of the study data and subject to the assumption of normality.

The objective of this study was to provide an assessment of the changes (from historical baseline) and trends of projected precipitation and temperature along with the trends in projected drought over California. This study extended beyond relevant previous studies in terms of: (1) focusing on the scale consistent with the water resources planning and management practices in the state; (2) using climate projections that reflects the latest climate science; and (3) applying the widely-used non-parametric Mann–Kendall approach in trend analysis. Compared to the traditional linear regression method, this method requires less assumption on data distribution and is less affected by the beginning and ending values of the study data. Specifically, the current study was built upon a previous study [14] that explored changes in historical precipitation, temperature, and drought in California. However, the current study differs from [14] in terms of study variables, study metrics, study method, study period, and study purpose. Particularly, this study aimed to offer insight into potential changes to California's hydroclimate on the scale meaningful for water resources management practices and to inform decision-makers in developing strategies to cope with these changes.

2. Materials and Methods

2.1. Study Area and Dataset

Different from the previous study [14] that looks at seven climatic divisions in California, the current study focuses on the 10 hydrologic regions (Figure 1 and Table 1) defined by the California Department of Water Resources (DWR) for operational water resources planning and management purposes [5]. These regions include four coastal regions (North Coast, San Francisco Bay, Central Coast, and South Coast), three Central Valley regions (Sacramento River, San Joaquin River, and Tulare

Lake), and three Eastern regions (North Lahontan, South Lahontan, and Colorado River). For each of these three categories (Coastal, Central Valley, and Eastern), climate tends to be drier towards the southern regions.

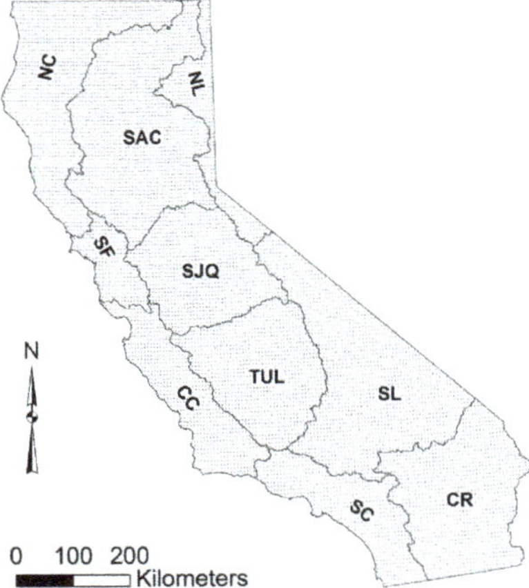

Figure 1. Ten hydrologic regions in California: North Coast (NC), San Francisco Bay (SF), Central Coast (CC), South Coast (SC), Sacramento River (SAC), San Joaquin River (SJQ), Tulare Lake (TUL), North Lahontan (NL), South Lahontan (SL), and Colorado River (CR). Dots represent the centroid points of individual climate projection grids (1/16th degree) located in each region.

Table 1. Geographic and climatic characteristics of study hydrologic regions.

ID	Region Name	Area (km^2)	Annual Precipitation (mm)	Annual Mean Temperature (°C)	Population (as of 2010; Million)
NC	North Coast	49,859	1390	9.3	0.81
SF	San Francisco Bay	11,535	641	14.3	6.35
CC	Central Coast	28,995	504	13.0	1.53
SC	South Coast	27,968	459	15.6	19.58
SAC	Sacramento River	69,750	925	11.4	2.98
SJQ	San Joaquin River	38,948	680	12.8	2.10
TUL	Tulare Lake	43,604	408	13.9	2.27
NL	North Lahontan	15,672	542	6.4	0.11
SL	South Lahontan	68,434	191	15.2	0.93
CR	Colorado River	51,103	127	20.2	0.75

The North Coast region contains the California Coast Ranges, the Klamath Mountains, and parts of the Modoc Plateau [5]. The eastern side of the region is mostly mountainous with crests around 1800 m (6000 ft) and a few more than 2400 m (8000 ft) in elevation. It is the wettest region in terms of annual precipitation received (1390 mm; Table 1). As such, the region is prone to flooding. Major floods were recorded in 1955, 1964, 1986, 1997, 2006, and 2017. The San Francisco Bay region is the smallest in size. It is bounded by the Pacific Ocean on the west and Coast Ranges on the east where the peaks are above 1200 m (4000 ft) in elevation. The region faces multiple water management challenges including an unreliable water supply, declining water quality and ecosystems, increasing flood risks, and threats posed by sea level rise to coastal areas. The Central Coast region is the most groundwater-dependent region. Groundwater supplies about 80% of its total water usage. The water management challenges of this region include managing groundwater quality and overdraft, sea water intrusion, and flood risks. The South Coast region is the most urbanized and populous region. It accounts for about 7% of the state's total area but accommodates more than half of the state's population. As a result, water supply is always a concern of local water managers. The region is also prone to flooding including debris flows and mud slides, particularly in areas where hillsides have been damaged by wildfires. It is the driest and warmest region in the coastal regions (Table 1).

Central Valley regions are the major water supply sources for the state, of which the Sacramento River region is the primary source. It is the largest and second wettest region (925 mm/year; Table 1) of all 10 hydrologic regions. It contributes a majority portion of the water supplied to the SWP and CVP. The region is bounded by Coast Ranges on the west and Sierra Nevada on the east. In this region, about one in three residents is exposed to a 500-year flood event. The region has approximately $65 billion of assets, 1.2 million acres of farmland, and over 340 sensitive species [5]. Major floods in the region normally originate from extreme atmospheric river events during the winter. The San Joaquin River region receives less precipitation than the Sacramento River region. It is also bordered by the Sierra Nevada on the east. However, Sierra Nevada watersheds in this region are higher in elevation, making them more dominated by snow compared to Sacramento River region watersheds. Floods in this region come from both winter rainfall and melting Sierra snowpack [5]. The Tulare Lake region is the driest in the Central Valley and one of the driest regions in the state. It is the largest agricultural region in the state heavily relying on groundwater and imported water supply. Groundwater pumping in this region accounts for more than 38% of the state's total annual groundwater extraction. The region is also prone to floods caused by winter rainfall and spring snowmelt.

The eastern regions are the least populous. The North Lahontan region accommodates approximately 0.3% of the state's population. It comprises arid high desert (1200–1500 m in elevation) in the north and the eastern slopes of the Sierra Nevada (up to 3750 m in elevation) in the central and southern portions. It is the coolest region in the state (Table 1). In contrast, the Colorado River region is the hottest. It is also the driest region, receiving only about one tenth of the precipitation received by the North Coast region. The Colorado River region is, however, also subject to flooding which threatens about 38% of its population. Different from all other regions, most flooding events occur from infrequent but high-intensity summer storms in this region. The South Lahontan region is the second driest region in the state. Precipitation for this region comes from both winter storm events and summer thunderstorms.

In general, California has a typical Mediterranean-like climate, with the summer (winter) being dry and warm (cool and wet). This is evident for the 10 regions on the monthly scale (Figure 2). Most of the precipitation occurs during the wet season (November to April). During that period, those regions receive 69% (Colorado River region) to 91% (Central Coast region) of their total annual precipitation. Statewide, 85% of annual precipitation occurs during the wet season. January normally observes the highest amount of precipitation while July is typically the driest month. Meanwhile, January is the coolest month while July has the highest average temperature. South Lahontan and Central Coast regions have the largest (22.1 °C) and smallest (10.1 °C) variations in monthly temperature, respectively. Across all regions, the Colorado River region is the driest and hottest. The North Lahontan region is the coolest and the North Coast region is the wettest. Those observations are consistent with values shown in Table 1.

This study looked at both the historical and projected precipitation, and maximum and minimum temperature data. The projections for 2020–2099 were based on climate model simulations from the Coupled Model Intercomparison Project Phase 5 (CMIP5) [32], which represents the current state of the climate science. Specifically, 20 individual projections from 10 Climate Circulation Models (GCMs) under two newly developed emission scenarios named Representative Concentration Pathways (RCP) 4.5 and RCP 8.5 [33] were selected for the analyses. These 10 GCMs (Table 2) were chosen by DWR Climate Change Technical Advisory Group and deemed as the most suitable for California climate and water resources assessment [34]. RCP 4.5 (RCP 8.5) assumes low (high) future greenhouse-gas concentrations. These projections were downscaled to a very high spatial resolution at 1/16th degree (approximately 6 by 6 km, or 3.75 by 3.75 miles) to better capture the spatial variability of the climate via the Localized Constructed Analogs (LOCA) method [35]. This dataset is made available for California's Fourth Climate Change Assessment (http://cal-adapt.org/). There are other ways of selecting representative GCMs models [36,37] for water planning analysis. However, they are beyond the scope of this study which exclusively used the GCMs recommended by the CCTAG. These 20 CCTAG-recommended projections have been applied in DWR's and the California Water Commission's planning activities including the Central Valley Flood Protection Plan [38] and the Water Storage Investment Program [39]. There is no consensus that some of those projections are more likely to occur than the remaining projections in the future. As a result, these projections are typically treated equally in planning activities. In this study, we looked at these 20 projections together. When looking at the mean of future projections on the annual scale, however, the 10 RCP 4.5 projections and 10 RCP 8.5 projections were analyzed separately.

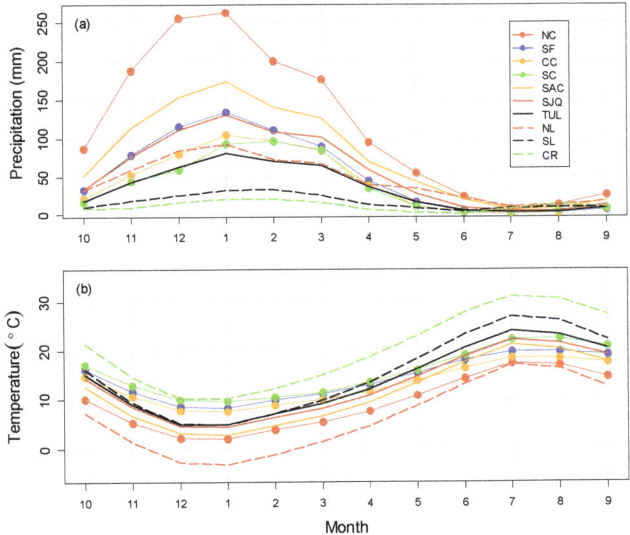

Figure 2. Long-term (1951–2013) mean monthly precipitation (**a**) and temperature (**b**) of 10 hydrologic regions.

Table 2. GCMs Selected for California Water Resources Planning [1].

Model ID	Model Name	Model Institution
1	ACCESS-1.0	Commonwealth Scientific and Industrial Research Organisation (CSIRO) and Bureau of Meteorology (BOM), Australia
2	CCSM4	National Center for Atmospheric Research
3	CESM1-BGC	National Science Foundation, Department of Energy, National Center for Atmospheric Research
4	CMCC-CMS	Centro Euro-Mediterraneo sui Cambiamenti Climatici
5	CNRM-CM5	Centre National de Recherches Météorologiques/Centre Européen de Recherche et de Formation Avancée en Calcul Scientifique
6	CanESM2	Canadian Centre for Climate Modeling and Analysis
7	GFDL-CM3	Geophysical Fluid Dynamics Laboratory
8	HadGEM2-CC	Met Office Hadley Centre
9	HadGEM2-ES	Met Office Hadley Centre/Instituto Nacional de Pesquisas Espaciais
10	MIROC5	Atmosphere and Ocean Research Institute (The University of Tokyo), National Institute for Environmental Studies, and Japan Agency for Marine-Earth Science and Technology

[1] Adapted from Tables 2–4 of [34].

The gridded historical observational dataset of these three variables on daily scale for water years 1951–2013 of Livneh et al. [40] (https://data.nodc.noaa.gov/) were employed as the historical baseline. The spatial resolution (1/16th degree) of this dataset is consistent with that of the LOCA-downscaled climate model projections. This dataset has been applied extensively in hydrologic modeling and drought assessment [41–44], and deemed as the best available historical data at this spatial resolution. In this study, both projected and historical datasets were aggregated from grid scale to (hydrologic) regional scale in the analyses presented below.

2.2. Study Method and Metrics

2.2.1. Difference from the Baseline

This study employed difference as a parsimonious metric to represent changes in future conditions from historical conditions. This is a standardized metric applied extensively in climate change related studies [29]. Specifically, the 40-year period, 1951–1990, was used as the historical baseline period. Compared to late 1990s and early 2000s, this period is relatively less impacted by anthropogenic climate change. Additionally, this 40-year window allows enough sample size to represent a wide range of natural variability in hydroclimatic variables. Similar studies have normally used 30-year periods [34]. Two 40-year future periods, mid-century (2020–2059) and late-century (2060–2099), were considered. Mean annual precipitation, and maximum and minimum temperature in the baseline period and future periods were computed and compared. Differences (from the baseline) were subsequently derived. Specifically, when looking at precipitation variables, the focus was on relative differences (i.e., percent different from the baseline); for temperature variables, absolute difference (in degree Celsius) was used.

In addition to annual precipitation and temperature, wet season precipitation and seasonal temperature were also applied as important indices in planning studies [45]. Wet season precipitation accounts for a majority portion of the annual precipitation. Seasonal temperature typically affects water supply and demand. For instance, spring temperature impacts snowmelt timing and amount. Summer temperature impacts evapotranspiration demand. Changes in wet season precipitation and seasonal temperature were also explored in this study.

2.2.2. Drought Index

Numerous drought indices have been developed for drought monitoring, assessment and prediction purposes [46–48]. Among these indices, the most widely used index might be the Standardized Precipitation Index (SPI) [49] because of its parsimonious (only requiring precipitation as input) and standardized (can be used across different spatial and temporal scales) nature. Despite its popularity, more and more studies noted that evapotranspiration also plays an important role in drought development [50–52]. This is particularly true in a warming climate for dry regions where evapotranspiration is an important component of the water budget. For instance, the most recent 2012–2015 California drought was a typical "warm drought" characterized by record-low precipitation and snowpack as well as record-high temperature [45,53–55]. As a result, SPI may not be the most appropriate index for drought analysis in California which contains many arid or semi-arid areas.

Most recently, based on the same concept employed in defining the SPI, Vicente-Serrano et al. [56] proposed a Standardized Precipitation-Evapotranspiration Index (SPEI). It first calculates the discrepancies between precipitation (P) and potential evapotranspiration (PET) on a monthly time scale (D = P − PET). Monthly discrepancies can be aggregated to other time scales (e.g., 3-month, 6-month, 12-month, among others) to calculate SPEI values at corresponding temporal scales. Next, a three-parameter Log-logistic distribution is selected to model the discrepancy time series. The probability distribution function of D is calculated according to the fitted Log-logistic distribution (F(x)).

Lastly, the SPEI value is determined as the standardized values of F(x) following the approximation of Abramowitz and Stegun [57]:

$$SPEI = W - \frac{C_0 + C_1 W + C_2 W^2}{1 + d_1 W + d_2 W^2 + d_3 W^3} \tag{1}$$

where $W = -2\ln(p)$; p is the probability of exceeding a determined D value; and C_0, C_1, C_2, d_1, d_2, and d_3 are preset constant coefficients. A positive (negative) SPEI value indicates wet (drought) conditions. Depending on the specific values, a drought event can be classified into different categories. Typically, a SPEI value less than -2 indicates extreme drought conditions. A value ranging from -2 to -1 denotes moderate drought conditions. A SPEI greater than -1 but less than 0 represents mild drought conditions.

SPEI has been shown to be a robust index. It compares favorably to other popular drought indices [58–63]. The PET is calculated using the Thornthwaite equation [64] which only requires temperature data as input. As such, the SPEI index implicitly considers the impact of temperature on drought situation, making it suitable in assessing drought conditions in future warming scenarios (represented by different model projections in the current study). For detailed explanations on the concept and calculation of the SPEI index, the readers are referred to [56]. The SPEI values on annual scale (SPEI-12), two-year scale (SPEI-24), three-year scale (SPEI-36), and four-year scale (SPEI-48) were chosen in this study. Drought occurs in California at those time scales regularly. It is meaningful to look at future drought at those scales for adaptive planning purpose.

Figure 3 exemplifies the SPEI-12 calculated for the three representative regions from each of the coastal, Central Valley, and eastern areas in the historical period: (1) the highly urbanized San Francisco Bay region; (2) the largest water supply source of the State: Sacramento River region; and (3) the driest Colorado River region. The San Francisco Bay region and the Sacramento River region have similar patterns due to their geographic proximity. The SPEI index for both regions well captures the 1983 and 1997 wet conditions as well as the 1976–1977, 1988–1992, 2007–2009, and the 2012–2013 droughts. The Colorado River region differs from those two regions in terms of annual precipitation (driest) and temperature (hottest). Long-duration droughts occur more frequently after 1990s in this region.

2.2.3. Trend Analysis

The methods applied in climatic and hydrological trend analysis are typically classified into two types: parametric and non-parametric [65,66]. The latter normally requires fewer assumptions (e.g., normality of study data) compared to the former. In reality, the assumptions on data distribution are difficult to satisfy. Therefore, the parametric methods are considered less robust than the non-parametric methods [66]. Among all non-parametric methods, the Mann–Kendall test (MKT) [67,68] has been applied extensively in the field of climatology and hydrology [14–16,45,69,70]. The approach first identifies the sign of each possible pair of data in the study time series, followed by the determination of the corresponding test statistic z. The null hypothesis (H0) assumes no significant monotonic trend in the time series while the alternative hypothesis suggests otherwise. The null hypothesis is rejected when $|Z| > Z_{1-\alpha/2}$, where $Z_{1-\alpha/2}$ is the probability of the standard normal distribution at a significance level of α. This study employed the MKT in assessing the significance of a trend and uses 0.05 as the significance level.

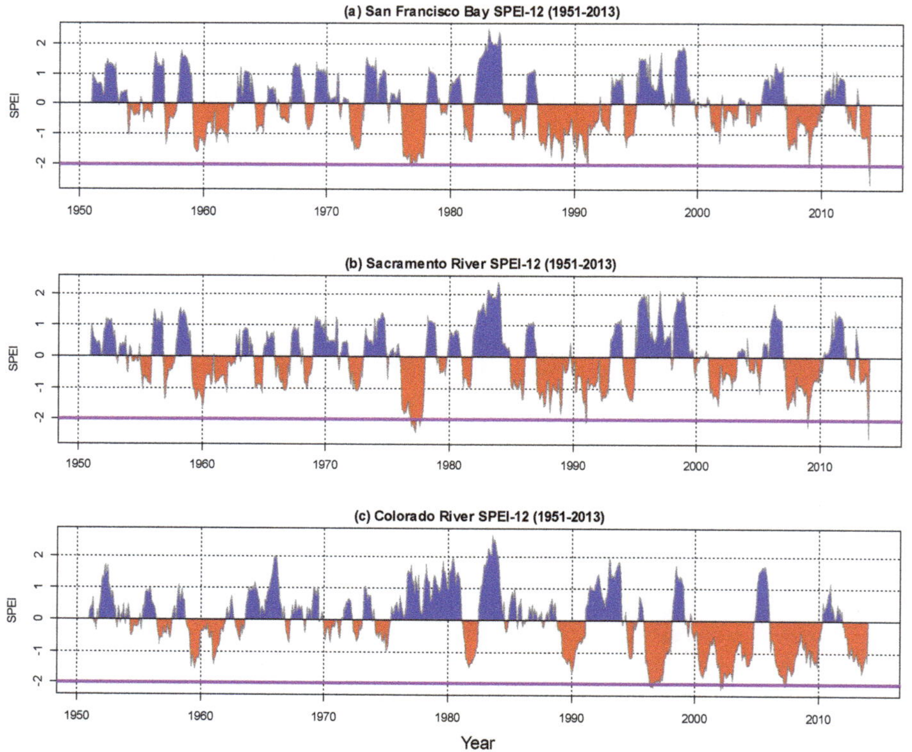

Figure 3. SPEI-12 of: (**a**) San Francisco Bay region; (**b**) Sacramento River region; and (**c**) Colorado River region during the historical period (1951–2013). Blue color indicates wet conditions; red color designates drought conditions. The purple line is the threshold below which extreme drought conditions exist.

This study further applied the non-parametric Theil–Sen approach (TSA) [71,72] to identify the slope of significant trends determined via the MKT. In this approach, the slope values (vector TS) of all data pairs are first calculated:

$$TS = \frac{V_i - V_j}{i - j} \; i = 1, 2, \ldots, n; j = 1, 2, \ldots, n; i > j \tag{2}$$

where n is the length of study record period; and V_i and V_j are time series values at time i and j, respectively ($i > j$). The median of TS is then used as overall slope of the trend identified for the study time series. A positive (negative) slope value represents an increasing (decreasing) trend. In this study, trend analysis is conducted in both historical (1951–2013) and future periods (2020–2099).

3. Results

3.1. Differences from the Baseline

3.1.1. Precipitation

Figure 4 shows the percent differences between historical precipitation and mean (of 10 individual RCP 4.5 projections) projected precipitation in mid-century (Figure 4a,b) and late-century (Figure 4c,d), respectively, on both the annual scale (Figure 4a,c) and during the wet season (Figure 4b,d). It is evident that all regions are expecting increases in precipitation during the wet season, with increases ranging from 2.8% (1.5%) to 9.8% (10.5%) in mid-century (late-century). This observation implies that future storms in the wet season would likely become more frequent, which is in line with the findings of previous studies [25,28]. On the annual scale, most regions are also projected to receive more precipitation, except for the driest two regions: South Lahontan and Colorado River. This suggests that those two regions are expecting much less precipitation in the dry season, although more precipitation is projected for them during the wet season. Typically, summer monsoons are a major contributor to dry season precipitation in these two regions [73,74]. This finding denotes that future monsoons over both regions are likely to become weaker or more sporadic. Across all regions, the San Francisco Bay and the South Coast generally have the highest and lowest increases in precipitation in late-century, respectively, on both temporal scales. This indicates that they are the most and least prone to changes in future storms during this period, respectively, yet they are not the wettest or driest regions. Comparing two future periods, the late-century period is generally expecting a higher increase in precipitation than the mid-century period except for the dry regions including Colorado River, South Lahontan, and South Coast.

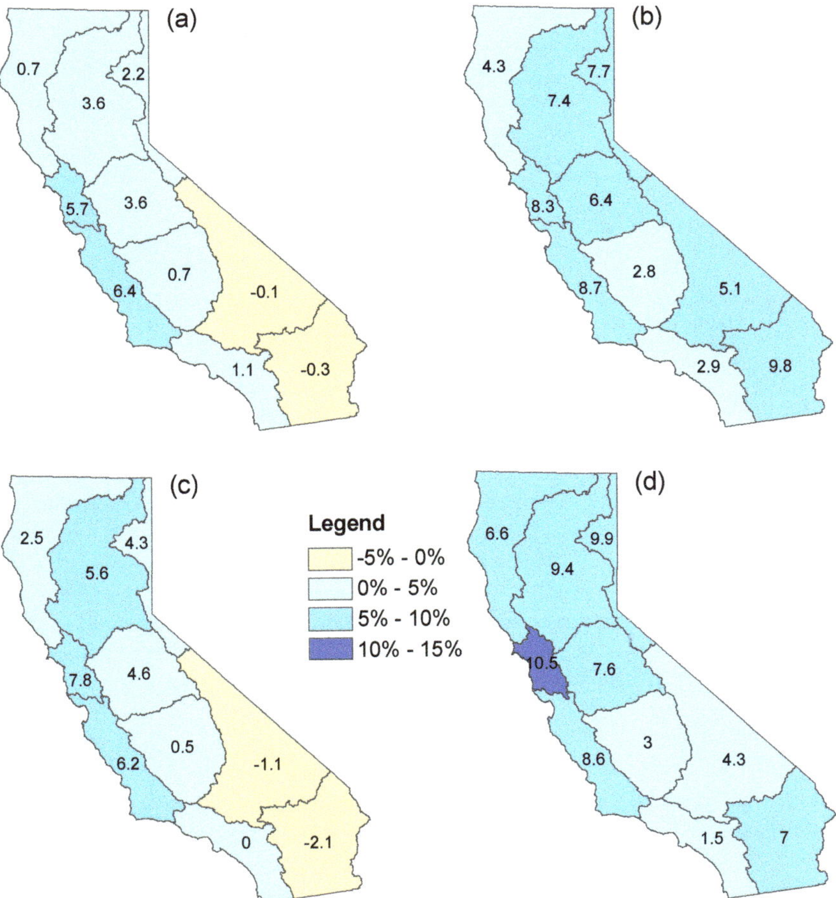

Figure 4. Percent differences (%) between historical and mean RCP 4.5 projections on: (**a**) annual precipitation in mid- century; (**b**) wet season precipitation in mid-century; (**c**) annual precipitation in late-century; and (**d**) wet season precipitation in late-century.

The differences between historical precipitation and mean RCP 8.5 precipitation projections are also explored (Table 3). Similar to what Figure 4 indicates, wet season precipitation is expected to increase in both mid-century and late-century across all regions. Increases are expected for annual precipitation for most regions except for three dry regions (i.e., Colorado River, South Lahontan, and South Coast) in mid-century and one region (i.e., Colorado River) in late-century. The increases in late-century are higher. Comparing annual precipitation and wet season precipitation, changes in the latter is more significant in terms of magnitude, which is in line with the RCP 4.5 results as illustrated in Figure 4. Comparing two future periods, changes in the late-century is more pronounced compared to those of the mid-century. Comparing differences of the mean RCP 4.5 projections from the historical baseline and that of the mean RCP 8.5 projections, the latter are more notable. Those are expected since the late-century (compared to mid-century) and the RCP 8.5 scenarios (compared to RCP 4.5 ones) are both expecting higher increases in temperature (Section 3.1.2). A warmer atmosphere can hold more water moisture, indicative of more water available for precipitation.

Table 3. Percent differences (%) between historical and mean RCP 8.5 projections on annual precipitation and wet season precipitation.

ID	Region Name	Annual Precipitation (%)		Wet Season Precipitation (%)	
		Mid-Century	Late-Century	Mid-Century	Late-Century
NC	North Coast	4.4	5.2	8.4	10.6
SF	San Francisco Bay	10.3	14.4	12.7	18.7
CC	Central Coast	7.4	12.8	9.6	16.0
SC	South Coast	−0.1	1.4	1.4	3.9
SAC	Sacramento River	7.6	9.0	11.4	14.2
SJQ	San Joaquin River	5.4	7.4	8.0	11.3
TUL	Tulare Lake	0.5	2.7	2.9	5.8
NL	North Lahontan	6.6	10.3	11.8	16.9
SL	South Lahontan	−0.5	2.4	3.8	7.7
CR	Colorado River	−2.3	−1.5	4.7	5.9

In addition to looking at the mean of PRC 4.5 and RCP 8.5 projections, individual projections are also investigated (Figure 5) to provide insights on the potential range of precipitation changes. Overall, on both temporal scales, there is no consensus that all projections show increases or decreases consistently for any region in mid-century or in late-century. This finding is also reported in previous studies using old climate projections [22,26–28]. The changes mostly range from −25% to 50%, with a few outliers showing more than 50% increases in precipitation. Those outliners come from a single wet climate model under the higher greenhouse-gas emission scenario (RCP 8.5). The variation range is generally larger for late-century (compared to mid-century) and dry regions (compared to wet regions). Additionally, wet season precipitation shows larger change ranges compared to annual precipitation. These results indicate more uncertainties in the projections for the dry regions, in the wet season, and in late-century.

3.1.2. Temperature

Mean annual maximum temperature and minimum temperature are examined in a similar way to the precipitation. The mean of 10 RCP 4.5 projections in two future periods are compared with their counterparts in the historical period (Figure 6). Increases are expected for both maximum and minimum temperature in both future periods across all regions. The eastern regions (NL, SL, and CR) are generally expecting more significant warming compared to other regions. This is likely because of their geographic location (away from the Pacific Ocean, lacking ocean regulation). In contrast, the Coastal regions normally have the least significant warming except for the South Coast region which has similar climate pattern as the dry Tulare Lake region. Comparing two future periods, late-century is expecting more warming consistently for all regions, which is not surprising given the accumulated effect of the greenhouse-gas emissions. The increases in minimum temperature and maximum temperature are generally comparable to each other. Statewide, the increases in the latter is slightly higher. Specifically, for maximum temperature, a 2.4 °C warming is projected statewide in the late-century versus 2.0 °C in mid-century. For minimum temperature, the statewide increases are expected to be 2.2 °C and 1.8 °C, respectively, in those two periods. This is somewhat different from previous studies which claimed that increases in minimum temperature are more pronounced [16], leading to smaller diurnal temperature ranges.

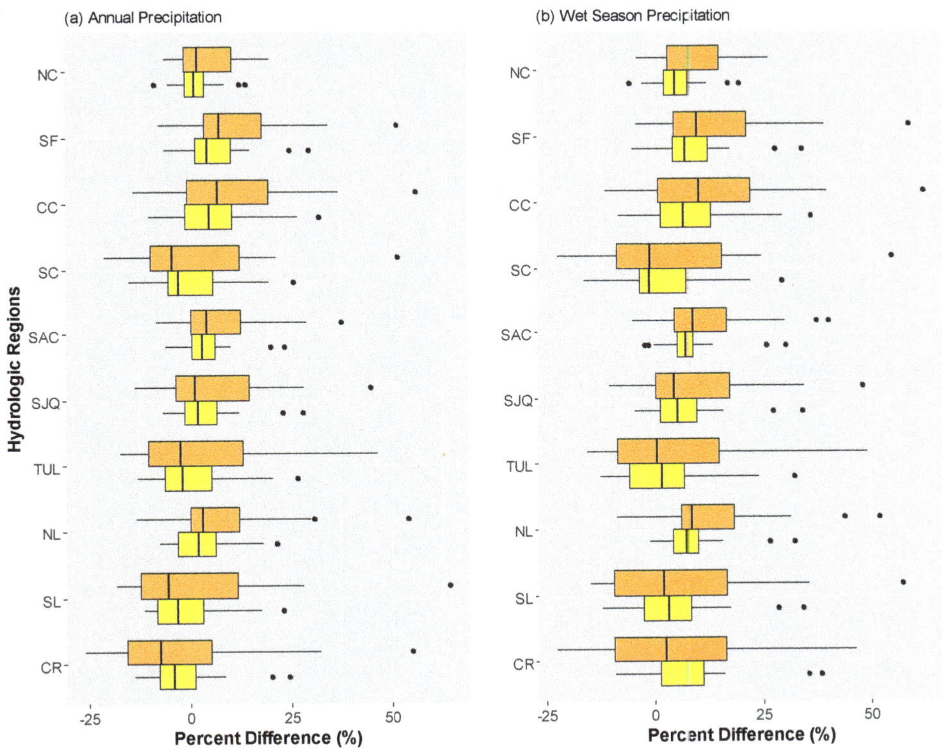

Figure 5. Box-and-whisker plots of percent differences (%) between historical and individual projections on: (**a**) annual precipitation; and (**b**) wet season precipitation. Yellow boxes represent mid-century results and orange boxes show late-century results.

In addition to the differences between mean RCP 4.5 projections and the historical baseline, the differences associated with the mean RCP 8.5 projections are also examined (Table 4). The messages are generally consistent with what the RCP 4.5 results (Figure 6) indicate. In general, warming (in both maximum and minimum temperature) is expected across all regions in both future periods. The inland eastern regions are projected to have the highest increases in temperature. The late-century is expecting more significant warming than the mid-century. Comparing RCP 4.5 and RCP 8.5 scenarios, warming of the latter is more pronounced in terms of increase amount. Specifically, for minimum temperature in the mid-century, RCP 8.5 scenario shows about 0.8 °C (for San Francisco Bay and Central Coast) to 1.1 °C (North Lahontan) warmer than the RCP 4.5 scenario; in the late-century, the range is from 1.9 °C (Central Coast) to 2.5 °C (North Lahontan). For maximum temperature, the differences between two scenarios are slightly higher than that of the minimum temperature.

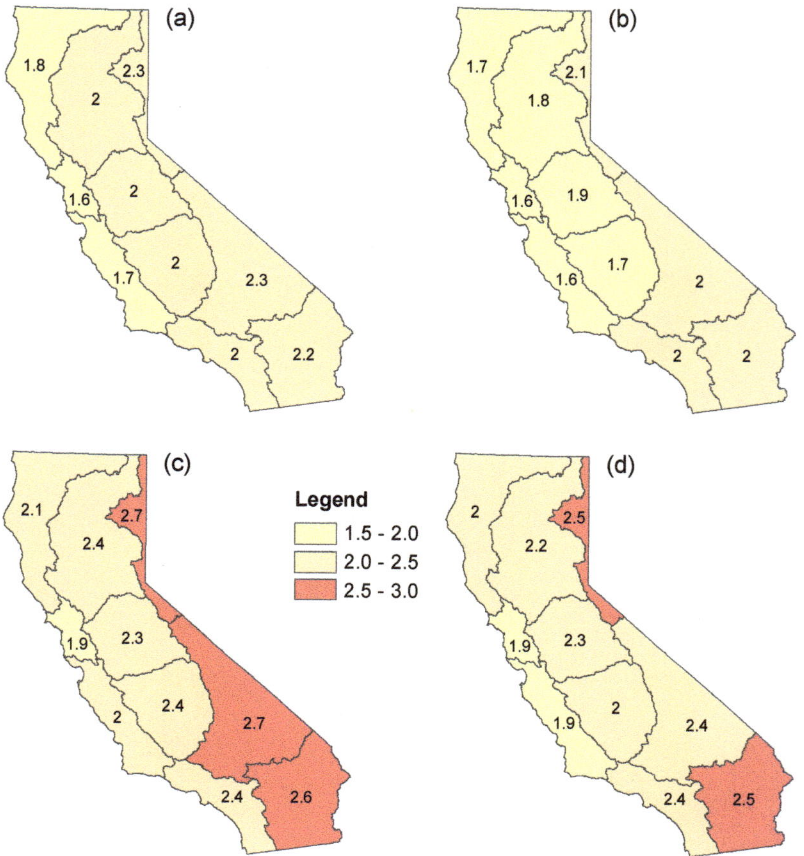

Figure 6. Differences (°C) between historical and mean RCP 4.5 projections on mean annual: (**a**) maximum temperature in mid-century; (**b**) minimum temperature in mid-century; (**c**) maximum temperature in late-century; and (**d**) minimum temperature in late-century.

Table 4. Differences (°C) between historical and mean RCP 8.5 projections on annual maximum and minimum temperature.

		Annual Tmax (°C)		Annual Tmin (°C)	
		Mid-Century	Late-Century	Mid-Century	Late-Century
NC	North Coast	2.7	4.3	2.5	4.1
SF	San Francisco Bay	2.4	3.8	2.5	4.0
CC	Central Coast	2.5	3.9	2.4	3.8
SC	South Coast	3.0	4.5	2.9	4.5
SAC	Sacramento River	3.0	4.7	2.7	4.4
SJQ	San Joaquin River	3.0	4.6	2.8	4.5
TUL	Tulare Lake	3.0	4.6	2.5	4.2
NL	North Lahontan	3.4	5.3	3.2	5.0
SL	South Lahontan	3.3	5.1	3.0	4.8
CR	Colorado River	3.2	4.9	3.0	4.9

Looking at individual projections on maximum (Figure 7a) and minimum temperature (Figure 7b), all of them show at least 1 °C warming. No projections indicate any decreases for any region, which is different from precipitation projections that have no such consensus. This is also reported in previous studies [30,75–79]. Comparing two future periods, higher increases are expected in the late-century. On average, increases in maximum temperature are generally higher than increases the minimum temperature, which is particularly true for the eastern regions. Those observations are consistent with what is noted in Figure 6. Similar to precipitation projections, the warming range of late-century is larger than that of mid-century across all regions. This indicates that climate models tend to disagree more with each other further into the future because of increasing uncertainty in climate model forcing.

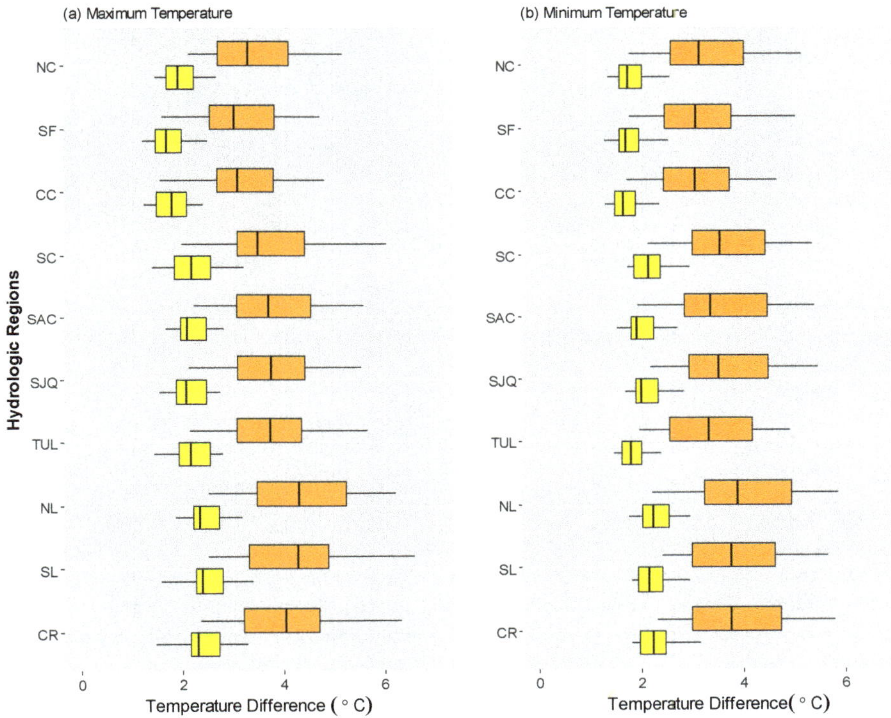

Figure 7. Box-and-whisker plots of differences (°C) between historical and individual projections on mean annual: (**a**) maximum temperature; and (**b**) minimum temperature. Yellow boxes represent mid-century results and orange boxes show late-century results.

At the seasonal scale, the mean projection in mid-century shows at least 1 °C warming in both maximum and minimum temperature across all seasons (Figure 8a). In comparison, at least 2.5 °C warming is expected in late-century (Figure 8b). The highest increases (2.9 °C and 5.0 °C in mid-century and late-century, respectively) are expected to occur in Summer maximum temperature in the coolest region, North Lahontan. Comparing different regions, the eastern regions are expecting higher increases in both minimum and maximum temperature than other regions. This is consistent with what Figure 6 illustrates on the annual scale. Looking at different seasons, fall and summer are expecting relatively higher warming than winter and spring. Particularly, summer is expecting the highest increases. Statewide, an amount of 2.4 °C and 2.5 °C warming is projected in mid-century in summer minimum and maximum temperature, respectively. In late-century, the corresponding increases in summer are expected to be 4.2 °C and 4.3 °C, respectively.

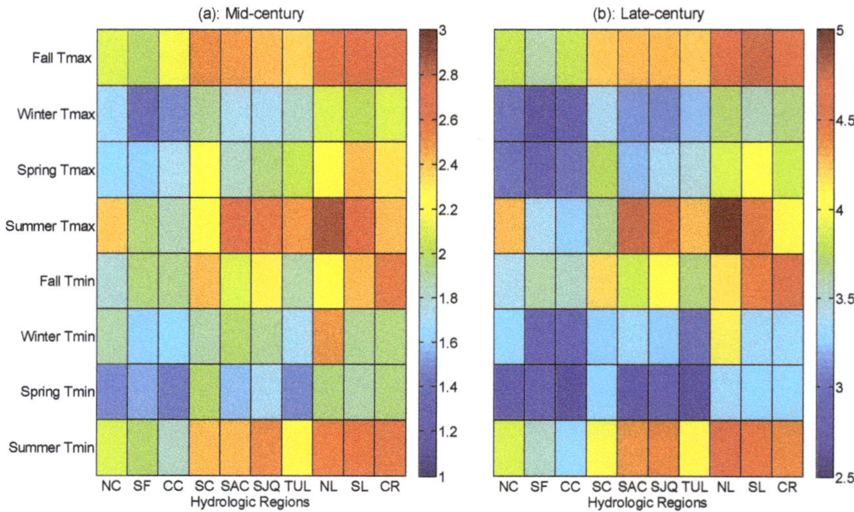

Figure 8. Differences (°C) between historical and mean (of all 20) projections on seasonal maximum temperature (Tmax) and minimum temperature (Tmin) in: (**a**) mid-century; and (**b**) late-century.

3.2. Trend Analysis

3.2.1. Precipitation

No significant trends are detected in historical annual and wet season precipitation for any study regions. Similar findings have also been reported in relevant previous studies [14]. During the projection period (2020–2099), a limited amount (no more than 15%) of model projections show significant trends (Table 5). For annual precipitation, only one projection (out of 20) has statistically significant trend for Sacramento River, South Coast, and Tulare Lake regions; three projections indicate significant trends in Central Coast and North Lahontan regions; for other regions, only two projections show significant trends. The slopes of those significant trends are all positive.

Table 5. Trend information of projected precipitation.

ID	Region Name	Number (Percent) of Projections with Significant Trend [1]		Range of Significant Trend Slope (mm/Year)
		Annual Precipitation	Wet Season Precipitation	
NC	North Coast	2 (10%)	3 (15%)	3.9~5.4
SF	San Francisco Bay	2 (10%)	2 (10%)	3.2~4.6
CC	Central Coast	3 (15%)	3 (15%)	−0.5~3.4
SC	South Coast	1 (5%)	1 (5%)	2.1~2.7
SAC	Sacramento River	1 (5%)	3 (15%)	−2.2~6.0
SJQ	San Joaquin River	2 (10%)	2 (10%)	2.8~5.0
TUL	Tulare Lake	2 (10%)	2 (10%)	1.7~3.2
NL	North Lahontan	3 (15%)	3 (15%)	1.2~5.3
SL	South Lahontan	2 (10%)	0 (0%)	0.8~1.9
CR	Colorado River	1 (5%)	0 (0%)	1.1

[1] Significance level 0.05.

For wet season precipitation, no projections show any significant trends for the driest two regions (Colorado River and South Lahontan). For San Francisco Bay, Central Coast, South Coast, San Joaquin River, Tulare Lake, and North Lahontan regions, the projections showing significant trends are exactly the same as those showing significant trends in annual precipitation. For the two wettest regions (North Coast and Sacramento River), three projections show significant changes. Different from annual precipitation, two projections on wet season precipitation (one for Central Coast region and the other for Sacramento River region) exhibit a decreasing tendency. Nevertheless, similar to the annual precipitation, no significant changes are expected in the majority of climate model projections on wet season precipitation through 2099.

3.2.2. Temperature

All 20 projections on mean annual maximum temperature (Figure 9a) and minimum temperature (Figure 9b) show significant increasing trends. On average, the increasing rates of the Central Valley regions (SAC, SJQ, and TUL) are fairly close to each other. The increasing rates of the coast regions (NC, SF, CC, and SC) and eastern regions (NL, SL, and CR) are slightly smaller and higher, respectively, compared to that of the Central Valley regions. Particularly, the median increasing rate in the coolest region, North Lahontan, is the highest among all regions in maximum temperature. This is mostly in line with what Figures 5 and 6 illustrate. In the historical period, both variables also exhibit increasing trends. However, for maximum temperature, only the trends for Central Coast, South Coast, San Francisco Bay, and South Lahontan regions are statistically significant at a significance level of 0.05. For minimum temperature, the trends of all regions except for Colorado River region are significant. Compared to historical trends, most projected trends have higher increasing rates. In general, the increasing trend is more significant in maximum temperature than in minimum temperature, implying that temperature range (difference between maximum and minimum temperature) is likely to increase. Comparing different regions, on average, the coastal regions (NC, SF, CC, and SC) tend to have the relatively smaller increasing rates while the eastern regions generally have the highest increasing rates. This is generally consistent with what has been observed in Figure 6.

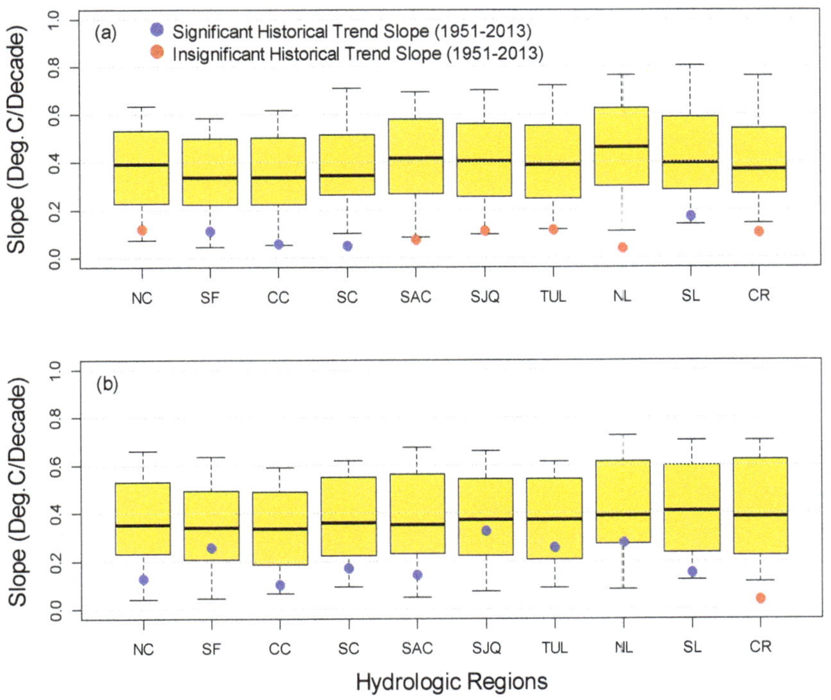

Figure 9. Box-and-whisker plots of trend slopes of historical (1951–2013) and projected (2020–2099) mean annual: (**a**) maximum temperature (Tmax); and (**b**) minimum temperature (Tmin) (at significance level 0.05).

Similar to those observed on the annual scale (Figure 9), not all regions have significant trends in historical maximum temperature and minimum temperature on the seasonal scale (Figure 10a). Specifically, fall and winter minimum temperature exhibits no statistically significant changes for any region. Furthermore, the Tulare Lake region does not observe any significant trends in its maximum or minimum temperature in any season. Comparing two temperature variables, maximum temperature shows significant increasing trend in most cases while minimum temperature only exhibits significant warming in a couple of seasons (spring and summer) for a few regions. In contrast, mean projections on seasonal maximum temperature and minimum temperature show significant warming trends consistently for all regions (Figure 10b). Warming in summer and fall is more pronounced than warming in two other seasons. Looking at different regions, the eastern regions generally have the highest increasing trend while the coastal regions have the smallest amount of increasing rate. Particularly, the coolest region, North Lahontan, has the most significant increasing tendency in both maximum temperature and minimum temperature. The region has the highest seasonal warming rate in both maximum temperature (0.52 °C/decade) and minimum temperature (0.51 °C/decade) in summer. Those observations are largely in line with what Figure 7 shows.

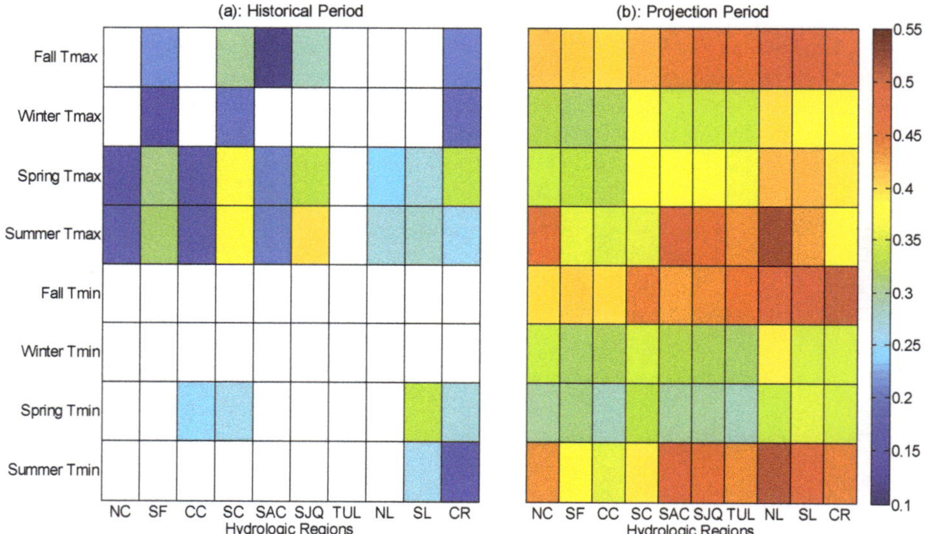

Figure 10. Trend slopes of (**a**) historical (1951–2013) and (**b**) projected (2020–2099) mean seasonal maximum temperature (Tmax) and minimum temperature (Tmin) (at significance level 0.05). Different colors mean different trend slopes (per decade). White color indicates no significant trends.

3.2.3. Drought Index

California is prone to drought, with examples being the 1976–1977, 1988–1992, 2007–2009, and 2012–2013 droughts [80]. While the occurrence and lasting period of drought events are difficult to predict decades in advance, the overall tendency (i.e., trend) of drought events can shed light on long-term drought response planning activities. This section looks at projected future drought conditions (represented by the SPEI index) at one- to four-year temporal scales which are relevant to our operational planning practices. Figure 11 shows trend slopes of SPEI-12, SPEI-24, SPEI-36, and SPEI-48 calculated from projected precipitation and temperature data, along with their counterparts in the historical period. On average, all regions are expecting a decreasing trend (negative slope value). This is particularly true for dry regions including the Colorado River, South Lahontan, and Tulare Lake. All 20 projections have a decreasing tendency consistently, indicating more severe droughts for those regions on the annual, two-year, three-year, and four-year scales. For other regions, there is no such consensus. However, the majority of projections show a decreasing trend. It should be highlighted that, for the wettest region, North Coast, most projections have a relatively milder decreasing trend compared to the historical baseline. This suggests that projected increase in precipitation over this region outweighs the effect of warming. For the coolest region, North Lahontan, the median trend slopes of projected SPEI values are generally around the historical trend slope values. This implies that projected precipitation increase in this region offsets the impact of warming. For other regions, most projections have a steeper decreasing trend compared to their historical counterparts, indicating that projected increases in precipitation are not sufficient to offset the effect of warming. Particularly, for the driest region, Colorado River, the decreasing rates of all 20 projections are higher than its baseline counterpart. This suggests that this region is the least resilient to warming and thus most prone to aridity (as represented by SPEI index) among all study regions.

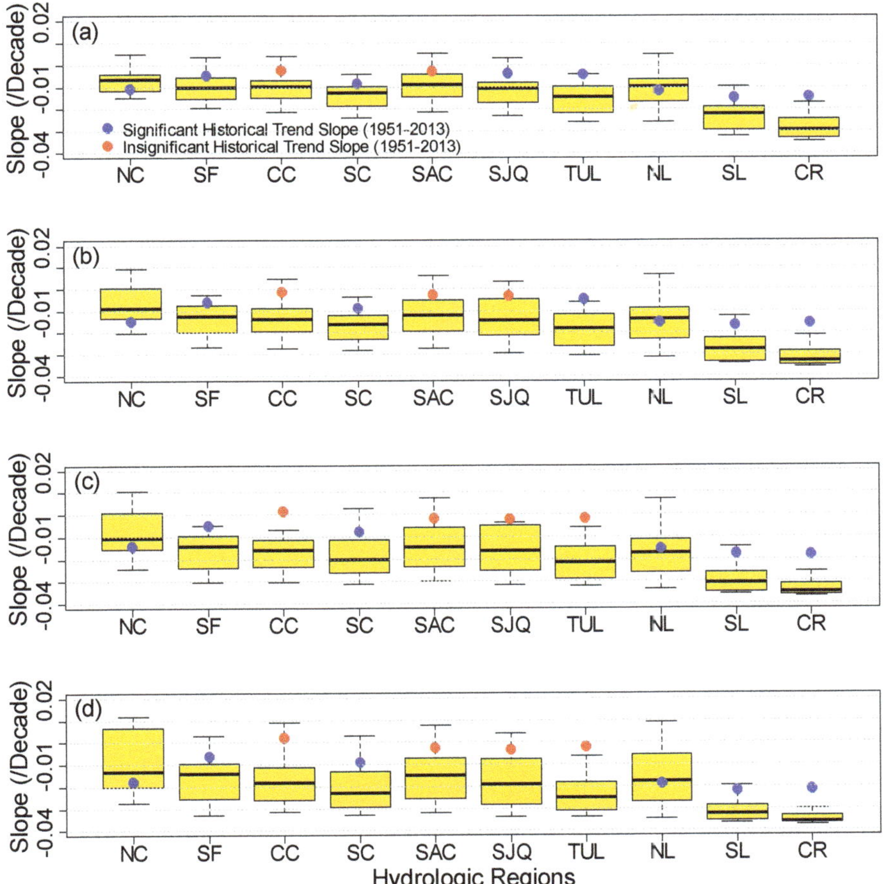

Figure 11. Box-and-whisker plots of significant trend slopes of: (**a**) SPEI-12; (**b**) SPEI-24; (**c**) SPEI-36; and (**d**) SPEI-48 during projection period (2020–2099) (at significance level 0.05). The slope information in historical period (1951–2013) is also shown.

It is worth noting that not all trends identified in the historical and projection periods are statistically significant at a significance level of 0.05. The Central Coast region and Sacramento

River region show no significant changes in SPEI values (Figure 11). Additionally, on three-year scale and four-year scale, all Central Valley regions (SAC, SJQ, and TUL) have no increasing or decreasing tendency in drought represented by SPEI during the historical period. In the projection period (Table 6), for the driest two regions, Colorado River and South Lahontan, all 20 projections show consistently significant trend in SPEI at the four time scales considered. This is no such consensus for other regions. However, the majority of projections still show significant trends. For instance, only one out of 20 (5%) projections exhibit insignificant changes in Tulare Lake region. For another relatively dry region, South Coast, all projections show significant trends in SPEI on the three-year scale (SPEI-36) and four-year scale (SPEI-48) while only one projection (5%) has insignificant trend on the annual scale (SPEI-12) and two-year scale (SPEI-24). The wettest region, North Coast, has the highest amount (25% to 30%) of projections that show no significant trend. For the second wettest region, Sacramento River, 15–20% of the projections indicate no significant trend. Overall, those projections agree more with each other on the increasing aridity in dry regions than in wet regions irrespective of the time scales investigated.

Table 6. Number (percent) of SPEI projections with insignificant trend [1] in the projection period.

ID	Region Name	SPEI-12	SPEI-24	SPEI-36	SPEI-48
NC	North Coast	6 (30%)	5 (25%)	5 (25%)	6 (30%)
SF	San Francisco Bay	5 (25%)	5 (25%)	5 (25%)	3 (15%)
CC	Central Coast	3 (15%)	3 (15%)	3 (15%)	2 (10%)
SC	South Coast	1 (5%)	1 (5%)	0 (0%)	0 (0%)
SAC	Sacramento River	4 (20%)	3 (15%)	4 (20%)	3 (15%)
SJQ	San Joaquin River	3 (15%)	2 (10%)	2 (10%)	2 (10%)
TUL	Tulare Lake	1 (5%)	1 (5%)	1 (5%)	1 (5%)
NL	North Lahontan	2 (10%)	3 (15%)	3 (15%)	0 (0%)
SL	South Lahontan	0 (0%)	0 (0%)	0 (0%)	0 (0%)
CR	Colorado River	0 (0%)	0 (0%)	0 (0%)	0 (0%)

[1] Significance level 0.05.

4. Discussion and Conclusions

This study investigated potential changes in future precipitation, temperature, and drought (as represented by SPEI) across 10 hydrologic regions defined by the California Department of Water Resources. The latest climate model projections on these variables through 2099 representing the state of the current climate science were applied for this purpose. Changes were explored in terms of differences from a historical baseline as well as the changing trend.

Results indicate that warming is expected across all regions in all temperature projections, particularly in late-century. There is no such consensus in precipitation, with projections ranging mostly from −25% to +50% different from the historical baseline. There is no statistically significant increasing or decreasing trend in historical precipitation as well as in the majority of the projections. However, on average, precipitation is expected to increase slightly for most regions. It should be noted that this finding is not completely in line with a previous study that indicates decreases in future California precipitation [81]. The major difference stems from the fact that different sets of data are applied in the two studies. Specifically, the current study focused on precipitation and temperature projections from 10 GCMs models (versus 42 models used by the previous study) that are deemed most appropriate for water resources planning studies in California. Compared to wet regions, dry regions are projected to have more severe drought conditions represented by SPEI. Those findings are generally consistent with what have been reported in previous studies [21,26,28,76]. A new finding of this study is that the coolest region, North Lahontan, tends to have the highest increases in both minimum and maximum temperature and a significant amount of increase in wet season precipitation, indicative of naturally increasing flood risk in this region. In another new finding, the warming in summer and fall (when water demand is typically high and precipitation is limited) is expected to be more significant than the warming in winter and spring

In general, the findings of this study are meaningful from both scientific and practical perspectives. From a scientific point of view, these findings provide useful information that can be utilized to improve the current flood and water supply forecasting models. For instance, the coolest region, North Lahontan, is expecting the most significant warming as well as increases in wet season precipitation. This region is largely impacted by snow because of its high elevation. These expected changes will most likely intensify regional rainfall (more precipitation comes as rainfall as warming elevates the snowline) and spring snowmelt, increasing flood risks in the future. This region needs to be closely monitored in the future, particularly near and above the current snowline. The current flood forecasting model uses a parameter to cap the maximum possible snowmelt rate [82]. To reflect the expected warming, this parameter needs to be increased accordingly to better model snowmelt. Taking one step further, the snow accumulation and snowmelt processes based on which the current forecasting model is developed are derived under the stationary assumption. In a non-stationary environment, these processes need to be revisited and updated accordingly as relevant new observations become available. Additionally, the current snowmelt model is temperature-index based. Snowmelt is a thermodynamic process driven more by radiation than temperature. Development and implementation of radiation-driven snowmelt model in operations are ongoing and will be reported in our future work.

From a practical standpoint, these findings can help inform water managers in making adaptive management plans. For instance, vulnerability assessment is typically the first step in developing any mitigation and adaptation strategies [83]. Corresponding adaptation strategies such as supply diversification or increased volume management capacity should be tailored for the characteristics of the regions and their particular impacts to a changing climate. All in all, this study has the potential to help decision-makers move from a reactive position of responding to hydroclimatic events as they happen to a pro-active position with region-specific strategies for improved water resources management in the future. These strategies facilitate improving the resilience of California's physical water framework and the preparedness of its institutional framework via investments (e.g., where, when, on what, and how much) in advance.

Despite its scientific and practical significance in guiding long-term strategical water resources planning, the study addressed temperature and precipitation changes at annual and seasonal scales at the hydrologic region scale. For time-sensitive and localized activities including emergency response and management, those changes at a finer temporal and spatial scale at which extreme events occur need to be explored. Extreme climatic indices (e.g., daily maximum precipitation, heat wave, etc.) with daily resolution at the watershed scale have been extracted from the 20 climate projections applied in this study. They will be analyzed and presented in a follow-up study. Furthermore, as opposed to precipitation and temperature, streamflow runoff is normally the variable directly used to inform real-time decision making (e.g., determination of reservoir release schedule). Those climate projections have been used as input to drive a distributed hydrologic model, the Variable Infiltration Capability model, to produce daily inflow projections through 2099 for major water supply reservoirs in California. Those flow data will be analyzed in terms of volume, variability, and frequency and reported in a companion study.

Acknowledgments: The authors would like to thank four anonymous reviewers for their valuable comments that largely helped improve the quality of this study. The authors would also like to thank their colleague Mahesh Gatuam and Jianzhong Wang for discussions on previous studies leading to the current work. Technical editing from Charlie Olivares is acknowledged. The authors also want to thank John Andrew, Prabhjot (Nicky) Sandhu, and Jamie Anderson for their management support on the study. Any findings, opinions, and conclusions expressed in this paper are solely the authors' and do not reflect the views or opinions of their employer.

Author Contributions: The study was conceived by the authors together. M.H. conducted the study and wrote the paper. A.S., E.L. and M.A. provided critical discussions.

Conflicts of Interest: The authors declare no conflict of interest.

References

1. Jones, J. California, a state of extremes: Management framework for present-day and future hydroclimate extremes. In *Water Policy and Planning in a Variable and Changing Climate*; Miller, K., Hamlet, A.F., Kenney, D.S., Redmond, K.T., Eds.; Taylor & Francis Group: Boca Raton, FL, USA, 2016; pp. 207–222.
2. Dettinger, M.D.; Ralph, F.M.; Das, T.; Neiman, P.J.; Cayan, D.R. Atmospheric rivers, floods and the water resources of California. *Water* **2011**, *3*, 445–478. [CrossRef]
3. U.S. Census Bureau. 2010 Census Summary File 1. Available online: https://www.census.gov/2010census/data/ (accessed on 1 August 2010).
4. Lund, J.R. Flood management in California. *Water* **2012**, *4*, 157–169. [CrossRef]
5. California Department of Water Resources. *California Water Plan Update 2013*; California Department of Water Resources: Sacramento, CA, USA, 2014.
6. Chung, F.; Kelly, K.; Guivetchi, K. Averting a California water crisis. *J. Water Resour. Plan. Manag.* **2002**, *128*, 237–239. [CrossRef]
7. Anderson, J.; Chung, F.; Anderson, M.; Brekke, L.; Easton, D.; Ejeta, M.; Peterson, R.; Snyder, R. Progress on incorporating climate change into management of California's water resources. *Clim. Chang.* **2008**, *87*, 91–108. [CrossRef]
8. Kapnick, S.; Hall, A. Observed climate–snowpack relationships in California and their implications for the future. *J. Clim.* **2010**, *23*, 3446–3456. [CrossRef]
9. McCabe, G.J.; Clark, M.P. Trends and variability in snowmelt runoff in the western United States. *J. Hydrometeorol.* **2005**, *6*, 476–482. [CrossRef]
10. Mote, P.W. Trends in snow water equivalent in the Pacific Northwest and their climatic causes. *Geophys. Res. Lett.* **2003**, *30*. [CrossRef]
11. Mote, P.W.; Hamlet, A.F.; Clark, M.P.; Lettenmaier, D.P. Declining mountain snowpack in western North America. *Bull. Am. Meteorol. Soc.* **2005**, *86*, 39–49. [CrossRef]
12. Stewart, I.T.; Cayan, D.R.; Dettinger, M.D. Changes in snowmelt runoff timing in western North America under a business as usual climate change scenario. *Clim. Chang.* **2004**, *62*, 217–232. [CrossRef]
13. Regonda, S.K.; Rajagopalan, B.; Clark, M.; Pitlick, J. Seasonal cycle shifts in hydroclimatology over the western United States. *J. Clim.* **2005**, *18*, 372–384. [CrossRef]
14. He, M.; Gautam, M. Variability and trends in precipitation, temperature and drought indices in the State of California. *Hydrology* **2016**, *3*, 14. [CrossRef]
15. He, M.; Russo, M.; Anderson, M. Predictability of seasonal streamflow in a changing climate in the Sierra Nevada. *Climate* **2016**, *4*, 57. [CrossRef]
16. He, M.; Russo, M.; Anderson, M.; Fickenscher, P.; Whitin, B.; Schwarz, A.; Lynn, E. Changes in extremes of temperature, precipitation, and runoff in California's Central Valley during 1949–2010. *Hydrology* **2017**, *5*, 1. [CrossRef]
17. Hatchett, B.J.; Daudert, B.; Garner, C.B.; Oakley, N.S.; Putnam, A.E.; White, A.B. Winter snow level rise in the northern Sierra Nevada from 2008 to 2017. *Water* **2017**, *9*, 899. [CrossRef]
18. Huppert, H.E.; Sparks, R.S.J. Extreme natural hazards: Population growth, globalization and environmental change. *Philos. Trans. A Math. Phys. Eng. Sci.* **2006**, *364*, 1875–1888. [CrossRef] [PubMed]
19. Cavallo, E.; Galiani, S.; Noy, I.; Pantano, J. Catastrophic natural disasters and economic growth. *Rev. Econ. Stat.* **2013**, *95*, 1549–1561. [CrossRef]
20. Hanak, E.; Lund, J.R. Adapting California's water management to climate change. *Clim. Chang.* **2012**, *111*, 17–44. [CrossRef]
21. Dettinger, M.; Anderson, J.; Anderson, M.; Brown, L.; Cayan, D.; Maurer, E. Climate change and the Delta. *San Fr. Estuary Watershed Sci.* **2016**, *14*, 1–26. [CrossRef]
22. Das, T.; Dettinger, M.D.; Cayan, D.R.; Hidalgo, H.G. Potential increase in floods in California's Sierra Nevada under future climate projections. *Clim. Chang.* **2011**, *109*, 71–94. [CrossRef]
23. Das, T.; Maurer, E.P.; Pierce, D.W.; Dettinger, M.D.; Cayan, D.R. Increases in flood magnitudes in California under warming climates. *J. Hydrol.* **2013**, *501*, 101–110. [CrossRef]
24. Sun, F.; Hall, A.; Schwartz, M.; Walton, D.B.; Berg, N. Twenty-first-century snowfall and snowpack changes over the southern California Mountains. *J. Clim.* **2016**, *29*, 91–110. [CrossRef]

25. Berg, N.; Hall, A. Increased interannual precipitation extremes over California under climate change. *J. Clim.* **2015**, *28*, 1–11. [CrossRef]
26. Tebaldi, C.; Hayhoe, K.; Arblaster, J.M.; Meehl, G.A. Going to the extremes. *Clim. Chang.* **2006**, *79*, 185–211. [CrossRef]
27. Wang, J.; Zhang, X. Downscaling and projection of winter extreme daily precipitation over North America. *J. Clim.* **2008**, *21*, 923–937. [CrossRef]
28. Yoon, J.-H.; Wang, S.S.; Gillies, R.R.; Kravitz, B.; Hipps, L.; Rasch, P.J. Increasing water cycle extremes in California and in relation to ENSO cycle under global warming. *Nat. Commun.* **2015**, *6*, 8657. [CrossRef] [PubMed]
29. Maurer, E.P. Uncertainty in hydrologic impacts of climate change in the Sierra Nevada, California, under two emissions scenarios. *Clim. Chang.* **2007**, *82*, 309–325. [CrossRef]
30. Cayan, D.R.; Maurer, E.P.; Dettinger, M.D.; Tyree, M.; Hayhoe, K. Climate change scenarios for the California region. *Clim. Chang.* **2008**, *87*, 21–42. [CrossRef]
31. Meehl, G.A.; Covey, C.; Taylor, K.E.; Delworth, T.; Stouffer, R.J.; Latif, M.; McAvaney, B.; Mitchell, J.F. The WCRP CMIP3 multimodel dataset: A new era in climate change research. *Bull. Am. Meteorol. Soc.* **2007**, *88*, 1383–1394. [CrossRef]
32. Taylor, K.E.; Stouffer, R.J.; Meehl, G.A. An overview of CMIP5 and the experiment design. *Bull. Am. Meteorol. Soc.* **2012**, *93*, 485–498. [CrossRef]
33. Van Vuuren, D.P.; Edmonds, J.; Kainuma, M.; Riahi, K.; Thomson, A.; Hibbard, K.; Hurtt, G.C.; Kram, T.; Krey, V.; Lamarque, J.-F. The representative concentration pathways: An overview. *Clim. Chang.* **2011**, *109*, 5. [CrossRef]
34. Climate Change Technical Advisory Group (CCTAG). *Perspectives and Guidance for Climate Change Analysis*; California Department of Water Resources: Sacramento, CA, USA, 2015.
35. Pierce, D.W.; Cayan, D.R.; Thrasher, B.L. Statistical downscaling using localized constructed analogs (LOCA). *J. Hydrometeorol.* **2014**, *15*, 2558–2585. [CrossRef]
36. Lutz, A.F.; ter Maat, H.W.; Biemans, H.; Shrestha, A.B.; Wester, P.; Immerzeel, W.W. Selecting representative climate models for climate change impact studies: An advanced envelope-based selection approach. *Int. J. Climatol.* **2016**, *36*, 3988–4005. [CrossRef]
37. Shrestha, N.K.; Wang, J. Modelling nitrous oxide (N2O) emission from soils using the soil and water assessment tool (SWAT). In Proceedings of the 2018 International SWAT Conference and Workshops, Chennai, India, 10–12 January 2018.
38. California Department of Water Resources. *2017 Central Valley Flood Protection Plan Update*; California Department of Water Resources: Sacramento, CA, USA, 2017.
39. California Water Commission. *Water Storage Investigation Program Technical Reference*; California Water Commission: Sacramento, CA, USA, 2017.
40. Livneh, B.; Bohn, T.J.; Pierce, D.W.; Munoz-Arriola, F.; Nijssen, B.; Vose, R.; Cayan, D.R.; Brekke, L. A spatially comprehensive, hydrometeorological data set for Mexico, the US, and southern Canada 1950–2013. *Sci. Data* **2015**, *2*, 150042. [CrossRef] [PubMed]
41. Livneh, B.; Hoerling, M.P. The physics of drought in the US central great plains. *J. Clim.* **2016**, *29*, 6783–6804. [CrossRef]
42. Bohn, T.J.; Vivoni, E.R. Process-based characterization of evapotranspiration sources over the North American monsoon region. *Water Res. Res.* **2016**, *52*, 358–384. [CrossRef]
43. Barnhart, T.B.; Molotch, N.P.; Livneh, B.; Harpold, A.A.; Knowles, J.F.; Schneider, D. Snowmelt rate dictates streamflow. *Geophys. Res. Lett.* **2016**, *43*, 8006–8016. [CrossRef]
44. Wi, S.; Ray, P.; Demaria, E.M.; Steinschneider, S.; Brown, C. A user-friendly software package for VIC hydrologic model development. *Environ. Modell. Softw.* **2017**, *98*, 35–53. [CrossRef]
45. He, M.; Russo, M.; Anderson, M. Hydroclimatic characteristics of the 2012–2015 California drought from an operational perspective. *Climate* **2017**, *5*, 5. [CrossRef]
46. Dai, A. Drought under global warming: A review. *WIREs Clim. Chang.* **2011**, *2*, 45–65. [CrossRef]
47. Heim, R.R., Jr. A review of twentieth-century drought indices used in the United States. *Bull. Am. Meteorol. Soc.* **2002**, *83*, 1149–1165. [CrossRef]
48. Keyantash, J.; Dracup, J.A. The quantification of drought: An evaluation of drought indices. *Bull. Am. Meteorol. Soc.* **2002**, *83*, 1167–1180. [CrossRef]

49. McKee, T.B.; Doesken, N.J.; Kleist, J. The relationship of drought frequency and duration to time scales. In Proceedings of the 8th Conference on Applied Climatology, Anaheim, CA, USA, 17–22 January 1993; American Meteorological Society: Boston, MA, USA; pp. 179–183.

50. Ciais, P.; Reichstein, M.; Viovy, N.; Granier, A.; Ogée, J.; Allard, V.; Aubinet, M.; Buchmann, N.; Bernhofer, C.; Carrara, A. Europe-wide reduction in primary productivity caused by the heat and drought in 2003. *Nature* **2005**, *437*, 529–533. [CrossRef] [PubMed]

51. Adams, H.D.; Guardiola-Claramonte, M.; Barron-Gafford, G.A.; Villegas, J.C.; Breshears, D.D.; Zou, C.B.; Troch, P.A.; Huxman, T.E. Temperature sensitivity of drought-induced tree mortality portends increased regional die-off under global-change-type drought. *Proc. Natl. Acad. Sci. USA* **2009**, *106*, 7063–7066. [CrossRef] [PubMed]

52. Breshears, D.D.; Cobb, N.S.; Rich, P.M.; Price, K.P.; Allen, C.D.; Balice, R.G.; Romme, W.H.; Kastens, J.H.; Floyd, M.L.; Belnap, J. Regional vegetation die-off in response to global-change-type drought. *Proc. Natl. Acad. Sci. USA* **2005**, *102*, 15144–15148. [CrossRef] [PubMed]

53. Swain, D.L. A tale of two California droughts: Lessons amidst record warmth and dryness in a region of complex physical and human geography. *Geophys. Res. Lett.* **2015**, *42*, 9999. [CrossRef]

54. Seager, R.; Hoerling, M.; Schubert, S.; Wang, H.; Lyon, B.; Kumar, A.; Nakamura, J.; Henderson, N. Causes of the 2011–2014 California drought. *J. Clim.* **2015**, *28*, 6997–7024. [CrossRef]

55. Wang, S.Y.; Hipps, L.; Gillies, R.R.; Yoon, J.H. Probable causes of the abnormal ridge accompanying the 2013–2014 California drought: ENSO precursor and anthropogenic warming footprint. *Geophys. Res. Lett.* **2014**, *41*, 3220–3226. [CrossRef]

56. Vicente-Serrano, S.M.; Beguería, S.; López-Moreno, J.I. A multiscalar drought index sensitive to global warming: The standardized precipitation evapotranspiration index. *J. Clim.* **2010**, *23*, 1696–1718. [CrossRef]

57. Abramowitz, M.; Stegun, I.A. Handbook of mathematical functions. *Appl. Math. Ser.* **1966**, *55*, 39. [CrossRef]

58. Beguería, S.; Vicente-Serrano, S.M.; Angulo-Martínez, M. A multiscalar global drought dataset: The speibase: A new gridded product for the analysis of drought variability and impacts. *Bull. Am. Meteorol. Soc.* **2010**, *91*, 1351–1356. [CrossRef]

59. Vicente-Serrano, S.M.; Beguería, S.; López-Moreno, J.I.; Angulo, M.; El Kenawy, A. A new global 0.5 gridded dataset (1901–2006) of a multiscalar drought index: Comparison with current drought index datasets based on the Palmer drought severity index. *J. Hydrometeorol.* **2010**, *11*, 1033–1043. [CrossRef]

60. Vicente-Serrano, S.M.; Van der Schrier, G.; Beguería, S.; Azorin-Molina, C.; Lopez-Moreno, J.-I. Contribution of precipitation and reference evapotranspiration to drought indices under different climates. *J. Hydrol.* **2015**, *526*, 42–54. [CrossRef]

61. Li, W.; Hou, M.; Chen, H.; Chen, X. Study on drought trend in south China based on standardized precipitation evapotranspiration index. *J. Nat. Disasters* **2012**, *21*, 84–90.

62. Beguería, S.; Vicente-Serrano, S.M.; Reig, F.; Latorre, B. Standardized precipitation evapotranspiration index (SPEI) revisited: Parameter fitting, evapotranspiration models, tools, datasets and drought monitoring. *Int. J. Climatol.* **2014**, *34*, 3001–3023. [CrossRef]

63. Banimahd, S.A.; Khalili, D. Factors influencing markov chains predictability characteristics, utilizing SPI, RDI, EDI and SPEI drought indices in different climatic zones. *Water Resour. Manag.* **2013**, *27*, 3911–3928. [CrossRef]

64. Thornthwaite, C.W. An approach toward a rational classification of climate. *Geogr. Rev.* **1948**, *38*, 55–94. [CrossRef]

65. Helsel, D.R.; Hirsch, R.M. *Statistical Methods in Water Resources*; Elsevier: New York, NY, USA, 1992; Volume 49.

66. Hirsch, R.M.; Helsel, D.; Cohn, T.; Gilroy, E. Statistical analysis of hydrologic data. *Handb. Hydrol.* **1993**, *17*, 11–55.

67. Mann, H. Non-parametric tests against trend. *Econometrica* **1945**, *13*, 245–259. [CrossRef]

68. Kendall, M.G. *Rank Correlation Methods*; Charles Griffin: London, UK, 1975.

69. Yue, S.; Pilon, P.; Cavadias, G. Power of the Mann–Kendall and Spearman's rho tests for detecting monotonic trends in hydrological series. *J. Hydrol.* **2002**, *259*, 254–271. [CrossRef]

70. Yue, S.; Pilon, P.; Phinney, B.; Cavadias, G. The influence of autocorrelation on the ability to detect trend in hydrological series. *Hydrol. Process.* **2002**, *16*, 1807–1829. [CrossRef]

71. Thiel, H. A rank-invariant method of linear and polynomial regression analysis, part 3. In *Proceedings of Koninalijke Nederlandse Akademie van Weinenschatpen A*; Royal Netherlands Academy of Arts and Sciences: Amsterdam, The Netherlands, 1950; pp. 1397–1412.

72. Sen, P.K. Estimates of the regression coefficient based on Kendall's tau. *J. Am. Stat. Assoc.* **1968**, *63*, 1379–1389. [CrossRef]

73. Adams, D.K.; Comrie, A.C. The North American monsoon. *Bull. Am. Meteorol. Soc.* **1997**, *78*, 2197–2213. [CrossRef]

74. Higgins, R.; Yao, Y.; Wang, X. Influence of the North American monsoon system on the US summer precipitation regime. *J. Clim.* **1997**, *10*, 2600–2622. [CrossRef]

75. Gutzler, D.S.; Robbins, T.O. Climate variability and projected change in the western United States: Regional downscaling and drought statistics. *Clim. Dyn.* **2011**, *37*, 835–849. [CrossRef]

76. Dettinger, M.D. Projections and downscaling of 21st century temperatures, precipitation, radiative fluxes and winds for the southwestern US, with focus on Lake Tahoe. *Clim. Chang* **2013**, *116*, 17–33. [CrossRef]

77. Elguindi, N.; Grundstein, A. An integrated approach to assessing 21st century climate change over the contiguous US using the NARCCAP RCM output. *Clim. Chang.* **2013**, *117*, 809–827. [CrossRef]

78. Scherer, M.; Diffenbaugh, N.S. Transient twenty-first century changes in daily-scale temperature extremes in the United States. *Clim. Dyn.* **2014**, *42*, 1383–1404. [CrossRef]

79. Ashfaq, M.; Bowling, L.C.; Cherkauer, K.; Pal, J.S.; Diffenbaugh, N.S. Influence of climate model biases and daily-scale temperature and precipitation events on hydrological impacts assessment: A case study of the United States. *J. Geophys. Res. Atmos.* **2010**, *115*. [CrossRef]

80. California Department of Water Resources. *California's Most Significant Droughts: Comparing Historical and Recent Conditions*; California Department of Water Resources: Sacramento, CA, USA, 2015; p. 136.

81. Kirtman, B.; Power, S.; Adedoyin, A.; Boer, G.; Bojariu, R.; Camilloni, I.; Doblas-Reyes, F.; Fiore, A.; Kimoto, M.; Meehl, G. Chapter 11—Near-term climate change: Projections and predictability. In *Climate Change 2013: The Physical Science Basis. IPCC Working Group I Contribution to Ar5*; IPCC, Ed.; Cambridge University Press: Cambridge, UK, 2013.

82. Anderson, E.A. *National Weather Service River Forecast System—sNow Accumulation and Ablation Model*; Technical Memorandum NWS HYDRO-17; NOAA: Silver Spring, MD, USA, 1973.

83. Andrew, J.T.; Sauquet, E. Climate change impacts and water management adaptation in two mediterranean-climate watersheds: Learning from the Durance and Sacramento rivers. *Water* **2016**, *9*, 126. [CrossRef]

MDPI
St. Alban-Anlage 66
4052 Basel
Switzerland
Tel. +41 61 683 77 34
Fax +41 61 302 89 18
www.mdpi.com

Climate Editorial Office
E-mail: climate@mdpi.com
www.mdpi.com/journal/climate